Computer-Aided Manufacturing

Third Edition

PRENTICE HALL INTERNATIONAL SERIES IN INDUSTRIAL AND SYSTEMS ENGINEERING

W. J. Fabrycky and J. H. Mize, Editors

AMOS AND SARCHET *Management for Engineers*

AMRINE, RITCHEY, MOODIE, AND KMEC *Manufacturing Organization and Management,* Sixth Edition

ANDREWS AND STALICK *Business Reengineering: The Survival Guide*

ASFAHL *Industrial Safety and Health Management,* Fifth Edition

BABCOCK AND MORSE *Managing Engineering and Technology,* Third Edition

BADIRU *Expert Systems Applications in Engineering and Manufacturing*

BADIRU AND PULAT *Comprehensive Project Management*

BANKS, CARSON, NELSON, AND NICOL *Discrete-Event System Simulation,* Fourth Edition

BLANCHARD *Logistics Engineering and Management,* Sixth Edition

BLANCHARD AND FABRYCKY *Systems Engineering and Analysis,* Fourth Edition

BUSSEY AND ESCHENBACH *The Economics Analysis of Industrial Projects,* Second Edition

BUZACOTT AND SHANTHIKUMAR *Stochastic Models of Manufacturing Systems*

CANADA AND SULLIVAN *Economics and Multi-Attribute Evaluation of Advanced Manufacturing Systems*

CANADA, SULLIVAN, AND KULONDA *Capital Investment Analysis for Engineering and Management,* Second Edition

CHANG AND WYSK *An Introduction to Automated Process Planning Systems*

CHANG, WYSK, AND WANG *Computer-Aided Manufacturing,* Third Edition

ELSAYED AND BOUCHER *Analysis and Control of Production Systems,* Second Edition

FABRYCKY AND BLANCHARD *Life-Cycle Cost and Economic Analysis*

FABRYCKY, THUESEN, AND VERMA *Economic Decision Analysis,* Third Edition

FISHWICK *Simulation Model Design and Execution: Building Digital Worlds*

FRANCIS, MCGINNIS, AND WHITE *Facility Layout and Location: An Analytical Approach,* Second Edition

GRAEDEL AND ALLENBY *Industrial Ecology,* Second Edition

HALL *Queuing Methods: For Services and Manufacturing*

HAMMER AND PRICE *Occupational Safety Management and Engineering,* Fifth Edition

HAZELRIGG *Systems Engineering: An Approach to Information-Based Design*

HUTCHINSON *An Integrated Approach to Logistics Management*

IGNIZIO AND CAVALIER *Linear Programming*

KROEMER, KROEMER, AND KROEMER-ELBERT *Ergonomics: Design for Ease and Efficiency,* Second Edition

KUSIAK *Intelligent Manufacturing Systems*

LANDERS, BROWN, FANT, MALSTROM, AND SCHMITT *Electronics Manufacturing Processes*

LEEMIS *Reliability: Probabilistic Models and Statistical Methods*

MUNDEL AND DANNER *Motion and Time Study: Improving Productivity,* Seventh Edition

OSTWALD *Engineering Cost Estimating,* Third Edition

PINEDO *Scheduling: Theory, Algorithms, and Systems,* Second Edition

PULAT *Fundamentals of Industrial Ergonomics*

PRASAD *Concurrent Engineering Fundamentals: Integrated Product and Process Organization*

SHTUB, BARD, AND GLOBERSON *Project Management: Processes, Methodologies, and Economics,* Second Edition

THUESEN AND FABRYCKY *Engineering Economy,* Ninth Edition

TURNER, MIZE, CASE, AND NAZEMETZ *Introduction to Industrial and Systems Engineering,* Third Edition

Computer-Aided Manufacturing

Third Edition

Tien-Chien Chang, Richard A. Wysk,
and Hsu-Pin (Ben) Wang

Upper Saddle River, New Jersey 07458

Library of Congress Cataloging-in-Publication Data

Chang, Tien-Chien, 1954-
    Computer-aided manufacturing / Tien-Chien Chang, Richard A. Wysk,
Hsu-Pin (Ben) Wang.—3rd ed.
      p. cm.
    Includes bibliographical references and index.
    ISBN 0-13-142919-1
    1. Computer integrated manufacturing systems.   2. CAD/CAM systems.
3. Flexible manufacturing systems.   I. Wysk, Richard A., 1948-
II. Wang, Hsu-Pin, 1954- III. Title.

TS155.6.C48   2005
670'.285—dc22

                                           2005045818

Vice President and Editorial Director, ECS: *Marcia Horton*
Executive Editor: *Eric Svendsen*
Associate Editor: *Dee Bernhard*
Executive Managing Editor: *Vince O'Brien*
Managing Editor: *David A. George*
Production Editor: *Craig Little*
Director of Creative Services: *Paul Belfanti*
Creative Director: *Jayne Conte*
Cover Designer: *Bruce Kenselaar*
Art Editor: *Greg Dulles*
Manufacturing Buyer: *Lisa McDowell*
Senior Marketing Manager: *Holly Stark*

PEARSON
Prentice
Hall
© 2006, 1998, 1991 Pearson Education, Inc.
Pearson Education, Inc.
Upper Saddle River, New Jersey 07458

Printed in the United States of America

10 9 8 7 6 5 4 3 2

ISBN 0-13-142919-1

Pearson Education Ltd., *London*
Pearson Education Australia Pty. Ltd., *Sydney*
Pearson Education Singapore, Pte. Ltd.
Pearson Education North Asia Ltd., *Hong Kong*
Pearson Education Canada Inc., *Toronto*
Pearson Educación de Mexico, S.A. de C.V.
Pearson Education—Japan, *Tokyo*
Pearson Education Malaysia, Pte. Ltd.
Pearson Education, Inc., *Upper Saddle River, New Jersey*

This book is dedicated to our wives.
Without their encouragement, it would not have been possible.

Contents

Preface

The paradigm of engineering is undergoing a major evolution throughout the world. The use of computers and the Internet has changed the way that we engineer and manufacture products. Among the recent trends in manufacturing are trends in which products are subject to a shorter product life, frequent design changes, small lot sizes, and small in-process inventory restrictions ("lean manufacturing"). The result of these trends is that today more than 90% of our products are manufactured in lots of less than 50 parts. These low lot quantities have eliminated many applications of dedicated production lines that were so effective in producing the inexpensive goods of the 1950 and 1960s.

The first step the nation employed to remain competitive with our international counterparts was the application of computer-aided design (CAD) and computer-aided manufacturing (CAM) to design and manufacture sophisticated products. Today, we routinely employ CAD to design products and flexible or programmable manufacturing systems to produce low- to medium-volume batch quantities. The Internet provides us with the connection to share design, marketing, and manufacturing information. We now look toward the advent of *distributed design and manufacturing using agile networking* as a means to produce products for the twenty-first century.

Employing numerical control (NC) and robotics in industry offers one potential solution to many manufacturing flexibility problems. This implementation, however, brings with it a variety of other problems. Robots and NC machines are designed to be flexible, self-contained, and capable of operating in both "stand-alone" and "integrated" manufacturing environs. Integrating this hardware into manageable systems has become a major focus of machine-tool makers and industry. Individual NC machines have also been made more versatile, more precise, more rigid and durable, and faster. More complex parts can be machined with higher accuracy and in less time. Time-compressed manufacturing technologies such as NC and rapid prototyping are being used more routinely to shorten product development cycles and to produce one-of-a-kind products. The benefit of all these new technologies cannot be achieved without the "communication networks" or an understanding of how these activities fit together. Today, it is not unusual for a design made thousands of miles away to be transferred to and realized in a remote site. Part programs and control instructions are downloaded from offices to machine controllers. Shop-floor operations can also be monitored either on-site or from afar. Manufacturing equipment has become part of the supply chain; capacity and availability are parameters used in planning and control of the entire chain. Further integration of the manufacturing component with design and business systems is also a key to our manufacturing success. These communication and control issues, coupled with a variety of sensing issues, are critical to the success of flexible automation in the United States.

This book focuses on the science, mathematics, and engineering of these new engineering methods. It is dedicated to making sure that the United States will remain the most efficient manufacturing nation in the world. The purpose of the book is to provide a comprehensive view of manufacturing, with a focus on design, automation, flexible automation, and the use of computers in manufacturing (CIM). Unlike other CIM books, this one attempts to provide a strong analytical science base and background in computer-aided manufacturing systems. The book is an excellent professional reference and also is an excellent text for CAM instruction.

We would like to thank the reviewers who provided feedback on the several drafts of this edition: Jeanette M. Garr, Youngstown State University; Nicholas G. Odrey, Lehigh University; Gary E. Rafe, The University of Toledo; Robert P. Van Til, Oakland University; and Gongyao (Jack) Zhou, Drexel University. We have also written an instructor solutions manual for this text. Copies are available either from your local Prentice Hall rep or by sending an email to engineering@prenhall.com

The book is written for advanced undergraduate and graduate courses. Each chapter covers general background, fundamental principles, and applications. Unlike most other manufacturing books on the market, it includes both descriptive information and analytical models. Whenever possible, MATLAB is used in examples. We do not assume that readers have a significant background beyond basic undergraduate engineering courses. However, the book does cover a very wide range of technologies and methodologies. Readers will gain in-depth and practical knowledge in CAM technologies.

Introduction to Manufacturing

Objective

Chapter 1 provides an overview of the development of manufacturing systems and machines. Beginning with an introduction to the product realization process, the chapter discusses the procedures for selecting, on the one hand, a manufacturing system and, on the other, processes for producing the products. The chapter also introduces computer-aided manufacturing (CAM). The historical development of CAM is examined as well. Finally, the chapter explains how the book is organized.

Outline

The wealth of a nation is commonly measured by its gross domestic product (GDP), or the per capita GDP of the nation. The GDP reflects the "standard of living" for industrial countries. The value of all the goods and services rendered by the entire nation in a calendar year, the GDP includes contributions from both private industry and government. Private industry is divided into the categories of agriculture, forestry, and fishing; mining; construction; manufacturing; transportation and public utilities; wholesale trade; retail trade; finance, insurance, and real estate; and services. All of these industry sectors contribute to the wealth of a nation. U.S. Department of Commerce statistics indicate that, in 2001, manufacturing sectors contributed $1.4923 trillion (1996 dollars) to the real GDP of the nation.[1]

This figure represents 16.2% of the entire U.S. GDP. Manufacturing also represents 14.8% of U.S. employment. Over the past 30 years, manufacturing's share of both GDP and of employment has dropped by more than 50%. This is a much bigger drop than that which has taken place in other major nations. For example, German manufacturing employment dropped 13% during the same period, and that country's manufacturing proportion of GDP fell from 36 to 23%. In Japan, manufacturing employment dropped by 6.5% and manufacturing's share of GDP fell from 33.5 to 21%.[2]

The statistics seem to tell a story of a declining and ever-marginalized manufacturing sector. However, the statistics also send us a signal of urgency and of a need for infrastructure transformation. If we still want to maintain or improve our standard of living, we must not ignore this signal. Whether we can maintain our standard of living depends on whether we can provide sufficient goods and services to the nation's population. Under the free-trade movement, goods and services are moved freely across international borders. "Trading" means the exchange of goods or services. One must have something that others want in order to trade. While a city or state may rely on tourism (service) dollars to trade food and goods, a large nation cannot afford to specialize in an area such as agriculture, manufacturing, or services in order to generate enough GDP to trade for other necessities. All world powers must compete on all fronts, adapt to changing environs, and find the right strategy to maintain their competitive edge in manufacturing. At the same time, they must recognize that, while the overall output of manufacturing should increase, employment will nonetheless inevitably decline.

Agriculture provides us a good example to follow. Today, agriculture accounts for 1.5% of employment and produces 1.7% of GDP. Low food prices keep the GDP contribution low. However, the 1.5% of employment engaged in agriculture produces enough food to feed this nation and many other nations in the world. A little more than a hundred years ago, 40% of the U.S. population was involved in farming. Agriculture produced 22% of the GDP. (The standard of living at the time was much lower than it is today.) Farming meant using animal power to plow the fields. Nowadays, riding on air-conditioned machinery guided by a global positioning system, a farmer can farm thousands of acres of land and produce a high yield. Improved crops, the precise delivery of fertilizer and insecticide, better weather forecasting, specialized farm equipment, and more made all that possible. The historical statistics show a clear trend: From 1900 to 2000, output per acre of corn increased from 28.1 bushels to 134.5 bushels. Soybean output increased from 10.9 bushels per acre to 36.7 bushels per acre. Rice production grew from 1,220 cwt (hundredweight, or 100 lb) per acre to 5,929 cwt per acre. Potato production improved from 52 cwt per acre to 348 cwt per acre. At the same

---

[1] http://www.bea.doc.gov/bea/dn2/gpox.htm, Bureau of Economic Analysis, U.S. Department of Commerce, NIPA data: Gross domestic product by industry.
[2] Michael R. Czinkota, "An Analysis of the Global Position of U.S. Manufacturing," *Thunderbird International Business Review*, 45 (5): 505–519 (September–October 2003).

time, the average farm size increased from 147 acres to 487 acres. Total farm machinery in the United States rose from 64,000 to 9,258,337 pieces of equipment. The mechanization of farming played an important role in agriculture productivity.

Manufacturing is no less important than agriculture, and, as with farming, we must learn to produce better products at lower cost. Automation—especially computer-aided manufacturing—is an answer. Manufacturing is an activity that is traditionally labor intensive. Before the era of automation, manufacturing industry employed 36% of the working population and produced about the same percentage of GDP as now. As manufacturing operations became more automated, labor productivity grew faster than the demand for workers. As a result, employment in manufacturing dropped. In the meantime, the demand for better quality, lower prices, and more variety continues to increase. The solution is not just to persist in finding lower-labor-cost regions of the world to produce the product, but to balance the need for low-cost labor with the development of high-technology intelligent and flexible production methods. Of course, we should never totally discount the importance of using lower-cost labor to reduce production costs, but we should keep in mind that it is often associated with lower quality products. To compete in the global market and still maintain the standard of living of this country, we must not ignore the importance of developing and deploying new manufacturing technology.

This book covers topics in computer-aided manufacturing ranging from design representation, to computer control of production machines, to machine-tool programming. It discusses the fundamentals of key technologies used in modern manufacturing facilities. It also tries to link various technologies and disclose the common threads among them. Finally, it shows how to integrate vertically the product realization process from design to manufacturing. The ultimate goal of the book is to provide readers with an in-depth knowledge of modern manufacturing technologies, so that they can help design and operate competitive manufacturing systems.

1.1 THE PRODUCT REALIZATION PROCESS

The purpose of manufacturing is to produce a product. The product can be a single component, such as a screw or a gear, or it can be a complex assembly, such as an airplane or a car. Regardless of the complexity of the product, all products go through a common development activity: the product realization process. The process begins with market demand for a certain product, followed by design, manufacturing, quality control, and, finally, product distribution. The ISO 9001:2000 standard defines the production realization process by dividing it into several stages (Figure 1.1). This classification evolved from a quality-control point of view. The purpose is to control the product realization procedure. In this book, we will take a different point of view—that of the engineering perspective. Although all development perspectives are equally important, the business process is intentionally ignored in the discussion that follows.

The engineering process for product realization begins with product specifications that are obtained from the customer requirements. Design synthesis is the process for converting the specifications into a product concept. For example, the specification for a subsonic aircraft might include functional requirements like a service ceiling of 35,000 feet, a cruise speed of 350 kts (knots, or nautical miles per hour), six

1 Planning of product realization

2 Customer-related processes

2.1 Determination of requirements related to the product
2.2 Review of requirements related to the product
2.3 Customer communication

3 Design and development

3.1 Design and development planning
3.2 Design and development inputs
3.3 Design and development outputs
3.4 Design and development review
3.5 Design and development verification
3.6 Design and development validation
3.7 Control of design and development changes

4 Purchasing

4.1 Purchasing process
4.2 Purchasing information
4.3 Verification of purchased product

5 Production and service provision

5.1 Control of production and service provision
5.2 Validation of processes for production and service provision
5.3 Identification and traceability
5.4 Customer property
5.5 Preservation of product

6 Control of monitoring and measuring devices

Figure 1.1   ISO 9001:2000, Section 7: The product
realization process.

passengers, and 300 pounds of cargo. The product concept could include a single jet, twin jets, a turboprop, a twin turboprop, a high wing, a low wing, a single tail, a twin tail, or something else. After a general concept is selected, conceptual and detailed geometric design can begin. Over the past 20 years, "Design for X" (where "X" could be "assembly," "manufacturing," "quality," etc.) has been lauded as the tool for better design. Be that as it may, the key is to take into consideration downstream processes during the design stage. For example, applying "Design for Manufacturing," one considers the ease of manufacturing in designing the overall shape and features of a product. Manufacturing rules, such as "It is hard to machine a flat-bottomed hole," are used to guide the designer through the process of creating an easier-to-manufacture design.

In order to share the design concept and document the design, a formal design representation is made. The traditional design representation is an "engineering drawing." Rules (ANSI and ISO standards) are applied to represent three-dimensional (3-D) geometries via two-dimensional (2-D) projections. Standard symbols are used to represent technological information such as dimensions, tolerances, and surface finishes. Since the 1980s, 3-D solid models have become more and more popular in replacing engineering drafting drawings. The 3-D solid models are better geometric representations for most artifacts designed. However, they still have their limitations, so they cannot completely supplant engineering drafting. (See Chapters 2 and 3.)

After an initial design is completed, engineering analysis is carried out. An assembly may need kinematic analysis to ensure that the motions of the parts will achieve the desired functions. For individual parts, stress and temperature properties under operational conditions may need to be determined. A wide range of analytical and computational tools, such as finite-element analysis software and kinematics simulation tools, are available. Some of these tools have been integrated into CAD software, and many others can import design data through a translator. The results, either in text or in graphical form, can be printed or displayed with the design model. On the basis of the results of the analysis, improvements to the design can be made to reduce the values of the targeted design parameters. Sometimes the entire shape of the design needs to be changed in order to satisfy design objectives. Using integrated software analysis tools shortens the design time and may allow for an increased number of design iterations in the same time span. In theory, the more iterations that are made, the better the final design will be. With the invention of rapid prototyping processes (Chapter 13), a physical prototype can be built quickly and economically. Some of the design analysis can be done or aided with the use of the physical prototype, instead of the designer's relying on software simulation to check gross fits or motion constraints for the parts. One can physically assemble parts together to test the ease of assembly of the product and the conditions under which it may be assembled. The technology is not quite ready, but when material property issues and achieving acceptable tolerances through rapid prototyping are met, it will be possible to put the prototype through physical product testing and observe physical changes to the prototype.

After the design objectives are satisfied through analysis, it is time to evaluate the design. It is especially important that several design alternatives be made. The purpose of design evaluation is to compare the cost and function of each alternative so that the best design can be selected. Even if there is only one design, the design should still be evaluated to determine whether all of its objectives have been accomplished and it is economical to produce.

Before the final design is built, a manufacturing plan must be prepared. Manufacturing planning begins with determining whether, and if so, how the part can be made and deciding whether it should be outsourced or made internally. If outsourcing is deemed as the best or only choice, the question of where to outsource the part must be resolved. The vendor must have the capability to manufacture the product and must charge a reasonable price. There are also supply-chain-related issues. The lowest-cost vendor is not always the best vendor. Consistency in quality and reliability in delivering the product are but two of the factors to consider. If the product is to be built in-house, then, a detailed manufacturing process plan has to be prepared. Such a plan consists of the operation-routing sequence used to produce the product, wherein each operation in the sequence includes the machine, tool, work-holding devices, operation parameters, etc., to be used. A detailed process plan document must be prepared for every part that will be manufactured in-house. This document may include texts with detailed instructions and setup or assembly drawings. Often, tools and work-holding devices have to be designed and built, and when they do, they also become products that are to be realized.

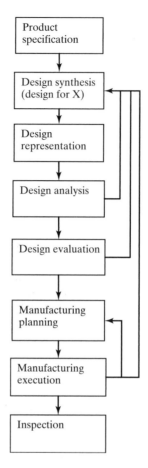

Figure 1.2 Engineering process for product realization.

Only after manufacturing process plans have been prepared can production begin. In Figure 1.2, the production process has two phases: "manufacturing execution" and "inspection." Lot size determination, master scheduling, daily scheduling, operation sequencing, and other processes have to be carried out in order to run the production system efficiently. "Lean manufacturing" principles are used to ensure that production is efficient. If the production facility is automated, material transportation and handling equipment must be laid out and programmed. Machine tools such as Computer Numerical Control (CNC) machines are also programmed and the programs verified. Normally, for CNC machine programming, CAD models of the product can be fed directly into the CAM package to generate part programs. In case robots are used for either production or handling of the parts, they can also be programmed with the CAD model. All of these operations can be very tedious and may require extreme precision. CAD/CAM tools are essential for success, however.

Finally, before the product is accepted, it must be inspected. Gauging and measuring operations, either manual or under computer control, are used to inspect the finished product. The entire product realization process can be long and tedious. For many new parts (especially precise parts with highly irregular geometries), numerous new fixtures and tools may be required to produce the product economically in large quantities. Since fixtures and tools must be more accurate than the part that they are used to manufacture, the time required for tool engineering can be significantly longer than the time required to design the part itself. The use of computers in many of the product realization steps helps to improve the product and speed up the process. CAM is a necessary tool in this regard.

1.2 THE EVOLUTION OF PRODUCT REALIZATION

Before we focus on the history of CAM, it is worthwhile to look at the historical development of manufacturing. We will begin by examining the major manufacturing milestones that took place during the course of human civilization (Figure 1.3). It is said that what differentiates human beings from other animals is our ability to use tools. Some animals do use tools to help them get their food. For example, once in a while an animal may pick up a rock and use it to break a nut. However, humans are the only species that knows how to *build* (i.e., manufacture) tools. Even during the Stone Age, humans learned how to fashion stone tools (such as knives made out of flint). Hand tools enable humans to make simple things. Extensions of our hands, hand tools also are much harder than our own fingers and resist impact and abrasion much better than we do. We, however, can self-heal cuts and breaks in our skin.

The next major improvement happened not long ago (ca. 1800). The Industrial Revolution was the result of the development of machine tools. It brought another major jump in the standard of living. A machine tool is defined as a machine that is used to build other machines. Machine tools added power and precision to manufacturing. With machine tools, harder materials can be processed at higher rates. As a result, humans can produce goods faster and more precisely. With the advent of machine

Category	Milestone
Skeleton	Hand tools: thousands of years to several thousands of years
Muscle	Machine tools: Industrial Revolution, 18th century, custom-made products
Smartness	Gauges: late 19th century, interchangeability
Resource	Manufacturing systems: early 20th century
Management	Modern management, transfer lines, pull technology
Nerve	Numerical control, robot (1950s, 1960s, 1970s), flexible manufacturing systems, sensors, vision
Brain	Computer-aided manufacturing, intelligent manufacturing

Figure 1.3   Manufacturing milestones.

tools, human productivity increased drastically and industrial goods replaced hand-made products. Today, machines make products that are both less expensive and of higher quality. Machine tools are used to build other machines and other machine tools. This amplifier effect was critical to industrial development: Without machine tools, the steam engine would not have been built. A human-made power source, the steam engine in turn made power available almost anywhere we choose. This ready availability of power further improves the usability of machine tools because machine shops do not have to be located beside rivers and streams in order to tap into water power.

With their newly invented machines, Europeans were able to expand their influence throughout the world. At the time of the Industrial Revolution, products were still custom made with manually operated machine tools. The interchangeability of parts, developed by Eli Whitney,[3] brought another major improvement to manufacturing. Ultimately, the combining of jigs and gauges that were developed for interchangeable manufacture aided in the realization of the production or assembly line and thence mass production in the 20th century. Now we work smarter and not just harder. Never before in human history had humanity enjoyed such an improvement in the standard of living than in the 20th century. Mass production and the scientific management of manufacturing together helped to produce more, better, and less expensive goods. Scientific management principles developed by F.W. Taylor [1911] and followed upon by Frank Gilbreth, Henry Gantt, and Charles Bedaux in the first quarter of the century made factories run more efficiently. Automated (mechanically controlled) machines and systems (e.g., transfer lines and screw machines) outproduced tens of hundreds of human workers. Mass-produced identical goods were plentiful and inexpensive. However, variety was limited, due to the high cost of changeover in the manufacturing system. As Henry Ford so aptly put it, "You can have any color Model T that you'd like—as long as it's black."

It was not until after World War II that another major breakthrough in manufacturing was observed: The invention of the computer and programmable automation led to another "Industrial Revolution." Today, manufacturing equipment is commonly controlled by digital devices through software programs. Such control starts the era of CAM. A detailed discussion of the history of CAM will be presented in the next section.

Figure 1.4 lists a few inventions in the area of machine tools. Compared with the thousands of years of human civilization, the history of machine tools is relative short. The development of CAM is especially recent. In our daily life, we all benefit from the results of these developments. Many modern-day conveniences, such as automobiles, airplanes, air conditioners, refrigerators, cell phones, and computers, are direct or indirect products of machine tools. All machines are made with the use of machine tools.

---

[3]Eli Whitney is commonly acknowledged as the inventor of interchangeable manufacturing. However, the concept of interchangeability of parts was first conceived by Thomas Jefferson upon his observation of French mechanic Honoré Blanc's work in 1785. Whitney sold the idea to the U.S. War Department in 1798 with an eye toward making muskets. It was not until 28 years later, however, in 1826, that John Hall finally succeeded in devising manufacturing methods, tools, and technologies for building rifles with truly interchangeable parts.

1750	Screw-driven lathe
1751	Slide lathe, first metal lathe
1770	Screw-cutting lathe
1775	Boring mill
1813	Interchangeability of parts: Simon North, horse pistols
1817	Planing machine
1845	Turret lathe
1847	Milling machine (Brown & Sharpe), making helical grooves by twist drill
1900	High-speed steel tooling (F. W. Taylor and M. White)
1946	ENIAC (Electronic Numerical Integrator and computer)

Figure 1.4   Important inventions in manufacturing.

To move our civilization forward, we cannot ignore the importance of manufacturing technology and machine tools.

1.3 CAM AND ITS HISTORICAL DEVELOPMENT

CAM can be defined as "the effective utilization of computers in manufacturing [Groover, 1987]." Computers may be used for direct manufacturing process control and monitoring or for indirect manufacturing operations support. Applications of direct manufacturing process control and monitoring include numerical-control (NC) machine tools, robotics, and manufacturing systems control (such as flexible manufacturing systems (FMS) and automated manufacturing cells). Applications of indirect manufacturing operations support include computer-aided facility planning and design, scheduling, manufacturing resource planning (MRP-II), enterprise resource planning (ERP), computer-aided process planning (CAPP), and computer-assisted part programming, among others. A narrow definition of CAM normally excludes scheduling, MRP, facility planning and design, and, sometimes, other functions.

Shortly after World War II, with increasing demand for more complex parts, NC machine tools were invented. NC replaced the need for the coordinated control of skilled machine operators, a skill that takes years to master. More scientific and technological developments have occurred since the 1950s than throughout all of the rest of human history. One of the most important of these developments was the invention of the digital computer. In discrete-product manufacturing, computers were essential to the development of NC, robotics, computer-aided design (CAD), CAM, and FMS. Today, new computer-based technologies enable us to produce small-batch products at low cost. Many human decision-making functions have been replaced or lent assistance by computers. Thus, computers further increase human productivity. Artificial intelligence (AI) and expert systems now provide our manufacturing systems with (limited) intelligence. Not only can computers replace manual labor, but they also can perform some mental processing. Figure 1.5 illustrates the short history of CAD and CAM.

SOFTWARE		HARDWARE
	1945	
		James T. Parsons proposed NC concept
	1950	MIT servo mechanism lab USAF NC milling m/c proj.
Part programs prepared manually		1st successful demo
MIT started APT development AI, Dartmouth Conference	1955	Automatic tool changer - IBM
		1st production skin-miller - G&L
LISP language APT language		Machining center - K&T
	1960	
SKETCHPAD Interactive computer graphics Coons patch, sculptured surfaces Ferguson's patch, sculptured surfaces		1st industrial robot CRT display Adaptive control - Bendix
UNISURF, Bezier's patch, sculptured surfaces CAD Drafting	1965	7,700 NCs installed
		CNC, DNC concept & mini-computers, PLC 1st DNC system
	1970	
Solid modeling development started		CAM, CAD/CAM
Build-1 solid modeler		Microcomputers FMS
	1975	
3-D CAD systems		
		Superminicomputers
PADL-1.0 solid modeler IGES graphics exchange standards Supercomputers	1980	
		Micro-based workstations GM MAP LAN standard
Solid modeler became commercialized PC-based CAD		
	1985	Computer vision
MAP, TOP LAN standards		
Expert CAD systems	1990	Automated factory In-situ sensing & control MEMS
Neural nets		
Virtual manufacturing	1995	
		Collaborative manufacturing
Enterprise manufacturing	2000	Nano manufacturing

Figure 1.5   History of CAD/CAM.

In pursuing the development of modern manufacturing, the Toyota Motor Company took a different approach to mass production. While U.S. industry followed the example of Henry Ford's assembly-line approach, Toyota started the concept of "lean manufacturing." (The term was coined later.) The company's executives believed that success in the post-mass-production era would be wrought by the economical production of small batches. They saw a mixture of zero inventory, flexible automation, and streamlined logistic support as the solution. They did not blindly apply technology for technology's sake, but tried to understand the process (operations) thoroughly before deciding what to do. The solution could be product redesign, process redesign, automation, improved human operation, or a combination of these approaches. Their success is well known, and the technologies warrant discussion.

From both the product demand and the manufacturing batch size, we also see a trend. Before the Industrial Revolution, demand was low and products were all custom made. Few products were available and their prices were uniformly high. Machine shops produced primarily small batches of a given product. Then, from the turn of the century to the 1950s, the world was transformed from a society with few industrial products for few people to abundant products for everybody. Demand for industrial products grew fast, and to respond to that demand, mass-production techniques were developed. Lower product costs further stimulated demand. Sometimes, the desire to have more variety led to the production of small batches again. Finally, in the 1980s, intense international competition mandated that products be made quickly and inventory be kept to a minimum. Small-batch dynamic production environs came into being. The production technology that has resulted produces better, less expensive products in response to rapid changes in demand. Because, however, demand drives the development of a new production technology, the survival of modern manufacturing industries is predicated on flexibility in the midst of automation. Automation provides good quality and low cost, and flexibility enables the company to adapt to changes in both the product and the demand for it. In pursuit of automation amidst flexibility, it seems obvious that the solution is to apply CAM at the shop floor. CAM includes a larger number of functions, ranging from FMS scheduling to machine control. In one application in which computers are totally integrated into manufacturing, the solution is *computer-integrated manufacturing* (CIM), which includes not only manufacturing functions, but also business and other engineering functions. CAM is part of the CIM solution; in this book, we focus on CAM technologies.

Today's new engineering thrusts include lean manufacturing, rapid prototyping, (micro-electromechanical systems (MEMS), and nanotechnologies. The first two are changes that are being made in the way we engineer products so that we can be more flexible to changes in designs and more responsive to customer needs or requests. *Lean manufacturing* is the systematic elimination of waste in manufacturing. Although the concept is relatively new, many of the tools used in lean manufacturing can be traced back to Fredrick Taylor and the Gilbreths at the turn of the 20th century. Lean manufacturing (無駄のない製造) is a Japanese method focused on the three "M's": **muda** (無駄), the Japanese word for no waste, **mura** (斑), inconsistency, and **muri** (無理), unreasonableness. Using methodologies developed in the Toyota Production System, such as "autonomation" (defined by Ohno[4] as a machine with a human touch, to eliminate problems) and "Kanban" (display cards adapting production scheduling to human beings), lean manufacturing makes modern manufacturing systems much more competitive. The same principles have been applied to service industries as well, with profound results.

Rapid prototyping refers to a broad set of engineering topics that are intended to produce the initial model of a new design. With many products, it is necessary to produce an initial physical item so that the fit, accessibility, and other properties of interrelated parts can be determined. Historically, production of the first item has occurred in a "tool or model shop," by highly skilled machinists and technicians. Rapid prototyping is the process whereby the initial model can be produced more quickly and inexpensively. Although rapid prototyping normally has to do with the fabrication of a physical product, it also can refer to the creation of software (especially control software). Among the processes that are used in the rapid prototyping of physical models is the layered deposition of material—for example, laser curing. Processes that use "layered methods" have been referred to as *3-D fabrication*, because layers of materials are deposited in 2-D form and then built up to make a 3-D object.

MEMS use semiconductor integrated-circuit manufacturing technology to build mechanical devices in the micron size range (1 micron = 0.000001 meter). These small products may be integrated with their electronic controllers in the same package. There are three basic building blocks in MEMS technology: the deposition of thin films of material on a substrate, the application of a patterned mask on top of the films by photolithographic imaging, and etching of the films selectively to the mask. So far, the most promising products of MEMS are biosensors (such as the DNA chip) and electro-mechanical sensors (such as the sensor in an air bag).

Nanotechnologies are a new set of processes that produce very small features and products (in the 1–100-nanometer range, where 1 nanometer = 0.000000001 meter). Nanotechnologies utilize processes that can be focused on atoms and molecules. Currently, many of nanotechnology studies focus on nanocoating or nanostructure materials, such as carbon tubes in the nanometer size range. Traditional discrete manufacturing processes do not play a role in manufacturing on that level.

Obviously, by no means can all CAM technologies be addressed in any depth in a single volume. Accordingly, we instead provide a concise, yet comprehensive, discussion of CAM technologies. Most technologies are linked through common product representations. Before we discuss manufacturing technologies, it is important to understand the input (i.e., the engineering design) into the manufacturing system. It turns out that both CAD and CAM share a set of common tools: coordinate transformations and kinematical representations. This tool set will be used throughout the book. In particular, NC part programming and layered manufacturing control are even more tightly integrated with CAD representations and algorithms. Some production equipment is also controlled by using similar principles. The basic control system architecture is almost the same, whether it be Computer Numerical Control (CNC) machine tool control, robot control, manufacturing cell control, or rapid-prototyping machine control.

---

[4]Taiichi Ohno, *Toyota Production System: Beyond Large-Scale Production* (Cambridge, MA: Productivity Press, 1988).

All share the same fundamental technologies. Because knowledge of one of these architectures helps us to understand the others, the best approach is to give the reader a sound understanding of each.

1.4 ORGANIZATION OF THE BOOK

The specific aim of this book is to provide a comprehensive view of CAM from a manufacturing engineering perspective. The main thrust of the volume is to impart an understanding of the principles of, and relationships among, the various components of CAM. The book is tailored to those wanting to obtain state-of-the-art knowledge of CAM. The specific objectives of the text are as follows:

- to provide in-depth knowledge of the various components of CAM systems
- to provide knowledge of state-of-the-art CAM technologies
- to develop skills in constructing and implementing CAM systems
- to develop skills in evaluating CAM technologies
- to examine the relationship between CAD and CAM

The book is divided into 15 chapters, each addressing a specific issue related to CAM. The following is a brief overview of the major focus of each chapter:

Chapter 1, "Introduction to Computer-Aided Manufacturing," covers the history of manufacturing and CAM, emphasizing the importance of CAM in modern manufacturing operations.

Chapter 2, "Engineering Product Specification," presents the principles underlying the specification of a product so that it can be uniquely interpreted. The fundamentals of design specification and tolerances are examined in detail. Basics of ASME Y14.5 are given, and single and multiple data are introduced.

Chapter 3, "Geometric Tolerancing," introduces tolerance stacking and the concept of tolerance management. Product interpretation and inspection using gauges are presented. Notations for maximum and minimum geometric size and location conditions are introduced.

Chapter 4, "Computer-Aided Design," introduces CAD and how it is used in product design. Basic CAD features and functions are presented, and information on CAD ranging from drafting to solid modeling is provided. Definitions and explanations of CAD terminology are offered.

Chapter 5, "Geometric Modeling," discusses the fundamentals of the geometric modeling that is the foundation of CAD. To understand CAD, one needs to know the mathematics and algorithms built into a CAD system. This chapter discusses computer graphics, 2-D and 3-D geometries, surface models, and solid models. It is intended to give readers a good background in the basics of geometric modeling.

Chapter 6, "Process Engineering," serves as an introduction to the remainder of the book and covers process planning, process capability, and issues related to the economies associated with selecting a machining process. In order to master the materials presented in this chapter, readers must understand engineering drawings and have a fundamental knowledge of various machining processes.

Chapter 7, "Tooling and Fixturing," presents the basics of tool engineering, an understanding of which is essential for manufacturing. Before any parts can be produced, tools and fixtures must be selected or designed. NC part programs are written on the basis of a knowledge of exact tool characteristics and the positioning of the workpiece on the machine table. A successful machining operation relies on the correct utilization of tools and fixtures.

Chapter 8, "Statistically Based Process Engineering," introduces process capabilities. The concepts of accuracy and precision are presented from a feature-based production point of view. The concept of statistically based decisions is introduced, and the cost of production as a function of product variability is addressed.

Chapter 9, "Fundamentals of Industrial Control," offers a general background on the topic. Readers are given an understanding of programmable logic controllers (PLCs) and NC. The discussion includes both mathematical analysis and an examination of how hardware is implemented. References are given for further reading. No background in control is assumed.

Chapter 10, "Programmable Logic Controllers," is an overview of the topic. The basic features of a PLC and its programming are discussed. The roles of PLCs in automated manufacturing are explored. Understanding this chapter requires knowledge of industrial control fundamentals.

Chapter 11, "Data Communications and LANs in Manufacturing," explains the importance of data communication for planning, monitoring, and control within a factory or between production sites. Since communication is a rapidly growing technology, the chapter explores various methods of communication.

Chapter 12, "Fundamentals of Numerical Control," introduces the fundamentals of NC in producing product geometries. Since machine tools in a CAM system are mostly NC machine tools, this chapter explains how NC machines are built. The focus of the chapter is on motion control of the machine tool, although other features of modern NC machine tools are also discussed.

Chapter 13, "Numerical Control Programming," builds on Chapter 11, which covered the hardware aspects of numerical control. The different levels of NC programming are introduced, and analytical geometry for part programming is discussed.

Chapter 14, "Rapid Prototyping," is an overview of rapid prototyping technologies and the basic principles they are based on. The capabilities and limitations of each process are discussed. General guidelines in part design and process selection are presented.

Chapter 15, "Robotics," provides information about the industrial use of robots. The chapter notes how industrial robots serve as an integral part of modern manufacturing systems and discusses the economic use of robots in such systems.

1.5 KEYWORDS

CAD	MEMS
CAM	Nanotechnology
Design for X	Numerical control
FMS	Product realization
GDP	Rapid prototyping
Lean manufacturing	Toyota Production System
Machine tool	

1.6 REVIEW QUESTIONS

1. What is CAM?
2. What major developments have come about in CAM technology?
3. What is the current trend in manufacturing? Describe it in terms of market changes, technological development (on both products and processes), managerial organization and shifts in philosophy, and social changes (brain generation, "the green movement," and so on).
4. List five direct and five indirect applications of computers in manufacturing.
5. Discuss how manufacturing affects our standard of living.
6. In the past four decades, we have seen manufacturing jobs moving from developed countries to developing countries. Do you believe that manufacturing jobs will always follow lower cost labor? Why?
7. CAM is embraced not only by developed countries but also by developing countries. Why is CAM important for the developing countries?

1.7 REFERENCES

Gilbreth, Frank and Lillian. *Applied Motion Study*. New York: Sturgis & Walton Co., 1917.

Gilbreth, Frank. *Motion Study*. New York: D. Van Nostrand Co., 1911.

Gilbreth, Frank. *Primer of Scientific Management*. New York: D. Van Nostrand Co., 1912.

Groover, M.P. *Automation, Production Systems, and Computer-Aided Manufacturing*. Englewood Cliffs, NJ: Prentice-Hall, 1987.

Hawke, D.F. *Nuts and Bolts of the Past: A History of American Technology, 1776–1860*. New York: Harper & Row, 1988.

Ketola, J. and Kathy Roberts. *ISO 9001:2000 in a Nutshell: A Concise Guide to the Revisions*, 2d ed. Chico, CA: Paton Press, 2000.

Koren, Y. *Computer Control of Manufacturing Systems*. New York: McGraw-Hill, 1983.

Ohno, Taiichi. *Toyota Production System: Beyond Large-Scale Production*. Cambridge, MA: Productivity Press, 1988.

Shingo, Shigeo. *A Study of the Toyota Production System from an Industrial Engineering Viewpoint*, tr. A. P. Dillon. Cambridge, MA: Productivity Press, 1989.

Smith, M. R. "Eli Whitney and the American System of Manufacturing," in *Technology in America: A History of Individuals and Ideas*, 2d ed., edited by C. W. Pursell. Cambridge, MA: MIT Press, 1990.

Taylor, Frederick Winslow. *The Principles of Scientific Management*. New York: Harper & Row, 1911.

Willey, Allan. "Technology Transition: An Historical Perspective," http://www.virtualschool.edu/

Chapter 2

Engineering Product Specification

Objective

Chapter 2 is an overview of the principles of specifying a product so that it can be uniquely interpreted. The emphasis of the chapter is mechanical product design. The fundamentals of design specification and tolerances are presented in detail. The objective of the chapter is to familiarize the student with the standards for representing mechanical components and provide the basis for the unique interpretation of product designs.

Outline

Before a product can be manufactured, it must be designed. The design process can be divided into five basic steps: (1) design conceptualization, (2) design synthesis, (3) design analysis, (4) design evaluation, and (5) design representation.[1] A design solution is conceptualized on the basis of the product requirements (functional requirements). The initial solution is usually rather broad, encompassing the general elements of the product, but lacking in specification detail. The synthesis step adds more details to the initial concept. At this stage, the geometry is laid out and design parameters and dimensions are assigned to the product. Steps 1 and 2 rely heavily on the creativity of the designer, and no scientific consensus exists regarding how to carry out these activities (perhaps with the exception of the axiomatic approach proposed by Suh [1982]).

During the first two steps, many design ideas are formed in the designer's head. This is the "brainstorming" stage, in which as many solutions are generated as is possible. These ideas are then sorted for further development. As the design takes on a more definite shape, a sketch is frequently used to help clarify the idea. When the design task is carried out by a group of people, a common understandable representation must be used in order for all involved to share in the development of the idea. The solution is then analyzed (e.g., via stress analysis) and evaluated (e.g., by a cost comparison) in order to identify a viable and, eventually, the best alternative. Before the design is released for manufacture, it is necessary to detail the design (including selecting standard components), determine dimensions and tolerances, determine special manufacturing notes, and produce the final draft. The design representation step includes both the rough sketch and the more detailed design layout.

In order for a part to be properly manufactured, a detailed part representation with information pertinent to manufacturing must be received before any manufacturing activity can begin. In this chapter, the various procedures used to specify a part design are discussed. The chapter begins with a general discussion of engineering design, followed by a description of how one goes about interpreting an engineering drawing. Conventional tolerances are discussed as well.

2.1 ENGINEERING DESIGN

Engineering design is the partial realization of a designer's concept. Today, a designer usually cannot directly transform a concept into a physical product. Instead, the designer conveys his or her idea to others through an alternative medium, such as an engineering drawing, and then the manufacturing engineer or machinist produces the design (Figure 2.1). Prior to the Industrial Revolution, if a farmer needed a tool, he normally went to a blacksmith and told the blacksmith the shape and size of the tool he wanted. Because most tools were simple and did not require significant accuracy, the blacksmith would get a pretty good picture of a hoe or a plow through the farmer's verbal description. If the blacksmith still did not understand, the farmer could sketch the tool on the dirt floor of the blacksmith shop.

As product requirements and designs became more complex, a picture became necessary to relate the information to others. Multiview orthographic drawings have long been adopted by engineers as the standard tool to represent a design. With such drawings, design information can be passed from the designer to others who are well

---

[1]The German VDI standard classifies the design process into four steps: clarification of the task, conceptual design, embodiment design, and detailed design.

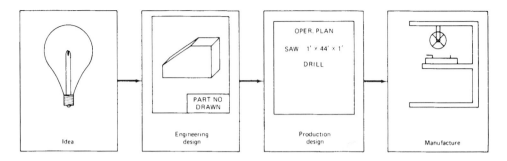

Figure 2.1   Evolution of a product.

trained in reading the drawings. The object a designer draws on paper can be interpreted and reconstructed in a viewer's mind. The ability to transform an object from one medium to another (e.g., from a 2-D, three-view drawing to a 3-D picture) and to convey an understanding of the rules of drawing are necessary to pass design information from the designer to others without error.

Several methods are available to represent an engineering drawing. The conventional method is drafting on paper with pen or pencil. Manual drafting is tedious, however, and requires a tremendous amount of patience and time. Computer-aided drafting (part of the CAD function) can improve drafting efficiency. The major objective of computer-aided drafting systems is to assist the draftsperson with tedious drawing and redrawing. Partially completed or completed drawings can be stored in a computer and retrieved when needed. Built-in symbols, templates, and tolerance tools greatly simplify the drafting task.

Although drafting is done in two dimensions, most CAD systems store drawings in a 3-D representation. Points (vertices), lines (edges), and curves are represented in (X, Y, Z) space. When a drawing is requested, a series of transformations is performed on the data, and the drawing is presented either in 2-D or 3-D perspective or in sectional views. The resultant representation can then be drawn physically by a printer or simply displayed on the computer monitor. These types of internal representations can be used not only for design drafting, but also for engineering analysis.

In the 1980s, the term "geometric modeling" became common in CAD. Geometric modeling is a technique for providing computer-compatible descriptions of the geometry of a part. Conventional CAD systems employ geometric models that use surface-oriented, 3-D representations of the part. Advances in CAD have provided bounded-shape models for such representations. In these models, individual surfaces are structured together to define the complete shell (boundary of the shape) of the part. Operations then can be applied to manipulate the shape. Chapter 4 introduces basic CAD concepts and systems; more advanced concepts on geometric modeling are discussed in Chapter 5.

2.2 DESIGN DRAFTING

Engineering drawing is an abstract universal language used to represent a designer's idea to others. It is the most accepted medium of communication in all phases of industrial and engineering work. Before today's multiview drawing standard was adopted, perspective drawings[2] were normally employed. Using perspective sketches (Figure 2.2), Leonardo da Vinci, the great master of art during the Renaissance, designed several machines and mechanical components (which still amaze contemporary designers).

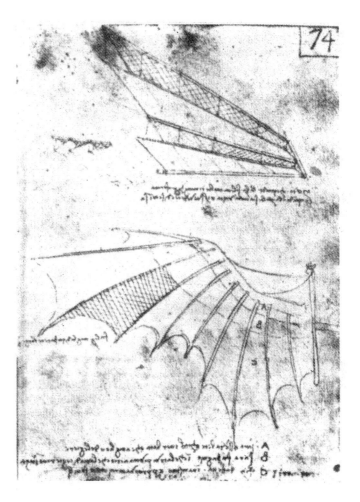

Figure 2.2   Idea sketch prepared by Leonardo da Vinci (1452–1519) (Courtesy of Biblioteque de l'Institut de France—Paris).

---

[2]A perspective drawing is a realistic 2-D image of a 3-D scene. The study of perspective projection goes back to the 15th-century artists Leon Battista Alberti, Albrecht Dürer, and, of course, Leonardo da Vinci, all of whom laid the aesthetic and empirical foundations for its subsequent analytical development. Orthographic projection, used in traditional engineering drawing, does not produce a convincing realistic image of an object. Perspective projection uses projection lines from the eye to the object and onto a projection plane. All projection lines come out from the same source. Orthographic projection uses parallel projection lines to project the object to the projection plane, under the assumption that the distance between the eye and the object is infinite.

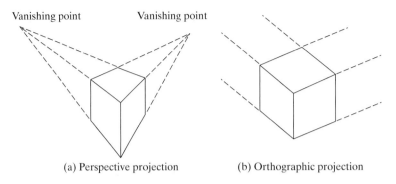

(a) Perspective projection (b) Orthographic projection

Figure 2.3 Perspective projection and orthographic projection.

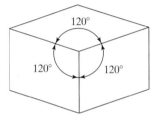

Figure 2.4 Isometric project axes.

Today, pictorial drawings are still employed to supplement other design representations. The basic engineering drawing that uses orthographic projection provides a complete and unambiguous representation of a part or product. Figure 2.3 illustrates the difference between perspective and orthographic projections. The perspective projection drawing in the figure uses two-point perspective. Drawings can be done with one-point, two-point, or three-point perspective. In the first two cases, the axis that is not drawn by perspective projection is drawn by orthographic projection. Most engineering drawings use isometric projection (Figure 2.4). Isometric projection uses orthographic projection to each projection plane. The projection planes are oriented in such a way that each of the axes is 120° apart. The reduction in length on each axes is proportionally the same.

2.2.1 Multiview Drawing

In today's modern manufacturing industry, several types of drawings are acceptable. The standard[3] is still the multiview drawing (Figure 2.5). A multiview drawing usually contains two or three views (front, top, and side). Each view is an orthographic projection

---

[3]The drafting standard in the United States is defined in ANSI Y14. The international standards ISO 128, 129, 3098, etc., also defines a standard for drafting.

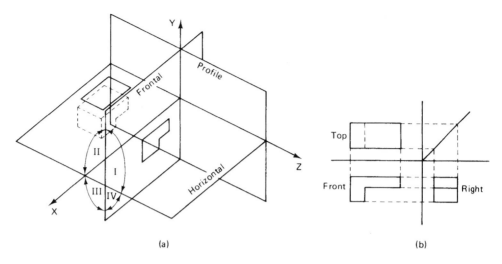

(a)   (b)

Figure 2.5   Multiview drawing of a bracket.

of a plane. In the United States and Canada, the third-angle projection is the system used. (See Figure 2.6.) In the figure, the four quadrants of the $Y–Z$ plane (called the I, II, III, and IV angles) are illustrated. For the third-angle projection, we always place the object in the third quadrant and project the object in three planes. This is done by projecting the object onto the frontal, horizontal, and profile planes. The projection on the frontal ($X–Y$) plane is fixed, and the image is called the *front view*. With the projected image, the horizontal ($X–Y$) plane is rotated 90° clockwise on the X-axis; the result is a *top view* of the object. The profile ($Y–Z$) plane is rotated 90° clockwise about the Y-axis to obtain a right-hand side view. The visible lines are drawn solid, and hidden lines (those edges which are either inside the object or on the side of the object opposite to the direction of projection) are drawn dashed.

With some training, one can reconstruct a 3-D object from the three views on the drawing. Figure 2.7 illustrates partial steps followed in reconstructing a 3-D geometry from the three-view drawing shown in Figure 2.6. Lines d, e, f, g, h, and i are used to reconstruct the slot on the upper right-hand side of the part. Initially, top (horizontal projection) and front (frontal projection) views are used to build the geometry, which is then refined as a side (profile projection) view is added.

2.2.2 Partial View

When a symmetrical object is drafted, two views are sufficient to represent it. (Hence, the third view is typically omitted.) A partial view can be used to substitute for one of the two views (Figure 2.8). Sectional and auxiliary views are also commonly used to present part detail. Sectional views, like that depicted Figure 2.8(a), are extremely useful in displaying the detailed design of a complicated internal configuration. If the section is symmetrical around a centerline, only one-half needs to be shown. The other half is typically shown only in outline. Casting designers often employ sectional views to "explode" details. When a major surface is inclined to all three projection planes, a

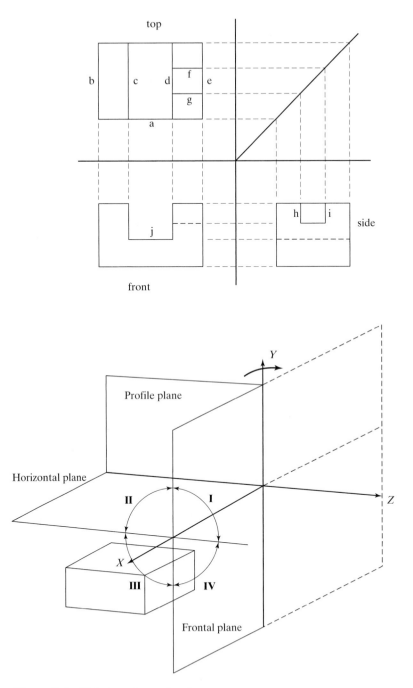

Figure 2.6   Third-angle projection.

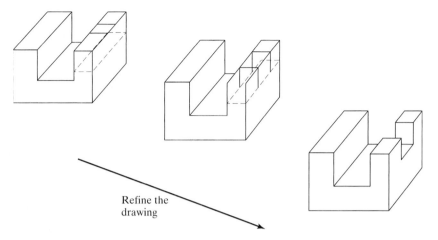

Figure 2.7   Reconstructing a 3-D geometry from the three-view drawing in Figure 2.6.

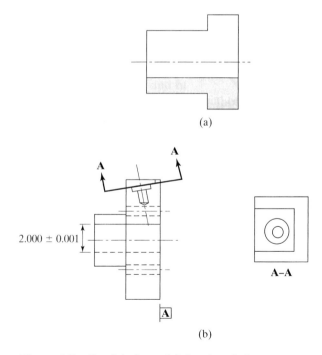

Figure 2.8   Partial views. (a) Sectional view; (b) auxiliary view in the A–A direction.

translation of the picture is seen, e.g., a circle becomes an ellipse. In that case, an auxiliary plane that is parallel to the major surface, like that in Figure 2.8(b), can be used to display an undistorted view.

2.2.3 Dimensioning and Tolerancing

A drawing is expected to convey a complete description of every detail of a part. In this regard, the dimensions are as important as the geometric information. In manufacturing, a drawing without dimensions is worth only as much as the paper on which it is drawn. Dimensions convey the required size, whereas tolerances convey the required precision.

According to American National Standards Institute (ANSI) standards, the following are the basic rules that should be observed in dimensioning any drawing:

1. Show enough dimensions so that the intended sizes and shapes can be determined without calculating or assuming any distances.
2. State each dimension clearly, so that it can be interpreted in only one way.
3. Show the dimensions between points, lines, or surfaces that have a necessary and specific relation to each other or that control the location of other components or mating parts.
4. Select and arrange dimensions to avoid accumulations of tolerances that may permit various interpretations and cause unsatisfactory mating of parts and failure in use.
5. Show each dimension only once.
6. Wherever possible, dimension each feature in the view in which it appears in profile and in which its true shape appears.
7. Wherever possible, specify dimensions so that those examining the drawing can make use of readily available materials, parts, tools, and gauges. Savings are often possible when drawings specify (a) commonly used materials in stock sizes, (b) parts generally recognized as commercially standard, (c) size that can be produced with standard tools and inspected with standard gauges, and (d) tolerances from accepted published standards.

Most designers are well aware of the impact of the mechanisms that they choose in creating a product. Frequent scrutiny is made of the number of parts required to achieve a function. Designers are also constantly concerned with the performance of the design. Unfortunately, tolerancing is often an afterthought of the process, even though the dimensioning and tolerancing of a part frequently presupposes information critical to the manufacture of the part. This information can affect the choice of process(es) to be used, tooling to be used, fixtures and fixture location(s), and machines required to produce a part.

2.2.4 Dimensioning

A basic dimension consists of two extension lines, a dimension line, and the dimension value (Figure 2.9). A tolerance value (see next section) often follows the dimension value. In dimensioning a circle, it is good practice to use either "dia." (added to the end

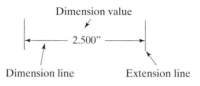

Figure 2.9   Specification of a dimension.

of the dimension value) or the symbol "Ø" (added in front of the dimension value). For the specification of a radius, an "R" is added in front of the dimension value. The basic rules of dimensioning can be summarized as follows:

1. Dimensions should be unambiguous and clearly and uniquely interpretable.
2. Dimensions should be complete, with none missing.
3. There should be no redundancy; each dimension should be shown only once.

The first two conditions are critical. If they are not satisfied, no correct parts can be produced on the basis of the drawing. The third condition is not as critical. If it is not followed, a cluttered drawing will result. As mentioned in the next section, redundancy introduces ambiguity in tolerances. Figure 2.10 shows a two-dimensional drawing with adequate dimensions. By contrast, the drawing in Figure 2.11 has missing height dimensions and a redundant width dimension.

2.2.5 Conventional Tolerancing

There are a few reasons that tolerances are added to the dimensions in a drawing. The most obvious one is to control the products that are produced. Since no manufacturing process is perfect, nominal dimensions (the dimension value shown in the

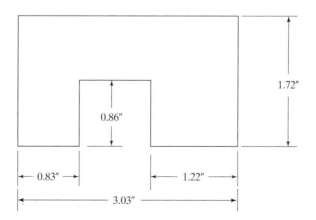

Figure 2.10   Adequate dimensions.

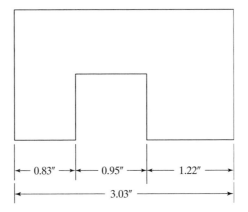

Figure 2.11   Redundant dimension on
X-axis and incomplete dimensions on
Y-axis.

drawing) cannot be achieved exactly. Without tolerances, one loses control over how accurate to make a feature, and, as a consequence, functional or assembly failure may result. The higher the quality desired in a product, the smaller the tolerance value must be. Tighter tolerances are translated into more careful production procedures and more rigorous inspection. There are two types of conventional tolerances: bilateral tolerance and unilateral tolerance (Figure 2.12). Unilateral tolerances, such as $1.00^{+0.00}_{-0.05}$, specify dimensional variation from the basic size (e.g., a decrease) in one direction; for example,

$$1.00^{+0.00}_{-0.05} = 0.95 \sim 1.00$$

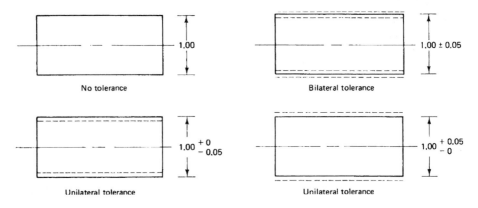

Figure 2.12   Bilateral and unilateral tolerances; dashed lines show the tolerance limits.

signifies that 0.95 is the *lower specification limit* (LSL) and 1.00 is the *upper specification limit* (USL).

The type of tolerance used for a feature depends on the application. For example, in designing a hole and a mating shaft, it might be easier to use unilateral tolerance. The dimension for both diameters could be the same. The hole might have a tolerance specification of $^{+0.01}_{-0.00}$ and the shaft $^{+0.00}_{-0.01}$. That way, the mating will always be a clearance fit. (The hole size is always bigger than the shaft size). The basic location from which most dimension lines originate is the reference location (also called the datum). For machining, the reference location provides the basis from which all other measurements are taken. By stating tolerances from a standard reference location, cumulative errors can be eliminated.

Since tolerances are attached to the dimension, the basic rules for specifying dimensions apply to tolerances. For some drawings, only critical dimensions have their tolerances specified; the tolerance for unspecified dimensions defaults to a larger value. Although it is common knowledge that a tighter tolerance (smaller value) improves the quality of the product and increases the manufacturing cost, the direct relationship between tolerance, on the one hand, and quality and cost, on the other, is not readily available. Chapter 6 discusses the manufacturing tolerance capability of several machining processes. The tolerance specified for a feature can be used as a constraint on the tolerance capability of a manufacturing process that will be used to make the feature. This is indeed part of the process planning function. It is only through process planning that manufacturing cost can be calculated.

While it is difficult to tell how a tolerance value will affect the performance of a part, it is easier to establish the influence of tolerance on assembly. Incorrect tolerances will cause assembly failure. For example, in the hole-and-shaft example mentioned before, suppose we change the tolerance for both the hole and the shaft to $^{+0.01}_{-0.01}$. Then the range of the hole diameter produced is $d \pm 0.01$, and so is the range of the shaft diameter produced. Half of the shafts produced have diameters greater than the nominal diameter d. This example is rather simple, since two dimensions (hole and shaft diameters) are compared directly. In many applications, the dimensional variation is the result of several dimensions and tolerances. In that case, tolerances are stacked together to find the total dimensional variations. The process is called *tolerance stacking*.

In Figure 2.13, the unspecified dimension and tolerance are inferred from the other three dimensions:

$$DIM = 0.80 + 1.00 + 1.20 = 3.00$$

$$Tol = \pm(0.01 + 0.01 + 0.01) = \pm0.03$$

Both the dimension and the tolerance values are said to stack up. Now let us look at another example (Figure 2.14). In this drawing, the total-part width is specified and the notch width is missing. The dimension for the notch width is easily calculated:

$$DIM = 3.00 - 0.80 - 1.20 = 1.00$$

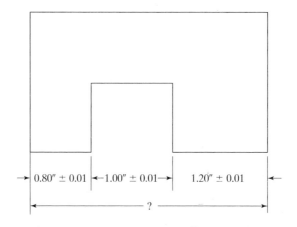

Figure 2.13   Tolerance-stacking example 1.

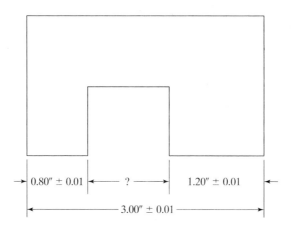

Figure 2.14   Tolerance-stacking example 2.

What about the tolerance? Should it be calculated as

$$\text{TOL} = \pm(0.01 - 0.01 - 0.01) = \pm(-0.01)?$$

Obviously, this is not correct. To find the tolerance, we need to find the maximum and the minimum width. The maximum notch width is the maximum total width minus the sum of the minimum widths of the other two dimensions. The same is true for the minimum notch width. Thus, we have

$$\text{Notch\_Width}_{\text{Max}} = 3.01 - 0.79 - 1.19 = 1.03$$

$$\text{Notch\_Width}_{\text{Min}} = 2.99 - 0.81 - 1.21 = 0.97$$

Therefore, the inferred dimension for the notch width is 0.97 ~ 1.03. Hence, the nominal dimension is (0.97 + 1.03)/2 = 1.00 and the tolerance is ±0.03. The latter is exactly the same as the inferred tolerance for the total width in Figure 2.13. One may conclude that tolerance is additive. That is to say, the inferred tolerance can be as large as the sum of all the other tolerances, or, put another way, the maximum total variation is the sum of the separate variations of the components.

2.2.6 Surface Finish

Most mechanical parts contain both working surfaces and nonworking surfaces. Working surfaces are surfaces for items such as bearings, pistons, and gear teeth, for which optimum performance may require control of the surface characteristics. Nonworking surfaces, such as the exterior walls of an engine block, crankcase, or differential housings, seldom require surface control. For surfaces that do require surface control, control surface symbols (to be discussed shortly) can be used. Although many designers frequently ignore or arbitrarily specify surface-finish requirements, these specifications can dictate secondary or tertiary machining requirements and significantly increase the cost of the product.

Figure 2.15 shows surface characteristics and Figure 2.16 shows the surface roughness symbol. In a number of design instances, the lay direction, waviness width, waviness height, and roughness width can be critical, especially for some parts-mating applications. For those applications, each of these attributes can be specified, and the interpretation of the specifications is straightforward. *Waviness* refers to surface variation on a relatively large scale, and *roughness* refers to the same variation on a smaller scale. Roughness is like the ripple on a wave. The *roughness width* is the width between successive peaks and valleys of roughness; it corresponds to the channel width that a production tool imparts on the workpiece surface. The *roughness width cutoff* is the largest spacing of irregularities, including the average roughness height, and corresponds to the minimum width of surface that yields the proper roughness value. The *waviness width* corresponds to the minimum length of workpiece used to evaluate surface waviness. The *lay* (Figure 2.17) is the orientation of the surface pattern and is determined by the manufacturing processes used. For example, =, ⊥ , and × are linear cutting marks that may be produced by planning, shaping, and grinding. C and R are circular cutting marks that may be produced by milling. *Flaws* are defects or irregularities that occur more or less at random over the surface. These defects can be anomalies such as cracks, blowholes, ridges, scratches, etc.

The *roughness height* is the roughness value as normally related to the surface finish. Recommended roughness heights are given in Table 2.1. An example of the use of control surface symbols is shown in Figure 2.18. Two methods are used to calculate the roughness height: root mean square (RMS) and arithmetic average (AA). AA surface roughness, more common today, is the average amount of irregularity above or below an assumed centerline. It is expressed in microinches (μ inch 0.000001 in.) or, in the metric system, in micrometers (μm = 0.000001 m). The variable R_a is used to denote the AA roughness value. Using Figure 2.19, one may define AA surface roughness as

$$R_a = \frac{\sum\limits_{i=1}^{n} y_i}{n} \tag{2.1}$$

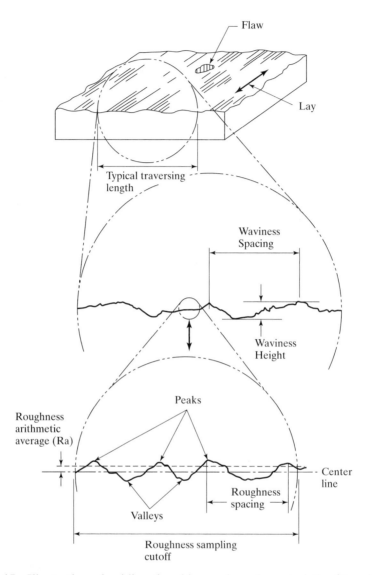

Figure 2.15 Illustration of unidirectional lay surface characteristics (Reprinted from ASME B46.1-1978, by permission of the American Society of Mechanical Engineers. All rights reserved).

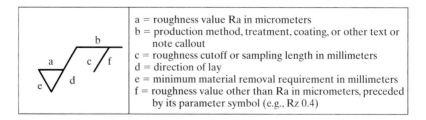

a = roughness value Ra in micrometers
b = production method, treatment, coating, or other text or note callout
c = roughness cutoff or sampling length in millimeters
d = direction of lay
e = minimum material removal requirement in millimeters
f = roughness value other than Ra in micrometers, preceded by its parameter symbol (e.g., Rz 0.4)

Figure 2.16 Surface roughness symbol.

Lay Symbol	Meaning	Example
=	Lay approximately parallel to the line representing the surface to which the symbol is applied.	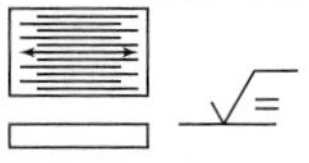
⊥	Lay approximately perpendicular to the line representing the surface to which the symbol is applied.	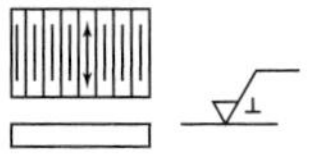
×	Lay angular in both directions to the line representing the surface to which the symbol is applied.	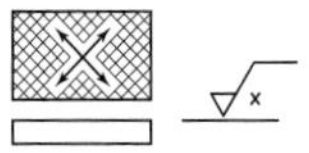
M	Multidirectional lay.	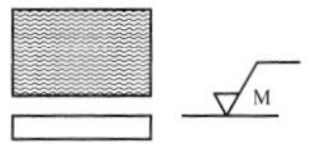
C	Lay approximately circular relative to the center of the surface to which the symbol is applied.	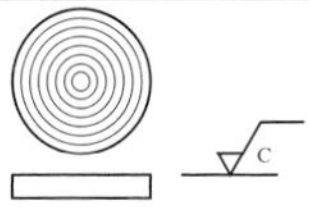
R	Lay approximately radial relative to the center of the surface to which the symbol is applied.	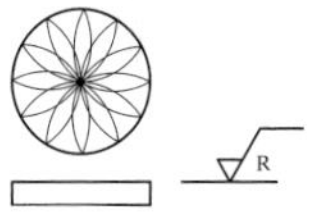
P	Lay particulate, nondirectional, or protuberant.	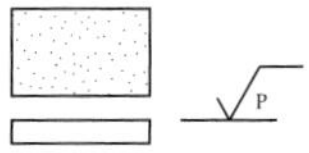

Figure 2.17   Lay symbols and examples.

RMS is calculated as follows:

$$\text{rms} = \sqrt{\frac{\sum_{i=1}^{n} y_i^2}{n}} \tag{2.2}$$

Conventional methods of dimensioning provide information concerning only size and surface condition. Thus, a component can be produced without a guarantee of interchangeability. For example, in Figure 2.20, both components (b) and (c) satisfy the dimension specified in (a); that is, the diameter of components (b) and (c) is 0.501 in. over the entire length of the component. Hence, although both (b) and (c) meet specifications, obviously, neither (b) nor (c) is desirable.

2.3 TOLERANCE GRAPH ANALYSIS

Tolerances are additive. One way to analyze the tolerance stacking problem is to use a *tolerance graph*, which can be defined by a set of nodes and arcs. Each node represents a datum (a reference surface, a line, a centerline, etc.). Each arc contains both the

TABLE 2.1 Recommended Height Values

Roughness Value (μin.)	Type of Surface	Purpose
1000	Extremely rough	Used for clearance surfaces only where good appearance is not required
500	Rough	Used where vibration, fatigue, and stress concentration are not critical and close tolerances are not required.
250	Medium	Most popular for general use where stress requirements and appearance are essential
125	Average smooth	Suitable for mating surfaces of parts held together by bolts and rivets with no motion between them
63	Better-than-average finish	For close fits or stressed parts except rotating shafts, axles, and parts subject to extreme vibration
32	Fine finish	Used where stress concentration is high and for such applications as bearings
16	Very fine finish	Used where smoothness is of primary importance, such as high-speed shaft bearings, heavily loaded bearings, and extreme tension members
8	Extremely fine finish produced by cylindrical grinding, honing, lapping, or buffing	Use for such parts as surfaces of cylinder
4	Superfine finish produced by honing, lapping, buffing, or polishing	Used on areas where packings and rings must slide across the surface where lubrication is not dependable

dimension and the tolerance. Arcs and nodes are taken directly from the drawing. The graph represents one dimension at a time. For example, a graph can be built for dimensions and tolerances in the X-coordinate direction, another can be built for the Y-coordinate direction, and still another can be built for the Z-coordinate direction. Figure 2.21 is an example of a tolerance graph. Each of A, B, C, D, and E is a datum. In the drawing, only the dimension and tolerance from datum D to datum E is not specified.

The graph is defined as follows:

$G(N,d,t)$

N: a set of reference lines, sequenced nodes (datums)

d: a set of dimensions, arcs

t: a set of tolerances, arcs

d_{ij}: dimension between references i and j, from left to right.

t_{ij}: tolerance between references i and j, from left to right.

Reference i is in front of reference j in the sequence.

To infer a missing dimension in the graph, a chain is formed. For dimensions whenever the direction of the index in variable d_{ij} is changed, the sign in front of the

 ✓	*Basic surface texture symbol.* Surface may be produced by any method, except when the bar or circle (Symbol b or d) is specified.
 ✓ (with bar)	*Material removal by machining is required.* The horizontal bar indicates that material removal by machining is required to produce the surface and material must be provided for that purpose.
 X✓	*Material removal allowance.* Value of X in millimeters defines the minimum material removal requirement.
 ✓ (with circle)	*Material removal prohibited.* The circle in the vee indicates that the surface must be produced by processes such as casting, forging, hot finishing, cold finishing, die casting, powder metallurgy and injection molding without subsequent removal of material.
 ✓—	*Surface texture symbol.* To be used when any surface texture values, production method, treatment, coating, or other text is specified above the horizontal line or to the right of the symbol. The surface may be produced by any method, except when the bar or circle (symbol b or d) is specified or when the method is specified above the horizontal line.
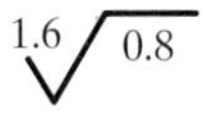 1.6 ✓ 0.8	The roughness average rating is placed at the left of the long leg, and the roughness cutoff rating or sampling length is placed at the right. The specification of only one rating for roughness average shall indicate the maximum value, but any lesser value shall be acceptable. Specify the roughness average in micrometers.
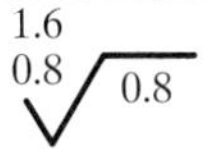 1.6 0.8 ✓ 0.8	The specification of maximum and minimum roughness average values indicates a permissible range of roughness. Specify in micrometers.
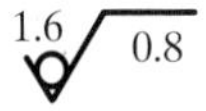 1.6 ✓ 0.8	Removal of material prohibited.
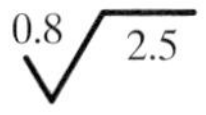 0.8 ✓ 2.5	The roughness sampling length or cutoff rating is placed below the horizontal extension and is mandatory in all cases when values are applied to the symbol. Specify in millimeters.
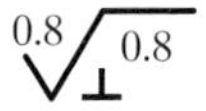 0.8 ✓⊥ 0.8	The lay designation is indicated by the lay symbol placed at the right of the long leg.

Figure 2.18   Examples of the surface texture symbol application.

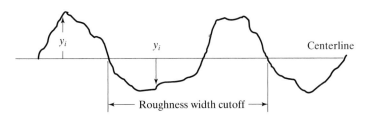

Figure 2.19   Roughness profile.

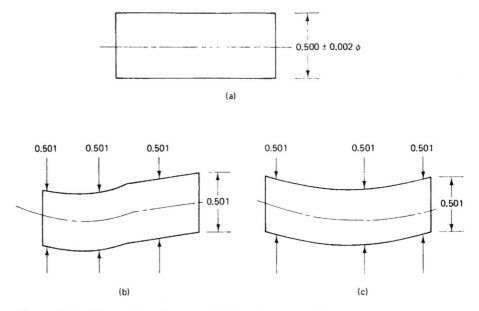

Figure 2.20   Illustration of some additional part conditions.

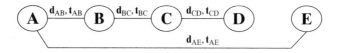

Figure 2.21   A tolerance graph.

dimension value is changed. For tolerances, the signs never change (in accordance with the additive property). For example, to find d_{DE}, we write

$$d_{DE} = d_{DC} + d_{CB} + d_{BA} + d_{AE}$$
$$= -d_{CD} - d_{BC} - d_{AB} + d_{AE} \tag{2.3}$$

Notice that the variables given in the graph are d_{CD}, d_{BC}, d_{AB}, and d_{AE}. For tolerance, we have

$$t_{DE} = t_{DC} + t_{CB} + t_{BA} + t_{AE}$$
$$= t_{CD} + t_{BC} + t_{AB} + t_{AE} \tag{2.4}$$

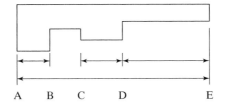

Figure 2.22   Example.

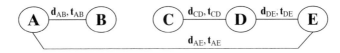

Figure 2.23   Solution.

Example

Convert the drawing shown in Figure 2.22 into a tolerance graph.

Solution

Five datums are identified in the figure, each represented by a node on the graph. Specifications between A–B, C–D, D–E, and A–E can be found in the drawing. Therefore, arcs between those nodes are added. (See Figure 2.23.)

The graph in Figure 2.23 can be used for several purposes. First, one can find the implied dimension and tolerance between any two datums. These are shown in the figure. The graph can also be used to check whether there is any over- or underspecification of dimensions and tolerances. Overspecification of dimensions, such as adding an arc between B and C, is not a major problem as long as the numbers add up. Overspecification of dimensions does, however, clutter the drawing. Overspecification of tolerances, by contrast, is not benign, as we shall see shortly. Underspecification of dimensions makes the drawing incomplete. Production cannot begin unless all dimensions either are given or can be inferred. Underspecification of tolerances is not as critical, since it is usually assumed that those dimensions whose tolerances are not specified are not critical dimensions. In general, a large tolerance value is used as default for those missing tolerances.

2.3.1 Overspecification

Overspecification exists if one or more cycles can be detected in the graph. For example, in Figure 2.24, all three segments of the dimension and tolerance are specified. No dimension needs to be inferred. The graph is shown in Figure 2.25.

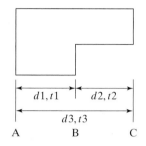

Figure 2.24   Overspecification of dimension and tolerance.

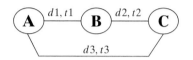

Figure 2.25   Tolerance graph for example in Figure 2.24.

From the graph, one can find dimensions between any pair of datums. The situation is totally different, however, where tolerances are concerned. For example, the tolerance between A and C is specified as $t3$. It can also be inferred by the following reasoning:

$$t_{AC} = t_{AB} + t_{BC}$$
$$= t1 + t2 \tag{2.5}$$

Just for argument's sake, let $t1 = t2 = 0.01$ and $t3 = 0.02$. Then the implied t_{AC} is the same as the specified t_{AC}. Now let us try to find the implied t_{AB}:

$$t_{AB} = t_{AC} + t_{BC}$$
$$= t3 + t2$$
$$= 0.02 + 0.01$$
$$= 0.03$$

The specified t_{AB} is 0.01. We see, then, that there is an inconsistency. Therefore, no redundant tolerance specification is allowed. The overspecified tolerance is shown as a cycle (a closed loop in the graph) in Figure 2.25.

2.3.2 Underspecification

Underspecification exists when one or more nodes are disconnected from the graph. The drawing in Figure 2.26 is underspecified. There are two missing specifications: between B and C and between D and E. One cannot infer both missing specifications at the same time. Such a product cannot be built, since d_{BC} and d_{DE} could be any values. The graph for the drawing is shown in Figure 2.27. Nodes C and D are disconnected from the rest of the graph. No chains can be formed between C–D and the other nodes, A, B, and E. In searching a tolerance graph, if a disconnected node is found, the drawing is underspecified.

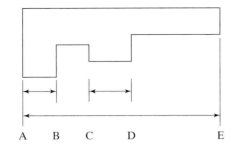

Figure 2.26   Underspecification.

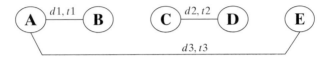

Figure 2.27   Tolerance graph for drawing in Figure 2.26.

If a drawing is neither over- nor underspecified, then it is said to be *properly specified*. In the figures shown, the under- and overspecification problems could be visually identified. In a more complex drawing, it is not always easy to spot this kind of problem. The tolerance graph approach allows validity checking (i.e., checking for proper specification) to be done automatically. Existing graph-theoretical algorithms can be used to identify cycles and disjointed nodes.

2.4 RELATIONSHIP BETWEEN PRODUCT AND PROCESS TOLERANCES

A product design is not complete until it is proven to be manufacturable. To evaluate the manufacturability of a design, knowledge of manufacturing processes must be applied. A good design engineer must have sufficient manufacturing knowledge so that the design geometry, dimensioning, and tolerancing can be carried out. In practice, manufacturing engineers often are assigned to the design team. Using the concept of design for manufacturing, or concurrent engineering, manufacturing engineers incorporate manufacturing capabilities into the design stage. When a design is deemed not manufacturable or difficult or expensive to manufacture, it should be modified. As long as functional constraints are not violated, alternatives can be developed and compared. In Chapter 6, we discuss process engineering, which deals with the basic capabilities of manufacturing processes. In the current section, we focus on the analysis of tolerance capabilities. We will assume that a manufacturing process plan (step-by-step instructions for how to produce the product) is available for the design. The question is whether the process plan can produce the part within the specified tolerances. If the answer is no, the process plan needs to be modified. If it is not possible to modify the process plan, the design tolerance may have to be changed. Only after the manufacturability of a design is verified can the design be declared complete.

Tolerance charting is a method used in industry for this purpose. A tolerance chart stacks up manufacturing process tolerances (process errors) for comparison with the design tolerance. Manufacturing process tolerance is normally prescribed at three- or six-sigma[4] error (depending on the quality program adopted) for a specific process. Information is generally collected through the inspection and quality control process. Any error greater than the quality limits is rejected through inspection. Only those design tolerances explicitly specified on the drawing are checked. With the process plan as a guide, tolerances on the sequence of operations resulting in the specific design tolerance are added together. This resultant process tolerance (called *tolerance stack-up*) is the expected value of the manufacturing tolerance. The manufacturing tolerance must be less than or equal to the specified tolerance.

There are two manufacturing errors to be considered: process error and setup error. Process error (or tolerance) is the result of machine inaccuracy (machine tolerance, assembly error, transducer error, and control resolution) and operation error (thermal error, tool deflection, etc.). Setup error is the result of setting up the workpiece in the fixture on the machine. Each time the workpiece is set up on the machine, an error is introduced. It is always a good practice to reduce the number of setups. On a machine, locators are used as a reference. The distances from the machine coordinate system to the locators are known. The workpiece reference (often the surface of the datum) is pushed against the locators. All positional control is based on the known locator position. The interface of the positioning part against a locator is normally assumed to be the location of the workpiece reference surface. The error between the true workpiece reference surface position and the assumed locator position is the setup error. No matter how small this misalignment might be, it is never zero. All surfaces on a workpiece machined under the same setup have the same setup error, which cancels when the distance between two surfaces machined in the same setup is measured. Cancellation occurs because the setup error for these surfaces has the same magnitude and direction. (It is either too long or too short.) This fact is important in constructing the tolerance chart.

Tolerance charting is a method for allocating process tolerances and verifying that the process sequence and machine selected can satisfy the design tolerance. The simple part shown in Figure 2.28 is used to illustrate how to construct a tolerance chart. In the drawing, a part is to be turned and both ends faced. Figure 2.29 shows the simplified process plan for the part.

Figure 2.30 shows a simple tolerance chart. On the top is the sketch of the part design. Overlaying the design is the sketch of the stock material (dashed line). A tolerance chart can be used to analyze one direction at a time (that is *X, Y,* or *Z*). Right below the sketch is the blueprint section. Each dimension line is shown in an entry in the blueprint section. The first entry is for the dimension and tolerance between A and B (d_{AB}, t_{AB}). The second entry is for C and D (d_{CD}, t_{CD}), and the third entry is for A

---

[4]Traditionally, the quality level is defined by $C_p \geq 1.33$, where $C_p = \dfrac{\text{tolerance range}}{6\sigma}$, in which σ is the standard deviation of the process. Since the $\pm$ tolerance value equals (tolerance range)/2, tolerance $\geq 3\sigma \times 1.33 = 4\sigma$. In the six-sigma quality program, developed by Motorola, the tolerance range is set at $2 \times 6\sigma$. However, it also allows a 1.5σ shift of the process center (the average) with respect to the nominal dimension. Therefore, the true quality level is equivalent to $(6 - 1.5)\sigma$.

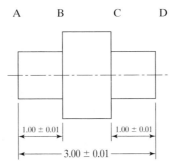

A   B   C   D

1.00 ± 0.01   1.00 ± 0.01

3.00 ± 0.01

Figure 2.28   Design specification.

Process Plan

1. Chuck on the left side, use A as reference.
   Cut C and D.
2. Turn the workpiece around, use D (a newly
   cut surface) as reference. Cut B and A.

Figure 2.29   Process plan.

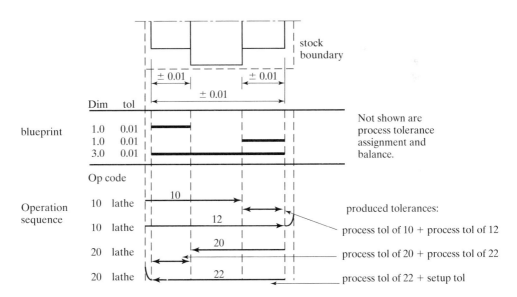

Figure 2.30   Tolerance chart (not shown are process tolerance assignment and balance).

and D (d_{AD}, t_{AD}). Both the dimension and tolerance are recorded, and a thick line connects the extension lines of corresponding surfaces from the drawing. Each thick line shows the dimension and tolerance between two surfaces. The next section shows the operation sequence. On the left-hand side is an operation plan. Operation 10 has two processes: Cut to datum C and cut to datum D. The thick arrow line marked "10" indicates that the first process uses the left-side stock surface as the reference. Process 12 uses the same reference surface. Processes 10 and 12 are said to share the same setup. Processes 20 and 22 use machined surface D as reference. They belong to the second setup. In a complete tolerance chart, two additional sections—the process tolerance assignment section and the process tolerance balance section—are listed below the operation sequence section. In the chart shown in Figure 2.30, this information is shown on the right-hand side.

Since the blueprint specifies t_{AB}, t_{CD}, and t_{AD}, we need to find the manufacturing tolerance for those specifications. CD is the result of processes 10 and 12 (the chain from C to the left-hand side of the stock and then from the left-hand side of the stock to D). The manufacturing tolerance is the sum of the process tolerances for processes 10 and 12. Similarly, we can find a manufacturing tolerance for AB that is the sum of the process tolerances for 20 and 22. For AD, the chain stops at D. Since D is located by a setup locator, the setup error must be added. Therefore, the manufacturing tolerance for AD is the process tolerance for 22, plus the setup error for setup 2. From this example, we can observe that setup 1 is not critical, since no specification is based on the setup surface.

Example

The process tolerance for all the processes discussed in Figure 2.30 is 0.005, and the setup error is 0.001. Can we produce the part from this process plan?

Solution

Let

t_{ij}^{M} represent the manufacturing tolerance for surfaces i and j.

t_{k}^{P} be the process tolerance for process k.

t_{l}^{S} be the setup error for setup l.

Then

$$t_{CD}^{M} = t_{10}^{P} + t_{12}^{P} = 0.005 + 0.005 = 0.01$$

$$t_{AB}^{M} = t_{20}^{P} + t_{22}^{P} = 0.005 + 0.005 = 0.01$$

$$t_{AD}^{M} = t_{22}^{P} + t_{2}^{A} = 0.005 + 0.001 = 0.006$$

All three manufacturing tolerances are less than or equal to the tolerance specified. Therefore, the part can be produced from this process plan.

2.5 INSPECTION AND MEASUREMENT

2.5.1 Inspection Errors

In order to detect errors, measure performance, and check materials and parts with standards, inspection and testing facilities are used. Tensile-strength-testing machines, hardness-testing machines, and other equipment are set up in laboratories that check materials received, made, or treated in the shop. Elaborate gauges and fixtures are designed to inspect complicated parts and assemblies. Complete sets of gauges and tools for measuring critical dimensions should be available to the operator as well as the inspector. Where controls are extensive, it is necessary to have a gauge section that checks, repairs, and adjusts gauges by using master gauges of the highest accuracy. Thus, duplicate and triplicate sets of gauges are sometimes necessary to maintain control.

Errors in measurement depend upon the method of measurement used and the accuracy of the measuring equipment, both of which are subject to errors. Measurement errors are the result of one or more of the following:

- Inherent errors in the measuring instrument
- Errors in the master gauge used to set the instrument
- Errors resulting from variation in temperature and from different coefficients of the linear expansion of the instrument and part being gauged
- Human errors on the part of the inspector

An engineer should not specify a dimension, characteristic, or function of a part or apparatus that cannot be measured. Fortunately, owing to the extensive development of inspection and testing apparatus, it is now possible to measure to a very high degree of accuracy. Instruments for measuring the roughness of surfaces are available, and through the use of oscilloscopes, spectrometers, X-rays, and other types of sensitive measuring equipment, the quality of products can be controlled. Some of these instruments, such as air gauges for measuring close dimensions, are rugged enough to use in production operations; thus, the operator has a means of controlling the quality of the part he or she is making.

Inspection operations, such as detecting surface defects and blowholes, can be a part of the operator's responsibility. Checking gauges can be built into jigs, fixtures, and equipment that can be used by the operator and inspector. Such gauges are especially valuable on medium-activity and complicated parts.

The ability to measure and control dimensions has progressed. A few years ago, it was said, "We can work to 0.001 in. and talk about holding to 0.0001 in." Now it can be said, "We work to 0.0001 in. and talk about holding to 0.00001-in. tolerance."

2.5.2 Quality Control for Surface Finishes

Surface finish is a function of the method used to produce a product (e.g., machining, honing, grinding, superfinishing, lapping, polishing, buffing, plating, and painting). In the control of surface finish, methods of measuring and comparing with samples are used. The discussion that follows will be confined to the control of surfaces that can be measured and designed on drawings. Some of the instruments used for measuring are the *profilometer* and the *Brush surface analyzer*. Both have a tracer or stylus that moves over the surface. The up-and-down movement is recorded on an electric meter or on paper or film that moves at a uniform rate as the measurement is taken.

In the past it has been difficult to duplicate, measure, and designate desired surfaces. The degree of surface roughness must be controlled and maintained, because it affects a number of characteristics:

- The life of the product, by controlling friction, abrasive wear, corrosion, and galling.
- The function of the product, by permitting the smooth surfaces of the parts to slide freely, fit properly, serve as bearings, reduce leakage, and rub against packings.
- The appearance of plated and other decorative surfaces.
- The safety of the part, by preventing stress concentration, fatigue, and notch sensitivity. For these reasons, airplane engine manufacturers are especially conscious of the value of controlling surface finish.
- The heat transfer, because smooth surfaces offer better heat conductivity.

Since it costs more to produce smooth surfaces, they should be specified only when desired. Progress has been made in advancing the methods of controlling surface finishes through the adoption of national standards.

2.5.3 Measurement or Evaluation?

For compliance with specified ratings, surfaces may be evaluated either by comparing their properties with those of specified reference surfaces or observational standards or by direct instrumental measurements. In many applications, these comparisons can be made by sight, feel, or instrument. In making comparisons, care should be exercised to avoid errors due to differences in material, contour, and type of operation represented by the reference surface and the work. In using instruments for comparison, or in direct measurement, care should be exercised to ensure that the specified quality or characteristic of the surface is measured. Unless otherwise specified, roughness measurements are taken across the lay of the surface or in the direction that gives the maximum value of the reading. The physical measurement of the roughness height value shall be the maximum *sustained* reading of a series of readings or the minimum *sustained* reading in case a minimum permissible value is specified also.

The physical measurement of the waviness height shall be the algebraic difference between the maximum and minimum readings of a dial gauge over a distance not exceeding 1 in. if no other definite waviness width is specified. The waviness height can also be determined by means of a straightedge. The recommended values for classifying roughness, in microinches, are as follows:

1	16	125
2	32	250
4	63	500
8		1000

The use of only one number or class to specify the height or width of irregularities shall indicate the maximum value. Any lesser degree or class on the actual surface of the part shall be satisfactory. When two numbers are used on the drawing or specification, they shall specify the maximum and minimum permissible values.

2.6 STATISTICAL QUALITY CONTROL

Statistical quality control is an analytical tool that can be used to evaluate machines, materials, and processes by observing their performance and variations thereof so that comparisons and reductions may be made to control the desired level of quality. Statistical quality control makes possible all of the following:

- Decreased inspection costs
- Reduced rejects, scrap, and rework
- A more uniform product
- Greater quality assurance
- The anticipation, and hence better control, of production trouble
- More efficient use of materials
- Rational setting of tolerances
- Better purchaser–vendor relationships
- Improved and concise reports to top management on the quality picture

In manufacturing, accuracy and precision are often used interchangeably. Unfortunately, such usage is incorrect and produces misleading results. Figure 2.31 illustrates the relationship between accuracy and precision. As can be seen from the figure, *accuracy* generally refers to the centroid, target dimension, or call-out. *Precision* generally refers to the scatter resulting from the process.

Statisticians became interested in the possibility of controlling quality by applying the laws of probability and statistics. For example, they knew that, in firing artillery weapons, the shots fall in a pattern around the target in spite of the greatest care in obtaining uniformity in the powder, projectile, gun, and accuracy of sighting. The observed pattern of shots was true to mathematical laws that had been worked out in other cases of probability. That is, statisticians could predict the number of shots that would hit the target, the number that would fall 5 ft away, 10 ft away, and 20 ft away. In studying a screw machine, they found similar data: The machine capable of producing uniform parts would make a certain quantity that would fall on the dimensions required, so many would be above or below 0.001, so many above or below 0.002, and so on. As the tool wore, the number below size decreased and the number above size increased. By studying this pattern, the observer could make the amount of correction needed to bring the machine back to its best performance (Figures 2.32 and 2.33).

In statistical quality control, observable features of a product or process are measured. A study of these measurements the usually determines the ability of the operator or equipment to produce the product within the desired specifications. The percentage gives the same story as if 100 percent of the products were inspected or tested. The size of sample is determined by the nature of the product and the standard of quality required.

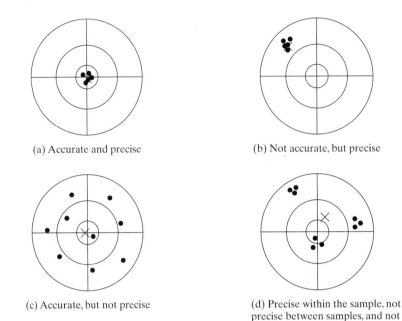

(a) Accurate and precise

(b) Not accurate, but precise

(c) Accurate, but not precise

(d) Precise within the sample, not precise between samples, and not accurate.

Figure 2.31 Accuracy versus precision in process. Dots in targets represent the locations of shots. Crosses represent the location of the average position of all shots.

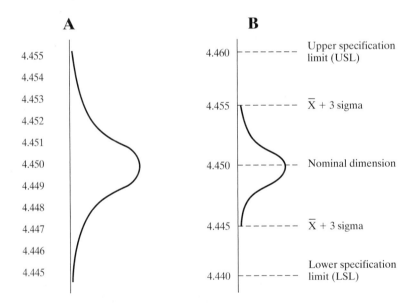

A

4.455
4.454
4.453
4.452
4.451
4.450
4.449
4.448
4.447
4.446
4.445

B

4.460 — — — — — Upper specification limit (USL)

4.455 — — — — — $\overline{X}$ + 3 sigma

4.450 — — — — — Nominal dimension

4.445 — — — — — $\overline{X}$ + 3 sigma

4.440 — — — — — Lower specification limit (LSL)

Figure 2.32 (a) A frequency curve of 50 measurements taken on a screw machine part. (b) Position of frequency curve of the process relative to the specification limits shows good statistical control.

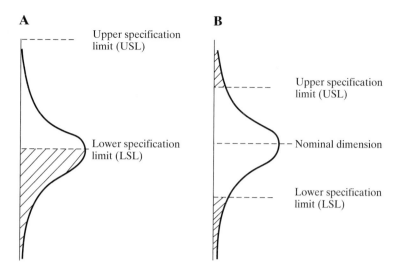

Figure 2.33 (a) Frequency curve of a process that is not centralized. (b) Frequency curve of a process that has too large a spread of variation.

The two statistics that the production–design engineer uses most frequently are the mean and the standard deviation. We define the *mean* of the set of numbers x_i as

$$\bar{x} = \frac{1}{n}\sum_{i=1}^{n} x_i \tag{2.6}$$

The *standard deviation* is a most useful measure of the extent to which an individual item may deviate from the expected mean. To employ the standard deviation, we must adopt confidence limits—that is, upper and lower limits representing the percentage of probability that the product will fall between them. No result can be absolutely assured and nothing is perfect, at least in production or anticipated sales. The standard deviation can be defined as the root-mean-square deviation, in that it is the square root of the mean of the squares of the deviations. The standard deviation of a sample of data is usually symbolized by s, the value of which is given by the equation

$$s = \sqrt{\frac{1}{n-1}\sum_{i=1}^{n}(x_i - \bar{x})^2} \tag{2.7}$$

The analysis of random errors of observation in gathering statistics is based on probability theory. The production–design engineer should be cognizant of the fact that the mean of a sample of data represents only an estimate of the mean of the parent distribution. Similarly, the standard deviation of the sample is an estimate of the standard deviation of the population. For most distributions, the precision of the estimate of both the mean and the standard deviation increases with the size of the sample of data taken. Therefore, the operator can select parts to check at random, plot the results, and determine the performance of his or her machine. Although the tool may wear gradually, as it approaches the limit, it can be adjusted to compensate for wear. If the tool is chipped and is producing defective parts, the sample immediately shows this and the entire lot will be rejected. Then 100% inspection of the lot will be necessary to salvage as many good parts as possible. When statistical control is established for an operation, the average size of the parts is plotted on one chart, and the range of size is plotted on another, for each set of samples. By observing the two charts, the operator can identify trends and know when the machine is exceeding its limits. Figure 2.34 illustrates how such a chart revealed the possibility of reducing limits of variation without increasing costs and resulted in a better product that assisted succeeding operations.

The patternmaker, mold maker, toolmaker, and designer have recently realized the value of making patterns, molds, and dies the same size for multiple patterns and cavities. For example, an electric-motor bracket hub varied to such an extent that a special machining chuck was required to hold the part. A study of deviations revealed that the patterns differed from each other. When they were made alike, the normal casting deviation permitted the use of a standard chuck and the reduction of material allowed for machining (Figure 2.35).

The production–design engineer will benefit by studying current articles and textbooks on statistical control. The subject has been reduced to "everyday" understandable terms that can be applied by quality-control personnel, shop supervisors, and engineers in any organization.

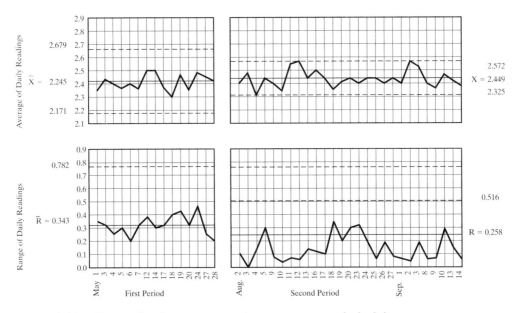

Figure 2.34   Chart reflecting average and range over a period of time.

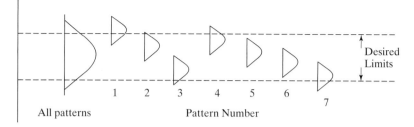

Figure 2.35   Frequency distribution of outside hub diameter.

2.7 MANUFACTURING RELIABILITY

Concepts of reliability need to be incorporated into the functional design of a product and, indeed, into all manufacturing operations in view of the long life requirements for many of today's products.

When we design for reliability, the product must withstand actual service conditions for a given period of time. We can prescribe the period, but we can only estimate the service conditions. Generally speaking, there are four types of failure a product is subject to while it is in service:

1. *Infant mortality.* This is the early failure of a product due to faulty material or manufacturing errors such as sharp corners, scratches, heavy undercuts, etc., that cause points of stress concentration. Sound quality-control procedures should minimize infant mortality.

2. *Chance failures.* These may be referred to as random or constant hazard failures—failures that take place during the product's life due to chance. Specific causes could be inadequate lubrication because of a clogged wick, leaking due to the development of excessive porosity, a burnout brought about by a surge of power, and the like. Chance failures are often mistakenly thought to be caused by wear-outs. (See shortly.)

3. *Abuse or misuse.* This kind of failure is due to uses of the design that go beyond the intended purpose of the design. Known excessive loads are typical examples.

4. *Wear-outs.* These failures are due to aging, fatigue, corrosion, etc., and usually progress until the product is retired because of inefficiency. Wear-out can usually be postponed by proper preventive maintenance.

Figure 2.36 illustrates a theoretical curve, which is seldom fulfilled in practice, of the relation in time and the magnitude of the failure rate of the preceding four types of failure. Through a well-conceived reliability and quality-control program, a reliability performance similar to that shown in Figure 2.37 can be achieved. Here, it can be seen that both the failure rate and the period of infant mortality have been reduced and the start of wear-out failures has been delayed.

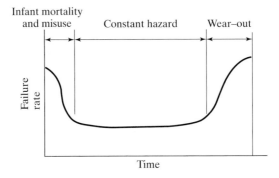

Figure 2.36   Mortality curve based on Robert Lusser's concepts.

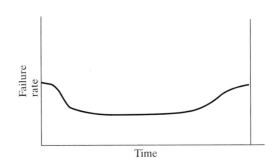

Figure 2.37   Mortality curve for a product designed and built on a well-conceived quality-control and reliability program.

2.8 SUMMARY

In this chapter, we have discussed the basics of engineering drawing. Various methods of representing drawings, along with drawing specifications, were discussed. We have long used some form of icons to represent designs. These icons are precise representations of the ideas we convey in engineering drawings. As we fully immerse ourselves into the era of computer-aided design, we seek to develop methods that can be automatically interpreted as well as compactly represented. We must be exceptionally careful and make sure that we fully understand the implications of our design and specification decisions. The chapter presented both the standards for representing engineering specifications and the interpretations of those specifications.

The cost of quality control and reliability is proportional to the level of reliability and accuracy desired in materials, processes, and functions of the product. The accuracy desired is based upon the judgment of management, which depends upon the advice of the sales, engineering, and manufacturing departments. The level of reliability established usually emanates from the customer. The ultimate decision as to acceptance or rejection rests with the quality-control and reliability department. In general, a company that is willing to maintain quality at considerable expense survives competition with companies that have lower quality standards. Over the long haul, the high-quality, reliable product wins because customers are satisfied. Through good engineering, a high-quality product can be obtained at less cost than the former poor-quality product produced under uncontrolled conditions. By going to the source of trouble and eliminating the cause, by providing the shop operators and members of the quality-control department with adequate equipment, by improving the maintenance of tools and machines, and by inspecting in the critical places, quality control has become an asset to industry.

Quality control and reliability departments are becoming effective in building quality into products with minimum overall costs of production, scrap, rework operation, and service adjustments. Goodwill is maintained by verifying the quality and reliability before the product is shipped. Years of control have brought about uniform materials and parts that are produced in enormous quantities. These materials make possible our mass-production lines that produce reliable products today.

2.9 KEYWORDS

Auxiliary view

Bilateral and unilateral tolerance

Datum

Dimension

Dimensional tolerance

Geometric tolerance

Manufacturing reliability

Multiview orthographic drawing

Nominal dimension

Orthographic projection

Partial view

Perspective projection

Quality control

Surface finish

Surface roughness: root mean square (RMS) and arithmetic average (AA)

Surface texture symbols

Third-angle projection

Tolerance chart

Tolerance graph

2.10 REVIEW QUESTIONS

2.1 What are the five major steps in a design process? Briefly explain each one.

2.2 Discuss how a design idea is represented in a designer's mind—that is, in the form of an equation, line drawing, etc.

2.3 What are the methods used in diameter inspection?

2.4 Why is it that 100 percent inspection seldom ensures that a shipment of parts will be 100 percent to specifications?

2.5 For what reasons is it advisable to control the surface roughness of a part?

2.6 What is meant by the root-mean-square deviation?

2.7 What is the value of 1 μin.?

2.8 Place the following processes in ascending order based on their surface roughness: cold rolled, die cast, plaster cast, hot rolled, and cold drawn.

2.9 What do we mean by statistical quality control?

2.10 Define "standard deviation."

2.11 What is a normal distribution?

2.12 Why is it desirable to plot the "range" of successive samples?

2.13 Why do chance failures plot as an approximate straight line when failure rate is plotted against time?

2.14 Show graphically the effect of a strong preventive maintenance program on the relation between failure rate and time.

2.11 REVIEW PROBLEMS

2.1 Prepare a three-view drawing in third-angle projection of the part shown in Figure 2.38.

2.2 Prepare an orthographic projection drawing of the part shown in Figure 2.38.

2.3 Prepare a two-point perspective projection drawing of the part shown in Figure 2.38.

2.4 Prepare a tolerance graph for the part drawing shown in Figure 2.39. Only dimensions along the X-coordinate are shown and considered in the problem. Is there any cycle or disconnect node?

2.5 Continuing with Problem 2.4, what are the implied distances between D–E, E–F, and C–E?

2.6 Again continuing with Problem 2.4, what are the implied tolerances between D–E, E–F, and C–E?

2.7 The process plan for the part in Figure 2.39 has only one setup that uses surface A as reference. Surfaces C, D, and F and holes B and E are machined under the same setup. The setup error is 0.001. The process error for C, D, and F is 0.002. The location process error for B and E is 0.003. Prepare a tolerance chart to verify whether the specifications can successfully yield a product.

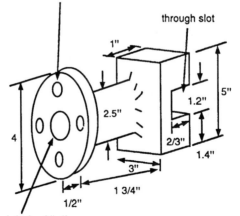

Figure 2.38

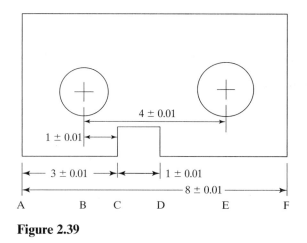

Figure 2.39

2.8 The weights of ceramic compacts were taken at random and recorded as follows:

Specimen Weights, kg				
1.24	1.31	1.29	1.34	1.26
1.32	1.41	1.37	1.28	1.31
1.40	1.34	1.29	1.32	1.30
1.31	1.26	1.28	1.34	1.31
1.25	1.29	1.34	1.31	1.32

Calculate the mean and standard deviation of the samples. Assume that the mean and standard deviation are representative of the population. Find the upper and lower control limits and plot an x control chart. Is the process in statistical control? Explain why or why not.

2.9 Given a process in statistical control with a population mean of 5.000 in. and a population standard deviation of 0.001 in., (a) determine the natural tolerance limits of the process and (b) if the specification limits are 5.001 ± 0.002 in., what percentage of the product is defective, assuming that the process output deviations are normally distributed?

2.10 Given a process with a population mean of 10.00 and a population standard deviation of unity, what can be said about the interval 10 ± 2 if the distribution is normal?

2.11 Determine the surface roughness values of the partial surface trace given in the following table and in Figure 2.40 in terms of the maximum peak-to-valley roughness, the arithmetic average (AA), and the root-mean-square (rms):

Point	Deviations from Mean, μ in.	Point	Deviations from Mean, μ in.
1	4	10	20
2	15	11	10
3	18	12	23
4	8	13	15
5	20	14	3
6	17	15	8
7	15	16	10
8	17	17	20
9	24	18	5

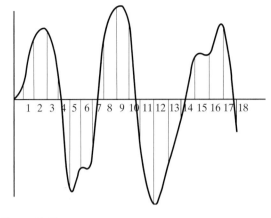

Figure 2.40

2.12 On an interference fit, the basic size of a hole is 3.5000 in. The interference between the shaft and the hole must be at least 0.0015 in. The tolerance on the shaft and the hole is 0.0009 in., using the basic hole system and unilateral tolerances. Divide the tolerances on each item into three groups so that the small shafts mate with the small holes, the medium shafts mate with the medium holes, etc. There will then be as nearly uniform as possible interference between the mating parts. This procedure is called *selective assembly* and is used when more precise metal fits are needed than can be obtained by conventional interchangeable manufacture. The hole diameter is $3.5000 \, {}^{+0.0000}_{-0.0009}$ and the shaft diameter is $3.5015 \, {}^{+0.0009}_{-0.0000}$. Do an experiment by randomly assembling hole and shaft within each of the three groups. What is the average interference for the entire batch? Now try to assemble hole and shaft without dividing the tolerances into groups. What is the average interference now?

Holes Produced:

3.5000	3.4991	3.4999	3.4992	3.4995
3.4992	3.4991	3.4993	3.5000	3.4998
3.4997	3.4992	3.4995	3.4999	3.4994
3.4993	3.4998	3.4991	3.4992	3.4998
3.4997	3.4991	3.5000	3.4991	3.4994

Shafts Produced:

3.5020	3.5015	3.5021	3.5023	3.5024
3.5023	3.5021	3.5024	3.5016	3.5019
3.5019	3.5022	3.5020	3.5022	3.5016
3.5020	3.5018	3.5017	3.5021	3.5020
3.5016	3.5017	3.5019	3.5018	3.5023

2.12 REFERENCES

ANSI, "Surface Texture." ANSI Standard B46.1-1978.

ANSI, for ANSI and ISO documents: http://global.ihs.com/

ASME Y14.36M-1996. Surface Texture Symbols.

Earle, James H. *Graphics for Engineers: With AutoCAD release 14 & 2000*. Upper Saddle River, NJ: Prentice Hall, 2000.

ISO 1302:1994. Technical Drawings: Method of Indicating Surface Texture.

ISO CD 1302. Geometrical Product Specifications (GPS): Indication of Surface Texture.

ISO, for ISO Standard: http://www.iso.ch/iso/en/ISOOnline.frontpage

Suh, N. P. "Qualitative and Quantitative Analysis of Design and Manufacturing Axioms," *CIRP Annals*, 31, (1982) 333–338.

Chapter 3

Geometric Tolerancing

Objective

Chapter 3 presents an overview of geometric tolerancing and inspection using gauges. The fundamentals of design tolerances and their interpretation are presented in detail. The basics of ASME Y14.5 ("Form Geometric Tolerance Specification") are presented. Single and multiple data are introduced.

Outline

3.1 BACKGROUND

In Chapter 2, we learned about engineering design and traditional tolerance specification. As products have gotten increasingly sophisticated and geometrically more complex, the need to specify regions of dimensional acceptability better has become more apparent. Traditional tolerances schemes are limited to translational (linear) accuracies that get appended to a variety of features. In many cases, bilaterial, unilaterial, and limiting tolerance specifications are adequate for specifying both location and feature size. Unfortunately, adding a translation zone or region to all features is not always the

54

best alternative. Often, in order to reduce the acceptable region, the tolerance specification must be decreased to ensure that mating components fit properly. The result can be a higher-than-necessary manufacturing cost. In this chapter, we will introduce geometric tolerances, how they are used, and how they are interpreted.

In 1973, the American National Standards Institute, Inc. (ANSI) introduced a system called "Geometric Dimensioning and Tolerancing (GD&T)" for specifying dimensional control. The system was intended to provide a standard for specifying and interpreting engineering drawings and was referred to as ANSI Y14.5-1973. In 1982, the standard was further enhanced and a new one ANSI Y14.5-1982 was born. In 1994, the standard was still further enhanced to include formal mathematical definitions of geometric dimensioning and tolerancing and became ASME Y14.5-1994.

3.2 GEOMETRIC TOLERANCES: ASME Y14.5

Geometric tolerancing specifies the tolerance of geometric characteristics, among which are the basic geometric characteristics as defined by the ASME Y14.5M 1994 standard:

Straightness	Perpendicularity (squareness)
Flatness	Angularity
Circularity (roundness)	Concentricity
Cylindricity	Runout
Profile	True position
Parallelism	

Symbols that represent these features are shown in Table 3.1. To specify the geometric tolerances, reference features—planes, lines, or surfaces—can be established. The geometric tolerance of a feature (a line, a surface, etc.) is specified in a feature control frame (Figure 3.1) consisting of (1) a tolerance symbol, (2) the tolerance value, (3) a modifier of the tolerance value, and (4) pertinent datum information. The diameter symbol (Figure 3.1) may be placed in front of the tolerance value to denote that the tolerance is applied to the diameter of the hole. The meaning of the modifier will be discussed later. In the fields following the modifier, there could be from zero to several datums (Figures 3.1 and 3.2). For tolerances such as straightness, flatness, roundness, and cylindricity, the tolerance is internal, so no external reference feature is needed. In this case, no datum is used in the feature control frame.

A datum is an idealized plane, surface, point or set of points, line, axis, or other source of information on an object. That is, datums are theoretical features that are assumed to be exact. From them, dimensions similar to the reference-location dimensions in the conventional drawing system can be established. In practice, imperfect surfaces on a part are generally treated as datums. Datums are used for geometric dimensioning and frequently imply information on the locations of fixturing surfaces. The correct (or in many cases the incorrect) use of datums can significantly affect the manufacturing cost of a part. Figure 3.3 illustrates the use of datums and the corresponding fixturing surfaces. The 3–2–1 principle is a way to guarantee that a workpiece is correctly located in three-dimensional space. Normally, three locators (points or, in

TABLE 3.1 Geometric tolerancing symbols [ASME Y14.5M-1994 GD&T (ISO 1101, geometric tolerancing; ISO 5458 positional tolerancing; ISO 5459 datums; and others)]

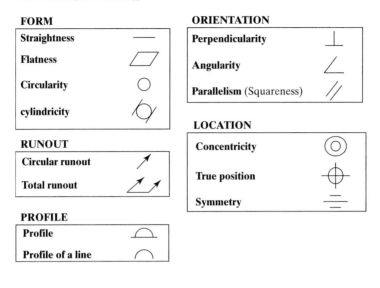

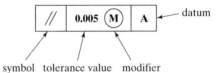

symbol   tolerance value   modifier

Figure 3.1   Feature control frame.

Figure 3.2   Datum symbol.

practice, locating pins) are used to locate the primary datum surface, two (again, points or, in practice, locating pins) for the secondary surface, and one for the tertiary surface. After the workpiece is correctly located, clamps are used on the opposing sides of the locators to immobilize it.

3.2.1 Symbol Modifiers

Symbol modifiers are used to clarify implied tolerances (Figure 3.4). There are three modifiers that are applied directly to the tolerance value: "maximum material condition" (MMC), "regardless of feature size" (RFS), and "least material condition" (LMC). RFS is the default; thus if no modifier symbol is present RFS is the intended

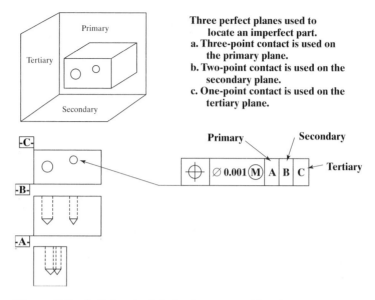

Three perfect planes used to
locate an imperfect part.
a. Three-point contact is used on
the primary plane.
b. Two-point contact is used on the
secondary plane.
c. One-point contact is used on the
tertiary plane.

Figure 3.3 3–2–1 principle in datum specification.

(M) **Maximum material condition**

Regardless of feature size

(L) **Least material condition**

(P) **Projected tolerance zone** ◀──── *Maintain critical wall thickness or*
critical location of features.

(∅) **Diametrical tolerance zone**

(T) **Tangent plane**

(F) **Free state**

MMC assembly
RFS (implied unless specified)
LMC less frequently used

Tolerances	Applicable modifiers
⊕ ⊥ // ∠	**MMC, RFS, LMC**
─	**MMC, RFS**
⌁ ⌁⌁ ◎	**RFS**

Figure 3.4 Modifiers and applicability.

call-out. MMC can be used to constrain both the feature location tolerance as well as the feature size (dimensional) tolerance. Used to maintain clearance and fit, MMC can be defined as the condition of a part feature in which the maximum amount of material is contained. For example, maximum shaft size and minimum hole size correspond to MMC. This concept is illustrated in Figure 3.5. LMC specifies the opposite of the maximum material condition. It is used for maintaining interference fit and, in

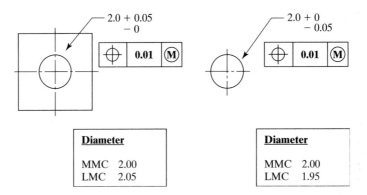

Figure 3.5   Maximum material diameter and least material diameter.

special cases, to restrict the minimum material in order to eliminate vibration in rotating components. MMC and LMC can be applied only when both of the following conditions hold:

1. Two or more features are interrelated with respect to the location or form (e.g., two holes or a flat plane and a hole). At least one of the features must refer to size.
2. MMC or LMC must directly reference a size feature.

When MMC or LMC is used to specify the tolerance of a hole or shaft, the modifier implies that the tolerance specified is constrained by the maximum or least material condition as well as some other dimensional feature(s). For MMC, the tolerance may increase when the actual feature size produced is larger (for a hole) or smaller (for a shaft) than the plan calls for. Because the increase in the tolerance is compensated for by the deviance of size in production, the final combined hole-size error and geometric tolerance error will still be larger than the anticipated smallest hole. Figure 3.6 illustrates the allowed tolerance under the hole size produced. The allowed tolerance is the actual acceptable tolerance limit; it varies as the size of the hole that is produced changes. The specified tolerance is the value.

The projected tolerance zone is used in the assembly of two parts with a gap between them. The assembly of an engine block with a cylinder head is a good example. A gasket is inserted in between the engine block and the cylinder head. In specifying the position of bolt holes on the cylinder head, the positional error is measured at a thickness of the gasket to ensure mating with the bolt on the engine block. Figure 3.7 illustrates the projected tolerance zone at a distance of 0.25″ over the top of the hole. The dimension above the feature control frame specifies 0.375″-diameter holes with UNC (Unified Coarse) 16-threads-per-inch screw threads. 2B stands for number-2 fit and internal thread.

In the sections that follow, the mathematical definitions of geometric tolerances are introduced. The gauge-based definitions are presented in the appendix. The math-based definition is more precise; however, for technologists with less mathematical education, the gauge-based definition is preferred.

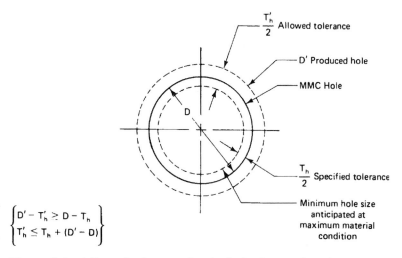

$$\begin{cases} D' - T'_h \geq D - T_h \\ T'_h \leq T_h + (D' - D) \end{cases}$$

Figure 3.6 Allowed tolerance for the hole size produced.

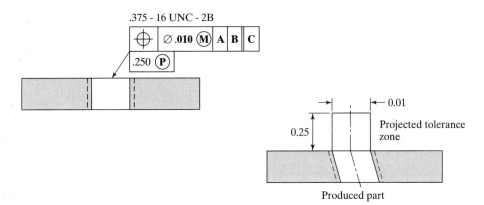

Figure 3.7 Projected tolerance zone.

3.2.2 Straightness [ASME Y14.5-1994]

Straightness is a condition wherein an element of a surface, or an axis, is a straight line. A straightness tolerance specifies a tolerance zone within which the considered element or derived median line must lie. A straightness tolerance is applied in the view in which the elements to be controlled are represented by a straight line.

Straightness of a Derived Median Line *Definition*: Straightness tolerance for the derived median line of a feature specifies that the derived median line must lie within some cylindrical zone whose diameter is the specified tolerance.

A straightness zone for a derived median line is a cylindrical volume consisting of all points $\vec{P}$ satisfying the condition

$$|\hat{T} \times (\vec{P} - \vec{A})| \leq \frac{t}{2} \tag{3.1}$$

where

$\hat{T}$ = direction vector of the straightness axis

$\vec{A}$ = position vector locating the straightness axis

t = diameter of the straightness tolerance zone (tolerance value)

This feature is illustrated in Figure 3.8 for a cylindrical surface and in Figure 3.9 for a planar surface.

Conformance. A feature conforms to a straightness tolerance t_0 if all points of the derived median line lie within some straightness zone as just defined, with $t = t_0$. That is, there exist $\hat{T}$ and $\vec{A}$ such that with $t = t_0$, all points of the derived median line are within the straightness zone.

Actual value. The actual value of straightness for the derived median line of a feature is the smallest straightness tolerance to which the derived median line will conform.

3.2.3 Straightness of Surface Line Elements

Definition. A straightness tolerance for the line elements of a feature specifies that each line element must lie in a zone bounded by two parallel lines that are separated by the specified tolerance and that are in the cutting plane defining the line element.

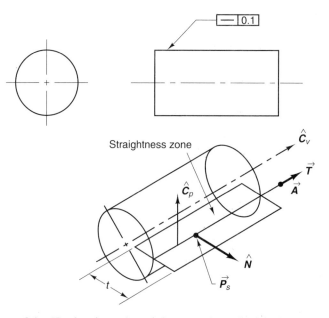

Figure 3.8 Evaluation of straightness of a cylindrical surface.

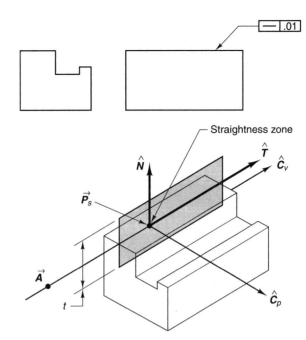

Figure 3.9   Evaluation of straightness of a planar surface.

A straightness zone for a surface line element is an area between parallel lines consisting of all points $\vec{P}$ satisfying the condition:

$$|\hat{T} \times (\vec{P} - \vec{A})| \le \frac{t}{2} \tag{3.2}$$

and

$$\hat{C}_p \cdot (\vec{P} - \vec{P}_s) = 0$$
$$\hat{C}_p \cdot (\vec{A} - \vec{P}_s) = 0$$
$$\hat{C}_p \cdot \hat{T} = 0$$

where

$\hat{T}$ = direction vector of the centerline of the straightness zone

$\vec{A}$ = position vector locating the centerline of the straightness zone

t = size of the straightness zone (separation between the parallel lines)

$\hat{C}_p$ = normal to the cutting plane, defined as being parallel to the cross product of the desired cutting vector and the mating surface normal at $\vec{P}_s$, in which

$\vec{P}_s$ = point on the surface, contained by the cutting plane

Figure 3.8 illustrates a straightness tolerance zone for surface line elements of a cylindrical feature. Figure 3.9 illustrates a straightness tolerance zone for surface line elements of a planar feature.

Conformance. A surface line element conforms to the straightness tolerance t_0 for a cutting plane if all points of the surface line element lie within some straightness zone as defined before, with $t = t_0$. That is, there exist $\hat{T}$ and $\vec{A}$ such that, with $t = t_0$, all points of the surface line element are within the straightness zone.

A surface conforms to the straightness tolerance t_0 if it conforms simultaneously for all toleranced surface line elements corresponding to some actual mating surface.

Actual value. The actual value of straightness for a surface is the smallest straightness tolerance to which the surface will conform.

3.2.4 Flatness

Flatness is the condition of a surface having all of its elements in one plane. A flatness tolerance specifies a tolerance zone defined by two parallel planes within which the surface must lie.

(a) *Definition.* A flatness tolerance specifies that all points of the surface must lie in some zone bounded by two parallel planes separated by the specified tolerance. A flatness zone is a volume consisting of all points $\vec{P}$ satisfying the condition

$$|\hat{T} \times (\vec{P} - \vec{A})| \leq \frac{t}{2} \tag{3.3}$$

where

$\hat{T}$ = direction vector of the parallel planes defining the flatness zone
$\vec{A}$ = position vector locating the midplane of the flatness zone
t = size of the straightness zone (separation between the parallel lines)

(b) *Conformance.* A feature conforms to a flatness tolerance t_0 if all points of the feature lie within some flatness zone as defined before, with $t = t_0$. That is, there exist $\hat{T}$ and $\vec{A}$ such that, with $t = t_0$, all points of the feature are within the flatness zone.

(c) *Actual value.* The actual value of flatness for a surface is the smallest flatness tolerance to which the surface will conform.

3.2.5 Circularity (Roundness)

Circularity is a condition of a surface wherein

(a) for a feature other than a sphere, all points of the surface intersected by any plane perpendicular to an axis are equidistant from that axis; and

(b) for a sphere, all points of the surface intersected by any plane passing through a common center are equidistant from that center.

A circularity tolerance specifies a tolerance zone bounded by two concentric circles within which each circular element of the surface must lie. A circularity tolerance applies independently at any plane just described in (a) and (b).

(a) *Definition.* A circularity tolerance specifies that all points of each circular element of the surface must lie in some zone bounded by two concentric circles whose radii differ by the specified tolerance. Circular elements are obtained by taking cross sections perpendicular to some spine. For a sphere, the spine is zero dimensional (a point); for a cylinder or cone, the spine is one dimensional (a simple, non-self-intersecting, tangent-continuous curve). The concentric circles defining the circularity zone are centered on, and in a plane perpendicular to, the spine.

A circularity zone at a given cross section is an annular area consisting of all points $\vec{P}$ satisfying the conditions

$$\vec{T} \cdot (\vec{P} - \vec{A}) = 0 \tag{3.4}$$

and

$$||\vec{P} - \vec{A}| - r| \leq \frac{t}{2} \tag{3.5}$$

where

$\hat{T}$ = a unit vector that is tangent to the spine at $\vec{A}$ for a cylinder or cone; a unit vector that points radially in all directions from $\vec{A}$ for a sphere

$\vec{A}$ = position vector locating a point on the spine

r = radial distance (which may vary between circular elements) from the spine to the center of the circularity zone ($r > 0$ for all circular elements)

t = size of the circularity zone

Figure 3.10 illustrates a circularity zone for a circular element of a cylindrical or conical feature.

(b) *Conformance.* A cylindrical or conical feature conforms to a circularity tolerance t_0 if there exists a one-dimensional spine such that, at each point $\vec{A}$ of the spine, the circular element perpendicular to the tangent vector $\hat{T}$ at $\vec{A}$ conforms to the circularity tolerance t_0. That is, for each circular element, there exist $\vec{A}$ and r such that, with $t = t_0$, all points of the circular element are within the circularity zone.

A spherical feature conforms to a circularity tolerance t_0 if there exists a point (a zero dimensional spine) such that each circular element in each cutting plane containing the point conforms to the circularity tolerance t_0. That is, for each circular element, there exist $\hat{T}$, r, and a common $\vec{A}$ such that, with $t = t_0$, all points of the circular element are within the circularity zone.

(c) *Actual value* The actual value of circularity for a feature is the smallest circularity tolerance to which the feature will conform.

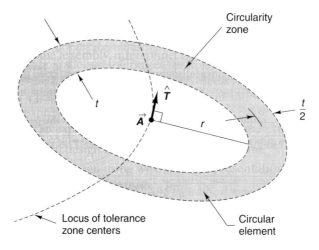

Figure 3.10   Illustration of a circularity tolerance zone for a cylindrical or conical feature.

3.2.6 Cylindricity

Cylindricity is a condition of a surface of revolution in which all points of the surface are equidistant from a common axis. A cylindricity tolerance specifies a tolerance zone bounded by two concentric cylinders within which the surface must lie. In the case of cylindricity, unlike that of circularity, the tolerance applies simultaneously to both circular and longitudinal elements of the surface (the entire surface). (*Note*: The cylindricity tolerance is a composite control of form that includes the circularity, straightness, and taper of a cylindrical feature.)

(a) *Definition.* A cylindricity tolerance specifies that all points of the surface must lie in some zone bounded by two coaxial cylinders whose radii differ by the specified tolerance.

A cylindricity zone is a volume between two coaxial cylinders consisting of all points $\vec{P}$ satisfying the condition

$$||\hat{T} \times (\vec{P} - \vec{A})| - r| \leq \frac{t}{2} \tag{3.6}$$

where

$\hat{T}$ = direction vector of the cylindricity axis
$\vec{A}$ = position vector locating the cylindricity axis
r = radial distance from the cylindricity axis to the center of the tolerance zone
t = size of the cylindricity zone

(b) *Conformance.* A feature conforms to a cylindricity tolerance t_0 if all points of the feature lie within some cylindricity zone as defined before, with $t = t_0$. That is, there exist $\hat{T}, \vec{A}$, and r such that, with $t = t_0$, all points of the feature are within the cylindricity zone.

(c) *Actual value.* The actual value of cylindricity for a surface is the smallest cylindricity tolerance to which the surface will conform.

3.2.7 Profile Control

A profile is the outline of an object in a given plane (a two-dimensional figure). Profiles are formed by projecting a three-dimensional figure onto a plane or taking cross sections through the figure. The elements of a profile are straight lines, arcs, and other curved lines. With profile tolerancing, the true profile may be defined by basic radii, basic angular dimensions, basic coordinate dimensions, basic size dimensions, undimensioned drawings, or formulas.

(a) *Definition.* A profile tolerance zone is an area (profile of a line) or a volume (profile of a surface) generated by offsetting each point on the nominal surface in a direction normal to the nominal surface at that point. For unilateral profile tolerances, the surface is offset totally in one direction or the other by an amount equal to the profile tolerance. For bilateral profile tolerances, the surface is offset in both directions by a combined amount equal to the profile tolerance. The offsets in each direction may or may not be disposed equally.

For a given point $\vec{P}_N$ on the nominal surface, there is a unit vector $\hat{N}$ normal to the nominal surface whose positive direction is arbitrary; this vector may point either into or out of the material. A profile tolerance t consists of the sum of two intermediate tolerances t_+ and t_- representing the amount of tolerance to be disposed in the positive and negative directions, respectively, of the surface normal $\hat{N}$ at $\vec{P}_N$. For unilateral profile tolerances, either t_+ or t_- equals zero; t_+ and t_- are always nonnegative numbers.

The contribution of the nominal surface point $\vec{P}_N$ toward the total tolerance zone is a line segment normal to the nominal surface and bounded by points at distances t_+ and t_- from $\vec{P}_N$. The profile tolerance zone is the union of the line segments obtained from each of the points on the nominal surface.

(b) *Conformance.* A surface conforms to a profile tolerance t_0 if all points $\vec{P}_S$ of the surface conform to either of the intermediate tolerances t_+ or t_- disposed about some corresponding point $\vec{P}_N$ on the nominal surface. A point $\vec{P}_S$ conforms to the intermediate tolerance t_+ if it is between $\vec{P}_N$ and $\vec{P}_N + \hat{N}_{t+}$. A point $\vec{P}_S$ conforms to the intermediate tolerance t_- if it is between $\vec{P}_N$ and $\vec{P}_N - \hat{N}_{t-}$. Mathematically, this is the condition that there exists some $\vec{P}_N$ on the nominal surface and some u, $-t_- \leq u \leq t_+$, for which $\vec{P}_S = \vec{P}_N + \hat{N}_u$.

(c) *Actual value.* For both unilateral and bilateral profile tolerances, two actual values are necessarily calculated: one for surface variations in the positive direction and one for surface variations in the negative direction. For each direction, the actual value of the profile is the smallest intermediate tolerance to which the surface conforms. Note that no single actual value may be calculated for comparison with the tolerance value in the feature control frame, except in the case of unilateral profile tolerances.

3.2.8 Orientation Tolerances

Angularity, parallelism, perpendicularity, and in some instances profile are orientation tolerances applicable to related features. These tolerances control the orientation of features to one another.

In specifying orientation tolerances to control angularity, parallelism, perpendicularity, and in some cases profile, the considered feature is related to one or more datum features. Relation to more than one datum feature is specified to stabilize the tolerance zone in more than one direction.

Tolerance zones are totals in value requiring an axis, or all elements of the considered surface to fall within this zone. Where it is a requirement to control only individual line elements of a surface, a qualifying notation, such as EACH ELEMENT or EACH RADIAL ELEMENT, is added to the drawing. This permits control of individual elements of the surface independently in relation to the datum and does not limit the total surface to an encompassing zone.

Where it is desired to control a feature surface established by the contacting points of that surface, the tangent plane symbol is added in the feature control frame after the stated tolerance.

Angularity is the condition of a surface or center plane or axis at a specified angle (other than 90°) from a datum plane or axis.

Parallelism is the condition of a surface or center plane, equidistant at all points from a datum plane or an axis, equidistant along its length from one or more datum planes or a datum axis.

Perpendicularity is the condition of a surface, center plane, or axis at a right angle to a datum plane or axis.

Mathematically, the equations describing angularity, parallelism, and perpendicularity are identical for a given orientation zone type when generalized in terms of the angle(s) between the tolerance zone and the related datum(s). Accordingly, the generic term *orientation* is used in place of angularity, parallelism, and perpendicularity in the definitions. See Appendix A.

An orientation zone is bounded by a pair of parallel planes, a cylindrical surface, or a pair of parallel lines. Each of these cases is defined separately below. If the tolerance value is preceded by the diameter symbol then the tolerance zone is a cylindrical volume; if the notation EACH ELEMENT or EACH RADIAL ELEMENT appears, then the tolerance zone is an area between parallel lines; in all other cases, the tolerance zone is a volume between parallel planes by default.

Planar Orientation Zone

(a) *Definition.* An orientation tolerance that is not preceded by the diameter symbol and that does not include the notation EACH ELEMENT or EACH RADIAL ELEMENT specifies that the toleranced surface, center plane, tangent plane, or axis must lie in a zone bounded by two parallel linear planes separated by the specified tolerance and basically oriented to the primary datum and, if specified, to the secondary datum as well.

A planar orientation zone is a volume consisting of all points $\vec{P}$ satisfying the condition:

$$|\hat{T} \cdot (\vec{P} - \vec{A})| \leq \frac{t}{2} \tag{3.7}$$

where

$\hat{T}$ = direction vector of the planar orientation zone

$\vec{A}$ = position vector locating the midplane of the planar orientation zone

t = size of the planar orientation zone (separation of the parallel planes)

The planar orientation zone is oriented such that, if $\hat{D}_1$ is the direction vector of the primary datum, then

$$|\hat{T} \cdot \hat{D}_1| = \begin{cases} |\cos \Theta| \text{ for a primary datum axis} \\ |\sin \Theta| \text{ for a primary datum plane} \end{cases} \tag{3.8}$$

where Θ is the basic angle between the primary datum and the direction vector of the planar orientation zone.

If a secondary datum is specified, the orientation zone is further restricted to be oriented relative to the direction vector $\hat{D}_2$, of the secondary datum by

$$|\hat{T} \cdot \hat{D}_2| = \begin{cases} |\cos \alpha| \text{ for a secondary datum axis} \\ |\sin \alpha| \text{ for a secondary datum plane} \end{cases} \tag{3.9}$$

where $\hat{T}$ is the normalized projection of $\hat{T}$ onto a plane normal to $\hat{D}_1$ and α is the basic angle between the secondary datum and $\hat{T}'$ $\hat{T}'$ is given by

$$\hat{T}' = \frac{\hat{T} - (\hat{T} \cdot \hat{D}_1)\hat{D}_1}{|\hat{T} - (\hat{T} \cdot \hat{D}_1)\hat{D}_1|} \tag{3.10}$$

Figure 3.11 shows the relationship of the tolerance zone direction vector to the primary and secondary datums. Figure 3.12 illustrates the projection of $\hat{T}$ onto the primary datum plane to form $\hat{T}'$.

(b) *Conformance.* A surface, center plane, tangent plane, or axis S conforms to an orientation tolerance t_0 if all points of S lie within some planar orientation zone as defined before, with $t = t_0$. That is, there exist $\hat{T}$ and $\vec{A}$ such that, with $t = t_0$, all points of S are within the planar orientation zone. Note that if the orientation tolerance refers to both a primary datum and a secondary datum, then $\hat{T}$ is fully determined.

(c) *Actual Value.* The actual value of orientation for S is the smallest orientation tolerance to which S will conform.

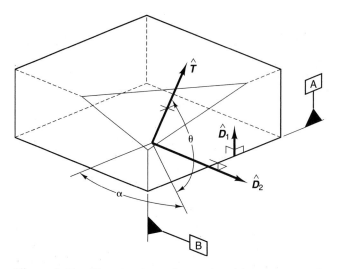

Figure 3.11   Planar orientation zone with primary and secondary datum planes specified.

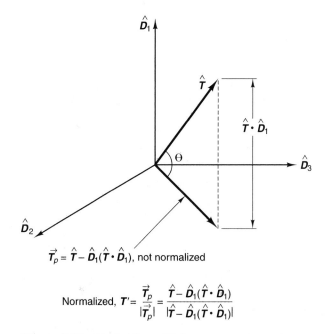

$$\vec{T}_p = \hat{T} - \hat{D}_1(\hat{T} \cdot \hat{D}_1), \text{ not normalized}$$

$$\text{Normalized, } \mathbf{T}' = \frac{\vec{T}_p}{|\vec{T}_p|} = \frac{\hat{T} - \hat{D}_1(\hat{T} \cdot \hat{D}_1)}{|\hat{T} - \hat{D}_1(\hat{T} \cdot \hat{D}_1)|}$$

Figure 3.12   Projection of tolerance vector onto primary datum plane.

3.2.9 Cylindrical Orientation Zone.

(a) *Definition.* An orientation tolerance that is preceded by the diameter symbol specifies that the toleranced axis must lie in a zone bounded by a cylinder with a diameter equal to the specified tolerance and whose axis is basically oriented to the primary datum. If secondary datum is specified, then the cylinder axis is basically oriented to that datum as well.

A cylindrical orientation zone is a volume consisting of all points $\vec{P}$ satisfying the condition

$$|\hat{T} \times (\vec{P} - \vec{A})| \leq \frac{t}{2} \tag{3.11}$$

where

$\hat{T}$ = direction vector of the axis of the cylindrical orientation zone

$\vec{A}$ = position vector locating the axis of the cylindrical orientation zone

t = diameter of the cylindrical orientation zone

The axis of the cylindrical orientation zone is oriented such that if $\hat{D}_1$ is the direction vector of the primary datum, then

$$|\hat{T} \cdot \hat{D}_1| = \begin{cases} |\cos \Theta| \text{ for a primary datum axis} \\ |\sin \Theta| \text{ for a primary datum plane} \end{cases} \tag{3.12}$$

where Θ is the basic angle between the primary datum and the direction vector of the axis of the cylindrical orientation zone.

If a secondary datum is specified, the orientation zone is further restricted to be oriented relative to the direction vector $\hat{D}_2$ of the secondary datum by

$$|\hat{T}' \cdot \hat{D}_2| = \begin{cases} |\cos \alpha| \text{ for a secondary datum axis} \\ |\sin \alpha| \text{ for a secondary datum plane} \end{cases} \tag{3.13}$$

where $\hat{T}'$ is the normalized projection of $\hat{T}$ onto a plane normal to $\hat{D}_1$ and α is the basic angle between the secondary datum and $\hat{T}' \cdot \hat{T}'$ is given by

$$\hat{T}' = \frac{\hat{T} - (\hat{T} \cdot \hat{D}_1)\hat{D}_1}{|\hat{T} - (\hat{T} \cdot \hat{D}_1)\hat{D}_1|} \tag{3.14}$$

Figure 3.13 illustrates a cylindrical orientation tolerance zone.

(b) *Conformance.* An axis S conforms to an orientation tolerance t_0 if all points of S lie within some cylindrical orientation zone as defined before, with $t = t_0$. That is, there exists $\hat{T}$ and $\vec{A}$ such that, with $t = t_0$, all points of S are within the orientation zone. Note that if the orientation tolerance refers to both a primary datum and a secondary datum, then $\hat{T}$ is fully determined.

(c) *Actual value.* The actual value of orientation for S is the smallest orientation tolerance to which S will conform.

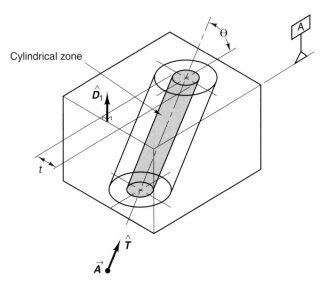

Figure 3.13   Orientation zone bounded by a cylinder with respect to a primary datum plane.

3.2.10 Linear Orientation Zone

(a) *Definition.* An orientation tolerance which includes the notation EACH ELE-MENT or EACH RADIAL ELEMENT specifies that each line element of the tol-eranced surface must lie in a zone bounded by two parallel lines that are (1) in the cutting plane defining the line element, (2) separated by the specified tolerance, and (3) basically oriented to the primary datum and, if specified, to the secondary datum as well.

   For a surface point $\vec{P}_S$, a linear orientation zone is an area consisting of all points $\vec{P}$ in a cutting plane of direction vector $\hat{C}_P$ that contains $\vec{P}_S$. The points $\vec{P}$ satisfy the conditions

$$\hat{C}_P \cdot (\vec{P} - \vec{P}_S) = 0 \tag{3.15}$$

and

$$|\hat{T} \times (\vec{P} - \vec{A})| \le \frac{t}{2} \tag{3.16}$$

where

$\hat{T}$ = direction vector of the centerline of the linear orientation zone

$\vec{A}$ = position vector locating the centerline of the linear orientation zone

$\vec{P}_S$ = point on S

$\hat{C}_P$ = normal to the cutting plane and basically oriented to the datum reference frame

t = size of the linear orientation zone (separation between the parallel lines)

The cutting plane is oriented to the primary datum by the constraint

$$\hat{C}_P \cdot \hat{D}_1 = 0 \tag{3.17}$$

If a secondary datum is specified, the cutting plane is further restricted to be oriented to the direction vector of the secondary datum $\hat{D}_2$ by the constraint

$$|\hat{C}_P \cdot \hat{D}_2| = |\cos \alpha| \text{ for a secondary datum axis}$$

$$|\hat{C}_P \cdot \hat{D}_2| = |\sin \alpha| \text{ for a secondary datum plane} \tag{3.18}$$

The position vector $\vec{A}$, which locates the centerline of the linear orientation zone, also locates the cutting plane through the following constraint:

$$\hat{C}_P \cdot (\vec{P}_S - \vec{A}) = 0 \tag{3.19}$$

If a primary or secondary datum axis is specified, and the toleranced feature in its nominal condition is rotationally symmetric about that datum axis, then the cutting planes are further restricted to contain the datum axis in accordance with the formula

$$\hat{C}_P \cdot (\vec{P}_S - \vec{B}) = 0 \tag{3.20}$$

where $\vec{B}$ is a position vector that locates the datum axis. Otherwise, the cutting planes are required to be parallel to one another.

The direction vector $\hat{T}$ of the centerline of the linear orientation zone is constrained to lie in the cutting plane by

$$\hat{C}_P \cdot \hat{T} = 0 \tag{3.21}$$

The centerline of the linear orientation zone is oriented such that, if $\hat{D}_1$ is the direction vector of the primary datum, then

$$|\hat{T} \cdot \hat{D}_1| = \begin{cases} |\cos \Theta| \text{ for a primary datum axis} \\ |\sin \Theta| \text{ for a primary datum plane} \end{cases} \tag{3.22}$$

where Θ is the basic angle between the primary datum and the direction vector of the linear orientation zone.

Figure 3.14 illustrates an orientation zone bounded by parallel lines on a cutting plane for a contoured surface.

(b) *Conformance.* A surface, center plane, or tangent plane S conforms to an orientation tolerance t_0 for a cutting plane $\hat{C}_P$ if all points of the intersection of S with $\hat{C}_P$ lie within some linear orientation zone as defined before, with $t = t_0$. That is, there exist $\hat{T}$ and $\vec{A}$ such that, with $t = t_0$, all points of S are within the linear orientation zone.

A surface S conforms to the orientation tolerance t_0 if it conforms simultaneously for all surface points and cutting planes $\hat{C}_P$.

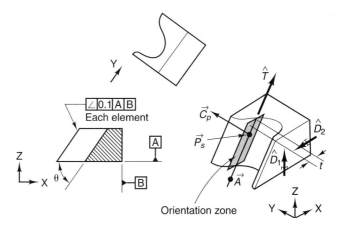

Figure 3.14 Orientation zone bounded by parallel lines.

Note that if the orientation tolerance refers to both a primary datum and a secondary datum, then $\hat{T}$ is fully determined.

(c) *Actual value.* The actual value of orientation for S is the smallest orientation tolerance to which S will conform.

3.2.11 Runout Tolerance

Runout is a composite tolerance used to control the functional relationship of one or more features of a part to a datum axis. The types of features controlled by runout tolerances include those surfaces constructed around a datum axis and those constructed at right angles to a datum axis.

Surfaces constructed around a datum axis are those surfaces which are either parallel to the datum axis or at some angle other than 90° to the datum axis. The mathematical definition of runout is necessarily separated into two subdefinitions: one for surfaces constructed around the datum axis and one for surfaces constructed at right angles to the datum axis. A feature may consist of surfaces constructed both around and at right angles to the datum axis. Separate mathematical definitions describe the controls imposed by a single runout tolerance on the distinct surfaces that make up such a feature. (Circular and total runout are handled later in the chapter.)

Evaluating runout (especially total runout) on tapered or contoured surfaces requires the establishment of actual mating normals. Nominal diameters and (as applicable) lengths, radii, and angles establish a cross-sectional *desired contour* having perfect form and orientation. The desired contour may be translated axially or radially (or both), but may not be tilted or scaled with respect to the datum axis. When a tolerance band is equally disposed about this contour and then revolved around the datum axis, a volumetric tolerance zone is generated.

Circular Runout
Surfaces Constructed at Right Angles to a Datum Axis

(a) *Definition.* The tolerance zone for each circular element on a surface constructed at right angles to a datum axis is generated by revolving a line segment about the datum axis. The line segment is parallel to the datum axis and is of length t_0, where t_0 is the specified tolerance. The resulting tolerance zone is the surface of a cylinder of height t_0.

For a surface point $\vec{P}_S$ a circular runout tolerance zone is the surface of a cylinder consisting of the set of points $\vec{P}$ satisfying the conditions

$$|\hat{D}_1 \times (\vec{P} - \vec{A})| = r \tag{3.23}$$

and

$$|\hat{D}_1 \cdot (\vec{P} - \vec{B})| \leq \frac{t}{2} \tag{3.24}$$

where

$\quad r =$ radial distance from $\vec{P}_S$ to the axis
$\quad \hat{D}_1 =$ direction vector of the datum axis
$\quad \vec{A} =$ position vector locating the datum axis
$\quad \vec{B} =$ position vector locating the center of the tolerance zone
$\quad t =$ size of the tolerance zone (height of the cylindrical surface)

(b) *Conformance.* The circular element through a surface point $\vec{P}_S$ conforms to the circular runout tolerance t_0 if all points of the element lie within some circular runout tolerance zone as defined before, with $t = t_0$. That is, there exists $\vec{B}$ such that, with $t = t_0$, all points of the surface element are within the circular runout zone.

A surface conforms to the circular runout tolerance if all circular surface elements conform.

(c) *Actual value.* The actual value of circular runout for a surface constructed at right angles to a datum axis is the smallest circular runout tolerance to which the surface will conform.

Surfaces Constructed around a Datum Axis

(a) *Definition.* The tolerance zone for each circular element on a surface constructed around a datum axis is generated by revolving a line segment about the datum axis. The line segment is normal to the desired surface and is of length t_0, where t_0 is the specified tolerance. Depending on the orientation of the surface with respect to the datum axis the resulting tolerance zone will be either a flat annular area or the surface of a truncated cone.

For a surface point $\vec{P}_S$, a datum axis $[\vec{A}, \hat{D}_1]$, and a given mating surface, a circular runout tolerance zone for a surface constructed around a datum axis consists of the set of points $\vec{P}$ satisfying the conditions:

$$\frac{\hat{D}_1 \cdot (\vec{P} - \vec{B})}{|\vec{P} - \vec{B}|} = \hat{D}_1 \cdot \hat{N} \tag{3.25}$$

$$||\vec{P} - \vec{B}| - d| \leq \frac{t}{2} \tag{3.26}$$

and

$$\hat{N}\,(\vec{P}_S - \vec{B}) > 0 \qquad\qquad (3.27)$$

where

$\hat{D}_1$ = direction vector of the datum axis
$\vec{A}$ = position vector locating the datum axis
$\hat{N}$ = surface normal at $\vec{P}_S$ determined from the mating surface
$\vec{B}$ = point of intersection of the datum axis and the line through $\vec{P}_S$ parallel to the direction vector $\hat{N}$
d = distance from $\vec{B}$ to the center of the tolerance zone as measured parallel to $\hat{N}$ ($d \geq t/2$)
t = size of the tolerance zone as measured parallel to $\hat{N}$

Figure 3.15 illustrates a circular runout tolerance zone on a noncylindrical surface of revolution.

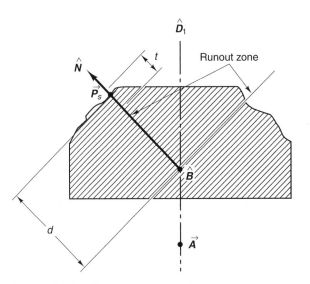

Figure 3.15 Circular runout tolerance zone.

(b) *Conformance.* The circular element through a surface point $\vec{P}_S$ conforms to the circular runout tolerance t_0 for a given mating surface if all points of the circular element lie within some circular runout tolerance zone as defined before, with $t = t_0$. That is, there exists d such that, with $t = t_0$, all points of the circular element are within the circular runout tolerance zone.

A surface conforms to a circular runout tolerance t_0 if all circular elements of the surface conform to the circular runout tolerance for the same mating surface.

(c) *Actual value.* The actual value of circular runout for a surface constructed around a datum axis is the smallest circular runout tolerance to which the surface will conform.

Total Runout
Surfaces Constructed at Right Angles to a Datum Axis

(a) *Definition.* A total runout tolerance for a surface constructed at right angles to a datum axis specifies that all points of the surface must lie in a zone bounded by two parallel planes perpendicular to the datum axis and separated by the specified tolerance.

For a surface constructed at right angles to a datum axis, a total runout tolerance zone is a volume consisting of the points $\vec{P}$ satisfying

$$|\hat{D}_1 \cdot (\vec{P} - \vec{B})| \leq \frac{t}{2} \tag{3.28}$$

where

$\hat{D}_1$ = direction vector of the datum axis
$\vec{B}$ = position vector locating the midplane of the tolerance zone
t = size of the tolerance zone (separation of the parallel planes)

(b) *Conformance.* A surface conforms to the total runout tolerance t_0 if all points of the surface lie within some total runout tolerance zone as defined before, with $t = t_0$. That is, there exists $\vec{B}$ such that, with $t = t_0$, all points of the surface are within the total runout zone.

(c) *Actual value.* The actual value of total runout for a surface constructed at right angles to a datum axis is the smallest total runout tolerance to which it will conform.

Surfaces Constructed around a Datum Axis

(a) *Definition.* A total runout tolerance zone for a surface constructed around a datum axis is a volume of revolution generated by revolving an area about the datum axis. This area is generated by moving a line segment of length t_0, where t_0 is the specified tolerance, along the desired contour, with the line segment kept normal to, and centered on, the desired contour at each point. The resulting tolerance zone is a volume between two surfaces of revolution separated by the specified tolerance.

Given a datum axis defined by the position vector $\vec{A}$ and the direction vector $\hat{D}_1$, let $\vec{B}$ be a point on the datum axis locating one end of the desired contour, and let r be the distance from the datum axis to the desired contour at point $\vec{B}$. Then, for a given $\vec{B}$ and r, let $C(\vec{B}, r)$ denote the desired contour. (*Note:* Points on this contour can be represented by $[d, r + f(d)]$, where d is the distance along the datum axis from $\vec{B}$.) For each possible $C(\vec{B}, r)$, a total runout tolerance zone is defined as the set of points $\vec{P}$ satisfying the condition

$$|\vec{P} - \vec{P}'| \leq \frac{t}{2} \tag{3.29}$$

where

$\vec{P}'$ = projection of $\vec{P}$ onto the surface generated by rotating $C(\vec{B}, r)$ about the datum axis

t = size of the tolerance zone, measured normal to the desired contour.

(b) *Conformance.* A surface conforms to a total runout tolerance t_0 if all points of the surface lie within some total runout tolerance zone as defined before, with $t = t_0$. That is, there exist $\vec{B}$ and r such that, with $t = t_0$, all points of the surface are within the total runout tolerance zone.

(c) *Actual value.* The actual value of total runout for a surface constructed around a datum axis is the smallest total runout tolerance to which the surface will conform.

3.2.12 Free-State Variation

Free-state variation is a term used to describe the distortion of a part after the removal of forces applied during manufacture. This kind of distortion is due principally to the weight and flexibility of the part and the release of internal stresses resulting from fabrication. A part of this kind—for example, a part with a very thin wall in proportion to its diameter—is referred to as a nonrigid part. In some cases, it may be required that the part meet its tolerance requirements while in the free state. In others, it may be necessary to simulate the mating part interface in order to verify individual or related feature tolerances. This is done by restraining the appropriate features. The restraining forces are those which would be exerted in the assembly or functioning of the part. However, if the dimensions and tolerances are met in the free state, it is usually not necessary to restrain the part, unless the effect of subsequent restraining forces on the concerned features could cause other features of the part to exceed specified limits.

3.3 INTERPRETING GEOMETRIC SPECIFICATIONS

Hole size and position   We will use the simple part shown in Figure 3.16 to illustrate how position and feature size interact. The part is first specified with the MMC modifier, where

X_M = largest value for x

Y_M = largest value for y

X'_M = smallest value for x

Y'_M = smallest value for y

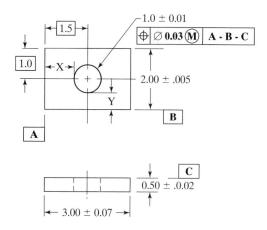

Figure 3.16   Part to illustrate true position.

The size of the hole feature can thus be from 0.99 to 1.01. The MMC diameter for the hole is for a size .99. For this example, we have the following table:

Ø (actual)	⊕ (tolerance)
.99	0.03
1.00	0.04
1.01	0.05

For D^* (the actual measured value), the position tolerance will vary with the hole diameter. If $D^* = 0.99$, then the allowable location error is contained in a circular region of diameter 0.03. If we seek to compute X_M, then the maximum size will correspond to a location residing as far right of datum A as allowable. In this case, we have $1.5 + .03/2 = 1.515$. Subtracting $D^*/2$ (0.99/2) from this yields $X_M = 1.020$. Interestingly, for any acceptable value of D and true position, X_M will remain the same, because of the way that MMCs are specified. X_M in this case will provide a maximum acceptable boundary for any acceptable diameter D^* and acceptable position tolerance zone. X'_M will provide the same acceptable boundary for the right side of the hole. $X'_M - X_M$ will yield the maximum diameter that will always assemble into the hole. The same calculations can be made for Y'_M and Y_M and will produce the same diameter.

The tolerances for the symmetric features (holes) are now specified with Form Geometry symbols (e.g., ⊕, ○, //). The hole features in the figure are specified as MMC entities. *This means that a virtual size for assembly is specified as the MMC for all of the part features, and this virtual size represents the minimum opening for all such labeled entities.* For female features, such as holes, the virtual size is specified as

$$\varepsilon_\nu = \varepsilon_s - \tau_s - \gamma_s \qquad (3.30)$$

where ε_v is the virtual size of the feature

ε_s is the nominal size specification of the feature
τ_s is the negative tolerance specification (for holes)
γ_s is the form geometric value (for holes)

For male features, such as a shaft, the virtual size is specified as

$$\varepsilon_v = \varepsilon_s + \tau_s + \gamma_s \tag{3.31}$$

For the hole shown in Figure 3.10, the virtual size is

$$\varepsilon_v = 1.000 - 0.01 - 0.03 = 0.96 \text{ inch}$$

Virtual-size information is useful for a variety of reasons. It can be used to determine how easily a part can be assembled, whether one part interferes with another, and more. The virtual size of the part in Figure 3.16 is shown in Figure 3.17. As can be seen, the virtual size is the maximum material condition for the entire part. It represents the silhouette for all possible combinations of acceptable parts stacked on top of each other. Therefore, it provides the condition under which all mating components can be assembled. This is a key feature for design as well as for inspecting components. It also relates only to parts specified with the MMC modifier.

Inspection equipment can be broadly classified as *general-purpose* or *special-purpose equipment*. To measure reasonably simple parts or very low-volume items, general-purpose inspection equipment is normally used. To inspect highly intricate or high-volume parts, special gauges are normally designed in order to reduce the amount of time required for the (calibration and) inspection process.

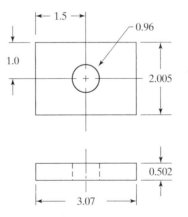

Figure 3.17   Virtual size of the part shown in Figure 3.16.

For the part shown in Figure 3.16, a single set of datums is used and the lone feature on the part is specified with an MMC modifier. Because of the foregoing interpretation, this means that a virtual part exists and that special gauges can be used to inspect the part, rather than having to use general-purpose gauges, such as micrometers or Vernier calipers. The use of special gauges can significantly reduce inspection time and improve the quality of the inspection process.

In order to inspect the part, "snap gauges" would be used to qualify the size requirements for the block portion of the bracket. The GO size is set to the largest acceptable size dimension, and the NOT GO size is set to the smallest acceptable size.

The snap gauge used to inspect the part height in Figure 3.16 is shown in Figure 3.18. A similar snap gauge for the width and depth would be used.

For the hole feature, a typical GO–NOT GO gauge can be used. The plug gauge to inspect the hole in the sample part is shown in Figure 3.19.

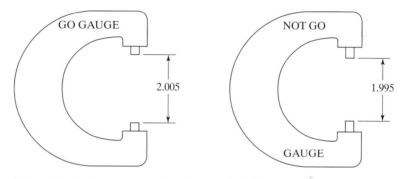

Figure 3.18   Snap gauges for the part height.

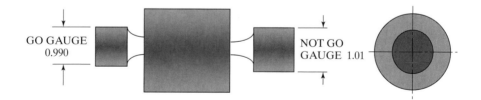

Figure 3.19   A typical plug gauge used to inspect a round hole.

In order to inspect feature location for the hole, a location gauge would be used. The size of the "pin" locator is determined from the formula

$$D_P = D_M - \oplus_M \text{ (for hole features)}$$

$$D_P = D_M + \oplus_M \text{ (for shaft features)}$$

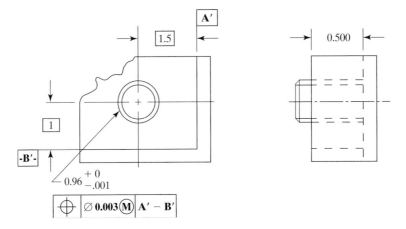

Figure 3.20   Location gauge for illustrative parts.

The location gauge for the part would look like the gauge shown in Figure 3.20. Note that the location planes correspond to datum surfaces –A- and –B-. The standard for gauge tolerance is that the gauge should be at least 10 times more precise than the part.

The inspection schema described in this section will work only for parts qualified with MMC. If the hole were specified for RFS or LMC, then conventional inspection equipment (micrometers and vernier calipers) would have to be used.

3.4 MULTIPLE PART FEATURES AND DATUMS

In the previous section, we looked at interpreting a part with a single hole feature. In this section, we will begin with a similar part with three holes and a single datum. We will then alter the specification of the part from one datum to two and discuss the why and hows of interpreting the part. Doing this has become more and more of an engineering requirement, because part datums are used to define critical relationships and dependencies that features have to other features.

We begin our discussion with the part shown in Figure 3.21. The part is a simple component with three hole features, all specified from a single part datum. The part is further specified with MMC modifiers. Each of the hole features can again be treated as independent features with one part coordinate system. Each of the holes is interpreted with respect to the part coordinate system or datums A–B–C. The part height, width, and depth would again be inspected with snap gauges (given that a high enough volume of parts was to be inspected). The holes would be inspected with plug gauges. This part of the inspection would ensure either that all of the part

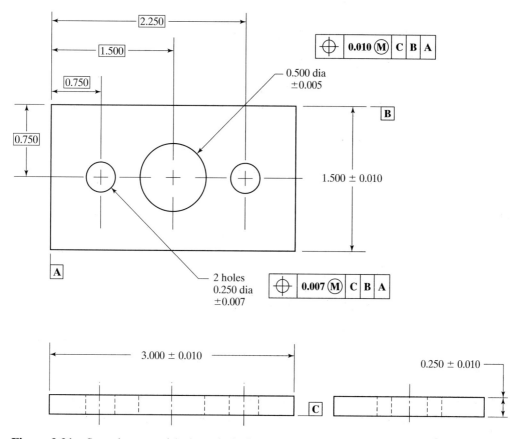

Figure 3.21   Sample part with three hole features.

sizes were within specification or that the part was defective. The last component to inspect is the location of each of the holes. Since there is only one datum and all of the features are called out with an MMC modifier, a virtual part exists and a location gauge can be used to inspect the hole locations. The diameter for the pins would again be calculated with the formula

$$D_P = D_M - \oplus_M$$

or

$$D_p\text{<large>} = 0.495 - 0.010 = 0.485$$

and

$$D_p<\text{small}> = 0.247 - 0.007 = 0.240$$

Since all three holes are called out from the same datum set, all three hole locations can be assessed with the use of a single location gauge. The gauge to inspect this part is shown in Figure 3.22.

A small variant in the part is shown in Figure 3.23. In the figure, the dimensions are all the same as before. The only difference is that a second datum set is used to define the locations of the two smaller holes (C–B–D) and datum D is qualified at MMC. The reason for qualifying a part like this could be that the larger hole might serve as an anchor for a part, with two smaller pins fitting into the other holes.

The size features for this part would again be inspected as they were for the previous part. The location requirements, however, will be very different. The large hole is called out with respect to the major part coordinates (A–B–C), so it will be inspected

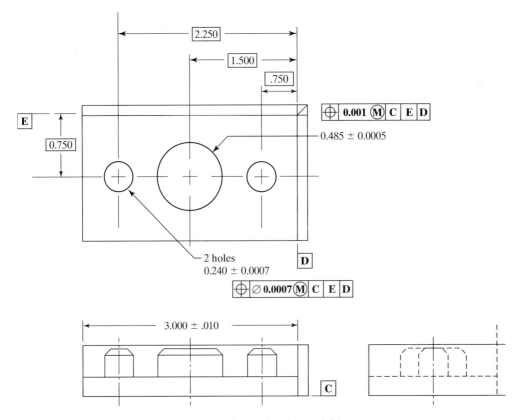

Figure 3.22   Location gauge for part shown in Figure 3.21.

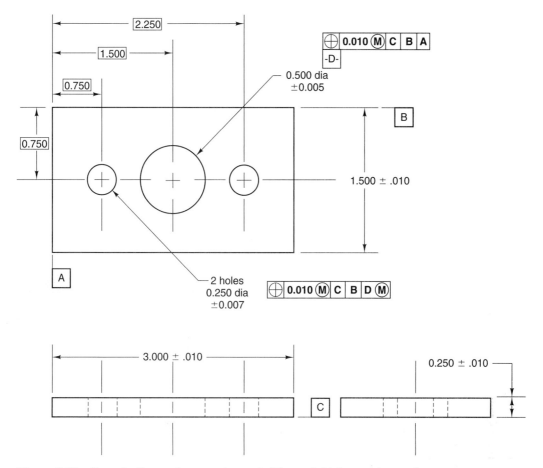

Figure 3.23   Part similar to the part shown in Figure 3.21, but with two datums.

in much the same manner as it was for the previous part. The location gauge for the large hole is shown in Figure 3.24. Since two datums are used, the locations of all of the holes cannot be qualified at the same time; a second location gauge will be necessary. The locations for the small holes are called out with respect to datum C–B–D. This datum will provide the locating surfaces for the smaller holes. The locations of the smaller holes will now be qualified in the original part A axis, using the large hole (datum feature D). The location gauge for the small holes is shown in Figure 3.25.

The gauge in Figure 3.24 is just like the previous gauges that were developed. This is because the large hole is called out with an MMC modifier and the hole is referenced from the general part coordinate system (datum). The gauge shown in Figure 3.25 is different in that, rather than using a plane to locate from the part's –A- datum, the large hole is used (at MMC) to position the small holes in this direction.

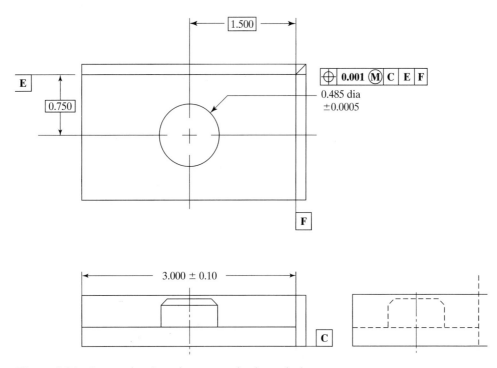

Figure 3.24   Inspection location gauge for large hole.

3.4.1 Multiple Datum Parts

Parts often have interfeature functionality, in that a component may be assembled over two or three features such that the performance is related to how one feature relates to another. In this case, multiple datums are used to characterize the functionality. The part shown in Figure 3.26 has two distinct datums, where the circular boss relates directly to the major part coordinate system. The hole on the boss requires a special location relationship with respect to the boss itself. In this case, the boss becomes a datum for the part ($-F$), and the hole on the boss is called out with respect to the boss at MMC. Often, this implies that the boss will be machined first and then the part will be refixtured with respect to the boss rather than the original part coordinate system. This part is another part that would require two different location gauges to fully qualify: first the gauge to qualify the boss and then a gauge to qualify the hole on the boss.

3.4.2 Limit and Fit

Since Eli Whitney first demonstrated the concept of interchangeable parts, fit has become a critical manufacturing issue. A tolerancing system for mating components was

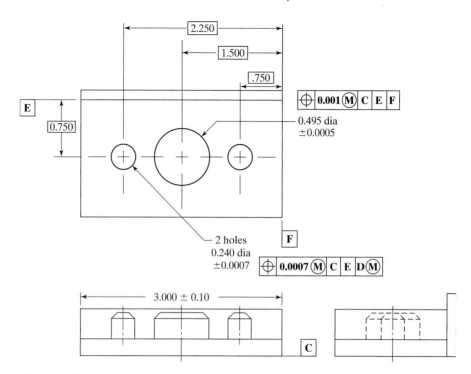

Figure 3.25   Inspection location gauge for smaller holes.

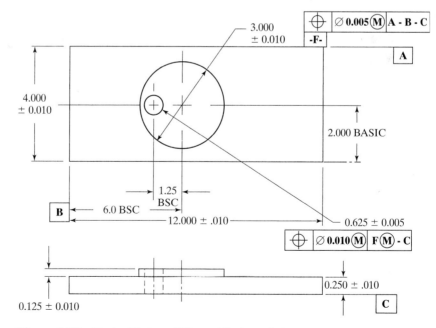

Figure 3.26   Part with two different feature datums.

developed as part of U.S. standards in 1967 [XUSAS B2.1X1967]. This standard has been adopted as part of ASME Y14.5. Figure 3.27 illustrates the recommended specifications for hole–shaft fit for various functionalities.

Three classes of fit are used to specify mating interaction: (1) clearance fits, (2) transitional fits, and (3) interference fits. Figure 3.27 illustrates this basic concept. As the name implies, a clearance fit indicates that a clearance remains between the shaft and the hole after they have been assembled, allowing the shaft to rotate or move about the major hole axis. The ANSI B4.1-1978 Standard uses nine subclasses of fit to describe hole–shaft fit. These subclasses range from RC1, a close sliding fit in which no perceivable play can be observed, to RC9, a loose running fit in which the shaft fits more loosely into the hole. The designer simply selects the subclass that best fits the needs of the part, knowing that the higher the specification subclass, the less expensive the manufacturing will be.

Transitional fits are normally used to specify tolerance for parts that are stationary. Location clearance fits (LC1 to LC11) are used for parts that are assembled together and that can be disassembled for service. The accuracy of these components is not exact. Transition location fits (LN1 to LN6) are specified when the location accuracy is of importance, but a smaller clearance or interference is acceptable.

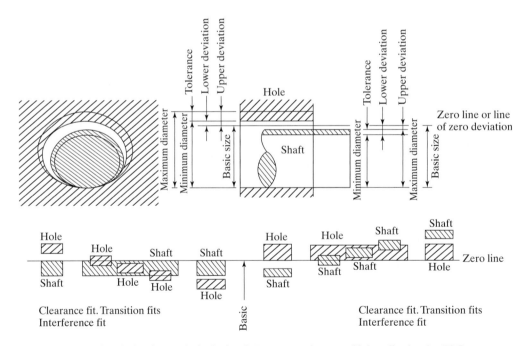

Figure 3.27   "Shaft-basis" and "hole-basis" systems for specifying fits in the ISO system. (By permission from *ASME Dimensioning and Tolerancing—Principles of Gauging and Fixturing Y14.43*, 2003, American Society of Mechanical Engineers, New York.)

Locational interference fits (LN1 to LN3) are specified when both rigidity and accuracy are required. Location interference fit parts can be assembled and disassembled, but not without special tooling (usually a shaft or a wheel puller) and considerable time. Other interference parts normally require special operations for their assembly. Tight drive fits (FN1) are used on parts requiring nominal assembly pressure. Force fits (FN5) are used for drive applications wherein the hole element is normally heated to expand the diameter prior to assembly. Tables that contain the specification for various classes of fit are available in most design handbooks.

3.5 SUMMARY

In this chapter, we presented the basics of ASME Y14.5, along with discussions on the use and interpretation of this standard. As engineering products become more complex and as tolerances get smaller and smaller, the need to describe the acceptable part variability more precisely becomes greater and greater. The chapter examined many simple parts to illustrate the basic concepts of form and location. Although the parts discussed were simple, they illustrate the same interpretation specifics that one would find in a more complicated part. Most times, part complexity is more related to the number of features on a part rather than to the interactions among these features.

3.6 KEYWORDS

ANSI	Least material condition
ASME	Limit and fit
Circularity	Maximum material condition
Cylindricity	Profile
Datum	Regardless feature size
Feature control frame	Runout
Flatness	Straightness
Gauge	Y14.5
GD&T	
ISO	

3.7 REVIEW QUESTIONS

3.1 Describe why ASME Y14.5 is important for product specification.

3.2 What is virtual size? What are the specification requirements for having a virtual size?

3.3 What are the different classes of fit? Describe specific instances when these classes would be used as part of a design.

3.4 What is a datum? Is it physically located on a part?

3.5 Why is it advantageous to use special plug and location gauges rather than micrometers and calipers? When is it possible from an economic point of view?

3.6 What conditions necessitate the use of gauges to measure both size and location?

3.8 REVIEW PROBLEMS

3.1 For the part shown in Figure 3.28, determine the maximum and minimum value of the dimension labeled X.

3.2 For the part shown in Figure 3.28, draw the virtual size (if there is one) as well as the gauging (if any can be used) that would be required for inspection of the part. Assume that the part is a high-volume product.

3.3 For the part shown in Figure 3.29, describe how the part would be inspected if it were to be produced in large quantities. What are the minimum and maximum acceptable values of X?

3.4 For the part shown in Figure 3.30, describe an efficient inspection schema if thousands of parts will be inspected.

3.5 For the part shown in Figure 3.31, develop an efficient inspection schema for inspecting thousands of components. Does the schema change if the modifier for position is changed from RFS to MMC?

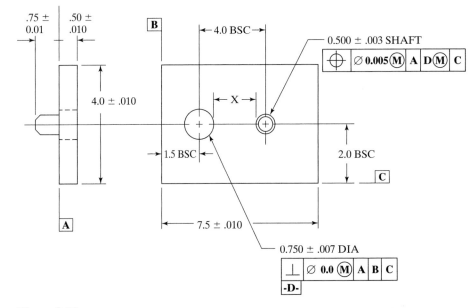

Figure 3.28

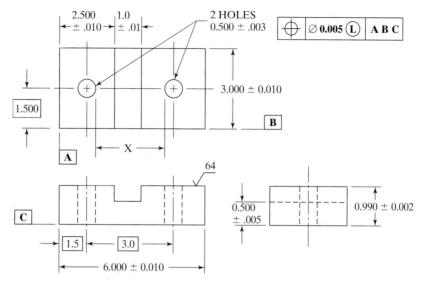

Figure 3.29

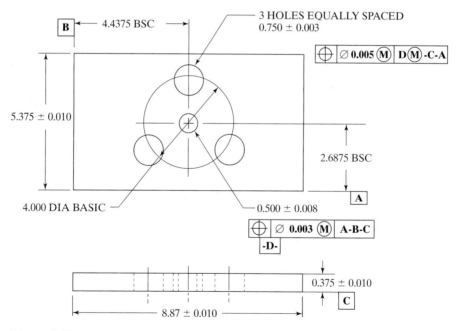

Figure 3.30

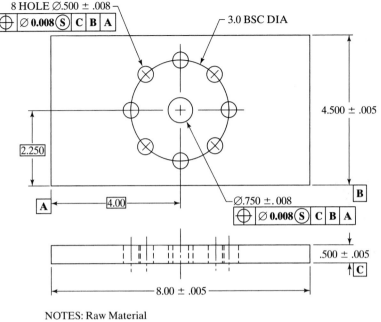

Figure 3.31

3.9 REFERENCES

ANSI, "Surface Texture." ANSI Standard B46.1-1978.

ANSI, for ANSI and ISO documents: http://global.ihs.com/

ASME Y14.36M-1996. Surface Texture Symbols.

ASME Y14.43 (2003) Dimensioning and Tolerancing Principles for Gauges and Fixtures.

Earle, James H. *Graphics for Engineers: With AutoCAD release 14 & 2000.* Upper Saddle River, NJ: Prentice Hall, 2000.

Foster, L. W. *Geo-Metrics III: The Application of Geometric Tolerancing Techniques.* Reading, MA: Addison-Wesley, 1994.

Gooldy, G. *Geometric Dimensioning and Tolerancing.* Englewood Cliffs, NJ: Prentice Hall, 1995.

ISO 1302:1994. Technical Drawings: Method of Indicating Surface Texture.

ISO CD 1302. Geometrical Product Specifications (GPS): Indication of Surface Texture.

ISO, for ISO Standard: http://www.iso.ch/iso/en/ISOOnline.frontpage

Suh, N. P. "Qualitative and Quantitative Analysis of Design and Manufacturing Axioms," *CIRP Annals*, 31 (1982), 333–338.

Voelcker, H. B. "A Current Perspective on Tolerancing and Metrology," *Manufacturing Review*, 6,4 (1993).

Voelcker, H. B., and A. A. G. Requicha "Geometric Modeling of Mechanical Parts and Processes," *Computer*, 10,12 (1977), 48–57.

APPENDIX: GAUGE-BASED GEOMETRIC TOLERANCE DEFINITIONS

STRAIGHTNESS
Tolerance zone between two straightness lines.

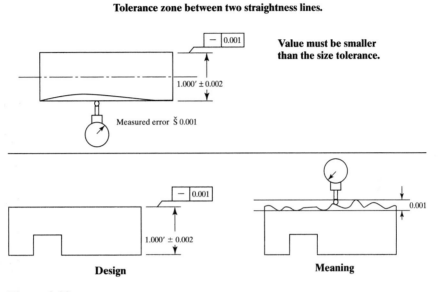

**Value must be smaller
than the size tolerance.**

Figure 3.32

FLATNESS
Tolerance zone defined by two parallel planes.

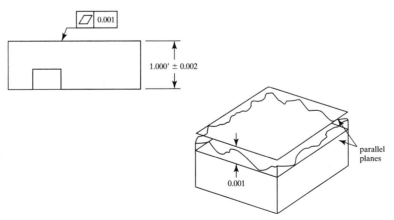

Figure 3.33

CIRCULARITY (ROUNDNESS)

a. Circle as a result of the intersection by any plane perpendicular to a common axis.

b. On a sphere, any plane passes through a common center.

Tolerance zone bounded by two concentric circles.

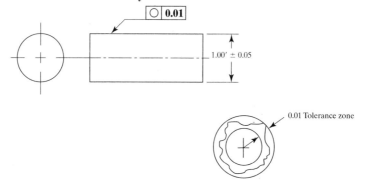

At any section along the cylinder

Figure 3.34

CYLINDRICITY

Tolerance zone bounded by two concentric cylinders within which the cylinder must lie.

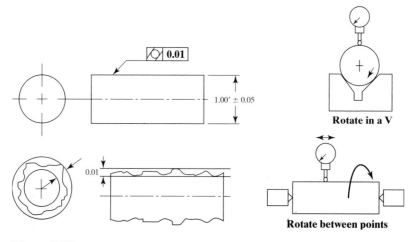

Figure 3.35

PERPENDICULARITY

A surface, median plane, or axis at a right angle to the datum plane or axis.

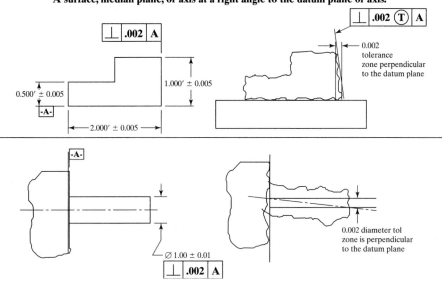

Figure 3.36

ANGULARITY

A surface or axis at a specified angle (other than 90°) from a datum plane or axis. Can have more than one datum.

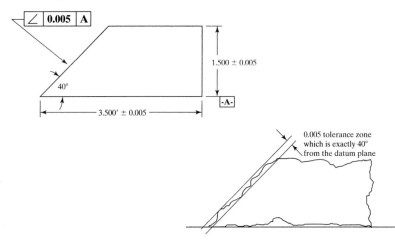

Figure 3.37

PARALLELISM

The condition of a surface equidistant at all points from a datum plane, or an axis equidistant along its length to a datum axis.

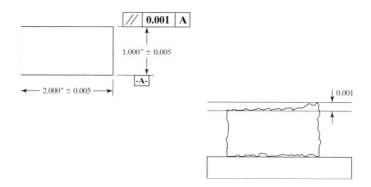

Figure 3.38

PROFILE

A uniform boundary along the true profile within which the elements of the surface must lie.

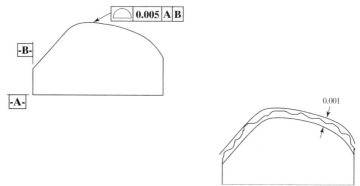

Figure 3.39

RUNOUT

A composite tolerance used to control the functional relationship of one or more features of a part to a datum axis. Circular runout controls the circular elements of a surface. As the part rotates 360° about the datum axis, the error must be within the tolerance limit.

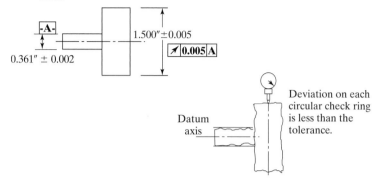

Figure 3.40

TOTAL RUNOUT

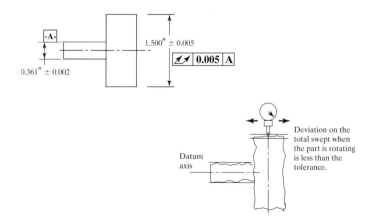

Figure 3.41

TRUE POSITION

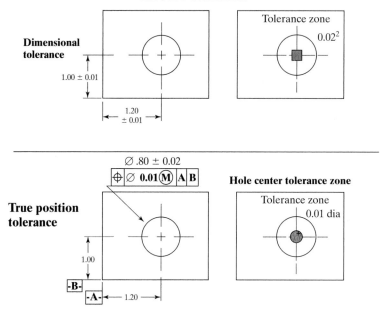

Figure 3.42

Chapter 4

Computer-Aided Design

Objective

Chapter 4 introduces computer-aided design (CAD) and how it is used in product design. Basic CAD features and functions are presented. Information from drafting to solid modeling is provided. Definitions and explanations of CAD terminologies are offered.

Outline

4.1 BACKGROUND

Computer-aided design (**CAD**) is the process of *utilizing computers to create and edit design **models** and **drawings***. CAD has been used in applications ranging from computer animation of movies, architectural design, and electric and electronic circuit design to fashion design, garment layout, and mechanical systems design. In most of these applications, CAD is used to model product geometries and to record some nongeometric design specifications. In applications in which the only output is an image, such as movie animation, games, and virtual-reality simulations, the design model is used for rendering or plotting. In most engineering applications, however, a physical artifact has to be realized from the design model. In that case, the design model serves as the interface between design and manufacturing. Since the main focus of this book is **computer-aided manufacturing** (**CAM**), and since CAM begins with a design model and ends with a physical artifact, CAD is critical to CAM. It is quite natural to use CAD to create a physical model of the product that will be used in manufacturing, instead of translating the product design from a paper drawing or text streams into a CAM system. As will be discussed in the next section, the development of CAD began in the 1960s. By the year 2000, most engineering designs were created with CAD systems. It is essential for a student of CAM to understand the basic features of CAD. The focus of this chapter, then, is to discuss those features; the next chapter examines the modeling of engineering geometry and the mathematics behind CAD systems.

4.2 A BRIEF HISTORY OF CAD

CAD is a product of the computer era. It originated from early computer graphic systems and persisted through the development of interactive computer graphics. In the 1950s, the U.S. Air Force's Semi Automatic Ground Environment (SAGE) air defense system, developed at the Massachusetts Institute of Technology's Lincoln Laboratory, pioneered the use of the **cathode-ray tube** (**CRT**) for computer displays. The SAGE project succeeded in linking CRT displays and computer operating systems. In 1960, Ivan Sutherland, at Lincoln, developed an interactive graphics input system called Sketchpad [Sutherland, 1963]. A CRT display and **light pen** input were used to interact with the system. This, coincidentally, happened at about the same time that numerical control (NC) and **Automatically Programmed Tool** (**APT**; see Chapters 13 and 14) first appeared. Later, X–Y plotters were used as the standard hard-copy output device for computer graphics. An interesting note is that an X–Y plotter has the same basic structure as an NC drilling machine, except that a pen is substituted for the tool on the NC spindle. CAD and CAM were being developed at the same time, from similar technologies and mathematics used by different groups.

In the 1960s, many major corporations, including General Motors, Ford, Boeing, and McDonnell-Douglas, began their own internal CAD projects. By the late 1960s, CAD companies such as Computervision and Applicon were founded, and the first commercial CAD systems became available. By the 1970s, the first generation of CAD systems had already populated many design departments.

In the beginning, CAD systems were no more than graphics editors with some built-in design symbols. The geometry available to the user was limited to lines, circular arcs, and the combination of the two. The development of free-form curves and surfaces, such as Coon's patch, Ferguson's patch, Bezier's patch, and the B-spline, enabled CAD systems to be used to represent and render sophisticated curves and surface designs. Soon, three-dimensional CAD systems were developed. Because a three-dimensional model (rendered via data abstraction in a computer) contains enough information for NC cutter-path programming, a linkage between CAD and NC could be developed. So-called turnkey **CAD/CAM** systems were developed on the basis of this concept and became popular in the 1970s and 1980s.

With the invention of three-dimensional solid modeling, the 1970s marked the beginning of a new era in CAD. In the past, three-dimensional wire-frame models represented an object only by its bounding edges. These representations can be ambiguous in the sense that several interpretations might be possible for a single model. There is also no way to find volumetric information with such a model. Solid models, by contrast, contain complete information; therefore, not only can they be used to produce engineering drawings, but engineering analysis can be performed on the same model as well. Solid modelers such as PADL-1 and PADL-2 [Voelcker and Requicha, 1977], Synthavision, BUILD-1 and BUILD-2 [Braid, 1973], COMPAC, EUCLID, GLIDE, and so on were developed in the 1970s. Later, many commercial systems and research systems were developed with the same methods. Several of these commercial systems were based on PADL and BUILD. Although powerful in representational properties, they contain many deficiencies. For example, such systems (1) have exorbitant computation and resource (memory) requirements, (2) use an unconventional way of modeling objects, and (3) lack an analytical tolerancing capability. These limitations have hindered CAD applications. It was not until the mid-1980s that solid modelers made their way into the design environment. Today, their use is as common as drafting and wire-frame model applications. Nearly all CAD packages have built-in solid modeling features.

Affordable CAD implementations on personal computers (PCs) have brought CAD to the masses. Originally, CAD was a tool used only by aerospace and other major industrial corporations. The introduction of PC CAD packages, such as Auto-CAD (Figure 4.1), VersaCAD, CADKEY, and so on, has made it possible for small companies and even individuals to own and use CAD systems. By 1988, more than 100,000 PC CAD packages had been sold. Today, PC-based solid modelers are both available and popular. Because rapid developments in microcomputers have enabled PCs to carry the heavy computational load necessary for solid modeling, virtually all solid modelers can now run on PCs, and the platform has become a nonissue. With the standard **graphics user interface** (**GUI**), CAD systems can be ported easily from one computer to another. Today, most major CAD systems are able to run on a variety of platforms; hence, most engineering designs are done with CAD systems. Add-ons to those systems, such as virtual reality, AI-based design, and network-based collaborative design systems, are available to engineers. Designers who are given these tools often go far beyond what traditional drafting can do.

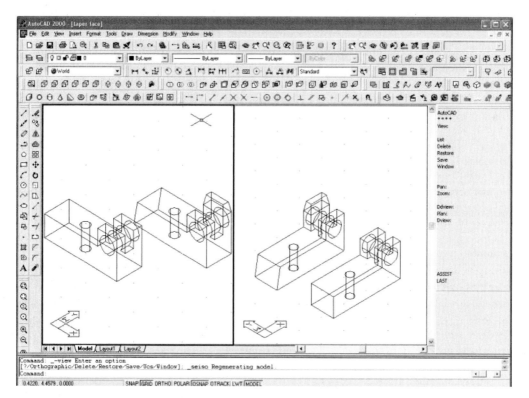

Figure 4.1   AutoCAD menu and icons.

4.3 DESIGN MODELING REQUIREMENTS

The purpose of design modeling is to capture the design concept in a sharable medium. Traditionally, this medium has been a piece of paper. Since engineering objects are mostly three dimensional, the model in the form of a drawing must capture all the details of the design on this two-dimensional medium. Engineering drafting standardizes the symbols, formats, and conventions used in drawing engineering objects. The purpose is to eliminate potential confusion due to different interpretations. Because early CAD drafting systems were a substitute for engineering drafting systems, the mechanical drafting functions were duplicated in software. Menus either display on the monitor or overlay on a digital tablet **drafting geometry** (line, circle, curve, etc.) and **drafting functions** (add, delete, modify, etc.). Being a software tool, CAD further expands the functionality of drafting to allow built-in standard symbols, semiautomatic dimensional tolerancing, and other desirable features. There are also many functions to help define geometry. As a result, a CAD system is more flexible than a manual drafting system and can produce more precise conventional engineering drawings.

While CAD drafting replaces paper-and-pencil based drafting with higher efficiency, 3-D CAD opens other possibilities in design modeling. Drafting produces a

2-D projection of the 3-D engineering object. Assuming that the main projection (top view) indicates the length and width of an object, we still require second or third projections (front view or side view, respectively) to obtain information on the height of the object. Only after comparing and relating the geometries (lines, circles, and curves) in multiple views can one discern the height information. As a matter of fact, all dimensional information has to be read from the dimension and tolerance symbols and annotations. 3-D CAD eliminates the need for multiple views to obtain the height information. All geometric entities are modeled in the 3-D space (X-, Y-, and Z-coordinates). With 3-D computer graphics, a 3-D CAD model can be rotated and viewed from any angle. Such a model is equally useful in manufacturing. NC part programming requires 3-D coordinates, which can be taken directly from the model. A 3-D CAD model is also called a **wire-frame** model, since it represents an object by its bounding edges. There is no explicit face (surface) or volume information in the model.

Although a 3-D CAD model contains richer information than a 2-D drafting model, it is still necessary to output drafting models. In order to generate a 2-D model from a 3-D model, a projection is used. Without faces, all edges will be projected on the projection plane. The correct drawing should include only the visible edges. By utilizing implied faces (adding a face to a closed loop of edges), **hidden-line removal** algorithms can remove the edges that are not supposed to be seen from a particular direction from which the object is viewed. This does not always work correctly, however, since faces cannot always be added correctly. Another complication of the 3-D model is adding dimensions and tolerances. Dimensions and tolerances in drafting are defined on 2-D projections. How to attach such information to the 3-D model in a way that will allow proper display on the 2-D projections is a practical problem.

A 3-D wire-frame model lacks face information and does not have the **volume property**. It is not a complete and unambiguous model. No correct section views can be guaranteed. It is also not possible to calculate the volume and mass of the object directly from the model. For applications such as rapid prototyping, which requires correct 2-D cross sections, none of the CAD models mentioned up to now can be used.

A **3-D solid model** is the answer to this new requirement. It models the entirety of a solid in the 3-D space, instead of just its edges or 2-D projections. While there are many ways a solid may be modeled, all of them have to represent the geometry of the object correctly and maintain the distinction between the inside and the outside of the object. A 3-D solid model is important for geometric representation, as it defines the *shape* of the object. The inside–outside distinction is unique to solid models. The inside and the outside are defined on the basis of the boundary of the object, a 3-D shell. In a **boundary representation** solid model, the shell is defined by a closed-loop set of faces. For example, for a cube without holes, the shell consists of six connected, but not intersecting, faces. (See Chapter 5 for further discussion.) One side of this shell is defined as inside and the other as outside. The solid cube is the inside; the rest of the universe is the outside. Of course, in this case we care only about the inside. If there is a hole drilled through the cube, the hole is represented by a single cylindrical face. Two of the six cube faces also get broken by the hole. Now the space inside the cylindrical face belongs to the outside of the solid.

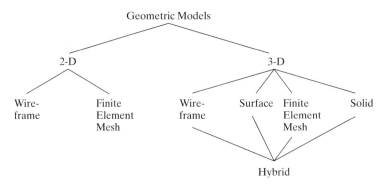

Figure 4.2   Classification of geometric models.

Other solid models that use the combination of intersecting solids to represent a bigger and more complex solid include **constructive solid geometry (CSG)** and **volume decomposition**. The geometry of the final object is defined by the building-block solids (**primitives**) and **operators** (**union, difference, intersection**). The information about the inside of the object is maintained using primitives and is modified by using operators.

Given the geometry and the inside–outside information, a solid model is said to have the volume property; that is, it can be sectioned, sliced, and projected, all the while maintaining the properties of a solid. It can thus satisfy the need of most engineering applications. Still, although a solid model is geometrically complete, it is not without shortcomings. For example, for many applications, including manufacturing, additional information, such as tolerances, are needed. Again, as with wire-frame models, the attachment of tolerances and annotations is not as straightforward as in drafting.

In conclusion, a design model needs the following information:

- Geometry
- Tolerances
- Volume (for some applications and available only in solid models)
- Annotation

Figure 4.2 shows one classification of geometric modeling techniques. In this chapter, we will discuss various geometric design methods. The finite-element mesh model, based on breaking the object into small cells of standard shapes (finite elements), is used only in engineering analysis. Hence, we shall not discuss it any further.

4.4 THE ARCHITECTURE OF CAD

A CAD system consists of three major parts:

1. Hardware: computer and input/output (I/O) devices
2. Operating system software
3. Application software: the CAD package

A wide range of hardware is used to support the software functions in CAD systems. The operating system software is the interface between the CAD application software and the hardware. It supervises the operation of the hardware and provides basic functions, including creating and removing operating tasks, controlling the progress of tasks, allocating hardware resources among tasks, and providing access to software resources such as files, editors, compilers, and utility programs. The operating system is important not only for CAD software, but also for non-CAD software.

The application software is the heart of a CAD system. It consists of programs that do 2-D and 3-D modeling, drafting, and engineering analysis. The functionality of a CAD system is built into the application software. It is the application software that makes one CAD package different front another. Application software is usually operating-system dependent. To transport a CAD system running on one operating system to another operating system is not as trivial as recompiling the software. Therefore, attention must be given to the operating system as well.

The general architecture of a CAD system is shown in Figure 4.3. The application software is at the top level and is used to manipulate the CAD model database. The graphics utility system performs coordinate transformation, windowing, and display control. Because several different I/O devices may be used, device drivers translate the data into and out of the specific data format utilized by each device. Device drivers also control the physical devices. The operating system is run in the background to coordinate the entire operation. Finally, a user interface links the human with the system.

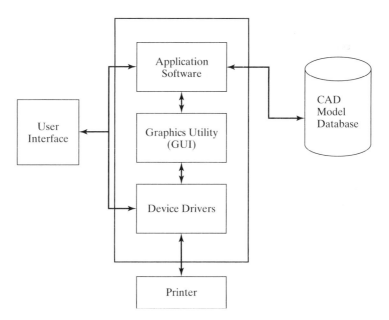

Figure 4.3   CAD system architecture.

4.5 DESIGN DRAFTING

Drafting deals with two-dimensional **geometric entities**. In CAD, the geometric entities are the point, line, circle, arc, polyline, spline, and area (rectangular, circular, etc.). Supplemental to the geometry are dimensions, tolerances, and annotations. Unlike mechanical drawings on paper, CAD also allows drafting be created in multiple **layers**. Each layer can be viewed as an overlying sheet of paper (Figure 4.4) that can be inserted or removed to display selected layers of entities. A user selects the current working layer to add, modify, or delete (geometric or nongeometric) entities.

To assist drawing precision, **grid** and **snap** tools may be used. A grid is laid on the screen or window at fixed X and Y distances (Figure 4.5). The grid can be either visible or invisible, and the distance between grid points can be set. When the drawing method is said to snap on the grid, one can position the drawing **cursor** only at the discrete grid points. (See Figure 4.5.) Better drawing control can thus be achieved.

To begin drawing geometry on a layer, first the layer is selected and then a built-in geometric entity is selected. Usually, there are many ways a geometric entity can be defined. For example, a line entity is defined by two endpoints that can in turn be defined in many ways:

1. Freehand drawing or snap to grid.
2. Points snap to endpoints, midpoint of another line.
3. Point snaps to the center or perimeter of a circle.
4. Point snaps to the intersection of two lines, a line and a circle, or two circles.

   The line can also be constrained:

1. Horizontal or vertical lines.
2. At an angle from the X-axis.
3. Parallel or perpendicular to another line.

Each CAD system has a different kind of interface that enables the user to make the appropriate selection and define the line. Figure 4.6 illustrates defining a line by snapping

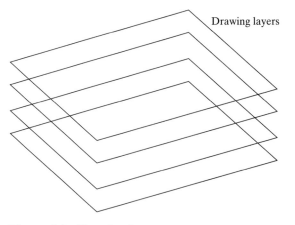

Drawing layers

Figure 4.4 Drawing layers.

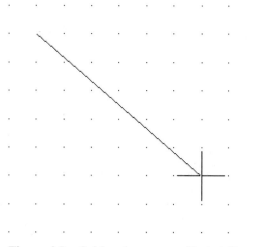

Figure 4.5   Grid and snap-on effect: A line is drawn.

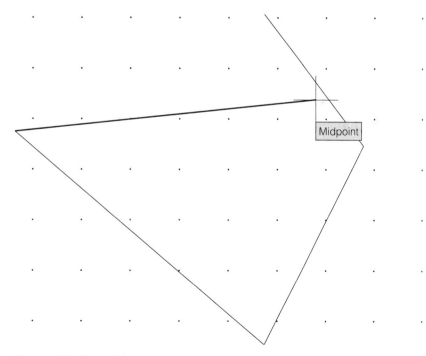

Figure 4.6   Defining the second point of a line by snapping on to the midpoint of another line.

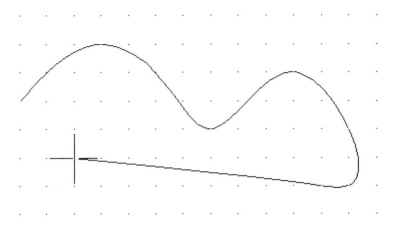

Figure 4.7   A spline.

onto the midpoint of another line. The triangle on the line highlights the midpoint. The dark line is a rubber line showing the line being defined.

As has been mentioned, there are multiple ways to define each geometric entity in CAD systems. Like the triangle, the point, circle, and circular arc need no explanation. A **polyline** consists of a sequence of lines and arcs. For example, the geometry in Figure 4.6 could be defined by each of the individual lines or by one polyline. A **spline** is a curve fitted to a set of points (Figure 4.7). There are several different types of spline curves. The simplest is one that passes through all of the given points. **Bezier's curve**, discussed in Chapter 5, is defined by curve segments, each of which is in turn defined by four control points. The curve begins at the first control point and ends at the last. Two middle control points pull and push the curve into shape, but do not touch it. (Actually, they define the tangent vectors at the beginning and the end of the curve.) Often, the user defines the points through which the curve will pass, and the CAD system computes the control points (those outside of the curve), using the curve-fitting algorithm. After a curve has been defined, the user can modify it with control points.

An **area** is enclosed by a closed loop of lines and arcs. An area can be shaded with **hatch** (Figure 4.8) and used to define cross sections that have been cut off.

A CAD system offers the user a set of tools with which to modify the geometry of an object. Commonly available tools include move, copy, rotate, trim, break, chamfer, and erase. Within each tool command are subcommands that allow the user better and more precise control of the operation. For example, copy can make a single copy, multiple copies in an array pattern, etc. Combined with the snap tool, copy makes precise drawings. The **move, copy**, and **erase** commands are fairly intuitive. **Trim** is used to cut away part of an object (a geometric entity), using another object as the reference. For example, in Figure 4.9, the circle is used as the cutting edge to trim the two lines. Later, the lines are used to trim the circle.

The **break** command is used to divide a line or other objects into two parts. Then the divided object can be manipulated separately. Finally, the **chamfer** (or **fillet**) command adds a chamfer (fillet) to the corner of an object (Figure 4.10).

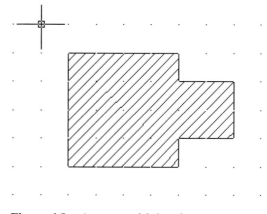

Figure 4.8   An area with hatch.

After the geometry is correctly defined, the next step is to add dimensions, tolerances, and annotations. Using built-in dimension tools, one can add dimensions and tolerances easily. A dimension has several parts: dimension and tolerance (±0.001) values, a dimension line with arrowheads, and two extension lines. Use snap to pick the

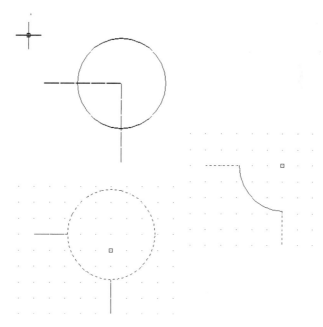

Figure 4.9   The trim command. The upper left drawing shows a circle and two lines. The bottom left drawing shows trimmed lines. Finally, the drawing on the right shows how the circle gets trimmed by two lines.

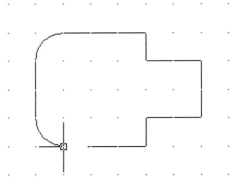

Figure 4.10   The chamfer command add two fillets to the corners.

feature (geometric entity) to be dimensioned; the dimension value is added automatically. Figure 4.11 is a simple drawing with dimensions and dimensional tolerances. The geometric tolerance (perpendicularity, as introduced in Chapter 3) and the datum "A" are added in the same manner. The dimension menu shows different ways dimensions and tolerances can be added to the drawing. The text "Center hole" over the hole diameter dimension is the annotation added to the dimension.

By using these geometric entities and tools, a complex 2-D drawing can be created step-by-step. The process is much more flexible than drawing on paper with a drafting board and drafting instruments. Editing the drawing is also easy. In manual drafting, one often uses templates to help draw standard components, such as bolts,

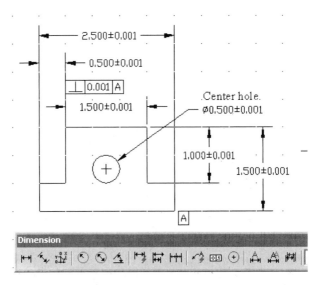

Figure 4.11   Dimensions, tolerances and menu for dimensioning.

screws, and gears. CAD systems, offer the user libraries of drafting symbols ranging from electronic components and machine parts to buildings and trees. Productivity increases far outpace the cost of the CAD system (software, hardware, and training).

4.6 3-D CAD AND SURFACE MODELING

A 3-D CAD model includes not only X- and Y-coordinates, but also the Z-coordinate. A simple 3-D model is a wire-frame model consisting only of lines, circles, and curves. For shapes such as the hull of a ship, the body of a car, or even the body of a hair blower, a **sculptured surface** (or **free-form surface**) is used. Such surfaces and many other analytic surfaces (e.g., a spherical surface, a conical surface, a surface of revolution) can't be represented by lines, circles, and curves. In the era of drafting, a sculptured surface was represented by its cross sections. At fixed intervals on the third axis (e.g., the Z-axis), cross-sectional drawings are made. The actual surface is interpolated from the set of drawings. Surface modeling allows the designer to model the sculptured surface directly.

4.6.1 3-D Wire-Frame Modeling

Even when using a 3-D CAD system, one is still working with a 2-D screen. One can work with either the ordinary projection (i.e., $X–Y$, $Y–Z$, and $X–Z$ views) or an orthogonal projection consisting of X, Y, and Z. Either kind of projection is represented by a **coordinate frame** (or a **coordinate system**) (Figure 4.12). The base coordinate frame is called the **world coordinate system** (WCS in AutoCAD). For the convenience of the designer, multiple **user coordinate systems** (UCSs) can be defined. A user coordinate system is defined on the basis of the world coordinate system. However, its origin can be translated (moved) and the system rotated (X-, Y-, and Z-axes not aligned with the X-, Y-, and Z-axes of the WCS). In Figure 4.12, two coordinate systems are displayed. In this case, the origins of both systems are the same. However, the UCS is rotated: The X-axis of the UCS is aligned with the Y-axis of the WCS, and the Y-axis of the UCS is aligned with the Z-axis of the WCS.

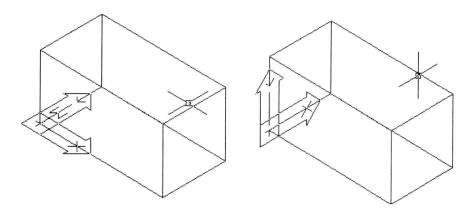

Figure 4.12   World and user coordinate systems.

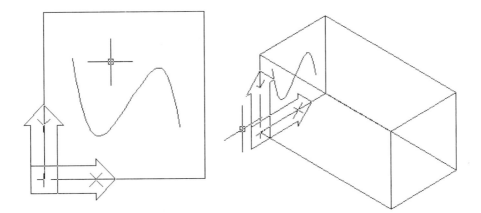

Figure 4.13 Drawing a curve on the UCS (left) and the result displayed in the orthogonal view (right).

The user may select either the WCS or the UCS for his or her work. In Figure 4.12, if the WCS is chosen, all the 2-D lines, circles, and curves will be defined on the bottom face of the rectangular box. If the UCS is chosen, the objects are drawn on the end of the box (Figure 4.13). To draw on the other face, one needs to define a new UCS on that face. Therefore, knowing how to use a coordinate system in 3-D modeling is an essential skill.

The drawings shown in Figures 4.12 and 4.13 are wire-frame models, represented by their edges only. In Figure 4.14, a second box with a dome is added to the original box. The second box is defined in the world coordinate system and rotated 10 degrees about the *X*-axis. A user coordinate system is then defined on the upper face (imaginary face) of the second box. The dome is defined in the user coordinate system and therefore is rotated with respect to the world coordinate system, too. In this drawing, the two boxes interfere with each other. All eight edges of each of the boxes are displayed. The dome is also represented by a wire mesh. The drawing is rather confusing, since, in the real world, no such view is possible.

4.6.2 Surface Modeling

There are several methods used in surface modeling. The simplest surface is a planar one. Other surfaces include a **ruled surface**, a **tabulated cylinder**, a **surface of revolution**, and **free-form surfaces**. A ruled surface is defined by two curves. The surface blends the two curves together (Figure 4.15). By contrast, a tabulated cylinder blends a curve and a generating vector (line segment) (Figure 4.16). The tabulated cylinder in the figure uses the same base curve as that shown in Figure 4.15. The vector is defined by the endpoints of the two curves in Figure 4.15. The result is a very different surface. A tabulated cylinder has a constant cross section in the direction of the generating vector. A surface of revolution is defined by a curve and an axis of revolution. In Figure 4.17, the surface is defined by the same curve as in the previous two examples. The axis of revolution is the *X*-axis. The rotation is from 0° to 180°. Complex axial symmetric objects, such as wineglasses, can be easily designed with the use of a surface of revolution (Figure 4.18).

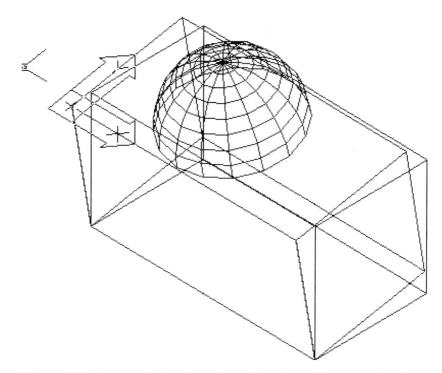

Figure 4.14   A second wire-frame box with a dome. The second box is tilted with respect to the first box.

Early surface design techniques from drafting days, such as **lofting**, were implemented in CAD systems for free-form surfaces. Lofting consists of meshing a set of cross-sectional (2-D) drawings into a surface. The mathematics for blending cross-sectional curves was already developed. Later, **parametric polynomial curves** and surfaces were developed for modeling free-form surfaces. Equations of some of those models are presented in Chapter 5.

A simple way to define a free-form surface is by using 3-D meshes. A mesh is defined by four corner points (vertices), each itself defined by its X-, Y-, and Z-coordinates. Figure 4.19 shows a surface defined by 4×4 meshes. Of course, such a surface is not smooth and is extremely tedious to define. A **Coon's surface** is defined by its edges (Figure 4.20). Four corner edge curves are blended together to form the surface. Coon's surface is named after its inventor. Other popular surfaces are Bezier's surface (invented by French engineer Dr. P. Bezier), the B-spline surface, and the Nonuniform rational B-spline (NURB) surface. These surfaces are defined by control points. (See Section 5.4).

A **bicubic**[1] Bezier's surface is defined by 16 control points (4 times 4). The surface is anchored at four corners, and the rest of the surface may not touch the control points (Figure 4.21). More examples can be found in Chapter 5. The shape of the curve is

---

[1]"Bi" means in two directions: t and s parameters, for example. "cubic" denotes the fact that the defining polynomial equation is a cubic curve—a curve up to the power of three ($t^1, t^2, t^3, s^1, s^2, s^3$).

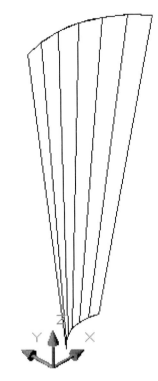

Figure 4.15 A ruled surface.

manipulated by the control points. More flexible than a Bezier's surface, a B-spline surface is defined by M times N control points. The order of the surface (say, cubic) is controlled by a separate variable. In the design of a large curved surface, a Bezier's surface requires multiple 4×4 patches. By contrast, one may define the entire surface with one B-spline surface. To add details to the surface, one just adds more control points.

A NURB surface is the most flexible free-form surface. Similar to a B-spline surface, it is defined by M times N control points. Each control point is also assigned a weight w_i. This weight can be considered analogous to gravity and the control points analogous to stars. The shape of the surface is formed by the pulling force of all of the control points. Without moving the control points, one can change the shape of the surface by changing the weights (producing stronger pulls on the affected control points). This feature lends an added flexibility to the surface. Mathematically, the weight is applied to the control point with rational-number coordinates. A rational number has the form X_i/h_i, where X_i is the X-coordinate of control point i and h_i equals $1/w_i$. It is the rational number from which the "R" derives in the name of the surface (NURB).

The "nonuniform" part of the name comes from another feature of the NURB surface: It means that the knot spacing associated with each control point is not a constant.

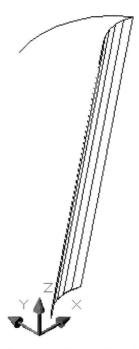

Figure 4.16   A tabulated cylinder. Use the same base curve as that for the ruled surface in Figure 4.14 to define the surface shown here.

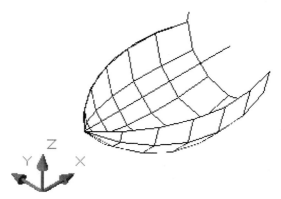

Figure 4.17   Surface of revolution.

The concept of a knot is not obvious physically; it is used in the mathematical model. (See Chapter 5.) Again, the shape of this kind of surface can be changed by varying the knot spacing. With control points, weights, and knot spacing, a NURB surface can model both free-form surfaces and analytic surfaces (such as the surface of a sphere or cone).

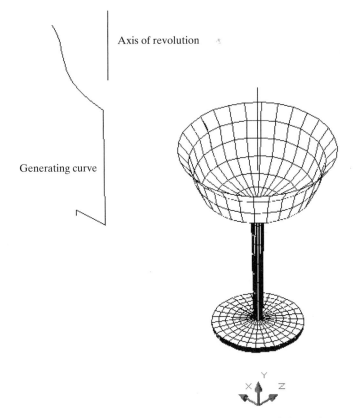

Axis of revolution

Generating curve

Figure 4.18   A glass, its generating curves, and the axis of revolution used.

4.7 PRODUCT DESIGN USING SOLID MODELS

Solid models provide "volume" information. A product designed by the use of a solid model is a complete and unambiguous geometric model. Mathematically, one can manipulate the model to project it into 2-D drawings, finding cross sections, calculating centers of gravity, computing moments of inertia, and the like. There are a few methods in building 3-D solids. **Constructive solid geometry** (CSG) is a technique that uses **Boolean operators** on solid primitives to build the desired solid. Internally, the solid is stored in a complex data structure (see Chapter 5) consisting of shells, faces, edges, and vertices. **Euler operators** can be used to manipulate individual data elements to change the solid. Another convenient way to build a solid is **sweeping**, the process of taking a 2-D closed area and extending it along a third dimension to form a solid. There are two types of sweeping operations: **linear sweep** (or **extrude**), and **rotational sweep** (or **revolve**).

4.7.1 Solid Primitives and Boolean Operators

Solid primitives are the basic building blocks of a solid model. They are simple solid shapes, such as a block, cylinder, sphere, wedge, torus, and cone (Figure 4.22). Each

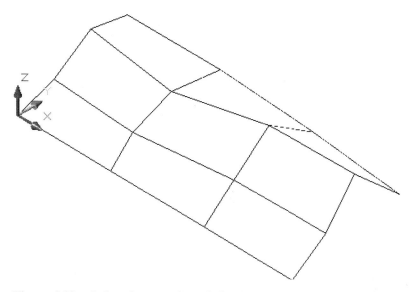

Figure 4.19   A free-form surface defined by 3-D meshes.

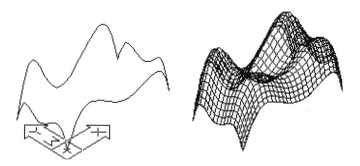

Figure 4.20   Coon's surface. On the left are four defining curves.

primitive is defined using parameters. For example, a block is defined by its width, length, and height. Using basic graphics operators such as move and rotate, the designer can position and orient solid primitives anywhere in the coordinate system. To build a complex solid object, Boolean operators are used to combine primitives (Figure 4.23). Boolean operators include union, subtraction, and intersection. Union operators join several solids together to form a single solid. The subtraction operator removes the intersection part of two solids from another solid. The intersection operator finds the intersection of two solids. Boolean operators cannot operate on nonsolids. Often, it is hard to tell whether an object in the design window is a solid or just a 3-D wire-frame model. Boolean operators do not work on wire-frame models. Boolean operators also fail if the solid objects being operated upon are disjoint.

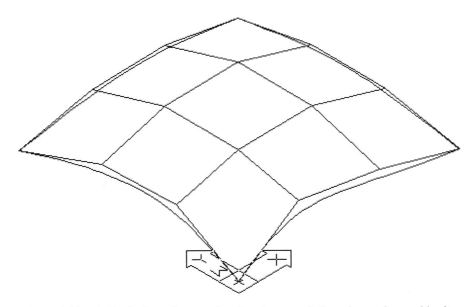

Figure 4.21   A Bezier's surface under the characteristic polygon formed by its control points.

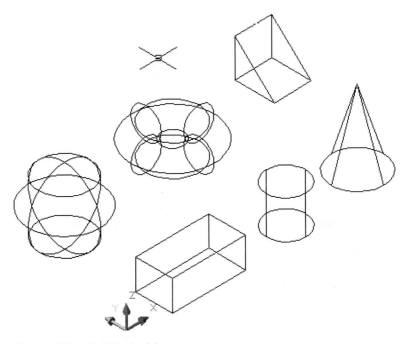

Figure 4.22   Solid primitives.

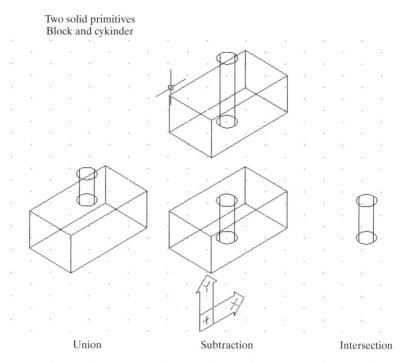

Figure 4.23   Boolean operators: union, subtraction, and intersection.

4.7.2 Euler Operators and Face and Edge Operators

Euler operators are used to manipulate the **topology** (see Chapter 5) of the solid. For the purpose of manipulating its representation, the solid is modeled with the use of a **boundary representation (B-rep)**. A simple B-rep is a hierarchical structure consisting of faces, a loop, edges, and vertices (Figure 4.24). The connections between these entities in the B-rep capture the topology of the solid object. While the face, edge, and vertex are geometrical entities, the loop is purely for storing topology. The **loop** is a closed loop of edges. For example, a rectangular face has a loop of four straight edges. A face with a hole in it has two loops: an inner loop for the hole and an outer loop for the face. When using Boolean operators to combine two solids, the designer employs a process called "**boundary evaluation**" to delete or otherwise modify the entities and connections of the two models and to merge the models. The graphics package then uses this boundary representation model to generate the desired display, as either a wire frame, a projection with hidden lines removed, or a shaded image. A process called **rendering** can add lighting, color, and texture to the faces and create realistic images of the quality of photographs.

Boolean operators do not operate on the B-rep elements (entities and connections) directly. To modify B-rep elements locally requires a different set of operators. Euler operators (named after the mathematician Euler) include **Mxyz and Kxyz**, where M indicates MAKE, K denotes KILL, and X, y, and z may represent a vertex, edge, face, loop, shell, or genus (hole). For example, MEV stands for "Make an edge

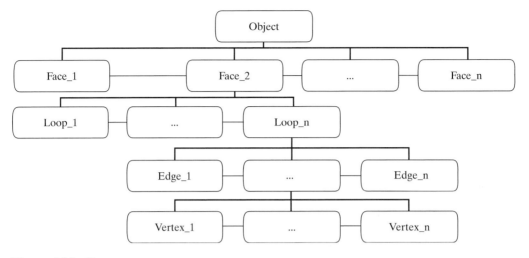

Figure 4.24   B-rep.

and a vertex," while KSG abbreviates "Kill a shell and a hole." Neither operator is intuitive to most designers. In CAD systems, a set of operators based on Euler operators allows designers to tweak faces, edges, and vertices. Note that tweaking is different from changing parameters, which does not affect the object's topology: The structure of the B-rep remains intact. By contrast, tweaking a face serves to extrude it, delete it, move it, etc. The structure of the B-rep changes accordingly.

In CAD systems such as AutoCAD, there are no explicit Euler operators. The set of operators related to those discussed so far might be called face and edge operators. Face operators include move, delete, copy, offset, extrude, taper, rotate, and color. Figure 4.25 illustrates two operations: face move and face taper. To move a face, that face is selected, a height is specified, and a taper angle is entered. The result is shown in the figure. Note that the original face is still in the model. AutoCAD uses ACIS solid modeler, a **nonmanifold** modeler. (See next section.) A solid is allowed to have an extra face in it; however, it is marked as internal and will not affect the computation of volume.

4.7.3 Manifold and Nonmanifold Models

Currently, the majority of 3-D solid modelers are based on either the CSG or the B-rep representation. CSG data input is the most popular. Often, many face types can be used in the solid model. The 3-D solid models discussed so far are called *manifold* (in particular, 2-manifold) models.[2] In a 3-D manifold model, the dimensionality is maintained. The B-rep of an object consists only of bounded faces, with no loose edge or face. Each

---

[2]"Manifold" is a term from topology. "2-manifold" or "two-dimensional manifold" describes objects that, when inspected locally, look like objects from Euclidean 2-space. For example, a ball is a 2-manifold object. When the ball is sufficiently larger than the observer (e.g., a human being on earth), the observer initially assumes that the ball is flat (a Euclidean two-dimensional plane). A torus, a box, a cylinder, and objects we use in real life are mostly 2-manifold objects.

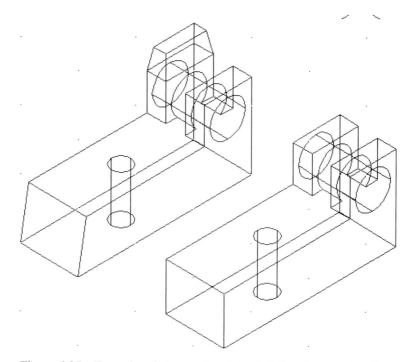

Figure 4.25   Front face is tapered and top left face is moved and side tapered.

edge is bounded by exactly two vertices and is adjacent to exactly two faces. Each vertex belongs to one disk. (Hence, traveling from one adjacent face to another will never cross the vertex itself.) One can easily verify this point by using vertex A in Figure 4.26. The manifold model does not allow any dangling faces and edges. There are also nonmanifold modelers, which allow additional faces and edges to exist in a solid model. Tracking the inside and outside, however, becomes more complex. A nonmanifold modeler such

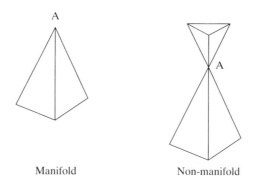

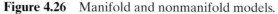

Figure 4.26   Manifold and nonmanifold models.

as ACIS (used in many CAD systems, including AutoCAD) allows much more flexibility, as more design intents can be saved in the design model.

Since a solid model is represented in its entirety, engineering analysis can be performed directly on the model, which, incidentally, provides a common linkage between design, analysis, and manufacturing. For applications such as **rapid prototyping** (layered manufacturing; see Chapter 14), a solid model is a must. A machine such as a stereolithography device takes as input a 3-D solid model in a special B-rep format (the model is a triangulated B-rep model) called STL file which means that every face is a triangular face. The machine controls a computer that slices the model into thin layers. A laser beam is used to draw the cross section of the layers and selectively cure the photopolymer. Parts can be built directly from this solid-model representation.

4.8 FEATURE-BASED MODELING

The data available in geometric models is maintained at a low level (e.g., faces, edges, vertices, etc., or solid primitives in a CSG model along with set operations). Often, the data are not directly suitable for performing analysis for downstream applications such as manufacturability analysis and process planning. These applications are not explicitly available in geometric models, and extraction of the entities needed for reasoning about the part is a nontrivial task.

Features are the next step in the evolution of CAD systems and product modeling techniques aimed at pushing the geometric models into product models that can be used in various stages of the product life cycle. In feature-based modelers, parts are viewed as sets of features that can best be described as a higher level entity imparting engineering significance to the geometry of the part or its assembly. A feature-based model can be viewed as consisting of two interrelated components: the feature model and an associated geometric model (Figure 4.27).

Several different approaches have been taken to create feature models.

4.8.1 Interactive Feature Definition

In interactive feature definition, a typical geometric model is created in the traditional manner, and features of interest are defined interactively later by picking the elements of the geometry and the topological entities that correspond to a feature. This approach

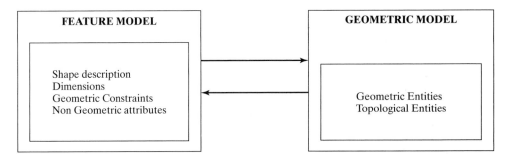

Figure 4.27   Feature model and its associated geometric model.

is not very popular for design purposes, but it does provide a means for establishing a feature library interactively and for creating user-defined features.

4.8.2 Automatic Feature Recognition

In automatic feature recognition, the geometric model is also created first, and computer programs are used to automatically recognize part features from the underlying geometry and topology of the part model. Algorithms capable of automatic feature recognition are based on recognizing machining volumes, as well as recognizing predefined features from the part geometry and topology. Automatic feature extraction still lacks robust recognition algorithms and has difficulty recognizing features in the presence of feature interactions. These procedures are often limited to classes of predefined features.

4.8.3 Design by Features

In **design by features**, the part is created by the addition of features, each of which changes the part geometry and adds information to the part model. Feature definitions are provided in feature libraries. The features to be used are instantiated by specifying the various parameters needed to completely specify the feature. The parameters may include dimensions, location parameters, constraints, and relationships. The manner in which the constraints are handled is different in different systems, and a further distinction can be made to partition the design-by-features approach into parametric and variational systems (discussed in the next section).

Since design by features is the most popular approach of the three, this section will discuss some of its aspects.

Features are often related to the process used to design, manufacture, or inspect parts and are often specific to the application. Various types of features can be used for modeling:

- *Geometric form features:* Geometric shapes that occur frequently in parts.
- *Tolerance features:* Features related to the specification of tolerances (e.g., datums).
- *Manufacturing features:* Features derived from the manufacturing process used (e.g., a hole, shaft, chamfer, slot, round, cut, protrusion, flange, rib, and shell).
- *Design construction features:* Features used to enable construction of the part (e.g., datum planes and coordinate systems and axes).
- *Assembly features:* Features describing relationships of various parts in an assembly (e.g., mating conditions, relative positions, fits, and kinematics relations).

The different features cannot all be defined a priori. Most modelers provide a diverse set of features for the application domains supported by the modeler. Most modelers also provide the capability of including user-defined features, and most support the creation of a user-defined feature library to customize the features for the user's particular domain.

Parts are stored as a series of features together with geometric information associated with each feature. The shape of the feature is defined by its dimensions, its geometric and topological entities, and its relationship to other features. Figure 4.28

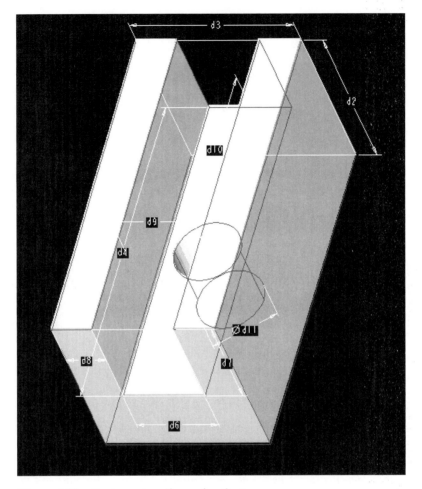

Figure 4.28   Part construction using features.

shows a part constructed from features. Initially, the design construction features (e.g., datum planes and coordinates systems) are defined, and then the first feature is created as an extrusion, followed by a cut feature, and a hole feature. The relationships between the features are defined through the dimensioning scheme used to locate the various features with respect to each other. By further defining relations between the various parameters, the design intent can be captured. For example, if the designer's intent is to maintain the hole in the center of the block, then relations defining the parameters used for placement of the hole enforce this design intent. The approach imposes a hierarchical ordering among the features, which is often represented in the form of parent–child relationships. Since the model is created one feature at a time, the creation of parent–child relationships often has an impact on editing of the features and on feature deletion. For example, in Figure 4.28, the hole feature is a child of the slot feature, because it is dimensioned from the edge of the slot. Deleting the slot feature will thus have an impact on the hole feature.

4.9 VARIATIONAL AND PARAMETRIC MODELING

Another problem associated with using only geometric models is their complete inability to capture the design intent. For example, if the designer requires that a hole be centered in a block, then in a pure geometric model the hole would be designed with the use of a cylinder primitive and a Boolean subtraction operation and would be placed in the center of the block. If the designer subsequently changed the dimensions of the block, the hole would no longer be centered.

This problem can be addressed by specifying the model in terms of various constraints between the higher level entities or features used to design the part, instead of specifying rigid dimensions. In the example of the previous paragraph, rather than explicitly specifying the dimensions of the hole from each face, we can specify the dimension as a constraint (or relationship) based on the size of the block. The hole is located at a distance $d9 = d3/2$ and $d10 = d4/2$ from the two edges of the part, where $d3$ and $d4$ are the dimensions of the block. In this case, the dimensions locating the hole are specified as relations based on the block's dimensions. Now if the designer changes the size of the block, the hole will remain centered in the block.

Since the part dimensions have specific variables called parameters associated with them, and since making changes to the parameters and subsequently reevaluating the constraints and relations can easily create new parts, these approaches are called *parametric* or *variational* models. The terms "parametric" and "variational" are often used interchangeably in the commercial CAD community; although fundamental technical differences exist between them, they support similar design processes, and, to the end user, the two types of systems are similar.

Variational/parametric modelers were developed to address two main design concerns: the inability of geometric modelers to capture a design's intent and the inability to support the concept of variational design. Eighty percent of all design tasks are adaptations and modifications of existing designs, and reusing existing drawings and making only small changes to them is important to support standardized parts and product families.

The flowchart shown in Figure 4.29 uses a variational/parametric modeler to illustrate the design process.

The basic technical difference between the parametric and the variational approach is the way in which the constraints are defined and solved. In the parametric approach, the construction is modeled as a sequence of assignments to model variables, where each assignment computes the values of those variables as a function of previously computed variables or original parameters. Hence, this approach dictates a sequential strategy for evaluating the constraint equations. To create variations of the product, the parameters are changed and the design construction sequence is reevaluated in creating each part. Variational systems, by contrast, employ a simultaneous solution of the set of governing equations representing the constraints.

Parametric models can be evaluated quickly, i.e., they do not require solving complex simultaneous equations. Equations are solved one at a time in a sequential manner. They cannot deal with constraints that are mutually coupled, however. Variational models can handle mutually coupled constraints, but are much slower and limited in their capability of handling models that have inconsistencies and are incompletely satisfied.

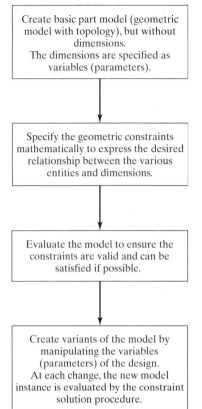

Figure 4.29 Design process using variational/parametric modeler.

4.10 CAD APPLICATIONS

Table 4.1 lists some important applications of CAD. No detailed explanation is given. Readers are encouraged to refer to other books on CAD and the Internet for further information.

4.11 DESIGN DATA EXCHANGE

Each CAD system has its own proprietary data format. To transfer data from one CAD system to another, translation software has to be written. When the number of CAD systems increases, the number of translators must also increase. Using a specialized translator is uneconomical and difficult to manage. The following equation shows the number N of two-way translators needed for n CAD systems to exchange data among them:

$$N = C\binom{n}{2} = \frac{n!}{2!(n-2)!}$$

TABLE 4.1   Some Important Applications of CAD

Area	Applications
Design	Assembly layout
	New-part design
	Standard part library
	Tolerance specification
	Interface and clearance specification
	Part relations in an assembly
Analysis	Interference checking
	Fit analysis
	Weight and balance
	Volume and area properties
	Structural analysis
	Kinematics analysis
	Tolerance stacking
Documentation	Drawing generation
	Technical illustrations
	Bill of materials
	Image rending
Manufacturing	Process planning
	NC part program generation
	NC part program verification
	NC machine simulation
	Inspection programming
	Robot programming and verification
	Factory layout
Management	Review and release
	Engineering changes
	Project control and monitoring
	Selection of standard parts and assemblies
	Design standards

The following table shows the number of CAD systems and the number of translators needed:

n	N
2	1
3	3
4	6
5	10
6	15
7	21
8	28

Data exchange happens not only between CAD systems, but also between CAD and CAM packages. A standard data exchange format (intermediate data format) will

Figure 4.30   CAD data exchange.

reduce the need for multiple translators. When such a data file is used, each CAD system needs only one preprocessor and one postprocessor (one set of translators). The preprocessor imports the data into the system, and the postprocessor outputs the file into standard format (Figure 4.30). Many CAD data-exchange formats have been developed by various countries and organizations. The most widely accepted formats are IGES and STEP. Both are international standards.

4.11.1 IGES

The International (originally "Initial") Graphics Exchange Standard (IGES)[3] was developed in the 1980s as the standard CAD data-exchange format. It was accepted as an ANSI American national standard and later as an ISO international standard. IGES was developed for 2-D, 3-D, and surface-modeling data exchange. Almost all CAD systems support IGES, which has the following general structure:

Flag section

Start

Global: pre- or postprocessor information

Directory entry: one for each entity, line type, color, and so on

Parameter data: parameters associated with each entity

Terminate

Geometric entities are coded in the directory entry in the form of numbers (e.g., 100, circular arc; 102, composite curve; and 110, line). Most 2-D and 3-D geometries, including NURB surfaces, are supported. Each entity has associated parameters stored in the parameter data section. For example, for a line entity, we could have the following entries:

Directory Data

   110 10

Parameter Data

   110, 1.0,1.0,0.0,2.0,2.0,0.0

The above line, called line_10, is defined from $(1.0, 1.0, 0.0)$ to $(2.0, 2.0, 0.0)$.

The purpose of IGES is to provide an intermediate data file between two CAD systems. Unfortunately, many CAD systems add features not covered by IGES, so, as a standard, it cannot catch up with new development in the marketplace. Also, certain entities in the original CAD model may be lost or incorrectly translated into the target CAD model. It is thus important to know the limitations of a system and to verify the results obtained.

---

[3]For the most recent IGES standard, see ANSI/USPRO 1996 [6].

4.11.2 Step

STEP (ISO 10303), the International Standard for the Exchange of Product Model Data, was a major effort originally undertaken by European countries under ISO. Parallel to the STEP effort was the Product Description Exchange Standard (PDES,[4] later changed to "for STEP") project in the United States. The international standard that resulted is a combined STEP–PDES standard (called STEP from this point on). The goal of STEP is to define a standard file format that includes all information necessary to describe a product from design to production (the product's life cycle). STEP also supports multiple-application domains (i.e., mechanical engineering, electronics, etc.) and contains product geometry, product functions, and a process plan for the product. The product geometry includes drawing, features, CSG, and B-rep and, like STEP itself, is arranged into application areas (mechanical, electrical, AEC, etc.). The standard is divided into categories. For example, the 01x series is concerned with definitional resources and the 02x series with implementation methods. The Application Protocols (APs) appear in the 2xx series, and the 5xx series is known as Application Interpreted Constructs (AICs). The following are standards published up to the year 2001 [Loffredo; Pratt, 2000]:

- Part 001. Overview (1994)
- Part 011. EXPRESS language (1994)
- Part 021. Clear text encoding of the exchange structure (1994)
- Part 022. Standard data access interface (1998)
- Part 023. Standard Data Access Interface (SDAI) C++ binding (2000)
- Part 027. SDAI Java binding (2000)
- Part 041. Product description and support (1994)
- Part 042. Geometric and topological representation (1994)
- Part 043. Representation structures (1994)
- Part 044. Product structure configuration (1994)
- Part 045. Materials (1998)
- Part 046. Visual presentation (1994)
- Part 047. Shape tolerances (1997)
- Part 049. Process Structure and Properties (1998)
- Part 101. Drafting (1994)
- Part 104. Finite-element analysis (2000)
- Part 105. Kinematics (1996)
- AP 201. Explicit draughting (1994)
- AP 202. Associative draughting (1996)
- AP 203. Configuration-controlled design (1994)
- AP 207. Sheet-metal die planning and design (1999)

---

[4]For most recent PDES standard, see reference ANSI/USPRO 1994 [7]. PDES is contained in 12-volume standards.

- AP 224. Mechanical product definition for process planning (1999)
- AP 225. Building elements using explicit shape representation (1999)
- AP 210. Electronic assembly, interconnection, and packaging design
- AP 212. Electrotechnical design and installation
- AP 227. Plant spatial configuration

Other, yet to be finished, standards include the following:

- AP 205. Mechanical design using surface representation
- AP 206. Mechanical design using wire-frame representation
- AP 207. Sheet-metal dies and blocks
- AP 208. Life-cycle product change process
- AP 213. Numerical control process plans for machined parts
- AP 218. Ship structures
- AP 221. Functional data and schematic representation for process plans
- AP 226 Shipbuilding mechanical systems

Of the preceding standards, AP203 is closest to IGES. A 600-page document covering more than geometry exchange, AP 203 includes assembly structure and configuration control information (e.g., part versioning, release status, authorization data, etc.). The standard covers a lot more than IGES and has taken hundreds of people since 1984 to develop it. So far, it is an ongoing project and has been implemented in most commercial CAD systems.

4.12 SUMMARY

In this chapter, CAD technology was introduced. A powerful tool for engineers, CAD has become an essential tool for every engineer and engineering student. It is important to understand the basic features of such a tool. While early CAD systems merely replaced the drafting board and drafting instruments, modern CAD systems are modeling and analysis tools. Knowledge of surface and solid modeling are important. A geometric modeling tool enables engineers to model, analyze, prototype, and produce the artifact. We have focused here on the concept of using CAD as a modeling tool. In the next chapter, the theory and mathematics of geometric modeling are presented. Readers will gain knowledge of how the system works, as opposed to how to use the system, the subject of the current chapter.

4.13 REVIEW QUESTIONS

4.1 Discuss how a design idea is represented in a designer's mind—that is, in the form of an equation, line drawing, and so on.

4.2 Discuss the advantages and disadvantages of using a 3-D solid model in mechanical part design.

4.3 Based on applications, what are the major classifications of CAD?

4.4 What are the differences between a wire-frame model and a solid model?

4.5 Why is a wire-frame model ambiguous and a solid model not?

4.6 What are parametric design and variational design?

4.7 What are the advantages and disadvantages of using a feature-based design system for mechanical part design?

4.8 What are the difficulties of including tolerance specifications in a solid model?

4.9 What is the progress of the STEP standard? How many more parts and APs have been released?

4.14 REVIEW PROBLEMS

4.1 Use a CAD system of your choice to make a three-view drawing of the part shown in Figure 4.31.

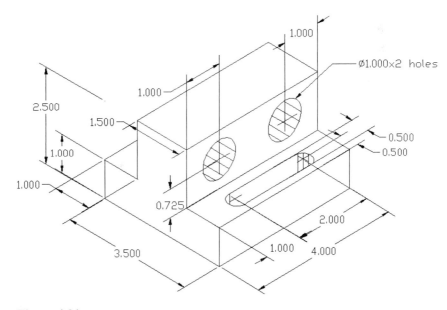

Figure 4.31

4.2 Using a 3-D wire-frame, model the part shown in Figure 4.32.

4.3 Use a solid modeler to model the part in Figure 4.33.

4.4 Use the profile defined in Figure 4.34 to design a solid model.

4.5 Use the profile defined in Figure 4.35 to design a surface model.

4.6 Use 2-D drafting to model the object shown in Figure 4.25. Since no dimensions are given, assign your own.

4.7 Use a 3-D solid model to model the object shown in Figure 4.25.

4.8 Model a pencil, using (a) 3-D drafting, (b) a wire-frame model, and (c) a 3-D solid model.

4.9 Use a surface model to model a toy car.

4.10 Use a surface model to model a coffee mug.

4.11 Model a cell phone, using (a) 3-D drafting, (b) a wire-frame model, and (c) a 3-D solid model.

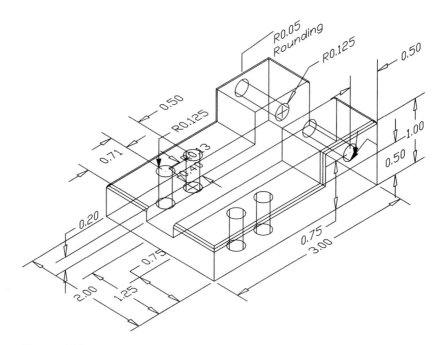

Figure 4.32

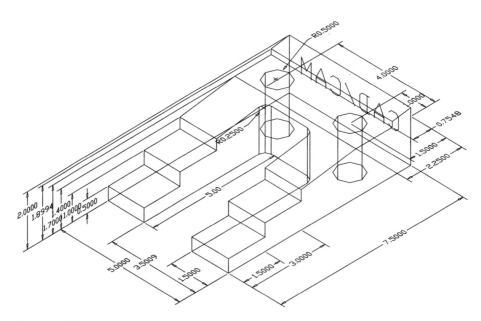

Figure 4.33

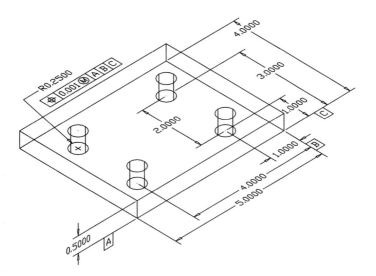

Figure 4.34

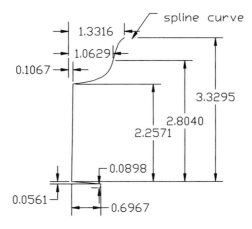

Figure 4.35

4.15 REFERENCES

ANSI/US PRO/IPO. *Product Data Exchange Using STEP (PDES)*, 12 volumes. US PRO, 1994.

ANSI/USPRO. *Initial Graphics Exchange Specification IGES 5.3*. US PRO, 1996.

Bezier, P. *Numerical Control—Mathematics and Applications*, trans. A. R. Forrest. London: John Wiley, 1972.

Braid, I.C. *Designing with Volumes* (technical report). Cambridge, U. K.: Cambridge University, CAD Group, 1973.

Burns, Marshall. *Automated Fabrication*. Englewood Cliffs, NJ: Prentice Hall PTR, 1993.

Chang, T. C., Wysk, R. A., and Wang, H.P. *Computer-Aided Manufacturing*, 2d ed. Upper Saddle River, NJ: Prentice Hall, 1998.

Das, Biman, and A.K. Sengupta. "Computer-Aided Human Modelling Programs for Work Station Design," *Ergonomics*, 1995, vol. 38, no. 9, 1958–1972.

ISO, for STEP standard document: http://www.iso.ch/iso/en/ISOOnline.openerpage

Jacobs, Paul F. Rapid Prototyping & Manufacturing: Fundamentals of Stereolithography. Dearborn, MI: SME, 1992.

Loffredo, D. "Fundamentals of STEP Implementation," STEP Tools, Inc.,
http://www.steptools.com

Mortenson, M.E. *Geometric Modeling*. New York: John Wiley, 1985.

"Ergonomic Modeling with MGPro," *Occupational Health and Safety*, October 1997.

Pratt, M.J. "Introduction to ISO 10303—the STEP Standard for Product Data Exchange," *ASME Journal of Computing and Information Science in Engineering*, November 2000.

Requicha, A. A. G., and H. B. Voelcker. "Solid Modeling: A Historical Summary and Contemporary Assessment," *IEEE Computer Graphics and Applications*, 2(2), 1982, 9–24.

Shah, Jami J., and Martti Mantyla. *Parametric and Feature Based CAD/CAM*. New York: John Wiley and Sons, 1995.

Sutherland, I. E. *SKETCHPAD: A Man–Machine Graphical Communication System*. Baltimore: Spartan Books, 1963.

Voelcker, H. B., and A. A. G. Requicha. "Geometric Modeling of Mechanical Parts and Processes," *Computer*, 10(12), 48–57.

Chapter 5

Geometric Modeling

Objective

CAD is a tool that is used to model the geometry of an artifact with computers. The purpose of a CAD system is to efficiently perform all of the required functions for geometric modeling. To understand CAD, one needs to know the mathematics and algorithms built into a CAD system. This chapter discusses computer graphics, 2-D and 3-D geometries, surface models, and solid models. It is intended to provide readers with a background on the basics of geometric modeling.

Outline

5.1 BACKGROUND

Today, CAD systems are used in the development of most engineering designs. To utilize a particular CAD system, one needs to know the syntax of the commands and the procedures used in the operation of the system. To develop a CAD system or applications based on it, knowledge of geometric modeling becomes crucial. Geometric modeling is the "physics" of the CAD system. How an object is represented internally in

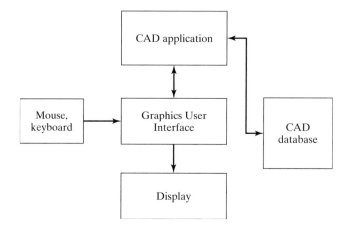

Figure 5.1   CAD architecture.

the computer and externally on the computer monitor dictates the limits of many applications. Also, mathematics is used in implementing operators that manipulate geometric entities. As engineers, we should not treat CAD as a black box; instead, we should understand the principles and mathematics built into the system. This chapter provides background on the fundamentals of geometric modeling. The concept of geometric modeling and the basic mathematics for its use are discussed. Implementing these models and operators requires a deep knowledge of computer programming. We make no attempt to discuss any actual computer implementation.

Figure 5.1 shows the architecture of a CAD system. The CAD application (CAD system) interacts with the user through a **Graphics User Interface** (GUI), a software package that handles both event-driven activities associated with input devices and display functions, such as windows, menus, dialogs, and 2-D and 3-D graphics. The CAD application calls functions in the GUI to accomplish the I/O and graphics interactions. At the same time, the CAD application handles the creation, modification, and manipulation of geometric models and the database that stores the model. Certain basic geometric functions are used in both GUI and CAD applications. Such functions include, but are not limited to, coordinate transformations. In the interests of providing a clear understanding of a CAD system, the basic geometric functions will be discussed first.

5.2 BASIC ELEMENTS OF COMPUTER GRAPHICS

All design models are set in a coordinate system. A 3-D **Cartesian coordinate system** is represented by three axes—X, Y, and Z—arranged in the manner of a right-hand coordinate system (Figure 5.2). Any point in the **Cartesian coordinate system** is represented by its coordinates (x, y, z). The origin of the coordinate system has coordinates $(0,0,0)$. A **polar coordinate system**, by contrast, represents a point by an angle and a distance. For example, in Figure 5.3, the point is represented by the distance l and the angle α. To convert from polar to Cartesian coordinates, we may use the formulas $x = l \cos \alpha$

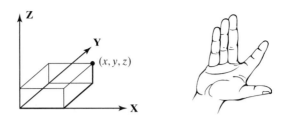

Figure 5.2   A right-hand Cartesian coordinate system.

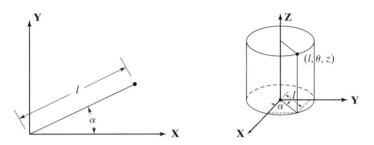

Figure 5.3   A point in a polar coordinate system.

and $y = l \sin \alpha$. For consistency and convenience, all design objects are stored in the Cartesian coordinate system—that is, coordinates x, y, and z. A point is customarily represented by a vector of the three coordinates: $\mathbf{P}_i = \begin{bmatrix} x_i \\ y_i \\ z_i \end{bmatrix}$. Vector $\mathbf{P}_i$ is a concise representation.

A cylindrical coordinate system (l, α, z)

$$x = l \cos \alpha$$
$$y = l \sin \alpha$$
$$z = z$$

In computer graphics, there are two coordinate systems: the **object coordinate system** and the **view coordinate system** (Figure 5.4). The object coordinate system is where the object is defined. The view coordinate system defines the viewing angle from which to observe the object. An object in the view coordinate system is at a position and orientation different from that of the same object in the object coordinate system (Figure 5.5). Viewing an object from different directions and distances can be achieved by moving the view coordinate system around. To do so, one uses a **coordinate system transformation**. Each object and view coordinate system is represented by **coordinate frames**. Let

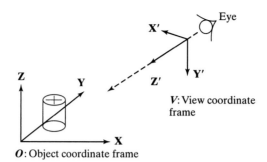

Figure 5.4   Object and view coordinate systems.

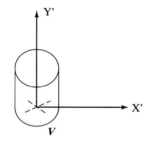

Figure 5.5   Object in Figure 5.4 as displayed in the view coordinate system.

O represent the object coordinate frame and **V** represent the view coordinate frame. In the beginning, both coordinate frames are collocated, and moving the object requires a transformation exactly the opposite of that required to move the view.

5.2.1 Coordinate Transformations

There are three major coordinate transformations: translation, rotation, and scaling.

 A. Translation.   Translation involves moving a point (or the view coordinate frame). Point **P** is moved to point **P′** by a distance Δ_x in the x direction, Δ_y in the y direction, and Δ_z in the z direction (Figure 5.6). This transformation can be expressed as

$$x' = x + \Delta_x$$

$$y' = y + \Delta_y$$

$$z' = z + \Delta_z \tag{5.1}$$

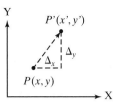

Figure 5.6   2-D translation.

or, in vector form,

$$\mathbf{P}' = \mathbf{P} + \begin{bmatrix} \Delta_x \\ \Delta_y \\ \Delta_z \end{bmatrix} \tag{5.2}$$

B. Rotation.   Rotation is defined by a rotation axis and a rotation angle. For ease of computation, the rotation axis is one of the axes of the coordinate system. Figure 5.7 illustrates rotation in 2-D. The rotational transformation equations are as follows:

$$x' = l\cos(\theta + \alpha) = l(\cos\theta\sin\alpha - \sin\theta\cos\alpha)$$
$$y' = l\sin(\theta + \alpha) = l(\sin\theta\sin\alpha + \cos\theta\cos\alpha) \tag{5.3}$$

$$l\cos\alpha = y$$
$$l\sin\alpha = x \tag{5.4}$$

$$x' = x\cos\theta - y\sin\theta$$
$$y' = x\sin\theta + y\cos\theta \tag{5.5}$$

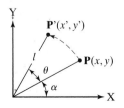

Figure 5.7   2-D rotation of a point.

Equation 5.3 can be written in matrix form as

$$\mathbf{P}' = \begin{bmatrix} x' \\ y' \end{bmatrix} = \mathbf{R}(z, \theta)\mathbf{P} = \begin{bmatrix} \cos\theta & -\sin\theta \\ \sin\theta & \cos\theta \end{bmatrix}\begin{bmatrix} x \\ y \end{bmatrix} \qquad (5.6)$$

The matrix $\mathbf{R}(z, \theta)$ is called a transformation matrix for 2-D rotation about the Z-axis. $\mathbf{R}(x, \theta)$ and $\mathbf{R}(y, \theta)$ can be found with the same procedure. In 3-D, the rotational transformation matrices have $3 \times 3 = 9$ elements:

$$\mathbf{R}(x, \theta) = \begin{bmatrix} 1 & 0 & 0 \\ 0 & \cos\theta & -\sin\theta \\ 0 & \sin\theta & \cos\theta \end{bmatrix} \qquad (5.7)$$

$$\mathbf{R}(y, \theta) = \begin{bmatrix} \cos\theta & 0 & \sin\theta \\ 0 & 1 & 0 \\ -\sin\theta & 0 & \cos\theta \end{bmatrix} \qquad (5.8)$$

$$\mathbf{R}(z, \theta) = \begin{bmatrix} \cos\theta & -\sin\theta & 0 \\ \sin\theta & \cos\theta & 0 \\ 0 & 0 & 1 \end{bmatrix} \qquad (5.9)$$

C. Scaling. Scaling is done on each axis. Let s_x, s_y, and s_z be three scaling factors on three axes. Then the transformation matrix is

$$\mathbf{P}' = S\mathbf{P} = \begin{bmatrix} s_x & s_y & s_z \end{bmatrix}\mathbf{P} \qquad (5.10)$$

To combine all three transformations into one matrix, a **homogeneous transformation** is used. A homogeneous transformation matrix is a 4×4 matrix. A point is represented by a 4-D homogeneous vector such as

$$\mathbf{P} = \begin{bmatrix} x \\ y \\ z \\ 1 \end{bmatrix}$$

Transformation matrices may be rewritten as

$$\mathbf{T_r} = \begin{bmatrix} 1 & 0 & 0 & \Delta_x \\ 0 & 1 & 0 & \Delta_y \\ 0 & 0 & 1 & \Delta_z \\ 0 & 0 & 0 & 1 \end{bmatrix} \qquad (5.11)$$

$$\mathbf{R}_x = \begin{bmatrix} 1 & 0 & 0 & 0 \\ 0 & C & -S & 0 \\ 0 & S & C & 0 \\ 0 & 0 & 0 & 1 \end{bmatrix} \tag{5.12}$$

$$\mathbf{R}_y = \begin{bmatrix} C & 0 & S & 0 \\ 0 & 1 & 0 & 0 \\ -S & 0 & C & 0 \\ 0 & 0 & 0 & 1 \end{bmatrix} \tag{5.13}$$

$$\mathbf{R}_z = \begin{bmatrix} C & -S & 0 & 0 \\ S & C & 0 & 0 \\ 0 & 0 & 1 & 0 \\ 0 & 0 & 0 & 1 \end{bmatrix} \tag{5.14}$$

$$\mathbf{S} = \begin{bmatrix} s_x & 0 & 0 & 0 \\ 0 & s_y & 0 & 0 \\ 0 & 0 & s_z & 0 \\ 0 & 0 & 0 & 1 \end{bmatrix} \tag{5.15}$$

where $C = \cos\theta$ and $S = \sin\theta$.

The foregoing transformation matrices are called object transformation matrices. Instead of transforming objects, one can transform the coordinate system. Moving an object in the positive X, Y, and Z directions is the same as transforming the coordinate system to the negative directions. This kind of transformation is called view transformation and is given by the following matrices:

$$T_r^v = \begin{bmatrix} 1 & 0 & 0 & -\Delta x \\ 0 & 1 & 0 & -\Delta y \\ 0 & 0 & 1 & -\Delta z \\ 0 & 0 & 0 & 1 \end{bmatrix}$$

$$R_x^v = \begin{bmatrix} 1 & 0 & 0 & 0 \\ 0 & C & S & 0 \\ 0 & -S & C & 0 \\ 0 & 0 & 0 & 1 \end{bmatrix}$$

$$R_y^v = \begin{bmatrix} C & 0 & -S & 0 \\ 0 & 1 & 0 & 0 \\ S & 0 & C & 0 \\ 0 & 0 & 0 & 1 \end{bmatrix}$$

$$R_z^v = \begin{bmatrix} C & S & 0 & 0 \\ -S & C & 0 & 0 \\ 0 & 0 & 1 & 0 \\ 0 & 0 & 0 & 1 \end{bmatrix}$$

Transformation matrices must be applied in the same order as the transformation. Otherwise the result will be incorrect.

Example 5.1.

Point $\mathbf{A} = [1 \quad 3 \quad 1 \quad 1]^T$ is translated by $\mathbf{T}_r = [1 \quad 1 \quad 1]^T$, rotated about the X-axis 90 degrees, and then rotated about Y-axis 45 degrees. What is $\mathbf{A}'$ in the object coordinate frame $\mathbf{O}$?

The relevant equations are as follows:

$$C_x = \cos 90 = 0$$

$$S_x = \sin 90 = 1$$

$$C_y = \cos 45 = 0.707$$

$$S_y = \sin 45 = 0.707$$

$$\mathbf{T}_r = \begin{bmatrix} 1 & 0 & 0 & 1 \\ 0 & 1 & 0 & 1 \\ 0 & 0 & 1 & 1 \\ 0 & 0 & 0 & 1 \end{bmatrix}$$

$$\mathbf{R}_x = \begin{bmatrix} 1 & 0 & 0 & 0 \\ 0 & C_x & -S_x & 0 \\ 0 & S_x & C_x & 0 \\ 0 & 0 & 0 & 1 \end{bmatrix} = \begin{bmatrix} 1 & 0 & 0 & 0 \\ 0 & 0 & -1 & 0 \\ 0 & 1 & 0 & 0 \\ 0 & 0 & 0 & 1 \end{bmatrix}$$

$$\mathbf{R}_y = \begin{bmatrix} C_y & 0 & S_y & 0 \\ 0 & 1 & 0 & 0 \\ -S_y & 0 & C_y & 0 \\ 0 & 0 & 0 & 1 \end{bmatrix} = \begin{bmatrix} 0.707 & 0 & 0.707 & 0 \\ 0 & 1 & 0 & 0 \\ -0.707 & 0 & 0.707 & 0 \\ 0 & 0 & 0 & 1 \end{bmatrix}$$

$$\mathbf{A}' = \mathbf{R}_y \mathbf{R}_x \mathbf{T}_r \mathbf{A}$$

$$= \begin{bmatrix} 0.707 & 0 & 0.707 & 0 \\ 0 & 1 & 0 & 0 \\ -0.707 & 0 & 0.707 & 0 \\ 0 & 0 & 0 & 1 \end{bmatrix} \begin{bmatrix} 1 & 0 & 0 & 0 \\ 0 & 0 & -1 & 0 \\ 0 & 1 & 0 & 0 \\ 0 & 0 & 0 & 1 \end{bmatrix} \begin{bmatrix} 1 & 0 & 0 & 1 \\ 0 & 1 & 0 & 1 \\ 0 & 0 & 1 & 1 \\ 0 & 0 & 0 & 1 \end{bmatrix} \begin{bmatrix} 1 \\ 3 \\ 1 \\ 1 \end{bmatrix}$$

$$= \begin{bmatrix} 4.242 \\ -2 \\ 1.414 \\ 1 \end{bmatrix}$$

Using Matlab to do the numerical computation yields

$\gg$ A = [1; 3; 1; 1];

$\gg$ Tr = [1 0 0 1; 0 1 0 1; 0 0 1 1; 0 0 0 1];

$\gg$ Rx = [1 0 0 0; 0 0 −1 0; 0 1 0 0; 0 0 0 1];

$\gg$ Ry = [0.707 0 0.707 0; 0 1 0 0; −0.707 0 0.707 0; 0 0 0 1];

Taking a step-by-step transformation, one can see how the point is moved in the coordinate system:

$\gg$ Tr*A

ans =

        2
        4
        2
        1

$\gg$ Rx*Tr*A

ans =

        2
       −2
        4
        1

$\gg$ Ry*Rx*Tr*A

ans =

      4.2420
     −2.0000
      1.4140
      1.0000

A' is located at (4.242, −2., 1.414)

Example 5.2.

The point does not move. The view coordinate frame V is translated from the current location by $\Delta_x = 1$, $\Delta_y = 1$, and $\Delta_z = 1$, rotated about its X-axis 90 degrees, and then rotated about its Y-axis 45 degrees. What are the coordinates of the point as viewed in V? Now the relevant equations are the following:

$$C_x = \cos(90) = 0$$

$$S_x = \sin(90) = 1$$

$$C_y = \cos(45) = 0.707$$

$$S_y = \sin(45) = 0.707$$

$$T_r^v = \begin{bmatrix} 1 & 0 & 0 & -1 \\ 0 & 1 & 0 & -1 \\ 0 & 0 & 1 & -1 \\ 0 & 0 & 0 & 1 \end{bmatrix}$$

$$R_x^v = \begin{bmatrix} 1 & 0 & 0 & 0 \\ 0 & 0 & 1 & 0 \\ 0 & -1 & 0 & 0 \\ 0 & 0 & 0 & 1 \end{bmatrix}$$

$$R_y^v = \begin{bmatrix} 0.707 & 0 & -0.707 & 0 \\ 0 & 1 & 0 & 0 \\ 0.707 & 0 & 0.707 & 0 \\ 0 & 0 & 0 & 1 \end{bmatrix}$$

We obtain

$$A' = R_y^v R_x^v T_r^v A$$

$$= \begin{bmatrix} 0.707 & 0 & -0.707 & 0 \\ 0 & 1 & 0 & 0 \\ 0.707 & 0 & 0.707 & 0 \\ 0 & 0 & 0 & 1 \end{bmatrix} \begin{bmatrix} 1 & 0 & 0 & 0 \\ 0 & 0 & 1 & 0 \\ 0 & -1 & 0 & 0 \\ 0 & 0 & 0 & 1 \end{bmatrix} \begin{bmatrix} 1 & 0 & 0 & -1 \\ 0 & 1 & 0 & -1 \\ 0 & 0 & 1 & -1 \\ 0 & 0 & 0 & 1 \end{bmatrix} \begin{bmatrix} 1 \\ 3 \\ 1 \\ 1 \end{bmatrix}$$

$$= \begin{bmatrix} 1.414 \\ 0 \\ -1.414 \\ 1 \end{bmatrix}$$

5.2.2 Projections and the Display

In general, the display is limited to 2-D rendering. A **view port** is the portion of the 2-D frame visible to the user. A view coordinate frame V is attached to the view port, which is defined by its limits: (vx_{min}, vx_{max}, vy_{min}, vy_{max}). The view port is projected to the display window on the terminal or on paper (Figure 5.8). The display window has its physical limits as well: (dx_{min}, dx_{max}, dy_{min}, dy_{max}).

To display something in the display window, first the 3-D object is projected onto the X–Y plane in the view coordinate frame. There are two types of projections: prospective and orthogonal. A **prospective projection** mimics what human eyes see. An object farther away from the viewpoint vanishes into a point (Figure 5.9). An **orthogonal projection** projects everything perpendicularly to the X–Y plane (Figure 5.10). It can be done by simply dropping the z-coordinate.

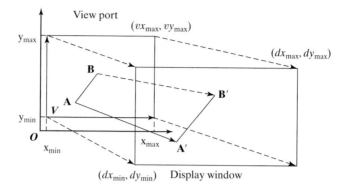

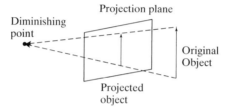

Figure 5.8   View port and display window.

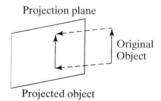

Figure 5.9   Prospective projection.

Figure 5.10   Orthogonal projection.

A view port sets a new coordinate frame and crops the projection. Finally, the contents in the view port are mapped onto the display window. To illustrate this point, let the line in Figure 5.8 be given by $A(1, 1, 2)$, $B(3, 3, 5)$. The view port limits are (0.2, 7, 0.4, 6). The projections of points A and B become $A(1, 1)$ and $B(3, 3)$. The Z-coordinate is dropped. The view coordinate frame V is translated to (0.2, 0.4). The view port does a view coordinate translation of $T(-0.2, -0.4)$. Therefore, point A becomes (0.8, 0.6) and point B becomes (2.8, 2.6). Then the view port is mapped to the display window. Suppose that the display window limits are (0, 800, 0, 600), in units of pixels. The next step in mapping is to scale the view coordinate system, using $s_x = 800/(7 - 0.2) = 117.65$, $s_y = 600/(6 - 0.4) = 107.14$. Point A's coordinates in the display window are $(117.65 \times 0.8, 107.14 \times 0.6) = (94, 64)$, and point B is $(117.65 \times 2.8, 107.14 \times 2.6) = (329, 279)$. Since d_x and d_y are both zero, there is no need to do another translation. The procedure is done.

The procedure of orthogonal projection is captured in the equation

$$\mathbf{P}' = \mathbf{T_{rd}} \mathbf{S} \mathbf{T_{rv}} \mathbf{P} \tag{5.16}$$

where

$$\mathbf{T_{rv}} = \begin{bmatrix} 1 & 0 & 0 & -vx_{min} \\ 0 & 1 & 0 & -vy_{min} \\ 0 & 0 & 1 & 0 \\ 0 & 0 & 0 & 1 \end{bmatrix} \tag{5.17}$$

$$\mathbf{S} = \begin{bmatrix} \dfrac{dx_{max} - dx_{min}}{vx_{max} - vx_{min}} & 0 & 0 & 0 \\ 0 & \dfrac{dy_{max} - dy_{min}}{vy_{max} - vy_{min}} & 0 & 0 \\ 0 & 0 & 0 & 0 \\ 0 & 0 & 0 & 1 \end{bmatrix} \tag{5.18}$$

and

$$\mathbf{T_{rd}} = \begin{bmatrix} 1 & 0 & 0 & -dx_{min} \\ 0 & 1 & 0 & -dy_{min} \\ 0 & 0 & 1 & 0 \\ 0 & 0 & 0 & 1 \end{bmatrix} \tag{5.19}$$

Notice that the z-coordinate is ignored.

5.3 3-D REPRESENTATIONS

Three-dimensional objects are represented in a 3-D Cartesian coordinate system. Basic 3-D objects are represented by lines, circles, and circular arcs. Although lines can be represented in many ways, the **canonical representation**[1] consists of two points. Other representations include a point and a vector and the intersection of two planes. A 3-D circle is defined by its center, radius, and a normal vector, which defines the plane in which the circle lies. Since 2-D circles are usually defined in the X–Y plane, the vector is omitted. A circular arc is represented by its center, starting point, direction, and endpoint, again, together with a normal vector. All the points used in defining 3-D geometries are defined in the 3-D Cartesian coordinate system.

---

[1] The canonical representation is the simplest form of representation of an object. While there are multiple representations of the same object, it is stored in the canonical form.

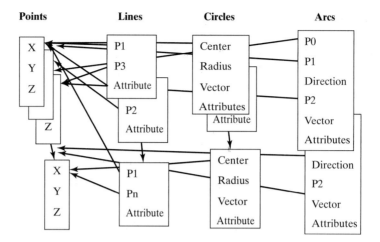

Figure 5.11   3-D CAD data structure.

As discussed in Chapter 4, there are many ways in which a user can interactively define 3-D objects. When they are defined directly following the canonical representation, the CAD system simply saves the data in a data structure for further manipulation. A simple data structure is shown in Figure 5.11; note that the line, circle, and arc entities are linked to the point entities through pointers. Parameters such as radius and direction are stored directly with the entity. Additional information (e.g., line type, color, line thickness) is stored as attributes. A practical CAD system has a more elaborate data structure, to accommodate properties such as layers, group, and polylines (groups of lines and curves). Their discussion is omitted here.

There are many methods used in defining geometric entities; yet, each type of entity is stored only in its canonical form. The CAD system has to use algorithms to convert the different definitions. For example, a point is defined by the intersection of two lines. The CAD system retrieves the definition of the lines and then calculates the intersection. For the geometric entities discussed so far, such calculations are all relatively simple, at the level of high school geometry. The problem becomes complicated when one tries to implement the derived equations into computer software. Without very careful thinking, the program will work in some cases, but fail in many more.

A simple example illustrates this point. A point is defined as the intersection of two lines. Suppose both lines are defined in 2-D. Line 1 is defined by $[(x_1, y_1)(x_2, y_2)]$ and line 2 is defined by $[(x_3, y_3)(x_4, y_4)]$. Then an equation for each line is

$$\frac{x - x_1}{y - y_1} = \frac{x_2 - x_1}{y_2 - y_1} \tag{5.20}$$

and

$$\frac{x - x_3}{y - y_3} = \frac{x_4 - x_3}{y_4 - y_3} \tag{5.21}$$

respectively.

The intersection point can be found by solving the two simultaneous equations. Let $a_1 = x_2 - x_1$, $b_1 = y_2 - y_1$, $a_2 = x_4 - x_3$, and $b_2 = y_4 - y_3$. Then Equations 5.20 and 5.21 can respectively be written as

$$x = x_1 + \frac{a_1}{b_1}(y - y_1) \qquad (5.22)$$

and

$$y = y_3 + \frac{b_2}{a_2}(x - x_3) \qquad (5.23)$$

Solving for y by substituting Equation 5.22 into Equation 5.23 yields

$$y = y_3 + \frac{b_2}{a_2}[(x_1 - x_3) + \frac{a_1}{b_1}(y - y_1)]$$

$$y = y_3 + \frac{b_2}{a_2}(x_1 - x_3) + \frac{a_1 b_2}{a_2 b_1}(y - y_1)$$

$$a_2 b_1 y = a_2 b_1 y_3 + b_1 b_2(x_1 - x_3) + a_1 b_2(y - y_1)$$

$$(a_2 b_1 - a_1 b_2)y = a_2 b_1 y_3 - a_1 b_1 y_1 + b_1 b_2(x_1 - x_3)$$

$$y = \frac{a_2 b_1 y_3 - a_1 b_1 y_1 + b_1 b_2(x_1 - x_3)}{a_2 b_1 - a_1 b_2} \qquad (5.24)$$

If Equation 5.24 is programmed into the software, special care must be taken for the case $a_2 b_1 - a_1 b_2 \approx 0$. When that approximation holds, a value of y will be equal to a real number divided by zero; in other words, y will be infinite. Note that the denominator does not have to equal zero exactly. A memory overflow error happens when the result of a mathematical computation is greater than the variable can store. In this case, the two lines are almost parallel to each other (Figure 5.12a).

If L_1 is horizontal, the $b_1 = 0$. In calculating the value of x from Equation 5.22, an overflow error will occur. If so, it does not mean that there is no solution; it simply means that the equation should not be used. If we modify Equations 5.20 and 5.21 to have y terms in the numerator and x terms in the denominator, the problem is solved. Of course, in doing so, we will find that the solution will not work for case c in the figure.

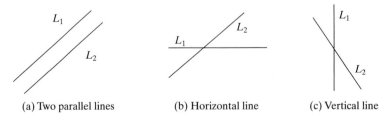

(a) Two parallel lines   (b) Horizontal line   (c) Vertical line

Figure 5.12   Three cases involving pairs of lines.

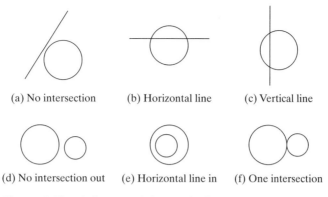

(a) No intersection   (b) Horizontal line   (c) Vertical line

(d) No intersection out   (e) Horizontal line in   (f) One intersection

Figure 5.13   A few special cases in finding intersections.

The same care has to be taken for other types of geometric computations. Figure 5.13 shows but a few of these cases.

5.4 CURVES AND CURVED SURFACES

Curves and curved surfaces are more difficult to handle than lines and circles. Curves might be limited to two dimensions or defined in three. Surfaces are always in three dimensions. Many surfaces are extensions of curves. In this section, we will discuss some basic curve and surface equations.

5.4.1 Curves

Curves can be classified as **analytical curves** and **free-form curves**. Analytical curves are those curves defined by an algebraic equation and its coefficients. For example, a 2-D parabola has the form $y^2 - 4ax = 0$. In order to define a parabolic shape, one selects the equation and the value of the coefficient a. In contrast, there is no algebraic formula that defines what the profile of a sports car is. The curve is drawn freehand by an artist (or stylist). A free-form curve is fitted to the shape that is of interest.

5.4.1.1. Analytical Curves   Analytical curves are represented either in **implicit polynomial** form or **explicit polynomial** form. A 2-D implicit function on the X–Y plane has the form $f(x, y) = 0$. For example, a line can be represented as $ax + by + c = 0$ and a circle as $(x - a)^2 + (y - b)^2 - r^2 = 0$. The variables a, b, c, r are coefficients. The most commonly used analytical curves are conic sections: the **ellipse**, the **parabola**, and the **hyperbola**. Conic sections are formed by slicing a right cone (Figures 5.14 to 5.16). A circle is also a conic section. Equations for conic sections are as follows:

a. Ellipse: $\dfrac{x^2}{a^2} + \dfrac{y^2}{b^2} - 1 = 0$

b. Parabola: $y^2 - 4ax = 0$

c. Hyperbola: $\dfrac{x^2}{a^2} - \dfrac{y^2}{b^2} - 1 = 0$

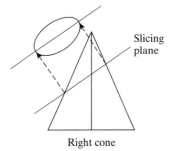
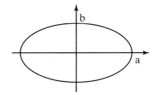

Right cone

Figure 5.14   Ellipse.

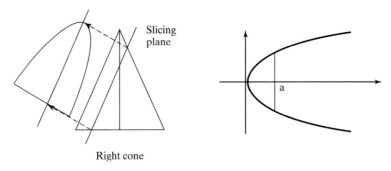

Right cone

Figure 5.15   Parabola.

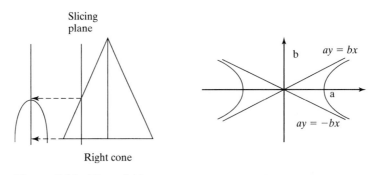

Right cone

Figure 5.16   Hyperbola.

Implicit curves have some convenient features. For example, it is simple to determine whether a point $P(x_1, y_1)$ is inside or outside of the region formed by the curve. Thus, for the curve $f(x, y) = 0$, a point (x_1, y_1) is inside the curve if $f(x_1, y_1) < 0$ and outside of the curve if $f(x_1, y_1) > 0$. The tangent and normal of the curve are also easy to find. The tangent line at $P(x_2, y_2)$ on the curve is

$$f_x(x_2, y_2)(x - x_2) + f_y(x_2, y_2)(y - y_2) = 0 \qquad (5.25)$$

where $f_x = \partial f / \partial x$ and $f_y = \partial f / \partial y$. The normal line is

$$f_y(x_2, y_2)(x - x_2) - f_x(x_2, y_2)(y - y_2) = 0 \qquad (5.26)$$

Explicit polynomial curves have the general form

$$y = f(x) = a + bx + cx^2 + \ldots$$

The highest exponent of the variable x is called the **degree of the polynomial** or the degree of the curve. The curve given by the preceding equation has a degree of two. (Ignore the ellipsis at the end of the equation.) An explicit curve can be converted into an implicit curve by transposing the right-hand side to the left-hand side: $y - f(x) = 0$. In addition to the nice properties implicit curves have, it is easy to trace such a curve in a sequential manner (incrementing the x value to find the y value). This property makes the curve attractive for plotting.

Example 5.3.

Find the tangent and the normal lines of the following curve at $P(1,2)$:

$$f(x, y) = y^2 - 4x = 0$$

Apply Equation 5.25 to find the equation for the tangent line:

$$f_x = \frac{\partial f}{\partial x} = -4; f_y = \frac{\partial f}{\partial y} = 2y$$

$$f_x(1, 2)(x - 1) + f_y(1, 2)(y - 2) = 0$$

$$-4(x - 1) + 4(y - 2) = 0$$

$$x - y = -1$$

Apply Equation 5.26 to obtain the equation for the normal line:

$$f_y(1, 2)(x - 1) - f_x(1, 2)(y - 2) = 0$$

$$4(x - 1) + 4(y - 2) = 0$$

$$x + y = 3$$

Figure 5.17 shows the normal and tangent lines to a curve.

Example 5.4.

Convert the implicit equation $y^2 - 4ax = 0$ to an explicit equation:

$$y^2 - 4ax = 0$$

$$y^2 = 4ax$$

$$y = 2\sqrt{ax}$$

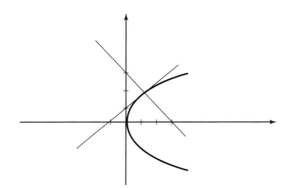

Figure 5.17   Normal and tangent lines.

5.4.1.2   Free-form Curves   Free-form curves are always represented by parametric polynomial equations. A parametric equation has the form $x = f(t)$, $y = g(t)$, $z = h(t)$. The variable t is a parameter with no units. To differentiate implicit and explicit equations from parametric equations, the former are called *nonparametric equations*. All three functions—$f(t)$, $g(t)$, and $h(t)$—are polynomials of the same degree. For most engineering applications, a third-degree polynomial is sufficient. Therefore, general parametric polynomial equations can be written as

$$x = a_x t^3 + b_x t^2 + c_x t + d_x$$

$$y = a_y t^3 + b_y t^2 + c_y t + d_y \tag{5.27}$$

$$z = a_z t^3 + b_z t^2 + c_z t + d_z$$

where a, b, c, and d are coefficients and the parameter t has a range of $0 \le t \le 1$.

To make the curve model concise, it can be written in vector form as

$$
\mathbf{r} =
\begin{bmatrix}
a_x & b_x & c_x & d_x \\
a_y & b_y & c_y & d_y \\
a_z & b_z & c_z & d_z
\end{bmatrix}
\begin{bmatrix}
t^3 \\
t^2 \\
t \\
1
\end{bmatrix}
=
[\mathbf{a} \quad \mathbf{b} \quad \mathbf{c} \quad \mathbf{d}]
\begin{bmatrix}
t^3 \\
t^2 \\
t \\
1
\end{bmatrix}
= \mathbf{AT}
\tag{5.28}
$$

where $\mathbf{r} = [x \quad y \quad z]^T$. $\mathbf{T}$ is called the *power basis vector* and $\mathbf{A}$ is the *coefficient matrix*.

The tangent property of a curve is important in designing the curve. The tangent can be derived easily:

$$
\dot{\mathbf{r}} = \frac{d\mathbf{r}}{dt} = \mathbf{A}\frac{d\mathbf{T}}{dt} = \mathbf{A}
\begin{bmatrix}
3t^2 \\
2t \\
1 \\
0
\end{bmatrix}
\tag{5.29}
$$

To design a curve is to determine its coefficients. There are several methods developed for this purpose.

a. Ferguson Curve   Ferguson [1964] introduced an easy way to define a curve. Since a long and complex curve is normally defined by several shorter curved segments, each curve segment has the form of Equation 5.28. To connect those curve segments together, one must be able to control the starting point, the endpoints, and the tangents at those two points. The reason for controlling the starting point and the endpoint is obvious, for that is the way to link the curve segments together. A curve connected in this manner has **first-degree continuity** (or **zeroth-order continuity**). When a curve is first-degree continuous, it is connected. There is no guarantee, however, that the curve is smooth. In Figure 5.18, curve segment *a* and curve segment *b* are connected at $\mathbf{r}_1^a = \mathbf{r}_0^b$, although there is a abrupt change in direction at the connecting point. To smooth the curve, the tangents at $\mathbf{r}_1^a$, $\mathbf{r}_0^b$ must be the same. (At least, the directions of the two tangent lines have to be the same.)

A Ferguson curve (Figure 5.19) is constructed by defining

1. Two endpoints: $\mathbf{P}_0$, $\mathbf{P}_1$
2. End tangent vectors: t_0, t_1

From the standard model in Equation 5.28, the following equations can be written:

$$\mathbf{P}_0 = \mathbf{r}(0); \ \mathbf{P}_1 = \mathbf{r}(1); \ \mathbf{t}_0 = \dot{\mathbf{r}}(0); \ \mathbf{t}_1 = \dot{\mathbf{r}}(1) \tag{5.30}$$

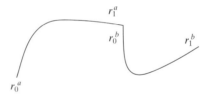

Figure 5.18   Two curves that are connected, but not smooth.

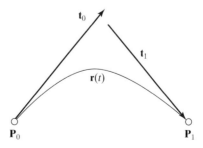

Figure 5.19   A Ferguson curve segment.

Substituting the values of the parameters into Equations 5.28 and 5.29 yields the following equations:

$$\mathbf{P}_0 = \mathbf{r}(0) = \begin{bmatrix} \mathbf{a} & \mathbf{b} & \mathbf{c} & \mathbf{d} \end{bmatrix} \begin{bmatrix} 0 \\ 0 \\ 0 \\ 1 \end{bmatrix} = \mathbf{d}$$

$$\mathbf{P}_1 = \mathbf{r}(1) = \begin{bmatrix} \mathbf{a} & \mathbf{b} & \mathbf{c} & \mathbf{d} \end{bmatrix} \begin{bmatrix} 1 \\ 1 \\ 1 \\ 1 \end{bmatrix} = \mathbf{a} + \mathbf{b} + \mathbf{c} + \mathbf{d}$$

$$\mathbf{t}_0 = \dot{\mathbf{r}}(0) = \begin{bmatrix} \mathbf{a} & \mathbf{b} & \mathbf{c} & \mathbf{d} \end{bmatrix} \begin{bmatrix} 0 \\ 0 \\ 1 \\ 0 \end{bmatrix} = \mathbf{c}$$

$$\mathbf{t}_1 = \dot{\mathbf{r}}(1) = \begin{bmatrix} \mathbf{a} & \mathbf{b} & \mathbf{c} & \mathbf{d} \end{bmatrix} \begin{bmatrix} 3 \\ 2 \\ 1 \\ 0 \end{bmatrix} = 3\mathbf{a} + 2\mathbf{b} + \mathbf{c}$$

From these four equations, we can easily solve for the unknowns $\mathbf{a}, \mathbf{b}, \mathbf{c}$, and $\mathbf{d}$:

$$\mathbf{a} = 2\mathbf{P}_0 - 2\mathbf{P}_1 + \mathbf{t}_0 + \mathbf{t}_1$$

$$\mathbf{a} = -3\mathbf{P}_0 + 3\mathbf{P}_1 - 2\mathbf{t}_0 - \mathbf{t}_1 \tag{5.31}$$

$$\mathbf{c} = \mathbf{t}_0$$

$$\mathbf{d} = \mathbf{P}_0$$

$$\mathbf{A} = \begin{bmatrix} \mathbf{a} & \mathbf{b} & \mathbf{c} & \mathbf{d} \end{bmatrix} = \begin{bmatrix} \mathbf{P}_0 & \mathbf{P}_1 & \mathbf{t}_0 & \mathbf{t}_1 \end{bmatrix} \begin{bmatrix} 2 & -3 & 0 & 1 \\ -2 & 3 & 0 & 0 \\ 1 & -2 & 1 & 0 \\ 1 & -1 & 0 & 0 \end{bmatrix} \tag{5.32}$$

$$\mathbf{r}(t) = \mathbf{AT} = \begin{bmatrix} \mathbf{P}_0 & \mathbf{P}_1 & \mathbf{t}_0 & \mathbf{t}_1 \end{bmatrix} \begin{bmatrix} 2 & -3 & 0 & 1 \\ -2 & 3 & 0 & 0 \\ 1 & -2 & 1 & 0 \\ 1 & -1 & 0 & 0 \end{bmatrix} \begin{bmatrix} t^3 \\ t^2 \\ t \\ 1 \end{bmatrix}$$

$$0 \le t \le 1 \tag{5.33}$$

Tangents to the curve can easily be found as well:

$$\dot{\mathbf{r}}(t) = \mathbf{A}\frac{d\mathbf{T}}{dt} = \begin{bmatrix} \mathbf{P}_0 & \mathbf{P}_1 & \mathbf{t}_0 & \mathbf{t}_1 \end{bmatrix} \begin{bmatrix} 2 & -3 & 0 & 1 \\ -2 & 3 & 0 & 0 \\ 1 & -2 & 1 & 0 \\ 1 & -1 & 0 & 0 \end{bmatrix} \begin{bmatrix} 3t^2 \\ 2t \\ 1 \\ 0 \end{bmatrix} \tag{5.34}$$

Example 5.5.

A Ferguson curve is defined by two points $P_0(0,0,0)$ and $P_1(4,0,0)$ and two tangents $t_0 = [1\ 4\ 0]$ and $t_1 = [1\ -4\ 0]$. What is the equation for the curve? Plot the curve. From Equation 5.33, the curve is

$$\mathbf{r}(t) = \begin{bmatrix} 0 & 4 & 1 & 1 \\ 0 & 0 & 4 & -4 \\ 0 & 0 & 0 & 0 \end{bmatrix} \begin{bmatrix} 2 & -3 & 0 & 1 \\ -2 & 3 & 0 & 0 \\ 1 & -2 & 1 & 0 \\ 1 & -1 & 0 & 0 \end{bmatrix} \begin{bmatrix} t^3 \\ t^2 \\ t \\ 1 \end{bmatrix}$$

$$= \begin{bmatrix} -6 & 9 & 1 & 0 \\ 0 & -4 & 4 & 0 \\ 0 & 0 & 0 & 0 \end{bmatrix} \begin{bmatrix} t^3 \\ t^2 \\ t \\ 1 \end{bmatrix}$$

$$= \begin{bmatrix} -6t^3 + 9t^2 + t \\ -4t^2 + 4t \\ 0 \end{bmatrix}$$

The curve can be plotted in MATLAB with the following commands:

```
t = 0:0.02:1;
p = [0 4 1 1; 0 0 4 −4; 0 0 0 0]
A = [2 −3 0 1; −2 3 0 0; 1 −2 1 0; 1 −1 0 0];
tt = [t.^3; t.^2; t; ones(size(t))];
r = p*A*tt;
plot(r(1,:),r(2,:))
```

Figure 5.20 shows the resulting curve.

Example 5.6.

What is the vector tangent to the curve of Example 5.5 at $t = 0.1$?

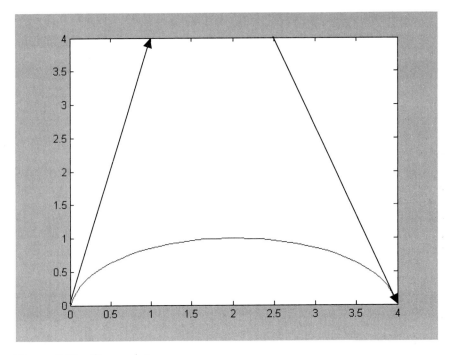

Figure 5.20 Curve plot.

Use Equation 5.34:

$$r(t) = \begin{bmatrix} 0 & 4 & 1 & 1 \\ 0 & 0 & 4 & -4 \\ 0 & 0 & 0 & 0 \end{bmatrix} \begin{bmatrix} 2 & -3 & 0 & 1 \\ -2 & 3 & 0 & 0 \\ 1 & -2 & 1 & 0 \\ 1 & -1 & 0 & 0 \end{bmatrix} \begin{bmatrix} 3t^2 \\ 2t \\ 1 \\ 0 \end{bmatrix}$$

$$= \begin{bmatrix} -6 & 9 & 1 & 0 \\ 0 & -4 & 4 & 0 \\ 0 & 0 & 0 & 0 \end{bmatrix} \begin{bmatrix} 3(.1)^2 \\ 2(.1) \\ 1 \\ 0 \end{bmatrix}$$

$$= \begin{bmatrix} 2.62 \\ 3.2 \\ 0 \end{bmatrix}$$

The same result can be obtained from the following MATLAB command:

```
>> [0 4 1 1; 0 0 4 -4; 0 0 0 0]*[2 -3 0 1; -2 3 0 0; 1 -2 1 0; 1 -1 0 0]*[3*.1^2; 2*.1; 1; 0]

   ans =

         2.6200
         3.2000
              0
```

Another way to represent the Ferguson curve is to expand the matrix:

$$r(t) = \begin{bmatrix} \mathbf{P}_0 & \mathbf{P}_1 & \mathbf{t}_0 & \mathbf{t}_1 \end{bmatrix} \begin{bmatrix} 2 & -3 & 0 & 1 \\ -2 & 3 & 0 & 0 \\ 1 & -2 & 1 & 0 \\ 1 & -1 & 0 & 0 \end{bmatrix} \begin{bmatrix} t^3 \\ t^2 \\ t \\ 1 \end{bmatrix}$$

$$= \begin{bmatrix} \mathbf{P}_0 & \mathbf{P}_1 & \mathbf{t}_0 & \mathbf{t}_1 \end{bmatrix} \begin{bmatrix} 2t^3 - 3t^2 + 1 \\ -2t^3 + 3t^2 \\ t^3 - 2t^2 + t \\ t^3 - t^2 \end{bmatrix} \tag{5.35}$$

$$= (2t^3 - 3t^2 + 1)\mathbf{P}_0 + (-2t^3 + 3t^2)\mathbf{P}_1 + (t^3 - 2t^2 + t)\mathbf{t}_0 + (t^3 - t^2)\mathbf{t}_1$$

$$= H_0^3(t)\mathbf{P}_0 + H_1^3(t)\mathbf{P}_1 + H_2^3(t)\mathbf{t}_0 + H_3^3(t)\mathbf{t}_1$$

$H_i^3(t); i = 0, 1, 2, 3$ are called **Hermite blending functions**. The index "3" indicates that the curve is cubic (of degree 3). The Hermite functions blend two points and two tangents together to form the curve.

---

b. Bezier Curve   One shortcoming of a Ferguson curve is that it is difficult to determine the tangents that will yield a desired shape. It is especially hard to decide the magnitudes of the vectors. To overcome this shortcoming, Dr. Pierre Bezier, working at the French automaker Renault, invented the Bezier curve in 1972. Since then, it has become the most popular curve design method used in graphics packages and CAD systems. A cubic Bezier curve is a cubic curve defined by four control points (Figure 5.21; Bezier's curve can be generalized to higher or lower order curves). A nice feature of Bezier's curve is that it is close to the polygon (**characteristic polygon**) formed by the control points. This property is important because it helps the designer relate the shape of the desired curve to the locations of the control points. The relationship between the control points and the curve is given by the equations

$$\mathbf{r}(0) = \mathbf{P}_0$$

$$\mathbf{r}(1) = \mathbf{P}_3 \tag{5.36}$$

$$\dot{\mathbf{r}}(0) = 3(\mathbf{P}_1 - \mathbf{P}_0)$$

$$\dot{\mathbf{r}}(1) = 3(\mathbf{P}_3 - \mathbf{P}_2)$$

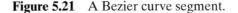

Figure 5.21   A Bezier curve segment.

where $\mathbf{P}_0$, $\mathbf{P}_1$, $\mathbf{P}_2$, $\mathbf{P}_3$ are the four (user-defined) control points.

Substituting Equation 5.36 into Equations 5.28 and 5.29 yields

$$\mathbf{P}_0 = \mathbf{r}(0) = \mathbf{d}$$

$$\mathbf{P}_3 = \mathbf{r}(1) = \mathbf{a} + \mathbf{b} + \mathbf{c} + \mathbf{d}$$

$$3(\mathbf{P}_1 - \mathbf{P}_0) = \dot{\mathbf{r}}(0) = \mathbf{c}$$

$$3(\mathbf{P}_3 - \mathbf{P}_2) = \dot{\mathbf{r}}(1) = 3\mathbf{a} + 2\mathbf{b} + \mathbf{c}$$

Therefore,

$$\mathbf{a} = \mathbf{P}_3 - 3\mathbf{P}_2 + 3\mathbf{P}_1 - \mathbf{P}_0$$

$$\mathbf{b} = 3\mathbf{P}_2 - 6\mathbf{P}_1 + 3\mathbf{P}_0$$

$$\mathbf{c} = 3\mathbf{P}_1 - 3\mathbf{P}_0$$

$$\mathbf{d} = \mathbf{P}_0$$

$$\mathbf{r}(t) = \begin{bmatrix} \mathbf{P}_0 & \mathbf{P}_1 & \mathbf{P}_2 & \mathbf{P}_3 \end{bmatrix} \begin{bmatrix} -1 & 3 & -3 & 1 \\ 3 & -6 & 3 & 0 \\ -3 & 3 & 0 & 0 \\ 1 & 0 & 0 & 0 \end{bmatrix} \begin{bmatrix} t^3 \\ t^2 \\ t \\ 1 \end{bmatrix}$$

$$0 \le t \le 1 \tag{5.37}$$

and the tangent vector is

$$\dot{\mathbf{r}}(t) = \begin{bmatrix} \mathbf{P}_0 & \mathbf{P}_1 & \mathbf{P}_2 & \mathbf{P}_3 \end{bmatrix} \begin{bmatrix} -1 & 3 & -3 & 1 \\ 3 & -6 & 3 & 0 \\ -3 & 3 & 0 & 0 \\ 1 & 0 & 0 & 0 \end{bmatrix} \begin{bmatrix} 3t^2 \\ 2t \\ 1 \\ 0 \end{bmatrix} \tag{5.38}$$

which can also be expressed in the polynomial form

$$\mathbf{r}(t) = \sum_{i=0}^{3} B_i^3(t)\mathbf{P}_i$$

$$0 \le t \le 1 \tag{5.39}$$

where $B_i^3(t)$ represents cubic Bernstein polynomials (or **Bernstein blending functions**). These polynomials blend the control points together to form the curve.

$$B_0^3(t) = (1 - t)^3$$

$$B_1^3(t) = 3t(1 - t)^2$$

$$B_2^3(t) = 3t^2(1 - t)$$

$$B_3^3(t) = t^3$$

which can be written as

$$B_i^3(t) = \frac{3!}{i!(3-i)!}t^i(1-t)^{3-i} \tag{5.40}$$

The function $B_i^3(t)$ is for the control point $\mathbf{P}_i$. The curve at a particular value of t is the sum of each control point multiplied by its corresponding blending-function value. Blending functions for cubic curves can be plotted as shown in Figure 5.22.

To generalize any Bezier curve to degree n, $n+1$ control points are required. Equation 5.39 then becomes

$$\mathbf{r}(t) = \sum_{i=0}^{n} B_i^n(t)\mathbf{P}_i \tag{5.41}$$

$$0 \le t \le 1$$

where

$$B_i^n(t) = \frac{n!}{i!(n-i)!}t^i(1-t)^{n-i}$$

From Equation 5.41, one can see that, at any location on the curve $\mathbf{r}(t)$ between $t = 0$ and $t = 1$, the blending-function values are nonzero. This means that, by changing any of the control points, the shape of the entire curve changes, except at the beginning point ($t = 0$) and the end point ($t = 1$). To maintain the zeroth-order (first-degree) continuity, the first point ($\mathbf{P}_0^2$) of the second curve segment must be the same as the fourth point ($\mathbf{P}_3^1$) of the first curve segment (Figure 5.23). For first-order continuity, points $\mathbf{P}_2^1$, $\mathbf{P}_3^1$, $\mathbf{P}_0^2$, and $\mathbf{P}_1^2$ must be collinear.

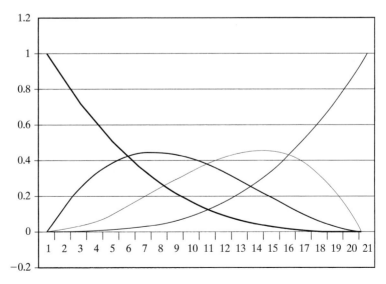

Figure 5.22 Cubic Bernstein blending functions.

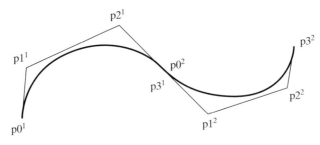

$p2^1$

$p3^2$

$p1^1$

$p0^2$

$p3^1$

$p2^2$

$p0^1$

$p1^2$

Figure 5.23   Two curve segments.

Example 5.7.

A Bezier curve is defined by four control points: (3,0,1), (4,0,4), (8,0,4), and (10,0,1). Find the equation of the curve.

From Equation 5.37,

$$
r(t) = \begin{bmatrix} 3 & 4 & 8 & 10 \\ 0 & 0 & 0 & 0 \\ 1 & 4 & 4 & 1 \end{bmatrix} \begin{bmatrix} -1 & 3 & -3 & 1 \\ 3 & -6 & 3 & 0 \\ -3 & 3 & 0 & 0 \\ 1 & 0 & 0 & 0 \end{bmatrix} \begin{bmatrix} t^3 \\ t^2 \\ t \\ 1 \end{bmatrix}
$$

$$
= \begin{bmatrix} -5 & 9 & 3 & 3 \\ 0 & 0 & 0 & 0 \\ 0 & -9 & 9 & 1 \end{bmatrix} \begin{bmatrix} t^3 \\ t^2 \\ t \\ 1 \end{bmatrix}
$$

$$
= \begin{bmatrix} -5t^3 + 9t^2 + 3t + 3 \\ 0 \\ -9t^2 + 9t + 1 \end{bmatrix}
$$

The curve can be plotted with the following MATLAB commands:

```
t = 0:0.02:1;
p = [3 4 8 10; 0 0 0 0; 1 4 4 1];
A = [-1 3 -3 1; 3 -6 3 0; -3 3 0 0; 1 0 0 0];
tt = [t.^3; t.^2; t; ones(size(t))];
r = p*A*tt;
plot(r(1,:),r(3,:))
```

Figure 5.24 shows the resulting output.

To plot a curve starting from $t = 0$, increment the t value by a small quantity Δt. Then, for each new t value, find $\mathbf{r}(t)$ and draw from the current position to the new $\mathbf{r}(t)$ position. Since the equation is simple, the computer can draw the curve very fast.

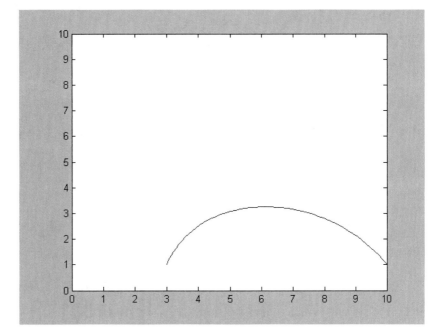

Figure 5.24   The resulting output.

c. B-spline Curve   B-spline curves are also very popular, since they offer more flexibility than Bezier curves. Like the latter, a B-spline curve is defined by a set of control points. This should come as no surprise, because a B-spline curve is a general case of Bezier's curve. The nonuniform B-spline presented in this section is defined as

$$\mathbf{r}(t) = \sum_{i=0}^{L} N_i^n(t)\mathbf{P}_i$$

$$t_i \leq t \leq t_{i+L}$$

(5.42)

where $N_i^n(t) = \dfrac{t - t_i}{t_{i+n-1} - t_i} N_i^{n-1}(t) + \dfrac{t_{i+n} - t}{t_{i+n} - t_{i+1}} N_{i+1}^{n-1}(t)$

$$N_i^1(t) = \begin{cases} 1 & t \in [t_i, t_{i+1}] \text{ and } t_i < t_{i+1} \\ 0 & \text{otherwise} \end{cases}$$

is a Cox–de Boor recursive function.

L is the number of control points and has nothing to do with the order of the curve, which is controlled by the parameter n. (The degree of the curve is $n - 1$; for example, a cubic curve has an n value of 4.) The t_i are called *knots* and are set by the user. Note the difference between t, which is a parameter variable, and the knots t_i, the effects of which can be seen in Figure 5.25. Compare this figure with Ferguson and Bezier curves, whose knot values are either 0 or 1 and cannot be altered.

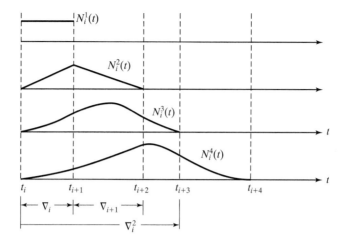

Figure 5.25 Nonuniform B-spline blending functions.

To simplify the algebra, a difference operator ∇ is introduced:

$$\nabla_i = (t_{i+1} - t_i) \tag{5.43}$$

$$\nabla_i^k = \nabla_i + \cdots + \nabla_{i+k-1} = (t_{i+k} - t_i) \tag{5.44}$$

The difference operator ∇ defines the knot span. A nonuniform B-spline is a B-spline that has a variable knot span. A uniform B-spline is a B-spline with a constant ∇ value. The blending functions of nonuniform B-spline are shown in Figure 5.25. For cubic curves, all four N functions are used. For a quadratic curve, only the first three are used. Figure 5.26 shows the blending functions of a cubic B-spline. Each blending function is associated with one control point. In this example, there are six control points.

A Bezier curve can be considered a special case of a nonuniform B-spline. By setting $L = 3, n = 4, t_{i-2} = t_{i-1} = t_i = 0$, and $t_{i+1} = t_{i+2} = t_{i+3} = 1$, the B-spline becomes a cubic Bezier curve.

d. NURB Curves The NURB curve is an extension of the B-spline. The difference between NURB and the nonuniform B-spline introduced in the previous section is the "rational" feature denoted by the "R" of "NURB." A **rational function** is defined by the ratio of two functions. A rational number is a number of the form a/b. A NURB curve has the form

$$\mathbf{R}(t) = \begin{bmatrix} x(t) \\ y(t) \\ z(t) \\ h(t) \end{bmatrix} \tag{5.45}$$

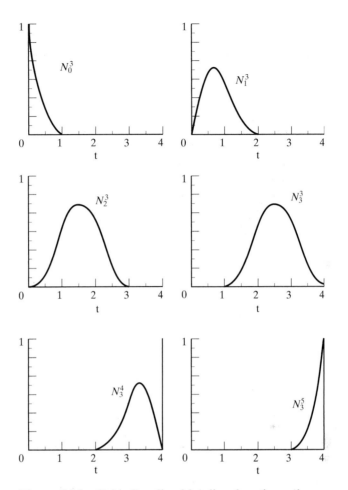

Figure 5.26   Cubic B-spline blending-function values.

$\mathbf{R}(t)$ is said to be a *homogeneous* vector. Since, in homogeneous coordinates, the last element is always 1, the homogeneous vector can be rewritten as

$$\mathbf{r}(t) = \begin{bmatrix} x(t)/h(t) \\ y(t)/h(t) \\ z(t)/h(t) \\ 1 \end{bmatrix} \qquad (5.46)$$

The transformation from $\mathbf{R}(t)$ to $\mathbf{r}(t)$ is called a **normalization**.

To obtain a NURB curve, homogeneous control points are used. Each control point is assigned a weight w_i. For example, a rational Bezier curve will add four weights, one for each control point, and will result in a curve with the form

$$\mathbf{Q}_i = \begin{bmatrix} w_i x_i \\ w_i y_i \\ w_i z_i \\ w_i \end{bmatrix} \tag{5.47}$$

the same form as Equation 5.45. Note that, instead of $\mathbf{P}_i$, $\mathbf{Q}_i$ is used in Equation 5.42. The result is a function $\mathbf{R}(t)$, which, when normalized, produces the curve $\mathbf{r}(t)$.

A rational curve introduces additional control over the shape of the curve. A NURB curve can represent both free-form curves and conical sections precisely. All of the curves discussed so far can be considered special cases of NURB. The additional controls make the equation quite complicated and computationally expensive to use. Because the equation is so complicated (assigning the locations of control points, as well as determining weights and knots), it is hard for users to take full advantage of its flexibility. In the real world, the user often defines a set of points the curve must pass through, and curve-fitting algorithms are used to obtain control points, weights, and knots. The user is able to modify the derived control points, etc., which makes the initial definition of the curve easier to understand.

5.4.2 Surfaces

Like curves, surfaces can be classified into analytical surfaces and free-form surfaces. Analytical surfaces include two groups. The first group is defined by nonparametric equations; examples are the plane, cylinder, sphere, cone, and torus. The second group is defined by parametric equations; examples are a surface of revolution, ruled surface, and tabulated cylinder. Free-form surfaces are defined by control points, as are free-form curves. Displaying a surface, however, is not as straightforward as displaying a curve. Drawing a curve basically involves connecting n equally spaced (either in Cartesian space or in parameter space) points on the curve. The value of n determines how precise the curve is interpolated. If the same strategy is used in drawing surfaces, $n \times m$ points are needed. When both n and m are large, the surface will be fully covered by curves and thus not discernible in shape. If n and m are small, then each curve will be coarse. Therefore, one usually draws sparsely positioned curves, and each one is interpolated with a high point count. Note that we are talking about line drawings here. Other ways to display surfaces are through shaded image and rendering, topics beyond the scope of this book.

5.4.2.1   Analytical Surfaces   In this section, equations for analytical surfaces are presented. These equations are used in representing analytical surfaces and to build applications.

a. Nonparametric surfaces   The simplest surface is a planar surface (see Figure 5.27), the equation of which has the general form

$$ax + by + cz = 1 \tag{5.48}$$

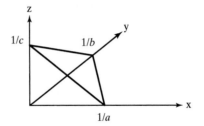

Figure 5.27   A planar surface.

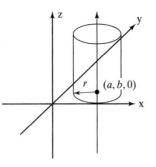

Figure 5.28   A right cylinder.

A cylindrical surface with axis on the Z-axis is basically a circle in the X–Y plane (see Figure 5.28):

$$(x - a)^2 + (y - b)^2 = r^2 \tag{5.49}$$

Alternatively, in parametric form with height h, we have

$$\begin{bmatrix} x \\ y \\ z \end{bmatrix} = \begin{bmatrix} r \cos \theta \\ r \sin \theta \\ z \end{bmatrix} \qquad 0 \le \theta < 2\pi, \;\; 0 \le z \le h \tag{5.50}$$

For a cylindrical surface in an arbitrary orientation, an axis vector is defined and the surface Equation 5.49 is rotated.

A sphere is a 3-D object defined by its center (x_0, y_0, z_0) and radius r:

$$\sqrt{(x - x_0)^2 + (y - y_0)^2 + (z - z_0)^2} = r^2 \tag{5.51}$$

The parametric form is

$$\begin{bmatrix} x \\ y \\ z \end{bmatrix} = \begin{bmatrix} r \cos \theta \sin \phi \\ r \sin \theta \sin \phi \\ r \cos \phi \end{bmatrix} \qquad 0 \le \theta < 2\pi, 0 \le \phi \le \pi \tag{5.52}$$

A right cone, as shown in Figures 5.14 to 5.16, forms a circle when cut on the X–Y plane. The circle is given by (x_0, y_0, r), and the cone has a height h. The equation is

$$\begin{bmatrix} x \\ y \\ z \end{bmatrix} = \begin{bmatrix} \dfrac{h - z}{h} r \cos \theta \\[2mm] \dfrac{h - z}{h} r \sin \theta \\[2mm] z \end{bmatrix} \qquad 0 \le z \le h, 0 \le \theta < 2\pi \qquad (5.53)$$

As with a cylinder, to have a different orientation, the cone can be transformed by rotating its axis.

Finally, a torus is defined by its center (x_0, y_0, z_0), the radius r of the meridian, and the distance c from the center of meridian to center of the torus (see Figure 5.29):

$$c - \sqrt{(x - x_0)^2 + (y - y_0)^2} + (z - z_0)^2 = r^2 \qquad (5.54)$$

The parametric form of the torus is

$$\begin{bmatrix} x \\ y \\ z \end{bmatrix} = \begin{bmatrix} (c + r \cos t) \cos s \\ (c + r \cos t) \sin s \\ r \sin t \end{bmatrix} \qquad 0 \le t, s < 2\pi \qquad (5.55)$$

b. Parametric surfaces **Tabulated cylinder.** A tabulated cylinder is a surface created with the use of a curve $\mathbf{r}(t)$ and a vector $\mathbf{V}$. (See Figure 5.30.) The equation is

$$\mathbf{r}(t, s) = \mathbf{r}(t) + s\mathbf{V} \qquad 0 \le t, s \le 1 \qquad (5.56)$$

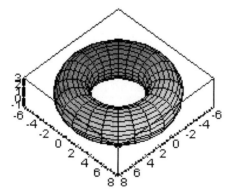

Figure 5.29   A torus centered at (1,1,1). The radius of the meridian = 2 and the distance from the center of the meridian to the center of the torus = 5.

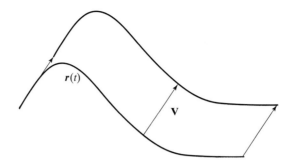

Figure 5.30 A tabulated cylinder.

The curve $\mathbf{r}(t)$ can be any parametric curve, even a line segment. A line segment can be written as $\mathbf{r}(t) = t\mathbf{P}_1 + (1 - t)\mathbf{P}_0$. For example, to create a planar surface, one may sweep a line:

$$\mathbf{r}(t, s) = [t\mathbf{P}_1 + (1 - t)\mathbf{P}_0] + s\mathbf{V} \qquad 0 \le t, s \le 1$$

Example 5.8.

Write the model for a cone with a base circle at (1,1), radius 1, and apex at (3,3,4). The circle has to be represented by a parametric form, so we have

$$\mathbf{r}(\theta) = \begin{bmatrix} r\cos\theta + a \\ r\sin\theta + b \\ 0 \end{bmatrix} = \begin{bmatrix} \cos\theta + 1 \\ \sin\theta + 1 \\ 0 \end{bmatrix} \qquad 0 \le \theta \le 2\pi$$

$$\mathbf{V} = \begin{bmatrix} 3-1 \\ 3-1 \\ 4-0 \end{bmatrix} = \begin{bmatrix} 2 \\ 2 \\ 4 \end{bmatrix}$$

$$\mathbf{r}(\theta, s) = \begin{bmatrix} \cos\theta + 1 \\ \sin\theta + 1 \\ 0 \end{bmatrix} + s\begin{bmatrix} 2 \\ 2 \\ 4 \end{bmatrix} = \begin{bmatrix} \cos\theta + 1 + 2s \\ \sin\theta + 1 + 2s \\ 4s \end{bmatrix} \qquad 0 \le s \le 1; 0 \le \theta \le 2\pi$$

Ruled surfaces.

A ruled surface is the result of blending two curves $\mathbf{r}_0(t)$ and $\mathbf{r}_1(t)$. (See Figure 5.31.) This kind of surface is defined as

$$\mathbf{r}(t, s) = \mathbf{r}_0(t) + s[\mathbf{r}_1(t) - \mathbf{r}_0(t)] \qquad 0 \le t, s \le 1 \qquad\qquad (5.57)$$

Equation 5.57 can also be written as

$$\mathbf{r}(t, s) = (1 - s)\mathbf{r}_0(t) + \mathbf{r}_1(t) \qquad 0 \le t, s \le 1$$

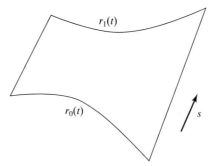

Figure 5.31   A ruled surface.

The two curves can be any parametric curves. In engineering products, ruled surfaces commonly create a surface between two curves. For example, the wing of an airplane can be defined by blending cross-sectional airfoil curves. Note that if $\mathbf{r}_0(t)$ degenerates into a vector, the ruled surface becomes a tabulated cylinder. One can simply replace $(1 - s)$ by s, since s has a range between zero and unity.

Example 5.9.

Find the ruled surface formed from the following two curves:

$$\mathbf{r}_0(t) = \begin{bmatrix} 3t^3 + 2t^2 + 4t \\ t^3 + 2t + 4 \\ t^2 + 4t \end{bmatrix} \qquad 0 \leq t \leq 1$$

$$\mathbf{r}_1(t) = \begin{bmatrix} 2t + 1 \\ 3t + 4 \\ t + 5 \end{bmatrix} \qquad 0 \leq t \leq 1$$

The ruled surface is given by

$$\mathbf{r}(t, s) = (1 - s)\begin{bmatrix} 3t^3 + 2t^2 + 4t \\ t^3 + 2t + 4 \\ t^2 + 4t \end{bmatrix} + \begin{bmatrix} 2t + 1 \\ 3t + 4 \\ t + 5 \end{bmatrix}$$

Combining the two terms, we obtain the following equation:

$$\mathbf{r}(t, s) = \begin{bmatrix} 3(1 - s)t^3 + 2(1 - s)t^2 + [4(1 - s) + 2]t + 1 \\ (1 - s)t^3 + [2(1 - s) + 3]t + 4(1 - s) + 4 \\ (1 - s)t^2 + [4(1 - s) + 1]t + 5 \end{bmatrix}$$

$$= \begin{bmatrix} 3(1 - s)t^3 + 2(1 - s)t^2 + (6 - 4s)t + 1 \\ (1 - s)t^3 + (5 - 2s)t + 8 - 4s \\ (1 - s)t^2 + (5 - 4s)t + 5 \end{bmatrix} \qquad 0 \leq t, s \leq 1$$

Note that the original equation is more concise.

Surface of revolution.

A surface of revolution is defined by a curve $\mathbf{r}(t)$ with rotational sweep around an axis. If a line is used instead of a curve, the result will be a cylindrical or conical surface. The equation for the surface shown in Figure 5.32 is

$$\mathbf{r}(t, \theta) = \begin{bmatrix} x(t) \cos \theta \\ x(t) \sin \theta \\ z(t) \end{bmatrix} \tag{5.58}$$

In the figure,

$$\mathbf{r}(t) = \begin{bmatrix} x(t) \\ 0 \\ z(t) \end{bmatrix}$$

c. Free-form surfaces   Free-form surfaces are based on free-form curves. The Bezier, B-spline, and NURB surfaces are formed from the Bezier, B-spline, and NURB curves, respectively. To blend points together to form a surface, two parameters t and s are needed:

$$\mathbf{r}(t, s) = \sum_i \sum_j B_i(t) B_j(s) \mathbf{P}_{ij} \qquad 0 \le t, s \le 1 \tag{5.59}$$

This equation may represent a surface patch, as in a Bezier surface (Figure 5.33), or a larger surface, using NURB. A surface patch must follow the necessary boundary conditions: zeroth- and first-order continuity.

In Equation 5.59, the blending functions $B_i(t)$ and $B_j(t)$ can be one of those discussed in the material on curve functions. A **bicubic Bezier surface**, as shown in Figure 5.34, is defined by $4 \times 4 = 16$ control points. Each point on the surface is the result of blending in all of the control points. Thus, the surface equation has 16 terms. Each term has a corresponding control point $\mathbf{P}_{ij}$ and combination of blending functions $B_i(t) B_j(s)$.

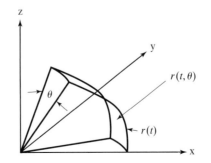

Figure 5.32   Surface of revolution.

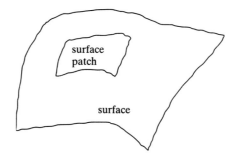

Figure 5.33 A surface patch on a larger surface.

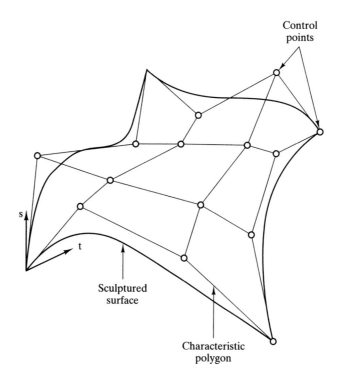

Figure 5.34 Characteristic polygon and surface of a bicubic Bezier surface.

The i and j values run from 0 to 3. The four corners of the surface are located at $\mathbf{P}_{0,0}$, $\mathbf{P}_{0,3}$, $\mathbf{P}_{3,0}$, and $\mathbf{P}_{3,3}$. Continuity and smoothness conditions must be satisfied. The tangent plane (instead of line) at any given point $\mathbf{r}(t,s)$ is defined by two **principal**

tangent vectors, one in the t direction and one in the s direction. The following equations are used to find the principal tangents:

$$\frac{\partial \mathbf{r}(t, s)}{\partial t} = \sum_i \sum_j \frac{dB_i(t)}{dt} B_j(s)\mathbf{P}_{ij} \tag{5.60}$$

$$\frac{\partial \mathbf{r}(t, s)}{\partial s} = \sum_i \sum_j B_i(t) \frac{dB_j(s)}{ds} \mathbf{P}_{ij} \tag{5.61}$$

The tangent plane is formed by these two vectors. The **surface normal** is

$$\mathbf{N}(t, s) = \frac{\partial \mathbf{r}(t, s)}{\partial t} \times \frac{\partial \mathbf{r}(t, s)}{\partial s} \tag{5.62}$$

Note that the normal vector could point either up or down. To change the vector from pointing in one direction to another, just multiply it by -1. Figure 5.35 shows the principal tangents and surface normal of a bicubic Bezier surface.

5.5 SOLID MODELS

In classifying a solid model, its **internal representation** is usually used. The internal representation is how a computer stores the model. It is different from the **external representation**, which is how the picture or image is displayed. There are six different types of solid internal representation schemes (see Figure 5.36):

1. Primitive instancing
2. Spatial occupancy enumeration (SOE)
3. Cell decomposition
4. Constructive solid geometry (CSG)
5. Boundary representation (B-rep)
6. Sweeping

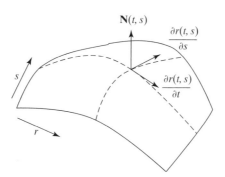

Figure 5.35   Principal tangents and surface normal.

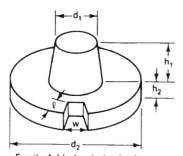

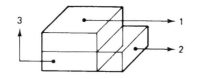

Family A (d_1, h_1, d_2, h_2, ℓ, w)

(a) Pure primitive instancing

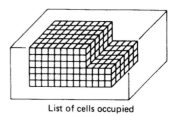

List of cells occupied

(b) Spatial occupancy enumeration

Cells 1, 2, and 3

(c) Cell decomposition

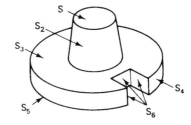

(d) CSG: ($\Delta \cdot$ dif $\cdot \Delta$) $\cdot$ un $\cdot$ (⬭ $\cdot$ dif $\cdot$ ▱)

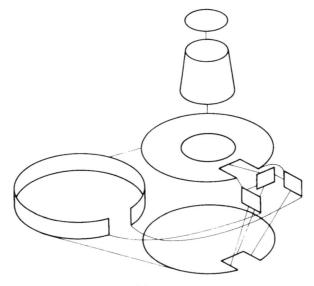

(e) Boundary:

Figure 5.36 Three-dimensional representations.

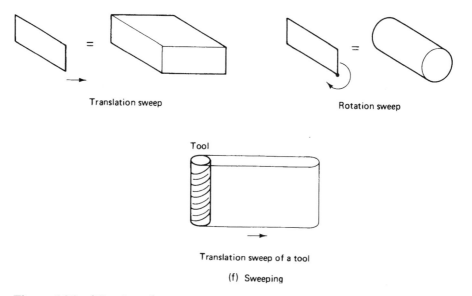

Translation sweep

Rotation sweep

Tool

Translation sweep of a tool

(f) Sweeping

Figure 5.36   (*Continued*).

Figure 5.36(a) shows pure primitive instancing. This scheme can be used to represent a family of objects. A composite model for the entire family of objects is parameterized, and features on the model can be turned on or off. The dimensions of each feature are set as variables (parameters). An object instance can be defined by assigning values to the parameters and can be used effectively to represent standard parts with different dimensions. Group Technology coding can also be considered an application of pure primitive instancing [Requicha, 1980]. The model is concise, since it has only a model name and a set of parameters. For this reason, it is difficult to do geometric operations directly on the pure primitive instancing model. Instead, the model must be converted into another model, such as a B-rep model, before any operation can be performed.

Spatial occupancy enumeration (SOE) records all spatial cells that are occupied by the object. (See Figure 5.36(b).) SOE is equivalent to storing the physical object in sections. To describe the object accurately, very small cells must be used. The massive memory required for even small objects makes this representation virtually useless in general engineering design. The memory needed equals the volume of the object divided by the volume of the cell. When many engineering applications require an accuracy of 0.001, the volume of each cell is 10^{-9} in^3. Even a small engineering object needs an enormous amount of storage.

Cell decomposition is a general class of spatial occupancy enumeration. In this difficult-to-create representation, a solid is decomposed into simple solid cells with no holes, whose interiors are pairwise disjoint (Figure 5.36(c)). A solid is the result of "gluing" component cells that satisfy certain boundary-matching conditions. Since cell size is a variable, it requires many fewer cells and thus less memory than that for SOE. Boolean operations are used to manipulate objects. Boolean operations can be difficult to perform.

Constructive solid geometry (CSG) is an excellent system for creating 3-D models. Using primitive shapes (see Figure 4.22) as building blocks, CSG employs Boolean set operators (union, difference, and intersection) to construct an object (Figure 5.37). A CSG model is represented by a tree structure (Figure 5.38). At its termini are primitives with dimensions (the size of the primitive) and coordinate transforms (the location and orientation of the primitive). At the nodes are Boolean operators. CSG models are not unique: The same object may be modeled with different primitives and different sequences of operations (Figure 5.39).

Boundary representations (B-reps) represent objects by their bounding faces. The object's faces are further divided and represented by edges and vertices (Figure 5.36(e)). A set of operators called Euler operators is available to build a B-rep from the ground up. To build a B-rep model by hand is tedious; therefore, most B-rep models are derived from a CSG model through boundary evaluation.

Sweeping is a powerful modeling tool for certain types of geometry. There are two types of sweeping: translation and rotation (Figure 5.36(f)). Translation sweeping of a rectangle produces a box, while rotation sweeping of the same rectangle produces a cylinder. Rotation sweeping can be used to create turned parts. In some design systems, an arbitrarily drawn face can be swept along either a line or a curve (translation sweeping), creating a highly complicated shape. One manufacturing application of sweeping is in NC cutting simulation: A volume to be removed can be represented by the sweeping of a tool.

Currently, the majority of 3-D solid modelers are based on either CSG or B-rep representations. CSG data input is the most popular. Often, many face types can be used in the solid model. In a 3-D manifold model, the dimensionality is retained. The B-rep of an object consists only of bounded faces, with no loose edge or face. Each edge is bounded by exactly two vertices and is adjacent to exactly two faces. Each vertex belongs to one disk. An object traveling from one adjacent face to another will never cross the vertex itself; see vertex A in Figure 5.40(i) and (ii). A manifold model does not allow any dangling faces or edges (Figure 5.40(iii)) and satisfies the Euler formula. A nonmanifold model allows additional faces and edges to exist in a solid model. Tracking the inside and the outside of the object then becomes more complex. A nonmanifold modeler, such as ACIS (a solid-modeler kernel used in many CAD systems, including AutoCAD R13 and up), enables more design intents to be saved in the design model.

Since solid-model representation is complete and unambiguous, engineering analysis can be performed directly on the model. It also provides a common linkage among design, analysis, and manufacturing. Today, the existence of solid modelers is essential to the automated manufacturing environment.

5.5.1 Topology

Topology represents the basic relationships of the geometric entities within an object. Each geometric entity, such as a point, a line, and a surface, has a corresponding topological element, such as a vertex, an edge, and a face. The topology of a solid is represented by an **object** that is bounded by several faces. Each face may be described by a surface equation (e.g., planar, conic, free form, etc.). A **face** is bounded by several

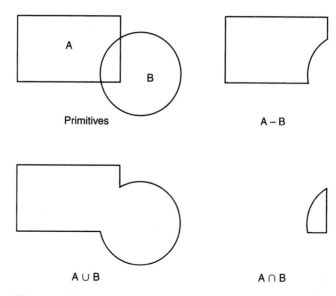

Figure 5.37   Boolean operators.

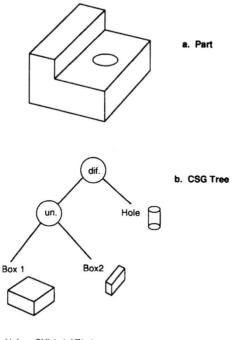

a. Part

b. CSG Tree

c. Instructions to construct part

Hole = CYL(...) AT(...)
Box1 = BLO(...) AT(...)
Box2 = BLO(...) AT(...)
Box = Box1 UN Box2
Part = Box DIF Hole

Figure 5.38   Example of CSG.

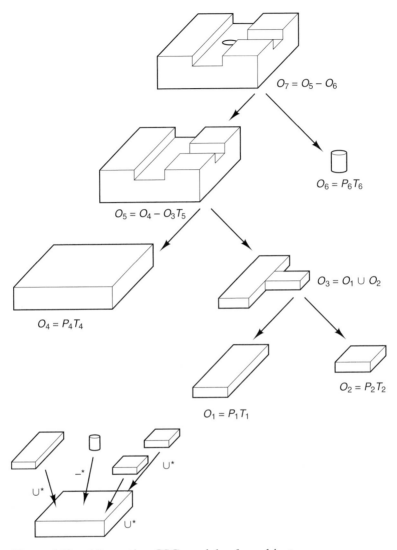

$O_7 = O_5 - O_6$

$O_6 = P_6 T_6$

$O_5 = O_4 - O_3 T_5$

$O_3 = O_1 \cup O_2$

$O_4 = P_4 T_4$

$O_2 = P_2 T_2$

$O_1 = P_1 T_1$

$\cup^*$

$-^*$

$\cup^*$

$\cup^*$

$\cup^*$

Figure 5.39   Alternative CSG models of an object.

edges. Each **edge** is given by an equation, of a line, a circle, or a curve. A face may have several disjointed sets of edges in it. Each set of edges is called a *loop* of edges. For example, a hole through a face will create an outer loop and an inner loop. The inner loop is the entrance of the hole. The **loop** is a topological element between face and edge. An edge is always bounded by two vertices. A **vertex** is represented by its counterpart in geometry, namely, a point. For example, in Figure 5.41, the topology of a tetrahedron is captured by the tree structure shown. The basic topological relationships remain the same for any tetrahedron, even though tetrahedrons can be of different sizes and may be oriented differently. Any object can be represented by (1) geometric entities (lines,

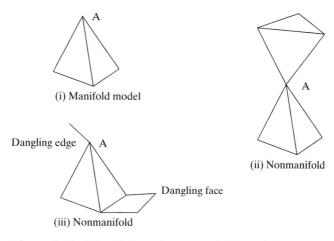

(i) Manifold model

Dangling edge

Dangling face

(iii) Nonmanifold

(ii) Nonmanifold

Figure 5.40   Manifold and nonmanifold models.

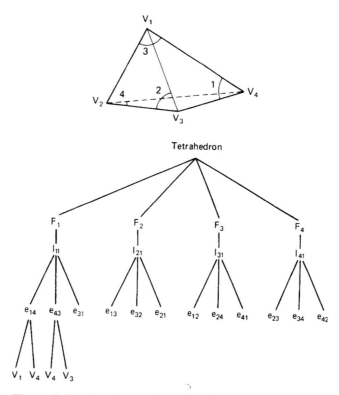

Tetrahedron

Figure 5.41   Topology of a tetrahedron.

curves, and surfaces), (2) topologic elements (vertices, edges, and faces), or (3) auxiliary information (color, tolerances, etc.).

Both geometry and topology are necessary in order to manipulate and display the object correctly. Some additional information may also be necessary to display the object or for other purposes (e.g., engineering analysis, manufacturing planning, programming, and so on). In three-dimensional solid models, the representation of the topology is critical. Given a data structure full of geometric entities (represented by coordinates, equations, and parameters) and topological elements (represented by the linkage in the data structure) how does one know if the structure represents a valid or an invalid solid? To answer this question, Euler's formula, developed in mathematical topology, may be used. Euler's formula for a solid without holes (passages) is

$$F - E + V = 2 \qquad (5.63)$$

where F is the number of faces, E is the number of edges, and V is the number of vertices.

When the object has a hole through it, a generalized Euler–Poincaré formula is applied:

$$V - E + F - (L - F) - 2(S - G) = 0 \qquad (5.64)$$

Here, L is the number of loops, S is the number of shells (for a manifold model, $S = 1$), and G is the number of genuses (holes).

Example 5.10.

The boundary model of the object shown in Figure 5.42 consists of six faces, eight vertices, and 12 edges. Applying Equation 5.63 yields

$$6 - 12 + 8 = 2$$

The object shown in Figure 5.43 has a square through hole at its center. It consists of 10 faces, 16 vertices, 24 edges, 12 loops (each face has 1 loop, and the top and the bottom faces have 2 loops—an inner and an outer one), 1 shell, and 1 genus. Applying Equation 5.64, we obtain

$$16 - 24 + 10 - (12 - 10) - 2(1 - 1) = 0$$

For an object with curved surfaces, the calculation is not as straightforward. For example, a cylinder is represented by two circles and a straight edge (Figure 5.44). The circle has one

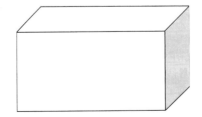

Figure 5.42   A parallel pipe (block) object.

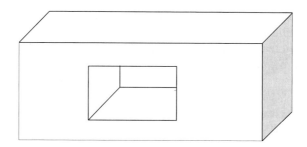

Figure 5.43   A block with a through hole.

Figure 5.44   A cylinder.

vertex that is both the beginning and the end vertex. The cylindrical face is bounded by two circular edges and one linear edge. There are two vertices, three edges (two circles plus a line), and three faces (two flat faces bounded by circles and one cylindrical face bounded by the edge). Thus, $3 - 3 + 2 = 2$.

Note that the Euler formula is only a necessary condition for a valid 3-D solid object and is valid for 2-manifold models. Every manifold object must satisfy the Euler formula. However, an object that satisfies the Euler formula may be nonmanifold (i.e., with a dangling face or edge). For example, a cube with a dangling rectangular face has 10 vertices, 15 edges, and 7 faces. Thus, $10 - 15 + 7 = 2$.

5.5.1.1 Loops   A loop represents an oriented set of close and connected edges. For a face with a hole or protrusion inside, there is only one outer loop. However, when there are holes or protrusions inside, the face has more than one loop (Figure 5.45). In a B-rep model, loops are represented explicitly. A proper B-rep model has a structure consisting of object–face–loop–edge–vertex (Figure 5.46).

By counting the number of inner loops an object has, one can estimate the number of holes and protrusions it has. This property is useful in feature recognition.

For a nonmanifold model, an empty shell within a solid is allowed. (A manifold model has only one shell.) The data structure includes the shell above the face. Some modelers also allow disjoint objects.

5.5.1.2 Euler Operators   Euler operators are used to edit a solid by adding or deleting vertices, edges, and faces. Euler operators ensure that the object being edited always satisfies the Euler–Poincare formula (Equation 5.64). There are two groups of such operators: the *Make* group and the *Kill* group. Euler operators are written as

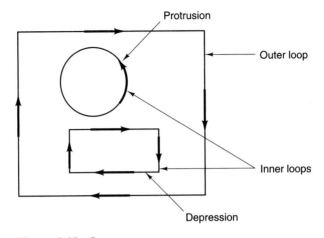

Figure 5.45   Loops.

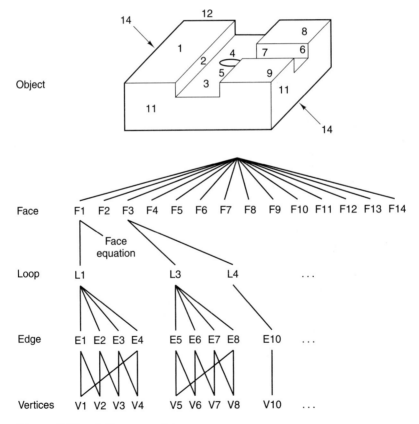

Figure 5.46   A B-rep model.

TABLE 5.1 Euler operators

Operator	Meaning	V	E	F	L	S	G
MEV	Make an edge and a vertex	+1	+1				
MFE	Make a face and an edge		+1	+1	+1		
MSFV	Make a shell, a face, and a vertex	+1		+1	+1	+1	
MSG	Make a shell and a hole					+1	+1
MEKL	Make an edge and kill a loop		+1		−1		
KEV	Kill an edge and a vertex	−1	−1				
KFE	Kill a face and an edge		−1	−1	−1		
KSFV	Kill a shell, a face, and a vertex	−1		−1	−1	−1	
KSG	Kill a shell and a hole					−1	−1
KEKL	Kill an edge and make a loop		+1		−1		

Mxyz and *Kxyz* for operations in the Make and Kill group, respectively, where *x, y,* and *z* are elements of the model (e.g., a vertex, an edge, a face, a loop, a shell and a genus). For example, **MEV** denotes the addition of an edge and a vertex. Euler operators keep the solid being edited topologically valid. Table 5.1 lists the Euler operators.

Example 5.11.

Use the Euler operator to change a tetrahedron into a cube.
Figure 5.47 illustrates a sequence of steps in which Euler operators are applied to change a tetrahedron into a cube. The first MEV operator stretches the right triangular face into a rectangular face. In the process, one edge and one vertex are added. The back face also gets changed into a face with four sides. This face may not be planar anymore. The next step modifies the left-hand triangular face into a rectangular face. The back face now has five sides. The following step modifies the bottom face. Then, an MFE operator is used to add a face and an edge to the back face. This new triangular face is modified into a rectangular face by an MEV operator. Finally, a top face is added with the use of MFE. When this face is added, so is an edge.

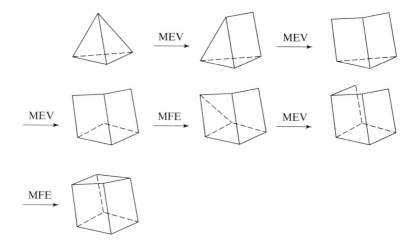

Figure 5.47 Turning a tetrahedron into a cube via Euler operators.

5.6 SUMMARY

Geometric modeling is the foundation for CAD. CAM functions such as NC part programming and rapid prototyping rely on a precise geometric model of the artifacts to be built. It is important to understand the geometric model itself. This chapter has presented an overview of geometric modeling. Among the topics discussed were how an engineering object is modeled by 2-D projections, a wire-frame model, a surface model, and a solid model. Equations for commonly used geometric entities were presented and discussed. The chapter should provide readers with a basic understanding of how CAD/CAM systems function. It is important that engineers understand not only how to operate devices, but also how things work. This is true for hardware, such as machines, as well as for the software we use in conducting our daily jobs.

5.7 KEYWORDS

Algebraic curves
Bezier curve
Boolean set operators
B-spline curve
Cartesian coordinate systems
Characteristic polygon
Control points
Coordinate system transformation
Edge
Euler operators
Explicit polynomial
Face
First-degree continuity
Free-form surfaces
Hermite blending functions
Hyperbola
Internal representation
Manifold
Nonparametric surfaces
NURB curves
Orthogonal projection
Parametric surfaces
Primitive instancing
Prospective projection
Rotation
Scaling
Surface normal
Sweeping
Topology
Vertex
View port

Bernstein blending functions
Bicubic Bezier surface
Boundary representation
Canonical representation
Cell decomposition
Constructive solid geometry
Coordinate frames
Degree of the polynomial
Ellipse
Euler–Poincare formula
External representation
Ferguson curve
Free-form curves
Graphics user interface
Homogeneous transformation
Implicit polynomial
Loops
Nonmanifold
Normalization
Object coordinate system
Parabola
Polar coordinate systems
Principal tangent
Rational function
Ruled surfaces
Spatial occupancy enumeration
Surface of revolution
Tabulated cylinder
Translation
View coordinate system
Zeroth-order continuity

5.8 REVIEW QUESTIONS

5.1   What are the differences between the view coordinate system and the object coordinate system? How are coordinate transformations handled in each?

5.2   In additional to computer graphics, can you think of any other applications that use a right-hand coordinate system?

5.3   What is the difference between homogeneous coordinates and ordinary coordinates? What are the advantages of homogeneous coordinates?

5.4   What curve results when a right cone is sliced perpendicular to its axis?

5.5   What will be the shape of the curves that result when a nonright cone (a cone whose axis is not perpendicular to the base circle) is sliced?

5.6   Why do engineering applications use the cubic curve instead of higher degree curves? Try to compare zeroth-degree, first-degree, second-degree, third-degree, and fourth-degree curves.

5.7   What is first-degree continuity? second-degree continuity? Why are they important in curve and surface design?

5.8   What are the advantages and disadvantages of using NURB in design?

5.9   What are the differences between manifold and nonmanifold modelers?

5.10   What is a dangling edge or face?

5.11   What's the difference between each of the following pairs of terms: face and surface; vertex and point; edge and line or curve?

5.9 REVIEW PROBLEMS

5.1   A point at $(1, 0, 0)$ is translated by a vector $[3\ 4\ 5]^T$. Then it is rotated $30°$ about the z-axis, followed by a rotation of $45°$ about the x-axis. What are the new coordinates of the point?

5.2   Repeat the previous problem, except do the x-axis rotation before the z-axis rotation.

5.3   A line is represented by two endpoints $(1, 0, 0)$ and $(2, 0, 0)$. The coordinate system is rotated $30°$ about the y-axis, and then the origin is moved to $(-1, -1, -1)$. What is the line in the new coordinate system?

5.4   The view port is defined by $x_{min} = 1$, $x_{max} = 3$, $y_{min} = 1$, $y_{max} = 4$. The corresponding display window is $dx_{min} = 0$, $dx_{max} = 1152$, $dy_{min} = 0$, $y_{max} = 864$. Two lines defined as (1.5, 1.5) to (3, 3) and (0, 0) to (5, 4) are drawn into the display window. What are the coordinates of the two lines in the display window? Notes that the view port clips one of the lines.

5.5   Based on the discussion in Section 5.3, develop algorithms and derive equations for finding the intersection of two 2-D lines. The algorithms must consider all possible cases. Try to implement the algorithms in a computer program.

5.6   Repeat the previous problem for a line and a circle (in 2-D).

5.7   Repeat the previous problem for two circles.

5.8   Coordinate transformation can be used to simplify computation in intersection problems. For example, in the line-and-circle intersection problem, if the line is vertical and the origin of the coordinate system is at the center of the circle, the problem is much simplified. In this case, the line equation is $x = a$ and the circle is given by $x^2 + y^2 = r^2$. Therefore, the intersections are $y = \pm\sqrt{r^2 - a^2}$. Develop an algorithm that calculates intersections for general cases.

5.9   Find the explicit representations of an ellipse, a parabola, and a hyperbola.

5.10   Find the parametric representation of a parabola. (*Hint:* An easy parameterization is to set $x = t$).

5.11   Find tangent and normal equations of a parabola.

5.12   Write a program to draw a Ferguson's curve. Try different values of Δt to obtain curves of different resolutions.

5.13   Write a program to draw a Bezier's curve. Try different values of Δt to obtain curves of different resolutions.

5.14   Write the blending functions for the following B-spline curve: $L = 5, n = 3, t_0 = 0, t_1 = 1,$ $t_2 = 3, t_3 = 4, t_4 = 5, t_5 = 6, t_6 = 8$. Plot $N_2^3(t)$.

5.15   Continue the previous problem. Let the control points be $p_0 = (0,0), p_1 = (1,1),$ $p_2 = (2,1), p_3 = (3,0), p_4 = (5,1),$ and $p_5 = (6,2)$. Let $n = 3$. Plot the B-spline curve.

5.16   A tabulated cylinder is defined by the formula

$$r(t) = \begin{bmatrix} 2t^2 + 2t + 1 \\ t^3 + 2t^2 + t + 4 \\ t + 1 \end{bmatrix}; N = \begin{bmatrix} 1 \\ 0 \\ 1 \end{bmatrix}$$

for $0 \leq t, s \leq 1$. Plot the surface.

5.17   What is the normal of the tabulated cylinder of problem 16 at $t = 0.5, s = 0.5$?

5.18   A ruled surface is defined by the two curves

$$r_0(t) = \begin{bmatrix} t^3 + t^2 + t - 4 \\ 2t^2 + 3t - 10 \\ t^3 + 2t^2 + t + 1 \end{bmatrix}, \quad r_1(t) = \begin{bmatrix} 3t^3 - 2t^2 + t - 1 \\ t^2 - 2t + 2 \\ 3t - 1 \end{bmatrix}, 0 \leq t, s \leq 1.$$

Plot the surface.

5.19   What is the normal of the ruled surface of problem 18 at $t = 0, s = 0$?

5.10 REFERENCES

Bezier, P. Numerical Control; Mathematics and Applications, trans. A.R. Forrest. London: John Wiley, 1972.

Braid, I. C. *Designing with Volumes*. Cambridge, U. K.: Cambridge University, CAD Group, 1973.

Choi, B. K. *Surface Modeling for CAD/CAM*. Amsterdam: Elsevier, 1991.

Farin, G. E. *NURB Curves and Surfaces: From Projective Geometry to Practical Use*. Wellesley, MA: A. K. Peters, 1995.

Farin, G. E., J. Hoschek, and M-S Kim. *Handbook of Computer Aided Geometric Design*. Boston: North-Holland/Elsevier, 2002.

Ferguson, J. C. "Multivariate Curve Interpolation," *J. of ACM*, 11(2): 221–228, 1964.

Gordon, W. J., and R. F. Riesenfeld. "B-Spline Curves and Surfaces," in R. E. Barnhill and R. F. Riesenfeld, eds., *Computer-Aided Geometric Design*. New York: Academic Press, 1974, pp. 95–126.

MathWorld. http://mathworld.wolfram.com/

LaCourse, D. E. *Handbook of Solid Modeling*. New York: McGraw-Hill, 1995.

Mantyla, M. *An Introduction to Solid Modeling*. Rockville, MD: Computer Science Press, 1986.

Mortenson, M. E. *Geometric Modeling*. New York: John Wiley, 1985.

Newman, W. M., and R. F. Sproull. *Principles for Interactive Computer Graphics*, 2d ed. New York: McGraw-Hill, 1979.

Patrikalakis, N. M., and T. Maekawa. *Shape Interrogation for Computer-Aided Design and Manufacturing*. Berlin and New York: Springer, 2002.

Requicha, A. A. G. "Representation of Rigid Solid: Theory, Methods, and System," *Computing Surveys*, 12(4), 1980, 437–464.

Requicha, A. A. G., and H. B. Voelcker. "Solid Modeling: A Historical Summary and Contemporary Assessment," *IEEE Computer Graphics and Applications*, 2(2), 1982, 9–24.

Chapter 6

Process Engineering

Objective

Chapter 6 serves as an introduction to process planning, process capability, and issues related to the economies of machining process selection, all dealt with in the remainder of this book. To master the materials presented in this chapter, readers must understand engineering drawings and have a fundamental knowledge of various machining processes.

Outline

Manufacturing is the activity that transforms raw materials into a finished product, typically specified on an engineering design (either a drawing on paper or a set of specifications in a CAD model). Selecting the manufacturing processes to transform the raw material into a finished part is based on matching the design requirements with the manufacturing process capabilities. Process capability involves historic and scientific knowledge of each process. Note that processes and machine tools are not equivalent. A process denotes a certain way an operation is carried out. For example, a drilling process denotes an operation that creates a hole by gradually removing material from a workpiece. There are many variations of drilling processes. A twist drill uses the two rotating chisel edges on the tool tip to cut the material while the flutes transport the cut material out of the hole being created. A paddle drill, on the other hand, employs no flutes for chip removal.

There are three levels of process capability: a universal level, a shop level, and a machine level. The universal-level process capability is our current ability to produce a product without regard to the individual shop or machines that perform the process. On the universal level, we say that a twist drill can produce a round hole with a certain accuracy. This statement is applicable to all twist drill processes; whether they are performed by company A or company B is not considered. Such universal-level process capability is an aggregate measure that is normally presented in handbooks and textbooks.

The second process-capability level is the shop level, at which additional processing detail is considered. Although one generally can say that a twist drill operation can produce a hole of a certain diameter with a certain accuracy, at the shop level we look at a specific machine or improved cutter requirements that can produce a much smaller hole or obtain better accuracy. This capability may not be achievable by other companies. Or, perhaps, the equipment is older and not as well maintained; thus, the capability is lessened. In either case, the shop-level capability represents the capability of the best machine in the shop. Shop-level capabilities are those published by companies for internal use.

The machine-level capability addresses the ability of a specific machine to achieve specific goals. Although the shop has certain capabilities, it does not necessarily follow that each machine in the shop has the same capability. For example, one may find an old milling machine in the same shop with a high-precision milling machine. Also, a sophisticated jig may be used to provide accurate location for a bolt-hole set. The capabilities of each machine are different, and so may be the capabilities of each setup. Statistical quality-control techniques usually are used to obtain accuracy capability data. Machine-level capability information is important in selecting the specific machine to perform a specific process.

Some important process-capability parameters are as follows:

1. The shapes and sizes a process can produce
2. The dimensions and geometric tolerances that can be obtained by various processes
3. The material removal rate
4. The relative cost of the process
5. Other cutting characteristics or constraints

Process capability does not necessarily imply that all process selection or machine selection is based on these parameters. However, the more information considered in selecting a process, the more complete will be the result. For conventional process planning (when human planners are used), all process-capability information comes either in the form of experience or in handbook tables and guides. A computer-aided process-planning system functions on the basis of this process-capability information. It is essential for a manufacturing engineer/process planner to know the capability of a process. In a computer-integrated manufacturing environment, process-capability information has to be captured and represented as computer-usable data so that it can be used for decision making.

Statistics-based process engineering is discussed in Chapter 7. Process capability and its representation methods are discussed in this chapter.

6.1 EXPERIENCE-BASED PLANNING

In everyday life and in many other domains, knowledge comes to us through the accumulation and application of experience. Most of our knowledge comes either from our own experience or is passed on to us by others and is based on their experience. This is also true in the context of manufacturing processes. If we go back to the example of the farmer and the blacksmith in Chapter 2, we can make a point of how experience works in a manufacturing environment. When the farmer verbally specified or drew a picture of the hoe he wanted, the blacksmith must have had a manufacturing method (process plan) in mind; otherwise he would not have been able to make the hoe. The process plan the blacksmith created was neither a written one nor a complete one. However, in his mind, he knew that he had some scrap iron in the back of his shop that he could use for the hoe. After retrieving the material, he heated it in his furnace until it was red hot (hot working). He then completed the forging cycle by hand and hammer. Had the farmer asked, "How did you know how to convert your scrap to my hoe?" the answer most assuredly would have been something like "I've been in the trade for 30 years—experience told me."

Even in this computer age, many activities still rely on the experience of process planners. Where do process planners obtain their experience? They get it from earlier training as machinists (most typical), from books, or from discussions with colleagues. This kind of information can be passed on from person to person and generation to generation. However, there are some problems associated with such a planning approach:

1. Gaining experience takes a significant amount of time.
2. Experience represents only approximate, not exact, knowledge.
3. Experience may not be directly applicable to new processes or new systems.

Because of these problems, we need to seek other ways to represent our process-capability knowledge base so that it can be preserved and installed as a decision-support system in a computer.

6.1.1 Machinist Handbooks

One way to store process-capability information is to print it in handbooks. This has long been a standard manufacturing practice. Process-capability information is usually presented in tables, illustrated in figures, or listed as guidelines. Large manufacturers typically prepare their own handbooks for internal use. Therefore, the knowledge is kept, but has traditionally been "proprietary." Handbooks can serve both as a reference and as a guide for process selection. Figure 6.1 represents some typical information on process capability (surface-finish ranges). The surface-finish chart shows the limiting extremes of several processes. For example, a flat surface of 8 μin. surface finish can be machined by grinding, polishing, and lapping. It can be rarely achieved by milling, yet a surface finish of 8 μin. is possible by using a finish milling cut. Other information, such

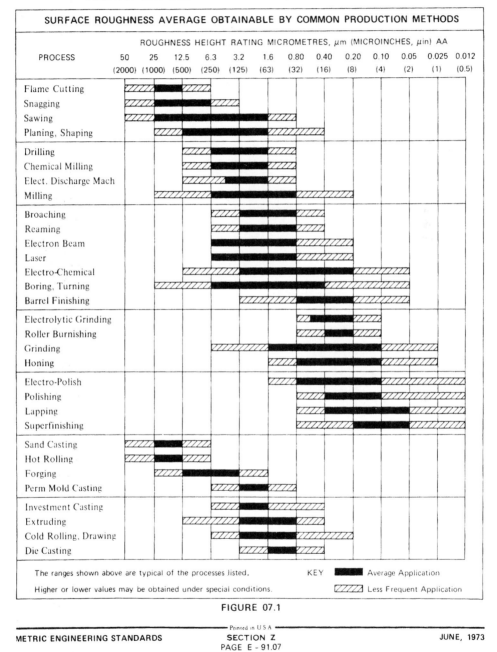

Figure 6.1 Surface-finish ranges. (Courtesy of General Motors Corporation [taken from GM Drafting Standards])

TABLE 6.1 Principles of Machining by Cutting, Abrasion, and Erosion

Classes (according to ISO) are grouped as *of accuracy* (columns 1 to 3 through 14 to 16) and *of surface quality* (columns 1 through 14).

Machining method	1 to 3	4	5	6	7	8	9	10	11	12	13	14 to 16	1	2	3	4	5	6	7	8	9	10	11	12	13	14
I. CHIP REMOVING PROCESSES																										
Turning				•	•	○	○	○	x	x	—	—		—	—	—	x	x	○	○	•					
Boring				•	•	○	○	○	x	x	—	—		—	—	—	x	x	○	○	•					
Drilling						•	○	x	x	—					—	—	x	x	○	•						
Reaming			•	•	○	○	x	x	—	—							—	x	○	○	•	•				
Peripheral milling				•	•	○	○	x	x	—	—				—	—	x	x	○	○	•	•				
Face milling				•	•	○	○	x	x	—	—				—	—	x	x	○	○	•	•				
Planing and shaping					•	○	○	○	x	x	—				—	—	x	x	○	○	•	•				
Broaching			•	•	○	○	x	x	—	—	—	—					—	x	○	○	•	•				
II. ABRASION PROCESSES																										
A. Using abrasive tools																										
Centre-type cylindrical grinding			•	○	○	x	x	—	—								—	x	x	○	•	•				
Centreless cylindrical grinding			•	○	x	x	x	—	—							—	x	x	○	○	•					
Internal grinding			•	○	○	x	x	—	—								—	x	x	○	•	•				
Surface grinding			•	○	○	x	x	—	—	—						—	x	x	○	○	•					
Abrasive belt grinding				•	○	○	x	x	—	—						—	x	x	○	○	•					
Surface honing		•	•	○	○	x	—	—										—	—	x	x	○	•			
Shaft and internal honing	•	•	•	○	○	x	—											—	—	x	x	○	○	•		
Superfinish	•	•	○	○	x	x	—												—	—	x	○	○	•	•	
B. Using loose abrasive																										
Lapping		•	•	○	○	x	x	—											—	—	x	○	○	•	•	
Mechanical polishing	•	•	○	○	x	x	—	—										—	—	x	x	○	○	•	•	
Vibratory and barrel finishing				•	•	•	○	○	x	x	—					—	—	x	x	○	•					
Abrasive-blast treatment				•	•	•	○	○	x	x	—	—				—	—	x	○	○	•	•				
Ultrasonic machining				•	•	○	○	x	x	—							—	x	○	○	•	•				

Accuracy	Surface quality
— rough	— rough
x fairly accurate	x fairly smooth
○ accurate	○ smooth
• highly accurate	• very smooth

Source: Courtesy of Peter Peregrinus, Limited, taken from J. Kaczmarek, *Principles of Machining by Cutting, Abrasion, and Erosion*, 1976.

as process accuracy, can be found in similar tables or charts. (See Table 6.1.) In the table, an accuracy class 10 for drilling is considered highly accurate, but for reaming, it is considered only fairly accurate. The information in these books, representing universal process capabilities, is highly general and not specific to any shop.

Some process-capability information is presented as listed guidelines or tables, so that process planners can follow general rules. Such guidelines are published in a shop to guide personnel in preparing process documents. These guidelines are not generally portable to other shops. We call such information *shop-level process-capability information*. For example, the following guidelines can be applied to produce holes:

I. Diameter ≤ 0.5 in.:	If the diameter is less than 0.5 in.
A. True position ≥ 0.010:	and true position greater than 0.010 in.
1. Tolerance > 0.010:	and the diametric tolerance is greater than 0.010 in.
Drill the hole	then drill.
2. Tolerance ≤ 0.010:	and the diametric tolerance is less than 0.010 in.
Drill and ream	then drill and ream.
B. True position ≤ 0.01:	and true position less than 0.01 in.
1. True position ≤ 0.01:	then drill and finish bore.
Drill, then finish bore	
2. Tolerance ≤ 0.002:	and the diametric tolerance is less than 0.002 in.
Drill, semifinish bore, then finish bore	then drill, semifinish and finish bore.
II. 0.05 in. < diameter ≤ 1.00:	
etc.	

6.2 PROCESS-CAPABILITY ANALYSIS

Let us now return to the chip-metal removal processes we mentioned in the first paragraph of this chapter.

Before any decision can be made (whether it be experience, table, or tree based), the information required to make the decision must be complete. This information includes a process-by-process breakdown of the following elements:

1. The shape(s) that a process can generate
2. Size considerations (boundaries of the tooling, machine tools, fixtures)
3. Tolerances (both dimensional and geometric)
4. The surface finish (as a limiting value or functional expression)
5. The cutting force involved
6. Power consumption.

TABLE 6.2   Shape Capability

Process	Shape Capability
Turning	External surfaces that can be generated by rotating a line or curve around an axis
Boring	Hole
Drilling	Hole
Reaming	Hole
Face milling	Flat surface
Peripheral milling	Flat surface, slot
Shaping and planing	Flat surface
End milling	Hole, flat surface, curved surface, slot
Grinding	Flat surface, hole, external cylindrical surface
Honing	Hole
Tapping	Internal thread

The first four elements are capabilities and the last two are limitations. The limitations form the constraints that the machinery is bounded by during processing. The shape elements imply or define the basic geometry producible by a process. (See Table 6.2). With little training, virtually anyone can develop a set of shapes that can be produced by a process. However, this feature is perhaps the most difficult to achieve with a computer geometry and topology, especially as regards sophisticated components. Many researchers are still working on the problem [Joshi and Chang, 1988]. A feasible alternative is to represent shape by a code. Human judgment can be used to identify machined surfaces and assign codes to them. The matching of shape capability and machined surface may not be automatic, but can be significantly simplified.

For internal machining processes (producing holes and so on), the size capability is constrained by the available tool size. For external machining, size is constrained by the available machine table size (machine cube). The other capabilities and limitations can be expressed mathematically in expressions that are straightforward to program on a computer if the "exact" equations and constraints can be found.

6.2.1 Process Boundaries

One way to represent process capabilities is to use process boundaries. A process boundary is interpreted as the limiting size, tolerances, and surface finish for a process. It is expressed as the best- (or worst-) case result of a process. Such a result can be obtained by careful control of the cutting conditions and process parameters (feed speed, depth of cut). Figure 6.2 is a typical process-boundary table.

Size boundaries are determined by available tool sizes or machine-table sizes. This limitation is purely a function of the production system. In a small custom machine shop, the largest chuck for the lathes may be 10 in. in diameter. Therefore, the upper part size boundary for turning is a 10-in.-diameter workpiece. However, in a shipyard, the largest lathe may be able to accommodate up to a 3-ft-diameter component. The upper boundary for turning would then be 3 ft instead of 10 in. Other process boundaries are also system dependent.

Boundary	Hole-processing processes	Plane-producing processes
Smallest tool size	S_n	S_s
Largest tool size	L_{fn}	L_s
Negative tolerance	$A_1(\text{Dia.})^{n_1} + B_1$	$A_1(\text{Dia.})^{n_1} + B_1$
Positive tolerance	$A_2(\text{Dia.})^{n_2} + B_2$	$A_2(\text{Dia.})^{n_2} + B_2$
Straightness (holes)	$A_3(\text{Len./Dia.})^{n_3} + B_3$	$A_3 = \dfrac{\text{depth of cut} \times \text{len.}}{\text{Dia.}} + B_3$
Parallelism	$A\left(\dfrac{\text{Len.}}{\text{Dia.}}\right) + B$	$A_5\left(\dfrac{\text{Len.}}{\text{Dia.}}\right) + B_5$
Roundness (holes); angularity (planes)	$R_n A_4$	$A_s A_4$
Depth limit	$D_n A_6 \cdot \text{Dia.}$	$A_6(\text{Dia.}) + B_6$
True position	A_7	
Surface finish	A_8	A_8

Figure 6.2   Process-boundary table [Wysk, 1977].

For instance, dimensional tolerance is affected by many variables (e.g., tool diameter, machine-tool accuracy, vibration, thermal effect). Several sources [Eary and Johnson, 1962; Trucks, 1974] have developed tables, charts, and functions that show tolerance dependency as a function of surface size and perhaps other variables, such as material or operational parameters. Scarr [1967] suggested that the tolerance for all hole-making processes can be expressed in the general form

$$\text{Tol} = A(D)^n + B \qquad (6.1)$$

where

$\text{Tol} =$ tolerance

$A =$ coefficient of the process

$n =$ exponent describing the process

$B =$ constant describing the best tolerance attainable by the process

$D =$ hole diameter

The diameter-tolerance capability is related to the tool diameter. Other factors that affect diameter tolerance are taken into account by the constant B. This general form also can be used to represent other processes. The values of the coefficients A, B, and n in the equation can be obtained through experimentation. A regression analysis can be carried but with the measurements taken from the process. The estimated value of the tolerance capability may not be precise, but it provides a first-cut estimation. Again, it is worth noting that A, B, and n are system dependent (i.e., no universal parameters can be found).

Straightness and parallelism tolerances for holes define the axial tolerance. Again, many factors affect these geometric tolerances. A theoretical prediction is not possible, so we need to find a reasonable empirical model. In pursuit of such a model, we would first like to find an expression that can estimate the major cause of error. Then, from the general form of such an expression, empirical data can be used to find the coefficients. The major cause of error on straightness and parallelism is tool deflection. A tool can be modeled most simply as a cantilever beam (Figure 6.3). The deflection of the beam is

$$\delta = \frac{Pl^3}{3EI} \tag{6.2}$$

where

δ = deflection of beam
E = modulus of elasticity
I = moment of inertia
l = beam length
P = force

and

$$I = \frac{\pi D^4}{64} \qquad \text{(for cylindrical beam or tool)} \tag{6.3}$$

$$\delta = \frac{64Pl^3}{3ED^4} = \frac{CP}{D}\left(\frac{l}{D}\right)^3 \tag{6.4}$$

where C is a constant.

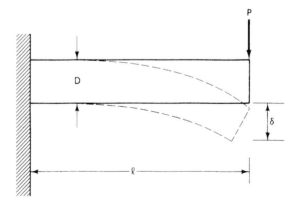

Figure 6.3   Cantilever beam.

Roughly, the error caused by deflection can be expressed as a coefficient multiplied by the length-to-diameter ratio raised to some exponential power. Because no machine is perfect, a constant is necessary to represent the effect of other errors. The final form of straightness and parallelism tolerances can be written as

$$\text{Tol} = A\left(\frac{l}{D}\right)^n + B \tag{6.5}$$

where A, n, and B are experimentally determined coefficients.

In surface production, flatness error is generally caused by the deflection of the tool and inaccuracy of the machine. Machining accuracy is a function of the machine-tool repeatability, backlash, deflection, distortion, and machine-spindle–tool alignment. If the machine is rigid, the tool can again be assumed to be a cantilever beam.

Cutting-force variables (width and depth of the cut, feed, number of cutter teeth) and the length and diameter of the cutter, as well as today's shop practices, all affect surface flatness. An empirical equation that has been used to estimate surface flatness is [Wysk, 1977]

$$\text{Flatness} = A\left(\frac{a_p l d}{w}\right)^n + B \tag{6.6}$$

where

$$a_p = \text{depth of cut}$$
$$l = \text{cutter length}$$
$$d = \text{cutter diameter}$$
$$w = \text{width of cut}$$

The roundness error of a hole-making process is a function of the machine rigidity, material used, tool geometry, and so on. Little quantifiable information is available. However, experimental results can be applied to find the boundaries for each process. The same is true for angularity and true position. Constant values or functional values can be used to model these characteristics.

Surface finish can be analytically expressed as a function of tool geometry and feed. The theoretical surface-finish heights for different tool geometries are shown in Figure 6.4. In the figure, h is the maximum height irregularity and f is the feed. The arithmetic mean value of surface finish, R_a, is defined as the sum of the areas above and below a line that divides these two areas equally. For a pointed tool, because $\triangle ABC = \triangle CDE$, we have the following equations:

$$AC = CE = \frac{f}{2} \tag{6.7}$$

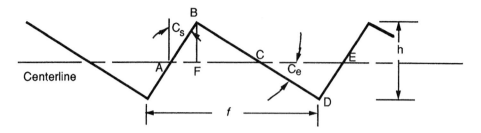

Figure 6.4   Arithmetic-mean surface finish.

Height of ΔABC = height of $\Delta CDE = \dfrac{h}{2}$

$$\overline{AF} = \frac{h}{2}\tan C_s \tag{6.8}$$

$$\overline{CF} = \frac{h}{2}\cot C_e \tag{6.9}$$

$$\overline{AF} + \overline{CF} = \frac{f}{2} = \frac{h}{2}(\tan C_s + \cot C_e) \tag{6.10}$$

$$h = \frac{f}{\tan C_s + \cot C_e} \tag{6.11}$$

$$\Delta ABC = \Delta CDE = \frac{(f/2)(h/2)}{2} = \frac{fh}{8} \tag{6.12}$$

$$\text{Total area} = \Delta ABC + \Delta CDE = \frac{fh}{4} \tag{6.13}$$

$$R_a = \frac{\text{total area}}{f} = \frac{h}{4} \tag{6.14}$$

$$= \frac{f}{4(\tan C_s + \cot C_e)} \tag{6.15}$$

For a round tool (of tip radius r) with a small feed,

$$R_a = \frac{0.0321 f^2}{r} \tag{6.16}$$

For a round tool with a large feed, Equation (6.16) also provides a reasonable approximation.

The preceding equations estimate the surface finish only under perfect cutting conditions; no consideration is given to other factors that affect cutting. Only the effect of feed is considered. Cook and Chanderamani [1964] showed that surface finish is affected by the cutting speed as well as the depth of the cut. Other factors, such as the

TABLE 6.3   Process-Boundary Data

	Twist drilling	Rough face milling
Length	$\leq 12.0D$	NA
D	$0.0625 \leq$	NA
D	≤ 2.0	NA
Tol +	$0.007D^{0.5} \leq$	$0.002 \leq$
Tol −	$0.007D^{0.5} + 0.003 \leq$	$0.002 \leq$
Straightness	$0.0005\left(\dfrac{l}{D}\right)^{3} + 0.002$	NA
Roundness	$0.004 \leq$	NA
Parallelism	$0.001\left(\dfrac{l}{D}\right)^{3} + 0.003$	$50 \leq$
True position	$0.008 \leq$	NA
Surface finish	$100 \leq$	$50 \leq$
Flatness	NA	$0.001 \leq$
Angularity	NA	$0.001 \leq$

sharpness of the cutter, vibration, and so on, also contribute to the roughness of the surface. The actual surface-finish value may be more than twice the theoretical value. The surface-finish boundary, however, should be the best attainable surface finish. Because feed and speed are control variables instead of process attributes, only a point estimator is used to predict the surface-finish capability. This point estimator can be taken from an existing data table, such as that shown in Figure 6.1.

Table 6.3 gives some sample process-boundary data taken from Wysk [1977]. As mentioned before, there is no universal process-boundary data; each machine tool has its own unique process capability.

6.3 BASIC MACHINING CALCULATIONS

Before examining the topic of process capability, it is worthwhile to discuss the basic calculations of machining. Our discussion is limited to drilling, turning, and milling. Calculations for other conventional machining processes can be derived therefrom.

6.3.1 Feed and Feed Rate

Feed f can be defined as the relative lateral movement between the tool and the workpiece during a machining operation. It corresponds to the thickness of the chip produced by the operation shown in Figure 6.5. In turning and drilling operations, feed is defined as the advancement of the cutter per revolution of the workpiece (turning) or tool (drilling). The typical unit is the inch per revolution (ipr). In milling, it is defined as the advancement of the cutter per cutter-tooth revolution, and the unit is inch per revolution per tooth. The feed rate is defined as the speed of feed; the unit is the inch per minute (ipm). Mathematically, the feed rate is

$$V_f = fnN \tag{6.17}$$

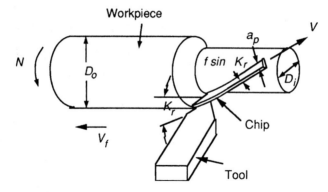

Figure 6.5   Chip formation.

where

> n = number of teeth in the cutter for milling; n = 1 for drilling and turning
> N = rotation speed of the cutter (drilling and milling) or workpiece (turning), in rpm

6.3.2 Cutting Speed

The cutting speed V can be defined as the maximum linear speed between the tool and the workpiece. The cutting speed for drilling, turning, and milling can be determined as a function of the workpiece or the tool diameter D and the rotation speed N and is given by

$$V = \frac{\pi D N}{12} \qquad (6.18)$$

where

> V = speed, feet per minute
> D = diameter, in.
> N = rotational speed, rpm

6.3.3 Depth of Cut

The depth of cut is determined by the width of the chip (Figures 6.5 and 6.6). During the roughing operation, the depth of cut is usually much greater than that of the finishing operation. For turning, it is one-half the difference between the inner and the outer diameters of the workpiece. That is,

$$a_p = \frac{D_o - D_i}{2} \qquad (6.19)$$

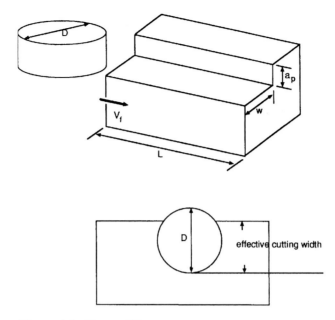

Figure 6.6   Face milling.

where

a_p = depth of cut, in.
D_o = outer diameter, in.
D_i = inner diameter, in.

6.3.4 Material-Removal Rate

The material-removal rate (MRR) is a measurement of how fast material is removed from a workpiece. It can be calculated by multiplying the cross-sectional area of the chip by the speed. A high material-removal rate (MRR) produces a short processing time, and a low MRR yields a long processing time. However, the MRR also affects the life of a cutter. The effect of MRR on tool life is discussed later. The unit of the metal-removal rate is usually expressed as cubic inches per minute. A different formula is used for different processes. For drilling, the cross-sectional area of the chip is $\pi D^2/4$. Hence,

$$\text{MRR} = \frac{\pi D^2}{4} V_f = \frac{\pi D^2}{4} f N \tag{6.20}$$

where

D = drill diameter, in.
V_f = feed rate, in. per min
f = feed, in. per revolution
N = rotational speed, rpm

Because

$$N = \frac{12V}{\pi D} \tag{6.21}$$

it follows that

$$\text{MRR} = \frac{\pi D^2}{4} f \frac{12V}{\pi D} = 3DfV \tag{6.22}$$

For turning, the chip width is $(D_o - D_i)/2$, where D_o and D_i are the outer and inner workpiece diameters, respectively. The cross-sectional area is $\pi(D_o^2 - D_i^2)/4$. Therefore, the material removal rate is

$$\text{MRR} = \frac{\pi(D_o^2 - D_i^2)}{4} V_f = \frac{\pi(D_o^2 - D_i^2)}{4} fN \tag{6.23}$$

$$N = \frac{12V}{\pi(D_o + D_i)/2} \tag{6.24}$$

$$\text{MRR} = 6(D_o - D_i)fV \tag{6.25}$$

For milling processes, the chip cross-sectional area is (Figure 6.6)

$$A = a_p w \tag{6.26}$$

$$\text{MRR} = a_p w V_f = a_p w f n N$$

$$N = \frac{12V}{\pi D} \tag{6.27}$$

$$\text{MRR} = \frac{12 a_p w n}{\pi D} fV \tag{6.28}$$

where

a_p = depth of cut, in.
w = width of cut, in.
n = number of teeth on the cutter
D = cutter diameter, in.
f = feed, in. per tooth per revolution
V = speed, feet per min

6.3.5 Machining Time

Machining time is the total amount of time it takes to finish processing a workpiece. Machining time is a function of workpiece size, depth of cut, feed, and speed. It can be calculated by dividing the tool-path length by the feed speed. The tool-path length is determined by the length of the workpiece, overtravel of the tool for clearance, and the number of passes required to clear the volume. For drilling, one-pass turning, and milling,

$$t_m = \frac{L + \Delta L}{V_f} \qquad (6.29)$$

where

t_m = machining time, min

L = hole depth, in.

ΔL = clearance height or overtravel, in.

V_f = feed speed, in. per min

For multiple-pass turning, the number of passes can be calculated as

$$n_p = \left| \frac{D_o - D_i}{2a_p} \right|^+ \qquad (6.30)$$

where

n_p = number of passes

D_o = raw-material diameter, in.

D_i = finished-part diameter, in.

a_p = depth of cut

$|x|^+$ = round off x to the next integer number

For milling processes,

$$n_p = \left| \frac{\Delta h}{a_p} \right|^+ \left| \frac{w}{\alpha D} \right|^+ \qquad (6.31)$$

where

Δh = total height of material to be removed, in.

w = workpiece width, in.

α = cutter overlapping factor = effective cutting width/tool diameter ≤ 1.0

D = cutter diameter, in.

Example 6.1.

A workpiece 3-in. in diameter is to be machined on a lathe to 2.7-in. in diameter. The total length of the workpiece is 10 in. It is recommended from the handbook that a feed of 0.01 ipr, a cutting speed of 200 fpm, and a maximum depth of cut of 0.1 in. be used. A 0.25-in. overtravel should be used for cutter clearance. How long will it take to finish the part?

Solution:

The spindle speed is

$$N = \frac{12 \times 200}{\pi 3} = 254 \text{ rpm}$$

so

$$V_f = 0.01 \times 254 = 2.54 \text{ ipm}$$

Because the required depth of cut is greater than the maximum allowable depth of cut, multiple passes are necessary:

$$n_p = \left| \frac{3 - 2.7}{2 \times 0.1} \right|^+ = 2 \text{ passes}$$

The time taken to finish the part is then

$$t_m = 2\left(\frac{10 + 0.25}{2.54} \right) = 8.07 \text{ min}$$

6.3.6 Tool Life

A tool's useful life can be ended by one of two mechanisms: erosion (wear) and breakage (catastrophic failure). The two major wear zones are crater wear (on the tool face) and flank wear (Figure 6.7). Crater wear is caused by the high temperatures generated on the rake face, combined with high shear stresses. Excessive crater wear weakens the cutting edge of the tool. Then the tool can no longer support the cutting force, and catastrophic tool failure is the result. Flank wear is a gradual wear that is the result of the frictional action between the tool flank and the workpiece. Catastrophic tool failure is usually quite unpredictable and a phenomenon that one tries to minimize; therefore, tool life is usually defined as the cut time a new tool undergoes before a certain flank

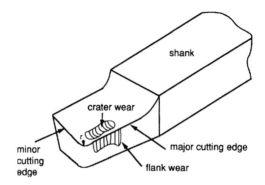

Figure 6.7   Cutter wear.

TABLE 6.4 Average Permissible Values of Lathe Tool (Flank) Wear

Tool-point material	Tool type	Workpiece material	Type of turning	Permissible values of flank wear, in mm	
				In dry turning	In turning with coolant
High-speed steel (HSS)	For external straight turning	Steel and malleable cast iron	Rough, medium accurate, and accurate	0.5–1.0 0.3–0.5	1.5–2.0
		Gray cast iron	Rough, medium accurate, and accurate	3–4	—
	Boring and undercutting tools	Steel and malleable cast iron	Rough, medium accurate, and accurate	0.3–0.5	1.5–2.0
		Gray cast iron	Rough, medium accurate, and accurate	1.5–2.0	—
	Cutoff and parting tools	Steel and malleable cast iron	Rough, medium accurate, and accurate	0.3–0.5	0.8–1.0
		Gray cast iron	Rough, medium accurate, and accurate	1.5–2.0	
	Form tools	Steel	Rough, medium accurate, and accurate	—	0.4–0.5
	Threading tools	Steel	Rough finish	— —	2.0 0.3
Sintered carbides	All types of tools	Steel	Rough, medium accurate, and accurate	0.8–1.0	—
		Gray cast iron	Rough and medium accurate ($s > 0.3$) ($s < 0.3$)	0.8–1.0 1.4–1.7	—
Sintered metal oxides	All types of tools	Steel and cast iron	Medium accurate and accurate	0.8	—

Courtesy of Peter Peregrinus, Limited, taken from J. Kaczmarek, *Principles of Machining by Cutting, Abrasion, and Erosion*, 1976.

wear is reached. Permissible values of flank wear have been recommended by the International Standards Organization [ISO, 1972] and appear in Table 6.4.

F.W. Taylor [1906] was the first to develop a generalized tool-life equation. He observed that, under the assumption of a constant feed, when tool life and cutting speed were plotted on a log–log graph the result was a straight line. Such a relationship can be expressed as

$$\frac{V}{V_r} = \left(\frac{t_r}{t}\right)^n \tag{6.32}$$

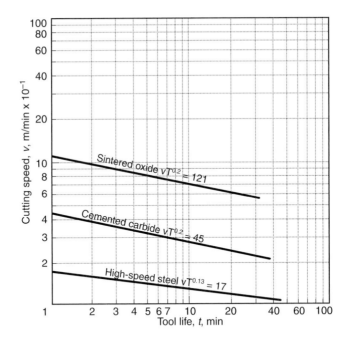

Figure 6.8 Log–log plot of tool life.

where

n = constant
V = cutting speed
t = tool life
V_r = reference cutting speed given tool life t_r

Figure 6.8 shows the results of tool-life experiments plotted on a log–log graph. Rearranging Equation (6.32), we obtain

$$t = \frac{t_r V_r^{1/n}}{V^{1/n}}$$

$$= \frac{C}{V^\alpha} \tag{6.33}$$

where

$\alpha = 1/n$
$C = t_r V_r^{1/n}$

Table 6.5 shows C and α for some metals, tool signatures, and sizes of cut.

TABLE 6.5 Some Typical Values for C and α in Tool-Life Equation $VT^n = C$

Tool	Work Material	f(ipr)	α_p (in.)	C	n
High C steel	Yellow brass	0.013	0.100	1400	0.1
HSS	Gray cast iron	0.026	0.050	800	0.1
HSS	SAE 1035 steel	0.013	0.050	610	0.11
HSS	SAE 1045 steel	0.013	0.100	850	0.11
HSS	SAE 3140 steel	0.013	0.100	810	0.16
HSS	SAE 4350 steel	0.013	0.100	360	0.11
HSS	SAE 4350 steel	0.026	0.100	215	0.11
HSS	Monel metal	0.013	0.100	790	0.08
HSS	Monel metal	0.026	0.150	590	0.07
T64 carbide	SAE 1040 steel	0.025	0.062	3720	0.16
T64 carbide	SAE 1060 steel	0.025	0.125	3070	0.17
T64 carbide	SAE 1060 steel	0.025	0.250	2600	0.17
T64 carbide	SAE 1060 steel	0.042	0.062	2375	0.16
T64 carbide	SAE 1060 steel	0.062	0.062	1860	0.16
High C steel	Bronze	0.013	0.100	1080	0.109
HSS	SAE B1112 steel	0.013	0.050	1040	0.109
Stellite	SAE 3240 steel	0.031	0.187	1000	0.19

Source: Adapted from Boston, 1951, p. 150.

Example 6.2.

When steel is cut with a high-speed steel (HSS) cutter, the tool life at a cutting speed of 100 sfpm (surface feet per minute) is 60 minutes; at 150 sfpm, it is 30 minutes. What is the tool-life equation of the cutter on the material?

Solution:

$$\frac{100}{150} = \left(\frac{30}{60}\right)^n$$

$$0.667 = (0.5)^n$$

$$n = \frac{\ln 0.667}{\ln 0.5} = \frac{-0.405}{-0.693} = 0.58$$

$$\frac{V}{150} = \left(\frac{30}{t}\right)^{0.58}$$

$$\left(\frac{V}{150}\right)^{1/0.58} = \frac{30}{t}$$

or

$$t = \frac{169,317}{V^{1.724}}$$

Later research confirmed that feed and depth of cut also contributed to the tool life. The result was an expanded Taylor tool-life equation of the form

$$t = \frac{\lambda C}{V^{\alpha_T} f^{\beta_T} a_p^{\gamma_T}}$$ (6.34)

where

λ, C = constants for a specific tool–workpiece combination

$\alpha_T, \beta_T, \gamma_T$ = exponents for a specific tool–workpiece combination

The influence of V on tool life is much greater than the influence of f. The depth of cut has the least influence on tool life. However, equation 6.34 is based on the flank wear of the tool. For HSS and carbide tools operating under normal conditions, flank wear is the major cause of tool deterioration. For ceramic tools, breakage is more significant than flank wear. A tool may fail before any significant flank wear can be detected. One must be also aware that a tool-life equation holds only when all cutting conditions, such as the tool geometry, material, workpiece material, cutting fluid, and so on, are held constant.

6.3.7 Machining Force and Power Requirements

Machining force and power requirements are not limiting values in process selection, but become important considerations in selecting process parameters (feed, speed, and depth of cut). Force and power are functions of process parameters. In general, when the same tool, machine, and workpiece material are used, the greater the volume of work material removed per unit time, the greater will be the power required. Reduced feed, speed, or depth of cut can decrease both the cutting force required and the power consumed. Because force and power are constrained by the machine output, it is necessary to know the power requirements of a cutting process as a function of the process parameters.

6.3.7.1 Machining Forces In orthogonal cutting, the resultant force, F_r, applied to the chip by the tool lies in a plane normal to the tool's cutting edge (Figure 6.9). F_c is the major cutting force and F_t is the thrust force. The cutting force can be expressed roughly as the product of the specific cutting resistance k_s and the cross-sectional area of undeformed chip; that is,

$$F_c = wh_c k_s$$ (6.35)

where

w = width of undeformed chip, in.

h_c = thickness of undeformed chip, in.

k_s = specific cutting resistance

The thrust force is just the cutting force multiplied by a coefficient b that is empirically determined by the tool geometry:

$$F_t = bF_c = bwh_c k_s$$ (6.36)

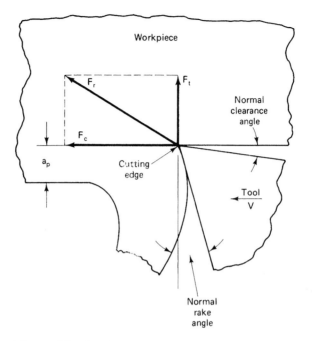

Figure 6.9   Cutting-force geometry.

Since the major cutting force is usually much greater than the thrust force, it is of more interest to us. Because the thickness of the chip, a_p, is determined by the feed and the width of the chip is determined by the depth of cut, equation (6.35) can be written as

$$F_c = K \cdot f^\alpha \cdot a_p^\beta \tag{6.37}$$

where

F_c = major cutting force, lbf
K = constant
f = feed, ipr
a_p = depth of cut, in.
α, β = coefficients

Through experiments, it has been shown that neither α nor β is unity. When f or a_p is plotted against F_c on a log–log graph, straight lines result.

6.3.7.2 Cutting Power   The cutting-power consumption P_c can be calculated as the following function of the product of the cutting speed v in fpm and the cutting force F_c in lbv:

$$P_c = \frac{F_c V}{33,000} \tag{6.38}$$

Example 6.3.

When a low-carbon steel is cut, it is found that the cutting-force equation is

$$F_c = 140{,}000 f^{0.8} a_p^{0.9}$$

If the selected cutting speed is 120 fpm, the feed is 0.01 ipr, and the depth of cut is 0.25 in., what is the required cutting horsepower? Assume that the machine efficiency is 0.85.

Solution:

$$F_c = 140{,}000(0.01)^{0.8}(0.25)^{0.9} = 1009 \text{ lbf}$$

$$P_m = \frac{1009 \times 120}{33{,}000} = 3.67 \text{ hp}$$

$$\text{Horsepower} = \frac{3.67}{0.85} = 4.310 \text{ hp at the machine}$$

The cutting-power consumption can also be estimated by multiplying a specific cutting energy in hp. min/in.^{3} by the metal-removal rate in in.^{3}/min:

$$P_c = p_s \text{ (MRR)} \tag{6.39}$$

Specific cutting energy is a function of the hardness of the workpiece material and the mean value of the undeformed chip thickness. Figure 6.10 shows the approximate values of the specific cutting energy p_s for various materials and operations. Undeformed chip thickness is $f \sin k_r$ for turning, $(f/2)\sin k_r$ for drilling, and V_f/Nn_t for face milling, where k_r is the major cutting edge angle, f is the feed, V_f is the feed rate, N is the rotation speed (rpm), and n_t is number of teeth on the cutter.

Equations for cutting force, power, surface finish, tool life, and machine time are summarized in Tables 6.6 and 6.7. The equations for F_c contain an empirically determined coefficient K_F to account for several factors that affect the cutting force. K_F must be determined for all tool–work-material combinations, process types, tool-wear conditions, workpiece hardnesses, tool geometries, and speeds.

6.3.8 Process Parameters

So far, we have discussed process capabilities, cutting force, and power. However, surface finish, force, and power constraints are directly affected by the process parameters of feed, speed, and depth of cut. Therefore, process selection becomes an iterative procedure: First, a process is selected, and then the machining parameters are adjusted to accommodate the system constraints. The selection of the machining parameters, however, affects the time and cost required to produce a component. These parameters are not arbitrary, nor are they constant for different operations. Process parameters are the basic control variables for a machining process. The earliest study on the economical selection of process parameters was conducted by F.W. Taylor [1906]. As a result of his effort and the later efforts of many others, machining data handbooks [Metcut Research Associates, 1980] and machinability data systems [Parsons, 1971] have been developed to recommend process parameters for efficient machining. These handbooks contain recommended feeds and speeds for different tooling, work-material, tool

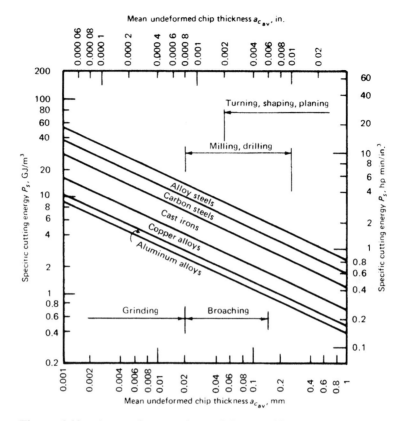

Figure 6.10   Approximate values of the specific cutting energy p_s for various materials and operations. (Reprinted from *Fundamentals of Metal Machining and Machine Tools*, G. Boothroyd, 1975, by permission of Hemisphere Publishing Corporation, formerly Scripta Book Company.)

diameter, depth of cut, and so on, combinations. The parameters recommended are good, but not necessarily the best or the most appropriate. An example is shown in Table 6.8, where the feeds and speeds for a 30-to-60 min tool life for HSS tools (for carbide insert, 1 to 2 h) are recommended.

Several machinability systems are currently marketed which recommend sets of parameters that optimize machining cost, time, or production rate or which simply retrieve data tables or calculated values. The FAST system [Parsons, 1971] is an example of a table-value retrieval system. An early General Electric (GE) system used a mathematical model developed by W. W. Gilbert [1950] to calculate one parameter, given all others. In Gilbert's model, speed, material, coolant, workpiece surface condition, tool geometry, flank wear, workpiece hardness, tool life, feed, depth of cut, and so on were represented as a mathematical function. A so-called machinability computer (a slide-rule-like device) using Gilbert's model was marketed by GE in the late 1950s [General Electric, 1957].

More recently (from 1960 to the present), machinability systems have been designed to run on computers to take advantage of their computational power and to

TABLE 6.6 Summary of Equations for Cutting Operations[a]

Operation	Machining Time t_m	Tool life t	Cutting force F_c	Power P_m	Surface finish R_a (4)
Turning	$\dfrac{l_w}{f \cdot n_w}$	$K_T \cdot v^{\alpha_T} \cdot f^{\beta_T} \cdot a_p^{\gamma_T}$	$K_F \cdot f^{\beta_F} \cdot a_p^{\gamma_F}$	$\dfrac{F_e \cdot v}{6120 n_m}$	$\dfrac{32 \cdot f^2}{r_e}$
Boring					
Facing	$\dfrac{D_m}{2 \cdot f \cdot n_w}$				
Parting	$\dfrac{b_w}{f \cdot n_t}$				
Shaping & planing	$\dfrac{l_w}{f \cdot n_t}$				
Drilling & reaming	$\dfrac{l_w}{f \cdot n_t}$	$K_T \cdot v^{\alpha_T} \cdot f^{\beta_T} \cdot a_p^{\gamma_T} \cdot D_t^{\delta_T}$	$K_F \cdot f^{\beta_F} \cdot a_p^{\gamma_F} \cdot D_t^{\delta_F}$ (2)	$\dfrac{M \cdot n_e}{9.74 \times 10^5 \eta_m}$	$\dfrac{64 \cdot 2 f^2}{D_t}$ (5)
Slab Milling	$\dfrac{l_w + k\sqrt{a(D_t - a)}}{f \cdot n_t}$ (1)		$K_F \cdot f^{\beta_F} \cdot a_c^{\gamma_F} \cdot D_t^{\delta_F} \cdot b_w \cdot z$		$\dfrac{64 \cdot 2 f^2}{D_t + e}$
Side & face milling		$K_T \cdot v^{\alpha_T} \cdot a_f^{\beta_T} \cdot a^{\gamma_T} \cdot D_t^{\delta_T} \cdot b_w^{???\cdot\varepsilon_T} \cdot z^{\zeta_T} \cdot \lambda_\beta^{\eta_T}$			
Face milling	$\dfrac{l_w + D_t}{f \cdot n_t}$		$K_F \cdot v^{\alpha_F} \cdot a_f^{\beta_F} \cdot a_p^{\gamma_F} \cdot d_t^{\delta_F} \cdot b_w^{\alpha_F} \cdot z^{\zeta_F}$	$\dfrac{F_e \cdot v}{6120 \eta_\mu}$	$K_a \cdot a_t^{14}$
Broaching	$\dfrac{l_t}{v}$	$K_T \cdot v^{\alpha_T} \cdot a_f^{\beta_T}$	$K_F \cdot a_f^{\beta_F} \cdot D_m \cdot z_c$ (3)		$\dots$

$$n_w = \frac{v}{nD_w} \qquad n_t = \frac{v}{nD_t}$$

(1) $k = \begin{cases} 1, & \text{case (a)} \\ 2, & \text{case (b)} \end{cases}$

(2) Same formula holds for torque M.

(3) Valid for broaching of round holes only.

(4) Valid for ideal conditions only; however, empirical formulas are available for specific cases in turning operations.

(5) Eccentricity e will be equal to zero in ideal conditions.

[a]From Armorego and Brown, 1968; Boothroyd, 1975; and Kaczmarek, 1976.
Source: Coelho [1980].

TABLE 6.7   Notation for Table 6.6

a, a_c, a_f, a_p	Depth of cut
b_w	Width of workpiece
D_m	Diameter of the machined surface
D_t	Diameter of the tool
f	Feed
K_F, K_T, K_R	Constants for cutting-force, tool-life, and surface-roughness empirical equations, respectively
l_t	Length of tool or broach
l_w	Length of surface to be machined
n_r	Frequency of reciprocation (strokes/min)
n_t	Tool-spindle speed (rpm)
n_w	Rotational frequency of the workpiece (rpm)
r_ε	Tool-nose radius
v	Cutting speed
z	Number of teeth on the cutting tool
z_c	Number of teeth cutting simultaneously in a tool
$\alpha_T, \beta_T, \gamma_T, \varepsilon_T, \eta_T, \delta_T$	Cutting speed, feed, depth of cut, tool diameter, machined surface width, number of teeth in the cutting tool, and tool cutting-edge inclination (L^0) exponents for cutting-force, surface-roughness, and tool-life equations, respectively
n_m	Overall efficiency of the machine-tool motor and drive systems
e	Tool cutting-edge inclination

include optimization models. For example, a later version of the GE data system contains optimum feed and speed for minimum cost and time using an unconstrained optimization scheme to identify the optimal tool life. Although most commercial machinability data systems use unconstrained optimization, several research papers have appeared that focus on constrained models. This class of models is discussed next.

6.4 PROCESS OPTIMIZATION

Before discussing process optimization, we must understand the basic tool-life equation. A faster metal-removal rate results in both a diminished tool life and reduced machining time. However, whenever a tool has been worn past some practical limit, it must be replaced. Therefore, there is a trade-off between increased machining rate and machine idle time. This trade-off results in frequent tool changes.

Machining optimization models can be classified as single-pass and multipass models. In a single-pass model, we assume that only one pass is needed to produce the required geometry. In that case, the depth of cut is fixed. In a multipass model, this assumption is relaxed and a_p also becomes a control variable. Any multipass model can be reconstructed into a single-pass model.

6.4.1 Single-Pass Model

In a single-pass model, processing time per component, t_{pr}, can be expressed as the sum of machining time t_m, material-handling time t_h, and tool change time t_t:

$$t_{pr} = t_m + t_h + t_t\left(\frac{t_m}{t}\right) \qquad (6.40)$$

TABLE 6.8 Example of Drilling Data

Material	Hardness, Bhn	Condition	Speed,[a] fpm m/min	Feed,[a] ipr mm/rev Nominal hole diameter								Tool material grade AISI or CISO
				$\frac{1}{16}$ in. 1.5 mm	$\frac{1}{8}$ in. 3 mm	$\frac{1}{4}$ in. 6 mm	$\frac{1}{2}$ in. 12 mm	$\frac{3}{4}$ in. 18 mm	1 in. 25 mm	$1\frac{1}{2}$ in. 35 mm	2 in. 50 mm	
Alloy steels, wrought high-carbon	50 R$_c$ to 52 R$_c$	Quenched and tempered	10	—	0.0005	0.001	0.002	0.002	0.003	0.003	0.004	T15, M42[b]
			3	—	0.013	0.025	0.050	0.050	0.075	0.075	0.102	S9, S11[b]
	52 R$_c$ to 54 R$_c$	Quenched and tempered	75	—	—	0.001	0.001	0.0015	—	—	—	C-2
			23	—	—	0.025	0.025	0.038	—	—	—	K10
	54 R$_c$ to 56 R$_c$	Quenched and tempered	60	—	—	0.001	0.001	0.0015	—	—	—	C-2
			18	—	—	0.025	0.025	0.038	—	—	—	K10
High-strength steels, wrought 300M 4340Si H11	225 to 300	Annealed	50	0.001	0.003	0.004	0.007	0.010	0.012	0.015	0.018	M10, M7, M1
			15	0.025	0.075	0.102	0.18	0.25	0.30	0.40	0.45	S2, S3

Material	Hardness	Condition	Speed									Tool material
4330V 98BV40 H13 4340 D6ac	300 to 350	Normalized	35	—	0.002	0.004	0.006	0.009	0.010	0.014	0.017	M10, M7, M1
			11	—	0.050	0.102	0.15	0.23	0.25	0.36	0.45	S2, S3
	350 to 400	Normalized	30	—	0.002	0.004	0.006	0.008	0.010	0.012	0.015	T15, M42[b]
			9	—	0.050	0.102	0.15	0.20	0.25	0.30	0.40	S9, S11[b]
	43 R_c to 48 R_c	Quenched and tempered	20	—	0.002	0.003	0.004	0.004	0.004	0.004	0.004	T15, M42[b]
			6	—	0.050	0.075	0.102	0.102	0.102	0.102	0.102	S9, S11[b]
	48 R_c to 50 R_c	Quenched and tempered	15	—	0.001	0.002	0.003	0.003	0.004	0.004	0.004	T15, M42[b]
			5	—	0.025	0.050	0.075	0.075	0.102	0.102	0.102	S9, S11[b]
	50 R_c to 52 R_c	Quenched and tempered	10	—	0.0005	0.001	0.002	0.002	0.003	0.003	0.004	T15, M42[b]
			3	—	0.013	0.025	0.050	0.050	0.075	0.075	0.102	S9, S11[b]
	52 R_c to 54 R_c	Quenched and tempered	75	—	—	0.001	0.001	0.0015	0.0015	0.0015	0.0015	C-2
			23	—	—	0.025	0.025	0.038	0.038	0.038	0.038	K10

[a] For holes more than two diameters deep, reduce speed and feed.

[b] Any premium HSS (T15, M33, M41–M47) or (S9, S10, S11, S12).

Source: From the *Machining Data Handbook*, 3d ed. (by permission of the Machinability Data Center, © 1980 by Metcut Research Associates, Inc.).

In this equation, t is the tool life and $(t_m/t)^{-1}$ represents the number of parts that can be produced before the tool requires changing.

Equations for t_m and t are found in Table 6.6. The depth of cut, a_p, in those equations is constant. The production cost per component can be written as

$$C_{pr} = \frac{C_b}{N_b} + C_m \left[t_m + t_h + \frac{t_m}{t} \left(t_t + \frac{C_r}{C_m} \right) \right] \tag{6.41}$$

where

C_b = setup cost for a batch

C_m = total machine and operator rate (including overhead)

C_r = tool cost (for an HSS tool, the cost of regrinding; for a tungsten carbide tool, TCT, steel, the cost of one insert cutting edge)

N_b = batch size

An optimization model for machining is as

$$\min t_{pr} \quad \text{(for time)} \tag{6.42a}$$

or

$$\min C_{pr} \quad \text{(for cost)} \tag{6.42b}$$

subject to the following constraints:

1. Spindle-speed constraint:

$$n_{min} < n_w < n_{max} \quad \text{(for workpiece)} \tag{6.43a}$$
$$n_{t, min} < n_t < n_{t, max} \quad \text{(for tool)} \tag{6.43b}$$

2. Feed constraint:

$$f_{min} < f < f_{max} \tag{6.44}$$

3. Cutting-force constraint:

$$F_c < F_{c, max} \tag{6.45}$$

4. Power constraint:

$$P_m < P_{max} \tag{6.46}$$

5. Surface-finish constraint:

$$R_a < R_{a, max} \tag{6.47}$$

Equations for n_w, n_t, F_c, P_m, and R_a are given in Table 6.6.

Many solution procedures can be used to solve the foregoing model. Berra and Barash [1968] and, later, Wysk [1977] used an iterative search procedure that approached optimum. Groover [1976] employed an "evolutionary operations" procedure somewhat similar to a Hooke–Jeeves search procedure. Hati and Rao [1976] applied a sequential unconstrained minimization technique [SUMT; Fiacco and McCormick, 1968] in conjunction with the Davidson–Fletcher–Powell (D–F–P) algorithm to solve the problem. Dynamic programming and other mathematical programming methods have also been used.

Example 6.4.

In a turning operation, the length of the part is 10 in., the diameter of the raw material is 3 in., and the diameter of the finished part is 2.9 in. The overtravel for the cut is 0.5 in. The recommended feed rate is 0.1 ipr. The time to load and unload a part is 2 min. The time for a cutter change is 3 min. The tool life can be expressed as $t = (1.7 \times 10^5)/V^{1.7}$.

a. What is the unconstrained minimum time-cutting speed?

b. The operator rate is \$20/h and the machine rate is \$15/h. The cutter costs \$3 each. What is the minimum cost-cutting speed? Ignore the batch size and the batch setup cost.

Solution:

(a)

$$\eta(\text{rpm}) = \frac{12V}{\pi D}$$

$$D = \frac{3 + 2.9}{2} \approx 3$$

$$\eta(\text{rpm}) = \frac{12V}{3\pi} = 1.273V$$

$$V_f = f \cdot \text{rpm} = 0.1 \cdot 1.273V = 0.1273V$$

$$t_m = \frac{L + \Delta L}{V_f} = \frac{10 + 0.5}{0.1273V} = \frac{82.48}{V}$$

$$t_{pr} = t_m + t_h + t_t \left(\frac{t_m}{t}\right)$$

$$= \frac{82.48}{V} + 2 + 3\left[\frac{82.48/V}{(1.7 \times 10^5)/V^{1.7}}\right]$$

$$= 82.48V^{-1} + 2 + 0.001456V^{0.7}$$

$$\frac{dt_{pr}}{dV} = -82.48V^{-2} + 0.0010192V^{-0.3} = 0$$

$$0.0010192V^{-0.3} = 82.48V^{-2}$$

$$V^{1.7} = 80{,}926$$

$$V^* = 769 \text{ sfpm}$$

$$t_{pr}^* = \frac{82.48}{769} + 2 + 0.001456(769)^{0.7}$$

$$= 2.26 \text{ min}$$

(b)

$$C_{pr} = \frac{C_b}{N_b} + C_m\left[t_m + t_h + \frac{t_m}{t}\left(t_t + \frac{C_t}{C_m}\right)\right]$$

$$= \frac{20 + 15}{60}\left\{82.48V^{-1} + 2 + 82.48V^{-1}\left(\frac{V^{1.7}}{1.7 \times 10^5}\right)\left[3 + \frac{3}{(20 + 15)/60}\right]\right\}$$

$$= 0.5833(82.48V^{-1} + 2 + 0.00395V^{0.7}) = 48.11V^{-1} + 1.1666 + 0.0023V^{0.7}$$

$$\frac{dC_{pr}}{dt} = -48.11V^{-2} + 0.00161V^{-0.3} = 0$$

$$V^{1.7} = 29,882$$

$$V^* = 428 \text{ sfpm}$$

$$C_{pr}^* = \frac{48.11}{428} + 1.1666 + 0.0023(428)^{0.7}$$

$$= 1.438$$

Example 6.5.[1]

This example presents the integer-transformation technique used in solving a milling-process optimization problem.

A low-carbon steel is machined by a milling process. The tool material is HSS. (See Figure 6.11.) The problem is to select the optimal parameters: n, d, R, Z, and V_p. The tool-change time is equal to 80 s, and the milling process is subject to the following constraints:

Machine power: $P_{max} = 7.5 \text{ kW}$

Cutting force: $F_{max} = 60 \text{ kg}$

Tool life: $T_{min} = 30 \text{ min}$

$T_{max} = 60 \text{ min}$

Spindle speed: $N_{min} = 1 \text{ rpm}$

$N_{max} = 1200 \text{ rpm}$

[1]This example is taken from Chang et al., 1982.

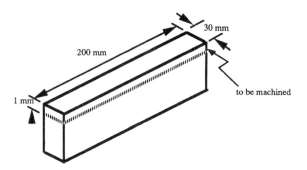

Figure 6.11   Workpiece for Example 6.5.

The tool-life equation [Wysk, 1977] can be stated as

$$T = \frac{C_v^3 (2R)^s}{V^\alpha (V_f/nZ)^\beta d^\gamma Z^{0.3} W^\delta}$$ (6.48)

where

$V =$ cutting velocity, m/min
$V_f =$ feed velocity, mm/s

For a milling operation using HSS tool and low-carbon steel material with ultimate tensile strength UTS $\cong$ 400 MN/m^2, the model parameters are

$C_v = 35.4,$ $a = 3.03,$ $b = 1.212,$ $\gamma = 0.909,$ $S = 1.364,$ $d = 0.303$

Equation (6.48) can be rewritten as

$$T = \frac{1.14 \times 10^5 R^{1.364} n^{1.212} Z^{0.909}}{V^{3.03} V_f^{1.212} d^{0.909} W^{0.303}}$$ (6.49)

and

$$R = \frac{25.6}{4} Z = 6.4Z$$ (6.50)

$$V = \frac{2\pi Rn}{1000} \times 60 = 2.413Zn$$ (6.51)

$$m = \frac{T \times 60}{T_c}$$ (6.52)

$$T_c = \frac{L}{V_f}$$ (6.53)

Substituting Equations (6.50) and (6.52) into Equation (6.49) yields

$$T = \frac{0.996 \times 10^5}{V_f^{1.212} d^{0.909} W^{0.303} n^{1.818} Z^{0.757}}$$

and including $W = 30$ mm, $d = 1$ mm, and $L = 200$ mm gives

$$T = \frac{3.55 \times 10^4}{V_f^{1.212} n^{1.818} Z^{0.757}} \tag{6.54}$$

Substituting Equations (6.54) and (6.53) into Equation (6.52) yields

$$m = \frac{1.066 \times 10^4}{V_f^{0.212} n^{1.818} Z^{0.757}}$$

and

$$V_f = \left(\frac{1.066 \times 10^4}{mn^{1.818} Z^{0.757}}\right) 4.717 \tag{6.55}$$

$$= \frac{9.98 \times 10^{18}}{m^{4.717} n^{8.576} Z^{3.571}}$$

The objective function is

$$T_T = T_c + T_t$$

or, in expanded form,

$$T_T = \frac{L}{60 V_f} + \frac{80}{60 m} \tag{6.56}$$

$$= \frac{m^{4.717} n^{8.576} Z^{3.571}}{2.99 \times 10^{18}} + \frac{1.333}{m}$$

The objective function is subject to the following constraints:

$$F = 1.46.6(2R)^{-1.1713} W^{1.1786} V^{-5557} d^{0.9128} \left(\frac{V_f^{0.7509}}{nZ}\right) \leq 60 \text{ kg} \tag{6.57}[2]$$

$$P = \frac{60 FV}{6010.2} \leq 7.5 \text{ kW} \tag{6.58}$$

$$30 \leq T \leq 60 \text{ min} \tag{6.59}$$

$$1 \leq n \leq 1200 \text{ rpm} \tag{6.60}$$

$$2 \leq Z \leq 24 \tag{6.61}$$

$$1 \leq m \tag{6.62}$$

Rewriting Equations (6.57) through (6.59) gives

$$F = 146.6(2 \times 6.4Z)^{-1.1713}(30)^{1.1796}(2.413nZ)^{-0.556}(1)^{0.9128}$$

$$\left(\frac{9.959 \times 10^{18}}{m^{4.717} n^{9.58} Z^{4.57}}\right)^{0.7509} = \frac{4.605 \times 10^{16}}{m^{3.542} n^{7.749} Z^{5.158}} \leq 60 \text{ kg} \tag{6.57a}$$

[2] Derived from Cincinnati Milling Machine Co. [1951].

$$P = \frac{1.1 \times 10^{15}}{m^{3.542} n^{6.749} Z^{5.158}} \le 7.5 \text{ kW} \qquad (6.58a)$$

$$30 = \frac{m^{5.717} n^{8.576} Z^{3.57}}{2.99 \times 10^{18}} \le 60 \qquad (6.59a)$$

The bounds of m can be found by the force constraint as

$$\frac{1.6 \times 10^4}{n^{2.188} Z^{1.456}} \le m \qquad (6.57b)$$

by the power constraint as

$$\frac{10^4}{n^{1.905} Z^{1.174}} \le m \qquad (6.58b)$$

and by the tool-life equation as

$$\frac{3.09 \times 10^3}{n^{1.5} Z^{0.624}} \le m \le \frac{3.48 \times 10^3}{n^{1.5} Z^{0.624}} \qquad (6.59b)$$

The bounds of n can be found by the machine characteristics:

$$1 \le n \le 1200$$

By Equation (6.57b) and the right-hand side of Equation (6.59b),

$$n \ge 2.47 Z^{-0.49}$$

By Equation (6.58b) and the right-hand side of Equation (6.59b),

$$n \ge 1.74 Z^{-0.29}$$

By the right-hand side of Equation (6.59b) with $m > 1$,

$$n \le 229.6 Z^{-0.416}$$

Because $W = 30$, it is preferred to have $2R > 30$; thus,

$$R = 6.4Z$$

$$Z \ge 2.34 \approx 3$$

By obtaining an optimum m, we have

$$\frac{\partial T_T}{\partial m} = \frac{m^{3.717} Z^{3.57} n^{8.576}}{6.34 \times 10^{17}} - \frac{1.333}{m^2} = 0$$

$$m^* = \frac{1367}{Z^{0.624} n^{1.5}}$$

$$\frac{\partial^2 T_T}{\partial m^2} > 0; \text{ consequently, } T_T \text{ is strictly convex}$$

Because m^* is always bounded by Equation (6.59b), we get

$$m^* = \left(\frac{3.09 \times 10^3}{Z^{0.624}n^{1.5}}\right)^{-} \qquad \text{(round off to the lowest integer)}$$

Therefore, Z and n should lie on their respective upper and lower bounds. In this case,

$$Z^* = 3 \quad \text{and} \quad n^* = 1$$

The results can be summarized as follows:

$$R = 19.2 \text{ mm}$$
$$Z = 3$$
$$W = 30 \text{ mm}$$
$$d = 1 \text{ mm}$$
$$V_f = 13.2 \text{ mm/min}$$
$$V = 7.239 \text{ m/min}$$
$$n = 1$$
$$m = 1556$$
$$T_T = 0.02 \text{ min}$$

6.4.2 Multipass Model

Multipass models also consider the depth of cut as a control variable. Let a_p be the height of material to be removed, and n the number of passes. Then the time required per component can be written as

$$t_{pr} = t_h + \sum_{i=1}^{n_p}\left[t_m^i + \left(\frac{t_m^i}{t}\right)t_t\right] \tag{6.63}$$

where t_m^i is the time required for machining pass i.

The cost per component is

$$C_{pr} = \frac{C_b}{N_b} + C_m t_h + \sum_{i=1}^{n_p}C_{pr}^i \tag{6.64}$$

where

$$C_{pr}^i = C_m\left[t_m^i + \frac{t_m^i}{t}\left(t_t + \frac{C_r}{C_m}\right)\right] \tag{6.65}$$

$$a_t = \sum_{i=1}^{n_p}a_p^i \tag{6.66}$$

and superscript i represents the ith pass.

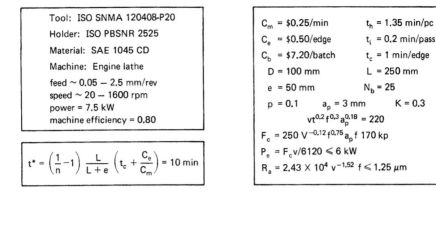

Tool: ISO SNMA 120408-P20

Holder: ISO PBSNR 2525

Material: SAE 1045 CD

Machine: Engine lathe

feed $\sim 0.05 - 2.5$ mm/rev
speed $\sim 20 - 1600$ rpm
power = 7.5 kW
machine efficiency = 0.80

C_m = \$0.25/min	t_h = 1.35 min/pc
C_e = \$0.50/edge	t_i = 0.2 min/pass
C_b = \$7.20/batch	t_c = 1 min/edge
D = 100 mm	L = 250 mm
e = 50 mm	N_b = 25
p = 0.1	a_p = 3 mm K = 0.3

$$vt^{0.2}f^{0.3}a_p^{0.18} = 220$$

$$F_c = 250 \, V^{-0.12} f^{0.75} a_p f \, 170 \, kp$$

$$P_e = F_c v/6120 \leqslant 6 \text{ kW}$$

$$R_a = 2.43 \times 10^4 \, v^{-1.52} \, f \leqslant 1.25 \, \mu m$$

$$t^* = \left(\frac{1}{n} - 1\right)\frac{L}{L+e}\left(t_c + \frac{C_e}{C_m}\right) = 10 \text{ min}$$

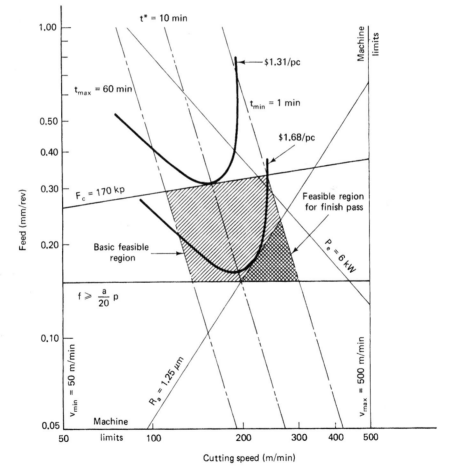

Figure 6.12   Contours of unit cost and the feasible region.

An additional constraint is also required in the formulation for depth-of-cut constraint:

$$a_{p,\,min} < a_p^i < a_{p,\,max} \tag{6.67}$$

The additional variable a_p makes the solution procedure more difficult than for a single-pass problem. In solving this class of problem, Challa and Berra [1976] used a modified Rosen gradient-search method. Philipson and Ravandaran [1978] and Subbarao and Jacobs [1978] both used goal programming to deal with the problem. Iwata et al. [1977] introduced a dynamic-programming procedure to solve a multistage machining-optimization problem. Hayes et al. [1981] and Chang et al. [1982] transformed certain variables, such as depth of cut, into the discrete domain. The number of passes then can be obtained iteratively by using a dynamic-programming procedure to optimize the feed and speed. There is no general solution method that can be used for all problems.

Figure 6.12 shows the contours of unit cost and the feasible region for a sample single-pass problem. In the figure, the bounding constraints as well as the objective function are plotted. The bounding constraint for the example is the surface finish. If the surface-finish constraint is relaxed, the force constraint becomes binding.

6.5 CONCLUDING REMARKS

There are two types of information that go into a process-planning system: design and process knowledge. Design representation was discussed in Chapter 2. In the current chapter, process knowledge was discussed. Process planning can be said to be a procedure that matches the knowledge of the processes with the requirements of the design. Process capability is the producibility knowledge for a process—the basic mechanism of the process-planning system. However, a logical structure is necessary to carry out the matching procedure. This chapter has provided both process capability and decision logic. In the next chapter, we discuss combining process planning with product design.

6.6 KEYWORDS

Cutting force
Cutting power
Cutting speed
Decision table and decision tree
Depth of cut
Feed
Flank wear
Machinist handbooks
Material-removal rate (MRR)
Process boundaries
Process capability
Process optimization
Process parameters
Process planning
Single-pass and multipass machining
Tool life

6.7 REVIEW QUESTIONS

6.1 It is desirable to produce a part with a surface finish of 3-3-μin. in AA. If the surface has already been machined to dimension, what process(es) are needed to finish the surface? (*Hint:* Use Figure 6.1.)

6.2 It can be seen from Figure 6.1 and Table 6.1 that the grinding process is more accurate, and can also produce a better surface finish, than the milling process. Why, then, aren't milling operations replaced by grinding operations? What factors should be considered before selecting a process?

6.3 Make a decision-table model for the following course-scheduling activity: Calculus is required before statistics can be taken, and two semesters of statistics are required before graduation. However, Professor Smith is the best statistics instructor, and unless taking her course postpones graduation, you will wait until she teaches the course.

6.4 Derive the ideal surface roughness for a tool with a round nose.

6.5 A hole 1 in. in diameter and 2.5 in. deep is to be drilled. The suggested feed is 0.05 ipr and the speed is 200 fpm. What are the feed speed, spindle rpm, MRR, and cutting time? Assume that the clearance height is 0.2 in.

6.6 A part 10 in. in diameter and 20 in. in length is to be turned down to 9.4 in. for the entire length. The suggested feed is 0.04 ipr and the speed is 450 fpm. The maximum allowable depth is 0.2 in. What are the feed speed, spindle rpm, MRR, and cutting time? Assume that the overtravel is 0.5 in.

6.8 REVIEW PROBLEMS

6.1 The end of the part described in Review Question 6.6 is to be faced. The spindle rpm is set so as not to produce a cutting speed higher than the suggested speed (450 fpm). What are the feed speed, spindle rpm, MRR, and cutting time?

6.2 A volume 10 in. long, 5 in. wide, and 0.5 in. thick is to be removed by a face-milling cutter that is 3 in. in diameter and has six teeth. The maximum depth of cut is 0.3 in. What are the feed speed, spindle rpm, and cutting time?

6.3 When a 0.5-in. HSS drill on SAE 3140 steel at a feed of 0.03 ipr is used, the tool-life equation can be expressed as

$$t = \frac{6.16 \times 10^{15}}{V^{6.25}}$$

The hole drilled is 2 in. deep. If the selected speed is 200 fpm, how many holes can be drilled before we need to change the drill?

6.4 When an HSS tool is used to turn a stainless-steel shaft, the cutting force can be predicted:

$$F_c = 151,000 \, f^{0.85} a_p^{0.96}$$

The recommended speed is 200 fpm and the feed is 0.03 ipr. The diameter of the workpiece is 3 in. and the final diameter is 2.8 in. The part is 4 in. long. A 10-hp motor is used to drive the machine, which has an efficiency of 0.90. What is the maximum depth of cut, based on the power-consumption constraint? How long will it take to finish the workpiece?

6.5 Assume that the workpiece in Review Question 6.6 is made of alloy steel. What is the required cutting power? Use Figure 6.10.

6.6 In Review Problem 6.3, if the part-loading/unloading time is 2 min and the tool-change time is 3 min, what is the minimum-time cutting speed? What is the cutting time per hole?

6.7 In Review Problem 6.6, the operator and machine rates are $30/h. Each drill bit costs $2. What is the minimum-cost cutting speed? What is the cost per hole?

6.8 Derive the general expression for the unconstrained minimum-time cutting speed and the maximum-cost cutting speed.

6.9 The rate of profit is defined as

$$P = \frac{S - C_{pr}}{t_{pr}}$$

Derive the expression for the maximum-profit-rate cutting speed V^*.

6.10 The processing time for single-pass model is given in Equation (6.40). The depth of cut, a_p, is a constant; therefore, the tool-life equation can be rewritten as

$$t = \frac{C'}{V^\alpha f^\beta}$$

Let the length of cut be L. If the maximum spindle speed is V_{max} and the maximum feed is f_{max}, what are the optimum cutting speed and feed?

6.11 Develop a method to solve the constrained optimization model for turning. Try to use the integer-transformation method illustrated in the milling optimization example.

6.12 Prepare a process plan for the part shown in Figure 3.9. First, assume that 10 parts will be made. What if 100,000 parts were made? How does this affect the process plan?

6.13 Find the general form of the tool-life equation for Example 6.4.

6.9 REFERENCES

ARMOREGO, E.J., and BROWN, R.H. *The Machining of Metals.* Englewood Cliffs, NJ: Prentice-Hall, 1968.

BERRA, P.B. *Investigation of Automated Planning and Optimization of Metal Working Processes.* Ph.D. thesis, Purdue University, West Lafayette, Indiana, 1968.

BERRA, P.B., and M.M. BARASH. *Investigation of Automated Plating and Optimization of Metal Working Processes.* Report 14. West Lafayette, IN: Purdue University, Purdue Laboratory for Applied Industrial Control, 1968.

BOOTHROYD, G. *Fundamentals of Metal Machining and Machine Tools.* New York: McGraw-Hill, 1975.

BOSTON, O.W. *Metal Processing.* New York: John Wiley, 1951.

CHALLA, K., and P.B. BERRA (1976). "Automated Planning and Optimization of Machining Procedures—a Systems Approach," *Computers and Industrial Engineering,* 1(1), 1976, 35–36.

CHANG, T.C., R.A. WYSK, R.P. DAVIS, and B. CHOI. "Milling Parameter Optimization through a Discrete Variable Transformation." *International Journal of Production Research,* 20(4), 1982, 507–516.

CINCINNATI MILLING MACHINE CO. *A Treatise on Milling and Milling Machines.* Cincinnati: CMM, 1951.

COELHO, P.L.F. *The Machining Economics Problem in a Probabilistic Manufacturing Environment.* Unpublished M.S. thesis, Virginia Polytechnic Institute and State University, Blacksburg, Virginia, 1980.

COOK, N.H., and K.L. CHANDERAMANI. "Investigation on the Nature of Surface Finish and Its Variation with Cutting Speed," *Journal of Engineering for Industry*, 86 (134), 1964, 134–140.

DEVOR, R.E., W.J. ZDEBLICK, V.A. TIPNIS, and S. BUESCHER. "Development of Mathematical Models for Process Planning of Machining Operations NAMRC—VII," in *Proceedings of the Sixth North American Metalworking Research Conference.* Dearborn, MI: Society of Manufacturing Engineers, 1978.

EARY, D.F., and G.E. JOHNSON. *Process Engineering for Manufacturing.* Englewood Cliffs, NJ: Prentice-Hall, 1962.

FIACCO, A.V., and G.P. McCORMICK. *Nonlinear Programming.* New York: John Wiley, 1968.

GENERAL ELECTRIC. *Operation Manual for the Carboloy Machinability Computer.* Manual MC-101-B. Detroit: General Electric Company, Metallurgical Processes Department, 1957.

GILBERT, W.W. "Economics of Machining," in *Machining: Theory and Practice.* Metals Park, OH: American Society for Metals, 1950.

GILBERT, W.W., and W.C. TRUCKENMILLER. "Nomograph for Determining Tool Life and Power when Turning with Single-Point Tools," *Mechanical Engineering*, 65, 893–898, 1943.

GROOVER, M.P. *A Survey on the Machinability of Metals.* SME Technical Paper, Series MR76-269. Dearborn, MI: Society of Manufacturing Engineers, 1976.

GROOVER, M.P. *Fundamentals of Modern Manufacturing: Materials, Processes, and Systems.* Upper Saddle River, NJ: Prentice Hall, 1996.

GROOVER, M.P., A.M. GUNDA, and R.J. JOHNSON. "Determination of Machining Conditions by a Self-Adaptive Procedure," in *Proceedings of the Fourth North American Metalworking Research Conference.* Dearborn, MI: Society of Manufacturing Engineers, 1976, pp. 267–271.

HATI, S., and S. RAO. "Determination of the Optimum Machining Conditions—Deterministic and Probabilistic Approaches," *Journal of Engineering for Industry*, 98(2), 1976, 354–359.

HAYES, G.M., JR., R.P. DAVIS, and R.A. WYSK. "A Dynamic Programming Approach to Machine Requirements Planning," *AIIE Transactions*, 13(2), 1981, 175–181.

ISO, (1972), Tool Life Testing with Single-Point Turning Tools, ISO 5th draft proposal, ISO/TC 29/WGG22 (Secretariat 37), Vol 91, March 1972.

IWATA, K., et al. "Optimization of Cutting Conditions for Multi-pass Operations Considering Probabilistic Nature in Machining Processes," *Journal of Engineering for Industry*, 99(2), 1977, 210–217.

JOSHI, S., and CHANG, T. "Graph-based Heuristics for Recognition of Machined Features from a 3D Solid Model," *Computer Aided Design*, 20(2), 1988, 58–66.

KACZMAREK, J. *Principles of Machining by Cutting, Abrasion, and Erosion*, trans. A. Voellnagel and E. Lepa. Stevenage, England: Peter Peregrinus, 1976.

KALPAKJIAN, S. *Manufacturing Engineering and Technology*, 3d ed. Reading, MA: Addison-Wesley, 1995.

LUDEMA, K.C., R.M. CADDELL, and A.G. ATKINS. *Manufacturing Engineering, Economics and Processes.* Englewood Cliffs, NJ: Prentice-Hall, 1987.

McDANIEL, H. *Decision Table Software—a Handbook.* Princeton, NJ: Brandon/Systems Press, 1970.

METCUT RESEARCH ASSOCIATES, INC. *Machining Data Handbook*, 3d ed. Cincinnati: Machinability Data Center, 1980.

NIEBEL, B.W., A. DRAPER, and R.A. WYSK (1989). *Modern Manufacturing Process Engineering.* New York: McGraw-Hill, 1989.

PARSONS, N.R., ed. *N/C Machinability Data Systems.* Dearborn, MI: Society of Manufacturing Engineers, 1971.

PHILIPSON, R.H., and A. RAVANDARAN. "Application of Goal Programming to Machinability Data Optimization," *Journal of Mechanical Design*, 100(2), 1978, 286–291.

SCARR, A.J.T. *Metrology and Precision Engineering.* London: McGraw-Hill, 1967.

SUBBARAO, P., and C. JACOBS. "Application of Nonlinear Goal Programming to Machine Variable Optimization," in *Proceedings of the Sixth North American Metalworking Research Conference.* Dearborn, MI: Society of Manufacturing Engineers, 1978, 298–303.

TAYLOR, F.W. "On the Art of Cutting Metals." *Transactions of the American Society of Mechanical Engineers*, 1906, 28, 31.

TIPNIS, V.A., M. FIELD, and M.Y. FRIEDMAN (1975). *Development and Use of Machinability Data for Process Planning Optimization.* SME Technical Paper, Series MD75-517. Chicago: 1975.

TRUCKS, H.E. *Designing for Economic Production.* Dearborn, MI: Society of Manufacturing Engineers, 1974.

WYSK, R.A. *Process Planning Systems*, MAPEC Module. West Lafayette, IN: Purdue University, School of Industrial Engineering, 1977.

Chapter 7

Tooling and Fixturing

Objective

This chapter gives readers a background on cutting tools, the machinability of work materials, and work-holding methods. This information is essential in selecting machining operations and in programming and operating NC machine tools.

Outline

7.1 INTRODUCTION

In order to process parts, certain conditions must be realized. For example, in order to assemble two items, one is normally held firmly by some type of clamping device (a vise, a hand, and so on) while the second item is moved to the required assembly position with the proper orientation. The items are then joined together (screwed, snapped, slid, and so on) to form an assembly. If the parts are to be screwed together, then some additional processing resources (tooling—in this case a wrench or screwdriver) may also be required. The methods and resources used to assist in processing can significantly affect the efficiency and profitability of a process. We refer to the methods and resources required to process a part as tool engineering.

The trend toward flexibility in manufacturing systems has brought a new focus to tool engineering. In the past, for high-volume, long-run production items, significant time and effort were expended on custom tooling and fixturing. It was not uncommon for several months of engineering and manufacturing to proceed the first production item. Much of today's emphasis on small-batch, custom manufacturing has led to the use of standard tooling and work-holding devices that can be used for a variety of products. This chapter will again focus on machining and will begin with a discussion of tools. An introduction to fixtures and fixture design will then be presented.

F. W. Taylor was an early pioneer of tool engineering. His 1906 paper still provides many principles used in industry today. Nonetheless, the structures and characteristics of modern machining facilities have changed a great deal. Higher powered, more rigid numerically controlled machines with multitool capacity and a variety of available cutting tools have added significant complexity to tool engineering. In Taylor's time, not many tool materials were available, and their performance was relatively easy for the operator to command. In the case of mass production, dedicated tests could be conducted to find the optimal performance. These circumstances are no longer globally true today. Beyond high-speed steel (HSS) and the original carbides, industry currently uses ceramics, cermets, cubic boron nitride (CBN), diamond tool materials, a variety of coatings, and ion and physical vapor-deposition (PVD) implementation techniques. These tools come with different geometries for different operations and various mounting configurations for many available toolholders. It is extremely difficult for a process planner or an operator to select the best tool geometry and the optimal application parameters. The situation in which a machinist grinds a tool from an HSS blank or brazed carbide is rare today. New tools and tool materials are developed routinely.

Over the years, thousands of person-hours and millions of research dollars have been spent to reveal the secrets of the metal-cutting process in hope of predicting the process performance for a given operation. Metallurgical, microstructural, thermal, and mechanical aspects of the process have been investigated. Although important progress has been made to improve the understanding of the metal-cutting process, the ultimate goal of predicting process performance from the setup and the material involved is still just a vision. It is unlikely that we will mathematically describe the process in general in the near future.

Despite the rapid development of other components in manufacturing systems, there has been little progress in tool management. This stagnation is due primarily to the number of variables involved and limited understanding of the processes. Direct and indirect performance measures of tool life are usually modeled as a function of the primary machining parameters of speed, feed, and depth of cut, and possibly some simplified tool geometry and workpiece variables, be the equation in Taylor's exponential form or some other form (e.g., polynomial). Nonetheless, the metal-cutting process takes place at high temperature and pressure; the chemical, physical, and mechanical phenomena are beyond what a few parameters can describe. Moreover, the process appears random in nature, so any tool-life model can give only approximate values for the average tool life. The variation in tool life is usually high, with the coefficient of variation as high as 0.3. Practitioners often indicate that tool-life consistency is more important than peak performance.

With the development of modern machining systems, new material-handling systems, and, especially, flexible manufacturing systems (FMSs), setup times have been greatly reduced while the percentage of time in cut has increased dramatically. In addition, the power and the rigidity of machine tools have improved tremendously. These new technologies upgrade system capability; however, capital investment in them is much higher. The utilization of equipment is crucial for improving system productivity. Due to the increasing percentage of time in cutting, high system cost, less human attendance, and relatively low knowledge of the increased complexity of the process, technical operation management is more and more critical for improving productivity in the machining industry.

7.2 TOOL CHARACTERISTICS

In metal cutting, a proper tool must be identified for any operation. The technical aspect of the decision increases with the increasing number of possible tool selections and workpiece materials. Determining factors in selecting tool material, geometry, and construction include setup characteristics, the material itself, the shape and size of the workpiece; the design requirement; the type of operation (roughing or finishing); and other factors. For a given setup-and-tool combination, machining parameters (feed, speed, and depth of cut) must also be selected. Although machining handbooks, which offer excellent application guides, are available (even in software form today), the selection of parameters is still no easy task. The suggested data in handbooks are based on isolated laboratory tests using standard specimens. Many restrictive factors, such as the rigidity of the setup, the quality of the machine tool, force, and power, are carefully controlled. The environment determined by the site, machine tool, and fixtures on the factory floor is often different from environments encountered in laboratory tests. Performance is strongly dependent on environment variables, so the suggested data can be used only as a guide or starting point. The data pertinent to a specific application environment must be collected and documented in order to produce reliable performance guidelines. Even when applied to similar conditions, the suggested parameters tend to be conservative.

A metal-cutting process is the removal of unwanted metal by a wedge-shaped tool. Material is removed from a workpiece in the form of chips so as to obtain a desired surface. The cutting tool is made of a harder substance than the workpiece is, and when forced against it, the edge of the tool deforms and removes a thin layer of material. A machine tool supplies the power, force, and necessary cutting-tool movement at the proper speed with certain degrees of freedom. Fixtures are used to hold the workpiece firmly, conveniently, and accurately.

In manufacturing, the term *machining* is used to cover all the chip-making processes—for instance, turning, milling, drilling, and boring. These operations are also classified as secondary processes, as opposed to the primary processes of forming, rolling, forging, and casting. Common features of cutting tools are discussed in the sections that follow.

7.2.1 Tool Geometry

In metal cutting, the chip represents the volume of plastically deformed material removed from the workpiece surface. The deformation process normally takes place at a very high rate. High strain rates occur in both the chip and the newly formed surface on the workpiece. A large amount of energy must be input to perform metal-cutting operations. The geometry of the tool point is one of the most important factors in determining the way in which the metal deforms, which in turn determines temperature, stress, and force on the tool. The geometry also affects the interaction between the tool surfaces and the work material. The fundamental geometric features of a turning tool can be characterized by the elements and parameters shown in Figure 7.1. The rake face is the surface on which the chip is formed. The highest temperature and stress on the cutting tool occurs on the rake face, which flows away from the workpiece. The clearance face, or flank, intersects the rake face to form the cutting edge. Due to elastic deformation on the finished workpiece surface, friction and rubbing take place at the region of the flank closest to the cutting edge. The cutting action at the area closest to the end cutting edge formed by the side end clearance face and the rake face is similar to that at the side clearance face, except that less cutting action takes place on that part of the tool point. In order to strengthen the tip of the tool and improve heat dissipation, a radius is used to connect the side clearance and end clearance faces. The nose radius also can improve the theoretical surface roughness. When cracking, rubbing, ploughing, chattering, and other effects in the process are very small, the theoretical surface roughness can be approximated by the side cutting angle, end cutting angle, nose radius, and feed rate. A large nose radius can generate a better surface finish.

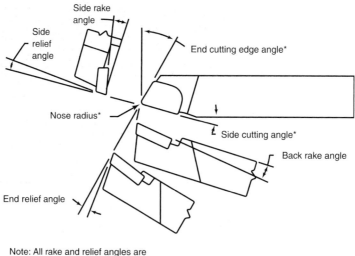

Note: All rake and relief angles are measured in normal direction.

Figure 7.1   Turning-tool geometry. (Based on Neibel, B.W., Alan B. Draper, and R.A. Wysk, *Modern Manufacturing Process Engineering*, McGraw-Hill, 1989. Reproduced with permission of McGraw-Hill Companies.)

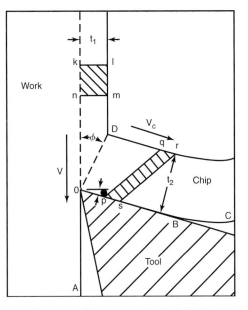

Figure 7.2   Metal-cutting diagram. (Based on Neibel, B.W., Alan B. Draper, and R.A. Wysk, *Modern Manufacturing Process Engineering*, Mc Graw-Hill, 1989. Reproduced with permission of McGraw-Hill Companies.)

The influence of tool geometry can be illustrated in a metal-cutting diagram, as shown in Figure 7.2. Deformation takes place primarily in three zones: the primary deformation zone, the secondary zone, and the clearance zone. The influence of fundamental tool geometry parameters can be addressed with respect to these three zones. The initial—and major—deformation takes place in the primary zone. The compressed material plastically sheared along a certain direction is determined primarily by properties of the workpiece material and the tool rake angle. The shear angle ϕ is an important characteristic of the metal-cutting process, determining the strain rate in the material removed. In general, the lower the shear angle, the higher is the strain and the higher is the plastic deformation. More heat, higher stress, and a larger force are induced. During the process, the square-shaped, shaded area *klmn* deforms into *pqrs*. The shear angle ϕ is related to the rake angle and the work material properties. For a given material, this relation can be approximated by

$$\tan \phi = \frac{R \cos \alpha}{1 - R \sin \alpha} \qquad (7.1)$$

where α is the rake angle and R is the chip thickness ratio t_2/t_1. When α is increased, R is reduced. As a result, ϕ is an increasing function of α. This means that a higher α value commonly results in a high shear angle, lower strain, lower energy input, and lower compression stress.

In the secondary zone, the deformed chip moves along the rake face under pressure to exit the cutting area. A small portion of the heat generated in the primary zone

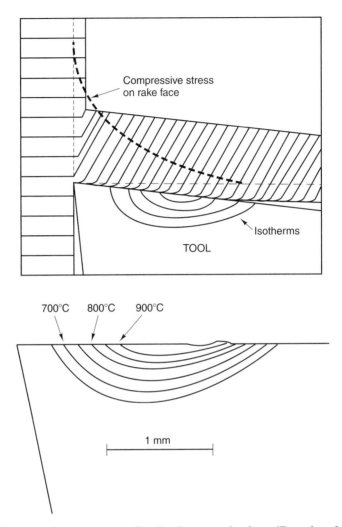

Figure 7.3   Stress and temperature distribution on rake face. (Based on Neibel, B.W., Alan B. Draper, and R.A. Wysk, *Modern Manufacturing Process Engineering,* Mc Graw-Hill, 1989. Reproduced with permission of McGraw-Hill Companies.)

and part of the heat generated by the interaction between the chip and the tool surface dissipate into the tool body, raising the temperature around the tool edge. Due to the high temperature and compression stress, the contact between the tool and work surface is nearly complete near the cutting edge. Sliding is virtually impossible under most cutting conditions. A flow zone is formed above the tool surface. (See Figure 7.3.) This zone is similar to the flow of fluid over a surface and has a very complicated nature. There is no simple relationship (as in the case of classical Coulomb friction) between the forces normal and parallel to the contact surface. The equivalent friction force is usually very high, which is (at least in part) why a restricted contact on the rake face (also called *land-edge chamfer*) can greatly reduce the force and temperature in the process. The temperature on the rake face can be so high that it is often the limiting factor in the material-removal rate. Overcoming this limitation is a major driving force in improving hot hardness and wear resistance of tool materials. Figure 7.3 shows the typical temperature contours that arise when low-carbon steels are cut with carbide tools. The high-temperature region is where the crater wear takes place. The temperature and stress are related to the rake angle. A high side rake angle allows a chip to move more freely on the rake face, so as to reduce contact length and shear stress. However, a high α value reduces the tool's strength, and the tool becomes less resistant to shock. A high α also reduces the tool's ability to dissipate heat, because a reduced included angle will have lower heat-transfer capacity. In certain operations, a slight change in the rake angle may weaken the tool edge. However, a dramatic increase in the rake angle may actually improve the process, because of a significant drop in cutting force, stress, and temperature.

In the clearance zone, the elastic- or plastic-deformed, newly formed surface on the workpiece recovers when it leaves the tool edge. The expanded surface pushes against the clearance face and creates friction between the two surfaces, resulting in flank wear. The length of contact, heat generated, and friction force decrease with the clearance angle. However, as in the case of the rake angle, an increased clearance angle can reduce the strength and heat-dissipation capability of the tool.

Other tool geometry parameters also have a significant influence on metal-cutting processes. The side cutting-edge angle (or lead angle) can change the force ratio along the X and Z directions (Figure 7.4). Varying the side cutting-edge angle should be considered in processing a part that lacks rigidity. The included angle formed by the clearance and back-clearance faces contributes to the strength of the tool and its heat-dissipation ability. It also affects surface finish when the nose radius is not large enough to form the machined surface completely. The size of the toolholder and insert can also affect the tool performance. The back rake angle can control the direction of chip flow. The orientation, affected by the rake and the inclination angle, determines either positive or negative entry in the milling operation. Figures 7.5 and 7.6 show the geometric features of milling and drilling tools. International Standards Organization (ISO) and

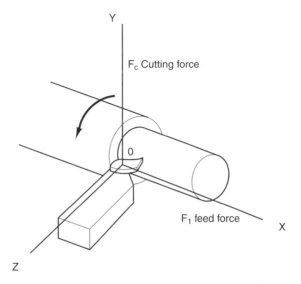

Figure 7.4   The cutting force in a turning operation.

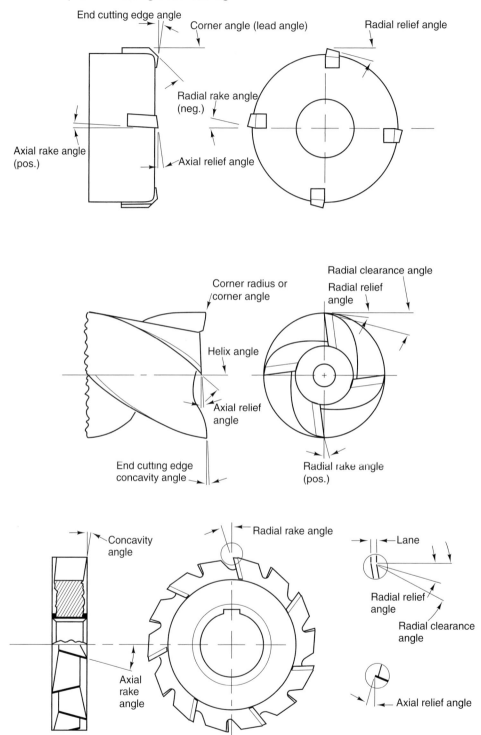

Figure 7.5 Geometry of milling tools.

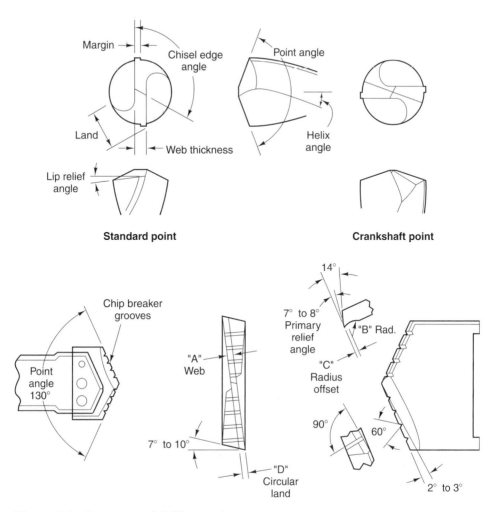

Figure 7.6   Geometry of drilling tools.

ANSI identification systems for toolholders and indexable inserts have been defined to classify these parameters. When standard toolholders and inserts are used, the geometry feature is determined by the combination selected.

7.2.2 Tool Material

The most commonly used cutting-tool materials in the machining industry today are HSS, carbides, ceramics, CBNs, and diamonds. These materials have different characteristics and are suitable for different cutting applications.

7.2.2.1. High-speed Steel.   The introduction of HSS was the first revolution in tool materials. Improved performance over carbon steel was made possible by the retention of hardness and compression strength at higher temperatures (up to 600°C). Although more advanced tool materials, including carbides, ceramics, CBNs, and diamonds, are harder and more wear resistant at higher temperatures, HSS is still one of the most important tool materials in use today, due to its higher strength and machinability. Its hardness at room temperature is close to that of carbides. These properties make HSS a top contender for high-volume, low-speed operations and interrupted operations. Higher rake angles, compared with those of carbides, ceramics, and CBNs, can be used in most processes. This is important for some low-machinability work materials, such as nickel-based high-temperature alloys, or for operations lacking rigidity. Because HSS can be machined in an annealed state, it is suitable for complex tool forms such as gear cutters, slotting tools, end mills, and twist drills.

Basically, there are three types of high-speed steels: tungsten, molybdenum, and ones that contain cobalt. (See Table 7.1.) Tungsten carbide steels, including T1 and T2 industrial grades, are similar to steels originally developed by Taylor. T2 is tougher than T1. The second type of HSS has higher strength due to added molybdenum. Grades M1, M2, M3, M7, and M10 are widely used in general-purpose operations and

TABLE 7.1   High-Speed Steels and their Applications

Type	Application
1. Tungsten High-Speed Steels	
T1	General purpose
T2	General purpose—higher strength
2. Molybdenum High-Speed Steels	
M1	General purpose
M2	General purpose
M3 Class 1	Fine-edge tools
M3 Class 2	Fine-edge tools
M4	Abrasion resistant
M7	Fine-edge tools—abrasion resistant
M10	General-purpose—high strength
3. High-Speed Steels Containing Cobalt	
M6	Heavy cuts—abrasion resistant
M30	Heavy cuts—abrasion resistant
M33	Heavy cuts—abrasion resistant
M34	Heavy cuts—abrasion resistant
M36	Heavy cuts—abrasion resistant
M41	Heavy cuts—abrasion resistant
M42	Heavy cuts—abrasion resistant
M43	Heavy cuts—abrasion resistant
M44	Heavy cuts—abrasion resistant
M46	Heavy cuts—abrasion resistant
M47	Heavy cuts—abrasion resistant
T4	Heavy cuts
T5	Heavy cuts—abrasion resistant
T6	Heavy cuts—abrasion resistant
T8	General-purpose—hard material
T15	Extreme abrasion resistant

fine-edge tools. The addition of cobalt to Type 3 HSS provides higher hot hardness and wear resistance. However, the resulting type of steel is difficult to grind and thus is more suitable for cutting hard work materials (with a hardness level of 350 BHN).

Both tungsten and molybdenum HSS can be hardened to 64–66 Rc and are recommended for the machining of materials with hardness up to 350 BHN. The heat treatment of the tungsten type is simpler than that of the molybdenum type, but the molybdenum type is tougher. Molybdenum's toughness is slightly reduced, though, and is advantageous only in machining harder materials if both performance and preparation are considered. Types M3 and M4 contain more vanadium, providing increased wear resistance, but make the material harder to grind.

New developments in HSS tools include the introduction of powder metallurgy processing and the use of coatings. Powder metallurgy has great potential for tools of larger size because it produces a uniform material structure consistently and economically. Chemical vapor deposition (CVD) coating of thin layers ($10\ \mu$m) of refractory titanium carbide (TiC), titanium nitride (TiN), hafnium nitride (HfN), alumina (Al_2O_3), and titanium carbonitride (TiCN) on HSS tools are widely adopted in commercial tools today. These coatings add high-temperature hardness, compressive strength, and wear resistance to very tough HSS substrates. Such coatings can also improve tool life by 300–400% or higher. Alternatively, tool life can be maintained when the cutting speed is increased by 25% or more. The surface finish also can be improved at a lower cutting speed, due to the elimination of the built-up edge. The speed limitation in the use of these coated tools is due primarily to the HSS substrate. At very high temperatures, HSS under the coating deforms and alters the cutting conditions.

7.2.2.2. Carbides. Carbides are tool materials made of very hard and fine grains of carbide powder that are sintered (bonded or cemented) together by a tougher metal, usually cobalt, nickel, or molybdenum. The most commonly used carbide substances are tungsten (WC), titanium (TiC), and tantalum (TaC). These cemented materials have good thermal conductivity (which is very important in dissipating heat in metal cutting) and a very fine structure.

Straight tungsten carbide was first introduced in the 1920s. It is still one of the most widely used tool materials today for cast irons and nonferrous metals. Tungsten carbide contains 4 to 12% cobalt by weight. The carbide grain size ranges from 0.5 to $10\ \mu$m. The hot hardness, wear resistance, and toughness are dependent on the composition and grain size. The hardness and compressive strength decrease with increasing cobalt content and grain size. However, the transverse rupture strength, a measure of toughness, increases with the cobalt content and grain size. The quality of the carbide particles and the sintering process are also very important to tool performance. There should be very few holes or nonmetallic inclusions in the material. The carbon content must be controlled within very tight tolerances.

Tungsten carbide has proven to be highly successful in cutting cast iron and nonferrous metals at much higher speeds than HSS tools can achieve. However, tungsten carbide is less effective in cutting steel, wherein the temperature on the tool surface is very high and diffusion wear dominates the wear process. The problem is that tungsten carbide does not offer high diffusion wear resistance, which results in greater crater wear on the rake face and a rapidly increasing wear land. This shortcoming led an investigation into carbides of the transition elements in Groups IV, V, and VI of the periodic table. The most successful additives have been tantalum, titanium, and niobium carbide, bonded with cobalt, nickel, or molybdenum. Tools made of these materials were used to cut steels beginning in the 1930s. The content of the bonding material, hardness, and grain size is in the same range as straight tungsten carbide. TiC and TaC content greatly improves diffusion wear resistance. However, it also reduces the toughness of the tool. For this reason, TiC and TaC content is kept relatively low, 4–20% for TiC and up to 20% for TaC. TaC is believed to cause less reduction in toughness and to increase high-temperature strength. However, it is much more expensive, because it is a rare metal. Also, these carbides reduce thermal conductivity, which is a drawback. The major advantage of adding TiC and TaC content is to attain high-temperature, compressive-strength, diffusion wear resistance. These tool materials can be used at a speed of up to three times faster than that of straight tungsten carbides in cutting steels. Even the addition of 5% TiC is very effective. The quality and performance of steel-grade carbides are affected by many more variables than are the quality and performance of straight tungsten. Researchers and manufacturers have been investigating composition, grain size, and sintering processes to make the best tools for certain operations. A very large number of commercial steel grades are available.

The ISO carbide identification system arranges the grades according to applications. The P series designates steel-cutting grades. The K series is for cast iron and non-ferrous-metal grades. The M series can be used in both cases. In America, industrial codes C1 through C8 are used to group manufacturers' products on the basis of their carbide content. C1 through C4 are straight tungsten carbide bonded with cobalt. Because straight tungsten carbide is recommended for cast iron and nonferrous metals, these metals are related to the K series in ISO identification. C5 through C8 are WC–TiC–TaC–Co types of carbides. These correspond to the ISO P series. C8 also includes TiC bonded with molybdenum or nickel. These tools are more wear resistant at high speed, but their reliability and consistency are not as good as those of WC–TiC–TaC–Co tools. In ISO groups, the series number increases in order of increasing hardness and decreasing toughness in each series. The relation is the same in industrial codes.

Each manufacturer offers a wide range of carbide grades for a variety of machining operations. It is the manufacturer's responsibility to assign an industrial code or ISO group to its product. There are considerable differences in composition, structure, and properties between grades offered by different manufacturers in each category. We also find that there are usually many manufacturers' grades that fall into a single ISO group or industrial code, although, in some cases, one grade may be recommended for several groups or codes. (See Table 7.2.)

As with HSS, the CVD coating techniques are used in carbide substrates. The result is an improved tool life by a factor of 2 to 3 or an improved speed by 25 to 50% (or both). Due to higher hot hardness of carbides, the potential of coating is better utilized than it is in the HSS substrate. In recent years, physical vapor deposition (PVD) has also been introduced. The PVD process takes place at a lower temperature (up to 500°C), which results in lower residual stress compared with that in the CVD process. (The range is from 850 to 1050°C.) The coated layer is thinner, providing a sharper

TABLE 7.2   Insert Codes from Several Toolmakers

Mfg.	C-1	C-2	C-3	C-4	C-5	C-6	C-7	C-8	Coated Turn Steel Al₂O₃	Coated Turn Steel TiN/TiC	Coated Turn Iron Al₂O₃	Coated Turn Iron TiN/TiC	Coated Mill Steel	Coated Mill Iron	Ceramics Turn Steel (Cold Press)	Ceramics Turn Iron (Hot Press)	Ceramics Mill Iron (Hot Press)	TIC Turn Steel	TIC Turn Iron	TIC Mill Iron	CBN Turn (>HRC45)	Ultra Fine End Mill
ADAMAS	B	A AM PWX	AM PWX		434 499 THER-MILL	499	548	Micro-cut Micro-Mill		CNC	Roxide	CNC								T80 T80 T50		
CARBOLOY	820 44A	883	895 905	999	370 373 395 390	370	350	120	570	550 515 516 518	570	545	523		570 518	570 518		210				
CARMET	CA3 CA12	CA443 CA4	CA7	CA8	CA740 CA730 CA720	CA720	CA711	CA704		CA740 CA-9740	CA-9443				CA-W	CA-8	CA-8					
DoALL	DO1 DO030	DO02 DO020	DO3	DO4	DO015 DO035	DO16 DO36	DO17 DO34	DO18		DO40 DO42 DO44	DO46				DO80	DO85	DO85					
EX-CELL-O	E8	E6 XLO29	E5 XLO28	E3	10A 8A 80S	8A 8A	6AX XLO61	6AX 50B		XL612 XL802	XL202		XL822	XL212						XL88 XL88 XL86		
FIRTH-STERLING	H91 H17 H	H21 HA HTA	HTA HF	HF	T04 T14 NTA TXH	T22 T24	T25 TXL		CC46	TC+4 HN-14 TC+ HN+		TC+4 HN-14 TC+ HN+								WF SO3		
GREENLEAF	G01 G10	G23 G25 G02	G23 G25 G30	G40	G52 G50	G53 G54	G70 G74		G1	TI5 T4 GA5 GA8	G1	GA2 TI2			GEM1 GEM9	GEM2 GEM3	GEM2 GEM3			G80		
INGERSOL		KM1	KM1		P25 Gx	P20	P10		CP1	CM3	CP1	CM3										
ISCAR	IC28	IC2	IC20	IC4	IC54 IC50 IC30M	IC50M IC70	IC70 IC78		IC848	IC656 IC757	IC848	IC424								IC100T IC80T IC60T		
KYOCERA															A65 A62	A65 A62	TC30 TC50					
KENNAMETAL	K1	K6 C735 K8	K68 K8	K11	K420 K21	K4M K2584	K45 K5M	K7M	KC910 KC950	KC810 KC850	KC910 KC950	KC210 KC850	KC250	KC250	K060	K090	K090	X165		KD120		
NEW-COMER	N10	N22 N25	N30	N40	N52 N55	N60	N70 N72	N80	NA02	NT5 NT8	NA02	NT2	NT55	NT25				N90 N95				
NTX															CX3	HC2	HC2	T3N T4N N20				
SANDVIK	H20	HM H20 H1P	H1P H10	H05	S6 SM30 S4	SM S2	S1P	S1P	GC415 GC435 GC015	GC-1025 GC135	GC415 GC435 GC015	GC315	GC315 GC120	GC320 GC310						C850		
SECO	HX	HX SU41	H13		S6 S80M	S25M S4 S2	S1F S10M	S1G	TP15	TP35 TP25	TP15 TX10		T15M									
VALENITE	VC1 VC111	VC2 VC24 VC28 VC27	VC3	VC4	V56 VC125 VC55 VC	VC8	VC7 VC76	VC8	VO1 VO5	VN5 VN8 V90 V99	VO1 VO5	VN2 VN8 V88 V91			V34	V32	V32	V83				
VR-WESSON	2A68 VR54	2A5 VR82	VR82 2A7	2A7	WS WM VR77	VR75 2B	WH VR73	VR71		850 860 870	630				VR97	VR100	VR100					
WALMET	WA1 WA59 WA110	WA2 WA35 WA107 WA69	CBD WA3	WA4	WA54 WA5 WA57	WA6 WA47	WA47 WA7 WA73	WA8		T3	T2											
WENDT-SONIS	CO22 CO12	CO2	CO3 CO23	CO4	CY17 CY12	CY16	CY14	CY31	918	714 715 716 U225	918	U222 027										

coated edge than CVD inserts offer. The thin layer also makes the tool life more predictable, a desirable outcome for tool management. However, the coated layer does not adhere to the substrate as strongly as CVD coatings do. Kennametal's KC710 is a PVD coated carbide suitable for interrupted operations that require a sharp edge.

7.2.2.3. Ceramics and Ultrahard Materials. Ceramic tool materials are high-hardness and high-melting-point refractory oxides. The first successful tool material of this kind is pure alumina. Ceramic tools are usually cold pressed (although hot pressing is also possible) and are white in color. Kennametal's K060 and Greenleaf's GEM9 are pure alumina ceramics. Alumina tools can be used at much higher speeds than carbides can. At feed rates of 0.25 mm/rev, there is no excessive wear for long periods cutting carbon steel at a speed as high as 6000 m/min. The material is, however, very brittle and can be used only for cast iron or steel with hardness lower than 32 Rc. A negative rake angle is always used, except in interrupted operations. Coolant cannot be applied because of low thermal shock resistance. Despite its limitations, alumina is the most economical ceramic tool material available.

Cermet, an alumina and titanium carbide composite, is a hot-pressed tool material. The added TiC increases the toughness of the tool. Cermet can be used for roughing and interrupted operations. Cermet can be used on steel and cast iron with hardness up to 65 Rc. Continuous coolant can be applied. Kennametal's K090 and Greenleaf's GM2 are two commercial examples of cermet tools.

A relatively new material, Sialon (Si–Al–O–N), is another ceramic, made of silicon nitride (Si_3N_4) plus alumina. This material has properties that lie between those of cermets and coated carbides and can be used for roughing cast iron at ceramic cutting speeds, with a feed rate normally associated with coated carbides. It also can be used for roughing nickel-based alloy at a higher speed (125–250 m/min) or a higher feed rate (0.1–0.3 min/rev) than those of either cermet or carbide inserts. This material has a very high hardness value, up to 94 Rc.

Another new ceramic material, WG-300, was developed by Greenleaf Corporation. It is a ceramic composite reinforced matrix with silicon carbide single crystals called "whiskers." Due to its high purity and lack of grain boundaries, the whiskers grow under carefully controlled conditions, approaching their theoretical maximum strength. Cemented with a fine-grain alumina matrix, the whiskers act like rods, adding strength to brittle alumina. In addition, the thermal conductivity and thermal shock resistance are enhanced, permitting the use of coolant in all operations. Substantial savings of up to 65 Rc can be achieved in the cutting of hardened steels, hard cast irons, and nickel-based aircraft-grade high-temperature alloy. The speed can be increased about 10 times more than speeds used for carbides. However, the increase in speed is not dramatic for commonly used, easy-to-cut materials.

Ceramic tools offer two advantages over carbides and HSS tools: a faster material-removal rate and high tip accuracy. However, their application in industry is still not widespread. One reason for this is that extra care must be exerted in the handling of these tools because of their brittle nature. Another reason is that, in some operations, the setup time constitutes a fixed portion of the total time, so that further reduction in the direct cutting time by using carbide tools may not have a significant impact on the tool's overall performance.

Cubic boron nitride (CBN) is a synthetic crystal not found in nature. CBN consists of two interpenetrating face-centered cubic lattices. The crystal, like diamond, has a rigid structure. Kennametal's KD120 polycrystalline cutting edge is composed of cubic boron nitride fused to carbide. The tool's hardness and chemical inertness at high temperatures permit the machining of hardened steel at speeds four to five times faster than coated carbides. This property is useful for machining tool steels, bearing steels, and chilled iron. It also can be used for difficult-to-cut, high-temperature alloys at speeds up to 300 m/min. Coolant can be applied for a more satisfactory performance.

The polycrystalline diamond tools are aggregates of randomly oriented small synthetic diamond crystals. These tools are extremely hard and wear resistant. Kennametal's KD100 belongs in this category. It can be applied to nonferrous and abrasive materials such as A390 aluminum, hybrid composites, and glass-fiber reinforced plastics.

CBN and diamond tool materials are very expensive. The cost of a tip can be 10 times that of a carbide or ceramic insert. Careful justification is necessary to take advantage of these tool materials.

7.2.2.4. Summary of Tool Materials.   In general, the tool materials listed here are in increasing order of hardness, wear resistance, and cost, and in decreasing order of toughness. In any application, trade-offs are necessary to find the best choice for a particular application.

7.3 MACHINABILITY OF WORK MATERIALS

The term *machinability* is a widely used term in both research and industry. In the *Metals Handbook* [American Society for Metals, 1985], it is defined as "the relative ease for a material to be machined." To the people engaged in a particular set of operations, machinability has a clear meaning, such as the number of components produced per hour or per tool, or the relative ease in achieving surface or dimensional specifications. However, the criteria employed for various tests can be different. A material with good machinability by one criterion may have poor machinability by another. Unlike the situation with most material properties, there is no generally accepted criterion for its measurement. The term tends to reflect the interests of the user. The most commonly used criteria in practice are as follows:

1. *Tool life:* The amount of material removed by a tool under standard conditions before the tool performance becomes unacceptable or the tool is worn by a standard amount. The main criterion for the ISO standard is basically tool life.
2. *Limiting material-removal rate:* This is often the criterion for ultrahard work materials.
3. *Surface finish achieved:* This is usually one of the dominant factors in the machining of ductile materials.
4. *Chip control:* In the machining of some ductile materials, this can be the most significant factor in certain operations.
5. *Force and power consumption:* The force can be the limiting factor when the setup lacks rigidity. In cutting "easy-to-machine" types of materials at a very high speed and high volume, the available power is often the limiting factor.

7.3.1 Machinability Tests

The machinability indexes in the *Metals Handbook* [American Society for Metals, 1985] and many other industry reports are the results of machinability tests. The long-term absolute machinability standard became available in 1977 [ISO 3685–1977]. Until then, the test conditions and criteria were determined by individual researchers, resulting in a vast amount of machinability data that were impossible to correlate for a cohesive body of data. The ISO standard test indicates the relative merit of two or more work-tool combinations for a range of cutting conditions. The test material should be mounted between centers or between the chuck and a center. It should have a length-to-diameter ratio of less than 10 to 1. The tool material should be HSS, P30, P10, K20, or K10. Four sets of machining conditions, intended to cover everything from light to heavy roughing operations, are recommended. At least four speeds should be used that ideally result in a tool life of 5, 10, 20, or 40 min. The tool failure criteria for HSS are as follows:

1. catastrophic tool failure (i.e., breakage)
2. 0.3-mm average flank-wear land width if flank wear is even
3. 0.6-mm maximum flank wear if flank wear is irregular, scratched, chipped, or badly grooved.

For carbide grades, the first criterion is a crater depth of $(0.06 + 0.3f)$ mm, where f is the feed in mm/rev. The criteria for ceramic tools are the same as those for HSS. Of the failure modes, flank wear is by far the most commonly used, except in the high-speed machining of cast iron, in which the most significant failure mode is cratering.

In addition to long machinability tests, many types of short-term tests are often conducted for a particular tool–material combination. Czaplicki [1962] proposed the following relationship between a cutting speed that results in a 60-min tool life and chemical content:

$$V_{60} = 161.5 - 141.4\% \text{ C} - 42.4\% \text{ Si} - 39.2\% \text{ Mn}$$
$$- 179.4\% \text{ P} + 121.4\% \text{ S (m/min)} \tag{7.2}$$

Boulger, Moorhead, and Govey [1951] determined the following relationship for the relative machinability index (MI; base 100):

$$\text{MI} = 146 - 400\% \text{ C} - 1500\% \text{ Si} + 200\% \text{ S} \tag{7.3}$$

Henkin and Datsko [1963] developed a relation using dimensional-analysis techniques based on the material's physical properties:

$$V_{60} \propto \frac{B}{LH_B}\sqrt{1 - \frac{A_r}{100}} \tag{7.4}$$

In this equation, B is the thermal conductivity, L is a characteristic length, H_B is the Brinell hardness, and A_r is the percentage reduction in a tensile test. Similar work by Janitzky [1944] yielded the expression

$$V_{60} \propto \frac{D}{H_B A_r} \tag{7.5}$$

where D is a constant that depends on the size of cut. However, the application of all these relationships is restricted to materials of the same type and thermal history. A more common empirical relation for machinability was developed by Gilbert and cited by Olson [1985]:

$$V = \frac{ABCDEFGPQ^{0.25}}{H^{1.72}T^n R^{0.16} f^{58} d^2} \tag{7.6}$$

The definitions of, and typical values for, the parameters in this equation are summarized in Table 7.3. Equation (7.6) is a very powerful relation, especially when adapted to a given system with known products.

TABLE 7.3   Parameters and Typical Values for Equation (7.6)

Parameter	Factor	Typical Values	
V	Cutting speed	Selected speed in ft/min	
A	Tool material	HSS: 180,000	Carbides: 300,000
		Ceramics: 1,500,000	
B	Coolant	Dry: 1.0	Cutting oils: 1.15
		Soluble oils: 1.25	
C	Material	Carbon steel: 0.8	Free-machining steel: 1.05
		Alloy steels: 1.1	
		Free-machining brass:	Cast irons: 0.75
		2.0 Magnesium	Aluminum alloys: 0.85
		alloys: 0.9	
D	Microstructure	Austenitic: 0.7	Most steel: 1.0
		Coarse spher. 1.4	
E	Surface	Sand cast: 0.7	Sand cast, blast: 0.75
		Heat treatment: 0.8–0.95	clean surface: 1.0
F	Tool type	Single-point turning:	Boring, milling: 1.0
		1.0 Drill, form	Reamers: 0.8
		tools: 0.7	
G	Tool profile	Sharp, 0° entering: 1.0	High radius: 1.5
		High entering: 1.5	
P	Tool material	HSS: 1.0	Carbides: 5.0
		Ceramics: 8.0	
Q	Flank wear	About 0.3 mm	Depends on tool material
H	Hardness BHN		
R	Number of	Single point: 1.0	Equal number for multiples
	points		
T	Tool life	A few minutes to a few	
		hundred minutes	
n	Taylor	HSS: 0.125	Carbides: 0.25
	expansion	Ceramics: 0.68	
f	Feed rate	inch/rev	
d	Depth of cut	inch	

In addition to these nonmachining tests, actual machining tests are conducted in practice. There are constant-pressure tests in which a constant feed force is maintained, accelerated tests in which a higher-than-normal speed is used to shorten test time, and so on.

Most machinability assessments are for single-point turning operations. Constant-pressure and other types of tests have also been carried out for various processes, including drilling operations. Tool failure results from either catastrophic failure or some measure of drill-tip wear. There is a high possibility, particularly in small-diameter drills, of drill breakage, which is influenced by drill length. Because there are a variety of milling operations, the chip equivalent concept [Colding, 1961] is often used to relate machinability to milling operations. However, as already indicated, machinability is highly dependent on process and cutting conditions. It is difficult to relate results from one process to another unless the processes involved are practically the same.

The machinability of several basic types of work materials under all criteria will be discussed shortly. The influence of some physical and mechanical properties of materials are as follows:

1. Material with high yield strength and work-hardening ability requires more power input, exerts higher compression stress, and, in general, generates higher temperatures on the tool surface.
2. In addition to requiring a high-energy input, machining ductile material results in a poor surface finish.
3. Material with high-fracture toughness tends to generate long chips that are hard to break.
4. Material with a high work-hardening capability requires more energy on the share plane. The tool may constantly cut against the work-hardened surface left by a previous cut.
5. Good thermal conductivity can reduce the temperature on the tool surface.
6. Material that tends to react chemically with the tool material at high temperatures can deteriorate tools.

7.3.2 Steels

Steels with very low carbon (commercially pure iron and steels with carbon content up to 0.15%C) have poor machinability by all criteria, because of their high ductility. A very low shear angle and large contact area on the rake face have been observed. The chips tend to adhere to the tool surface and may become hard to break, which often causes trouble. The high strain generates more heat, which requires more energy input and results in a higher temperature on the tool surface. Surface finish is also difficult to control because of rubbing.

The added alloys to low-carbon steel improve machinability in certain instances, depending on which elements are added and in what quantity. The increased carbon content decreases the ductility of steel, so that the energy consumption and required force are reduced. A more significant improvement can be achieved in surface finish. Although the heat that is generated diminishes, the compression stress on the rake does not change much, because the contact area is also reduced. The highest tool-temperature point on the rake face moves closer to the tool edge. The added carbon content also improves the work-material's strength and hardness. For steels with a carbon content of greater than 0.30%, the power input and tool-surface temperature increases for steel of lesser carbon content for the same machining conditions. The surface finish improves as the carbon content goes up to 0.35%, but if the carbon content goes above 0.35%, the quality of the surface finish begins to go down again.

Other alloys added to low-alloy steels (manganese, chromium, and so on) increase the strength and hardness of steel. In general, the tool wear rate increases with alloy content, but other machining characteristics remain unchanged. Machinability is also strongly affected by heat treatment. As a general rule, the material should be treated to the minimum hardness requirement.

One important problem in cutting steel is variations in tool life. When steel is cut with steel-grade carbides at high speeds, major variation in tool life is attributed to the nonmetallic inclusions that attach to the rake face and form an unstable glassy barrier. Although the phenomenon is not as pronounced in WC–Co or HSS tools, these tool materials do not cut steel effectively. Thus arises the need for free-machining steels.

7.3.3 Free-Machining Steels

Free-machining steels are typically alloys of steel and sulfur, lead, or some other suitable alloying agent. The addition of 0.1 to 0.3% S or 0.1 to 0.35% Pb or a small amount of Bi (bismuth), Se (selenium), Te (tellurium), and P (phosphorus) can greatly reduce force, power input, tool-surface temperature, and tool wear rate. Surface finish and chip control can also be improved. Most importantly, the tool performance is more consistent, because the tool is less sensitive to a detailed heat history. The difference in tool life may be very large for material with standard specifications, but quite different for non-free-machining steels. However, the mechanical properties of free-machining steels are not as good as those of regular steel, and they cost more. A compromise can be made among material cost, operation cost, and product performance. The manganese content of these steels must be high enough to ensure that all the sulfur present is in the form of MnS.

7.3.4 Stainless Steels

Stainless steels have three major types of microstructure: austenitic, ferritic, and martensitic. All of these materials have higher tensile strength and a greater spread between yield and fracture strength than do low-alloy steels. The energy input and the temperature on the tool surface are also higher than in ordinary steels. Due to their high alloy content, the stainless steels contain abrasive carbide phases. Both of these characteristics produce faster tool wear.

In addition to exhibiting the preceding properties, austenitic stainless steels possess strong work-hardening capability and low thermal conductivity, severely reducing the machinability measures for all criteria. Not only will higher temperatures be encountered, but the chips tend to bond to the tool surface and are hard to break. Specifically, problems arise when the tool edge is cutting into a work-hardened surface left by a previously machined surface. Because of this, a sharp tool, a reasonable feed rate, and a reasonable depth of cut are recommended to avoid excessive wear caused by continuously cutting a work-hardened surface. In order to improve the generally low machinability of austenitic stainless steels, sulfur, selenium, and tellurium are added to reduce ductility. When this is done, the resulting steels are called free-machining stainless steels. They are more expensive, and the additives reduce their resistance to corrosion slightly.

7.3.5 Cast Irons

Flake graphite and spheroidal graphite cast irons have good machinability with respect to all criteria. The graphite flakes and spheres initiate fracture on the shear plane at frequent intervals. The process has a low tool wear rate, a high MRR, low force, and low power consumption. A good surface finish can be easily achieved. The chips fall in very short segments for flake graphite and easy-to-break longer segments for spheroidal graphite. Tool life decreases mainly with hardness. In recent years, there have been more applications for ceramic tools run at a very high speed. Due to the high production rate and high surface finish achieved, additional grinding operations can be eliminated. Another ceramic tool application on cast iron is for chilled irons with hardness of 430 HV at speeds up to 50 m/min. CBN can also be used for cast-iron hardness ranging from 55–58 Rc (600–650 HV) at speeds up to 80 m/min. A higher clearance angle is recommended for spheroidal graphite cast iron to eliminate extremely ductile flow-zone material from clinging to the flank.

7.3.6 Nickel-Based Alloys

Nickel-based alloys are among the most difficult materials to machine, because of their very strong work-hardening capability and hard abrasive-carbide phases. At much lower speeds than cutting steels, the tool temperature can reach the point at which plastic deformation and diffusion can take place. Because of this work hardening, the feed rate is very important. When it is too low, the tool is continuously cutting through work-hardened material generated by a previous cut. By contrast, at high feed rates, even if the theoretical surface finish is acceptable, the stresses on the tool may be too high and can cause catastrophic tool failure. As a compromise between two extremes, a feed rate of 0.18 to 0.25 mm is recommended. In order to eliminate rubbing effects, a positive rake should be used. WC–Co grades of carbide can be used at speeds up to 60 m/min. Steel-cutting-grade tools fail more rapidly, and even coatings have shown little advantage. However, in recent years, ceramics, CBN, and Sialon tools have been found to be more effective in cutting nickel-based, high-temperature alloys. The surface speed can be as high as 250 m/min.

7.3.7 Aluminum Alloys

Pure aluminum is highly ductile. The chip tends to adhere to the tool surface, where it becomes stringy and can be hard to break. It is difficult to achieve a good surface finish, especially at low cutting speeds. However, aluminum alloys have good machinability in almost all criteria. Cast-aluminum alloys with silicon as the main alloying element are the most important casting group. These alloys contain abrasive silicon particles, which can reduce tool life and therefore are more economically machined at lower cutting speeds and feeds than other types of aluminum alloys. The addition of copper can improve not only the material's strength, but tool life as well. Because ductility is reduced, the chips are easier to control. The aluminum–magnesium and aluminum–zinc–magnesium alloys all have good machinability. The cutting speed can go up to 300 m/min for HSS tools and 2000 m/min for WC–Co carbides. The machinability of wrought aluminum can be improved by the addition of low-melting-point insoluble metals, tin, bismuth, and lead. The flank wear is most pronounced in cutting aluminum alloys and is usually the measure of tool failure.

7.3.8 Copper and Its Alloys

Pure copper is similar to other pure metals and has poor machinability. Unlike other pure metals, however, copper with a very low alloy content is widely used in electronic components and fittings. The cutting speed of these small-size components are usually limited by the spindle speed (up to 140–220 m/min). A built-up edge does not occur in this type of application. The tool forces are very high due to the large contact area on the rake face and the low shear angle. However, surface finish and chip control can become a problem. For this reason, high-conductivity coppers are regarded as one of the most difficult materials to machine. In drilling deep holes, for example, the forces are often high enough to break the drill. The addition of lead, sulfur, and tellurium improves the machinability after cold working. For most operations, the important concern is chip control.

7.4 MACHINING OPERATIONS

"Machining" is used as a general term to cover several operations. Generally, turning is the process in which a tool cuts a rotating workpiece held in the chuck. (See shortly.) Turning can be further classified as facing, cutoff, contouring, and so on. Figure 7.7 shows common turning operations. These operations can be performed in a conventional lathe, turret lathe, screw machine, or turning center. In turning operations, the workpiece is held to a rotating spindle. Standard workholding devices called "chucks" are used to locate and support the workpiece. In a milling operation, a rotating tool is placed in contact with a stationary workpiece. Milling operations are performed on vertical and horizontal milling machines, gear-cutting machines, and machining centers. Milling operations can be classified as face milling, end milling, and slotting. Figure 7.8 shows these operations. Important concepts in milling are conventional milling (or up milling) and down milling. In conventional milling, the feed direction is against the cutting force. Constant pressure can be maintained, so backlash has no major effect on the process. However, the cutter usually starts by cutting a very thin layer of material. As explained earlier, the cutter can virtually rub against the workpiece surface, reducing both tool life and cutting stability. Nevertheless, modern machines have an antibacklash mechanism in the drive system. Up milling introduces no such trouble in the process. The advantage to conventional milling is more pronounced in materials with high ductility and work-hardening capability. Milling is an interrupted operation that usually demands higher tool toughness, a honed edge, or negative land on the rake face. Drilling, another machining operation, is a process in which a fixed-diameter cutting tool is fed into a workpiece. The nominal size of the hole is the same as the nominal size of the tool. Drilling can be performed on a lathe, a milling machine, or a drilling machine. Drilling and turning operations are illustrated in Figure 7.7.

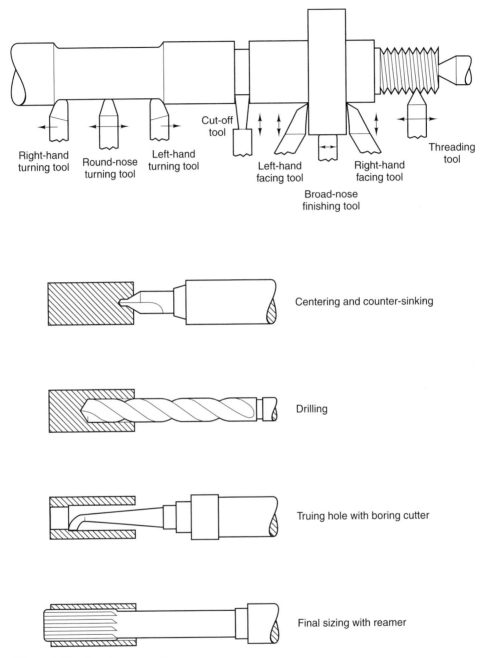

Right-hand turning tool

Round-nose turning tool

Left-hand turning tool

Cut-off tool

Left-hand facing tool

Right-hand facing tool

Broad-nose finishing tool

Threading tool

Centering and counter-sinking

Drilling

Truing hole with boring cutter

Final sizing with reamer

Figure 7.7   Turning operations.

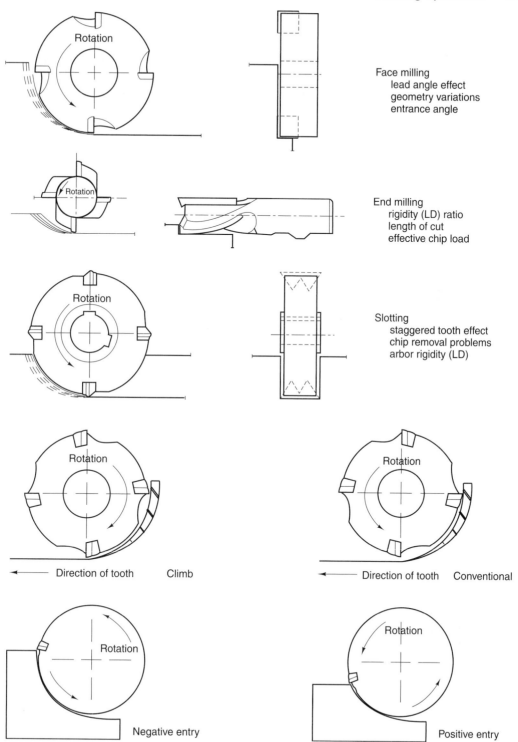

Figure 7.8   Milling operations.

7.5 WORKPIECE HOLDING PRINCIPLES

Fixtures are used to locate and constrain a workpiece in machining and other manufacturing operations. To ensure that the workpiece is produced according to the specified shape, dimensions, and tolerances, it is essential that it be appropriately located and clamped onto the machine tool. The configuration of a machining fixture depends not only on workpiece characteristics, but also on the sequence of machining operations, the magnitude and orientation of the expected cutting forces, the capabilities of the machine tool, and cost considerations. A fixture could be specially designed and fabricated for each workpiece (a dedicated fixture), or a fixture could be constructed from standardized fixturing elements chosen from a catalog (a modular fixture). Usually, it is expensive to design and build a fixture for each individual type of workpiece, so it is often considered satisfactory to select readily available fixturing elements from a catalog. Each fixturing element has a specific function, and a number of elements can be combined to build a complete fixture.

The traditional approach to designing fixtures for prismatic parts has been for the human designer to look at the workpiece drawing and to analyze the geometrical features from the viewpoint of obtaining the desired orientation and restricting the necessary degrees of freedom. A fixture designed from these initial considerations is further modified to conform to the machining sequence and to the configuration of the machine tool on which the part is manufactured. Other external issues, such as the mechanism for loading and unloading the workpiece (human–robot), setup times, chip disposal, and so on, will also influence the fixture design process. The ability to come up with a feasible solution will depend on the designer's experience, the designer's ability to recall fixture designs for similar workpieces, his or her knowledge of material-removal operations, and the workpiece's material properties. Obtaining a suitable design in this manner can be called *nonalgorithmic*, because it involves trial and error.

7.5.1 Fixtures and Jigs

General principles in the practice of fixture design are more or less agreed upon; however, a systematic structured design procedure does not exist. The relationship among the chosen locating and clamping scheme and the machining sequence, process parameters, and workpiece characteristics has not been adequately analyzed. A unified treatment of these issues is the key to building fixtures. We next survey some common principles relating to workholding and present the associated terminology.

In general, a workholding device serves three primary functions: location, clamping, and support. As regards *location*, the workpiece has to be correctly positioned with respect to the tool in order to maintain the specified tolerances. With respect to *clamping*, the position of the workpiece must be maintained while it is being subjected to cutting forces. Finally, in terms of *support*, the deflection of the workpiece due to the tool and the clamping forces must be minimized. Total workpiece control involves both linear equilibrium (a balance of forces) and rotational equilibrium (a balance of moments). The correct placement of locators, supports, and clamps enables this equilibrium to be achieved. In addition to the three primary functions, fixtures may perform the operations of centralizing and guiding. Where appropriate, as found in specialized fixtures called jigs, a special guiding system leads the tool to its precise position relative

Function	Symbol	Function	Symbol
Load			Guide
Locate			Longitudinal spacing
Centralize			Circular spacing
Position check			Measure
Clamp			Unclamp
Machine			Remove

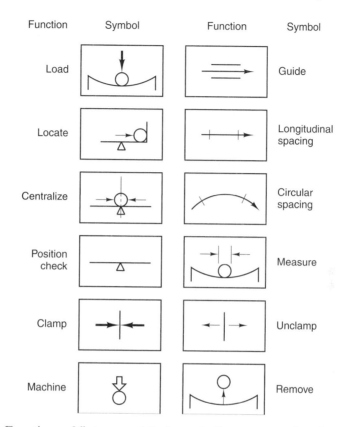

Figure 7.9   Functions of fixtures and their symbolic representation. (Based on Weck and Bibring. *Handbook of Machine Tools, Vol. 1*, 1984. Copyright John Wiley and Sons, Ltd. Reproduced with permission.)

to the work. An abstract description of the individual functions of a fixture is shown in Figure 7.9. To guarantee the exact relationship between the tool and the work, four important relationships, or "couplings," must be controlled [Weck and Bibring, 1984]. As shown in Figure 7.10, these are the relationships between (1) tool and toolholder, (2) toolholder and machine, (3) workpiece and clamp, and (4) clamp and machine. Among other criteria, certain practical considerations that indicate a good design are as follows:

1. Locating and clamping methods should reduce the idle time of the machine tool to a minimum.

2. The configuration of the locators and clamps should not interfere with the swept volume of the cutter. Avoiding collisions between the cutting tool and fixturing elements is imperative for the safety of the operator and to prevent damage to the machine tool or cutter.

3. Adequate clearance, in the form of channel ways, should be provided to allow for good chip clearance. This implies that awkward corners, wherein chips tend to

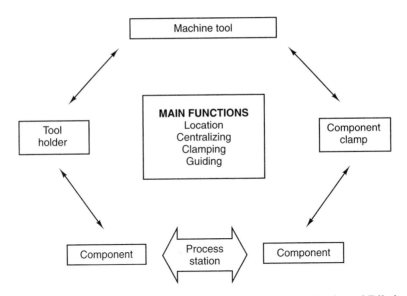

Figure 7.10   Four key fixturing relationships. (Based on Weck and Bibring. *Handbook of Machine Tools, Vol. 1*, 1984. Copyright John Wiley and Sons, Ltd. Reproduced with permission.)

collect, should be avoided. Similar considerations dictate the ease with which coolants will be able to access the cutting edges.

4. The design should be robust in order to withstand intermittent cutting and avoid vibration effects.

5. The design should be foolproof; that is, within reason, it should be impossible to insert the workpiece incorrectly into the fixture.

Table 7.4 summarizes the general considerations in fixture design.

7.5.2 Location

Location establishes a desired relationship between the workpiece and the fixture, which in turn establishes the relationship between the workpiece and the cutting tool. Weck and Bibring [1984] define "locating" as using faces of the component as reference planes. A free body in space has six degrees of freedom: a linear and rotational motion for each of the X-, Y-, and Z-axes. These six basic motions can occur in two directions each, for a total of 12 degrees of motion. Usually, locators will eliminate as many degrees of freedom as are possible, while still being able to perform the operation with the required accuracy. The most common method of location is 3–2–1, or the six-points principle. The first plane, which usually has the largest surface area, establishes the primary locating plane (the 3-plane) and is located by three points (Figure 7.11(a)). The surface with the next-largest amount of area generally establishes the secondary locating plane (the 2-plane) and is located with two locators (Figure 7.11(b)). The final locator is placed on the tertiary plane (the 1-plane), to complete the location of the part (Figure 7.11(c)). The 3-plane restricts two rotational

TABLE 7.4   Fixture Design Considerations

1. Locating considerations
    (a) Radial   (b) Concentric
    (c) From surfaces   (d) From points
    (e) Other

2. Positioning considerations (relation to tool and orientation in the fixture)
    (a) Indexing (linear and circular)   (b) Rotating
    (c) Sliding   (d) Tilting

3. Clamping considerations
    (a) Rapidity   (b) Amount of clamping forces
    (c) Direction of clamping forces   (d) Actuation (manual, power)

4. Supporting considerations
    (a) Relation to tool forces   (b) Relation to clamping pressure
    (c) Relation to thin-walled sections of workpiece

5. Loading considerations (including manual lifting and sliding; hoisting; unloading chutes, magazines)
    (a) Rapidity   (b) Ease
    (c) Safety

6. Coolant considerations
    (a) Direction

7. Chip considerations
    (a) Accumulation   (b) Disposal

Source: Wilson and Holt [1962].

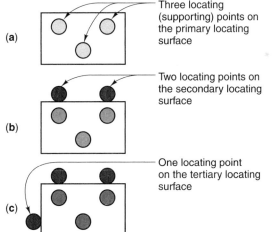

Figure 7.11   3–2–1 location: (a) 3-plane, (b) 2-plane, and (c) 1-plane [Hoffman, 1987]

motions and one linear motion, the 2-plane restricts one rotational and one linear motion, and the 1-plane restricts one linear motion.

The type of location is governed by the type of feature and the number of faces being machined. Locating arrangements for different production requirements are shown in Figure 7.12. To machine one face, control of dimension *a* is required; hence,

Partial location   Location   Full location

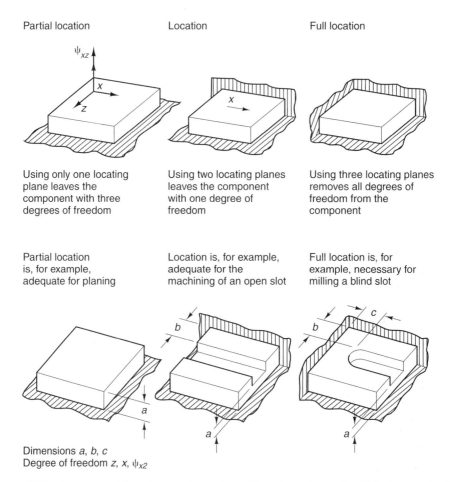

Using only one locating plane leaves the component with three degrees of freedom

Using two locating planes leaves the component with one degree of freedom

Using three locating planes removes all degrees of freedom from the component

Partial location is, for example, adequate for planing

Location is, for example, adequate for the machining of an open slot

Full location is, for example, necessary for milling a blind slot

Dimensions a, b, c
Degree of freedom z, x, ψ_{x2}

Figure 7.12 Degrees of freedom and number of locating planes for differing production requirements[a]

only one locating plane is necessary. Two locating planes are required for machining an open slot, as dimensions a and b need to be controlled. Full location (three planes) is necessary for milling a blind slot, as dimensions a, b, and c need to be controlled. The 3–2–1 principle is ideal for prismatic or rectangularly shaped workpieces only. For simple cylindrical or conical shapes, just five locators are needed, because one rotation is stopped purely by friction (Figure 7.13).

By the act of location, a machinist establishes a relationship between the features of the workpiece that are being machined in the given setup and other features being used as a reference, in order to locate the workpiece at the desired position in the machine-tool

[a]This material has been reproduced from Darvashi, A.R. and K. F. Gill, "Knowledge Representation Database for the Development of a Fixture Design Expert System," *Proceedings of the Institution of Mechanical Engineers, Part B, Journal of Management and Engineering Manufacture*, Volume 202: No B1, p 41, Fig. 5. Used with permission of the Council of the Institution of Mechanical Engineers.

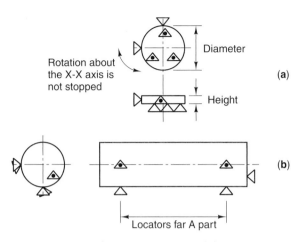

Figure 7.13   Workpiece control, (a) short cylinder, (b) long cylinder. (Based on Eary, D.F. & G.E. Johnson, *Process Engineering for Manufacturing*, 1962. Reprinted by permission of Pearson Education, Inc., Upper Saddle River, NJ.)

coordinate system. To determine the reference features or surfaces of the workpiece, it is vital to examine how the workpiece dimensions are specified on the blueprint. Incorrect placement of locators will cause the workpiece to go out of tolerance. It is important to (1) select the correct surfaces for placement of the locators and (2) position the locators correctly on the surface selected.

An excess of locators exists when more than six locators are provided for a prismatic workpiece in a single fixture. Four points of location, all on one surface, allow a workpiece to be clamped onto slightly different planes, which may be enough to throw the workpiece out of tolerance. A locator directly opposite another locator is also harmful, because the distance between the two locators may not be large enough to allow for size variation of the workpiece.

Location may involve centering. Whereas locating normally brings one surface of the workpiece into proper position relative to the fixture, centering is applied to two surfaces, usually locating a plane or an axis within the part. A *centralizer* is a combination of fixed and movable locators that provide positive contact and pressure without clearance. Typical combinations of locators and centralizers are shown in Figure 7.14. Double centering is accomplished whenever an axis is located. Locating requirements are achieved by providing plane location, concentric location, or radial location (Figure 7.15). In general, the following principles are commonly applicable:

1. Stability of the workpiece is best when the locators have the largest overall distance between them. This also diminishes the effect of surface irregularities, dirt, and locator wear on the position of the workpiece. However, locator spacing cannot be unduly large, in order to minimize workpiece deflection.

2. The center of gravity of the workpiece should be as low as possible and close to the centroid of the three locators in the 3–2–1 system.

3. Good dimensional control is achieved when locators are placed on one of the two surfaces on which the dimension is shown on the part drawing. When the drawing

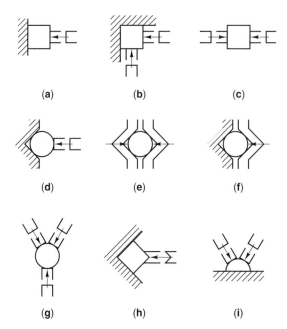

(a)   (b)   (c)

(d)   (e)   (f)

(g)   (h)   (i)

Figure 7.14   Centering operations: (a) single defined, not centered; (b) double defined, not centered; (c) single centered; (d) single centered; (e) double centered; (f) single centered; (g) double centered; (h) single centered; (i) single centered. (Based on Henriksen, E.K., *Jig and Fixture Design Manual.* New York: Industrial Press, 1973.)

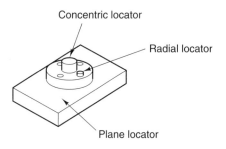

Figure 7.15   Three locating methods [Hoffman, 1987].

specifies close tolerances on parallelism, perpendicularity, or concentricity, more than one locator must be placed on one of the two surfaces to which the tolerance applies [Eary and Johnson, 1962].

4. Locating surfaces should not be larger than is necessary for proper support and wear. It is hard to keep larger surfaces clean and free of chips, the presence of which could cause inaccuracies. If possible, the locating surfaces should be completely covered by the workpiece, so that chips cannot fall on the locating points.

5. Ideally, locating surfaces should be fixed. Movable surfaces should be used for clamping only.

6. Buttons and pins—rather than flat surfaces—are preferable for locating, as they are easier to keep clean and afford easier adjustment for wear. Pins from previously drilled holes should be hardened and ground and have sufficient clearance for chips and burrs. These pins should be located from only one diameter when used with counterbored holes.

7. Foolproofing the locating arrangement is desired, so that it is not possible to place the part in an improper position [Hamner, 1982].

7.5.3 Supporting and Clamping

Once the workpiece is positioned by the locating arrangement, further control of the workpiece is necessary. The function of any clamping device is to apply and maintain a sufficient counteracting holding force to a workpiece while it is being machined. The workpiece may deflect, within its elastic limit, due to the cutting forces, clamping forces, or its own weight. Excessive clamping and cutting forces may also cause distortion of the workpiece—that is, deflection beyond the elastic limit. As defined by Eary and Johnson [1962], a support is a device to limit or stop the deflection of a workpiece. Supports are either fixed, adjustable, or equalizing. The fixed support is placed away from the locating plane, against the locators, so that the workpiece will not contact the support.

Under the influence of cutting forces, the workpiece is permitted to deflect and contact the fixed support, thus limiting deflection. An adjustable support is suitable for the uneven surfaces that are typical of castings. This type of support is kept away from the workpiece during location, but is adjusted to contact the workpiece after it has been clamped. Equalizing supports are connected units in which the depression of one point causes the other points to rise and maintain contact.

The clamping arrangement must force the workpiece to make contact with all locators if it is not already in contact with them. Clamps must also maintain the workpiece contact at all locators, in spite of cutting-force variations, inertial forces, dead weight, and vibrations in the machine–fixture–tool–workpiece (MFTW) system. To ensure correct clamping of the workpiece, the following points may be considered:

1. The design must permit the clamps to act against the locators, so as to minimize the deflection due to clamping forces. For geometries in which clamping forces cannot be applied opposite the locators, supports may be used to control the deflection caused by the clamping forces. A recent method of clamping the workpiece is the self-adapting fixture element (SAFE), which uses flat contact areas of hardened steel balls that are free to swivel within their sockets [Smith, 1982; Drake, 1984; Kuznetsov, 1986a]. The balls automatically accommodate irregularities in the workpiece surface and provide contoured support without causing distortion.

2. The clamping scheme should be such that maximum cutting forces are directed toward the solid part of the fixture body and not toward the clamp.

3. Cutting forces should be absorbed by a fixed locator and its support and not by friction between the work and the clamp.

4. The surface quality of the workpiece may dictate placing the clamps on noncritical surfaces.

Power clamping, which uses hydraulic or air-hydraulic-actuated components to provide the clamping force, is widely used ["Fast Clamping," 1975]. A number of U.S. companies provide off-the-shelf components for power clamping systems [Carr Lane, 1988; De-Sta-Co, 1988; Owatonna Tool Co., 1988; Vektek, 1988]. For a description of the various clamp designs used for NC machines, see Kuznetsov [1975].

7.6 PART SETUP OR ORIENTATION

The features of a workpiece are machined in a certain sequence, as given in its process plan. These features (pockets, slots, holes, and so on) are located on different sides (faces) of the workpiece. Usually, not all the features can be accessed (machined) while the workpiece is in a given orientation. (An *orientation*, or *setup*, refers to a unique locating, supporting, and clamping configuration). Every machining operation has a fixturing configuration that is best for the operation, but may not be a practical solution. One would like to find a subset of such feasible fixturing configurations whereby all the operations can be performed within those configurations. Therefore, it is necessary that the machining operations be grouped into setups. Grouping depends on the geometry of the individual features and the availability of cutting tools. Each setup will have a so-called primary positioning face, usually the 3-plane, on which the workpiece rests, and also the remaining positioning planes (the 2-plane and 1-plane). In addition, a setup will also have a unique clamping scheme. Thus, each orientation requires a separate fixture. Because changing from one setup to another involves unclamping or removing the workpiece from one fixture and locating or clamping it into another fixture, the number of setups (fixtures) should be minimized to get the best accuracy.

The face of the workpiece, used as reference for any machining operation in a given setup, is sometimes called the *machining reference face*. In some cases, the machining reference face could also serve to maintain the stability of the part. Inui et al. [1985, 1987] have given some guidelines on how the candidates for the machining reference face may initially be selected. Subsequently, the authors' system evaluates candidate faces for stability of the machined part when it is resting on any one of those faces. Candidates with stability not good enough to support the part are eliminated from consideration. It is recommended that the workpiece be supported by a face with as few bounding faces as possible. Other resting faces of the workpiece are determined in the same manner.

A number of rules for part orientation planning have been proposed and formalized [Ferreira et al., 1985; Ferreira and Liu, 1988]. One example of such a rule, as incorporated in a rule-based system by the authors, is "A workpiece is stable in an orientation if the vertical projection of its center of gravity passes through the convex hull of its support (base) face." A convex hull is obtained by connecting the outermost vertices of a part. A number of "basic" and "optimizing" objectives are identified to search for solutions in a given situation. Depending on the particular manufacturing environment, the objectives can change. For example, in large-scale production, a maximum number of operations should be performed in one setting. However, ease of workpiece restraint is the primary objective in small-batch sizes. For machining to tight tolerance specifications, the attainable accuracy will dictate the orientation of the workpiece. Stability in the resting position, against gravity, becomes important for large and heavy workpieces.

Guidelines for selecting the setup, based on orientation and tolerance relationships, have been proposed by Boerma and Kals [1988, 1989]. Their procedure incorporates the dual objectives of (1) reducing the number of "critical" tolerances in the geometrical relations between features belonging to the different setups and (2) keeping the number of setups to a minimum. In order to compare the significance of the different kinds of tolerances in the relations between features, a nondimensional "tolerance factor" is introduced. It is assumed that the smallest tolerance factor determines the maximum permissible rotation and translation errors of the part during fixturing. The two features, related by the smallest tolerance factor, determine the two orientations of the setup base. The third orientation is selected on the basis of the tolerance-factor relationship of a third feature with either of the first two. The procedure is somewhat simplistic because it ignores all tolerance relationships except tolerances of position, parallelism, and perpendicularity.

7.7 FIXTURING FOR NC MACHINING

For a fixture to be cost effective, it has to be used for large-batch sizes or be adaptable to different part geometries. Welded fabrications are commonly used as fixtures in production shops, but such custom-built fixtures are relatively expensive and can be limited to a single application. To eliminate the need for single-purpose fixtures, the use of standardized fixtures is quite popular. Tombstone tooling blocks, angle plates, parallels, V-blocks, riser blocks, and subplates are the basic components of standardized fixtures [Gouldson, 1982; Boyes, 1986]. A tooling block or a baseplate can be mounted precisely on the machine table. These components have a predetermined grid pattern of tapped holes to accept studs, clamps, and other fixture components, as shown in Figure 7.16. The built-up fixture can be removed and replaced exactly in the same position on one or more machines. There is also the provision of a tram hole, which is a bushing in the fixture base, at some known distance from the part location point. The tram hole establishes that point on the fixture from which all part dimensions are located.

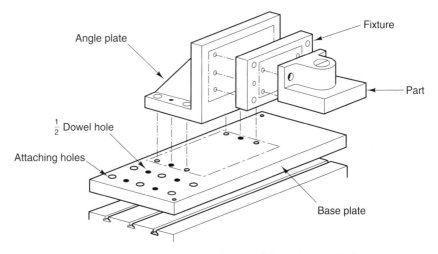

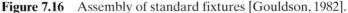

Figure 7.16   Assembly of standard fixtures [Gouldson, 1982].

Subplates can aid in reducing setup time, because they are usually designed and manufactured for the customer's machining center. A subplate needs to be aligned with the machine axes, by using an indicator in the machine spindle, only during the initial installation. Once the subplate is locked into position, the indicator is not used again. Each modular component of the fixture, which is dowel pinned and screwed to the subplate, is accurately aligned with the machine axes.

The design of the fixture also depends on whether it is a first-operation fixture or a second-operation fixture [Gouldson, 1982]. A first-operation fixture is used for a workpiece on which no machining has yet been performed. Such a fixture may be constructed with the use of adjustable supports and locators in conjunction with a predetermined target point or line. If possible, a premachining operation is carried out on a conventional machine to establish a datum surface or locating holes. A second-operation fixture is used where some machining has taken place, so that reference surfaces or holes exist that position the part with respect to the fixture and the fixture with respect to the machine table. Such a fixture is comparatively easier to design because an adequate reference surface is already available. Case studies on fixtures designed for NC machine tools can be found in Hatschek [1977].

A flexible manufacturing system (FMS) involves a group of computer numerical control (CNC) machine tools that perform a number of operations and manufacture many different parts. A material-handling system, such as an industrial robot, can transfer parts from one machine tool to another, provided the part geometry is not complicated. For heavier components and complex geometries, pallets are used. Fixtures are mounted on pallets that travel from one machine tool to another. With the work material already loaded in the fixtures, the pallets are held in the magazines of flexible manufacturing modules (FMMs) or in a central buffer store, from where they are delivered to the machines in the desired sequence [Karyakin, 1985]. Multisided or cubic fixturing is frequently used when machine uptime is important and the workpieces are relatively simple [Kellock, 1986]. In an FMS, 2 to 12 components of the same type or different types can be loaded at one time for palletized transportation. This kind of environment precludes the use of special-purpose fixtures from a cost viewpoint, because a large number of such dedicated fixtures would be required. This lack has led to extensive research in adaptable, or flexible, fixturing.

A flexible (adaptable) fixture has been defined as a single device that holds parts of various shapes and sizes that are subjected to the wide variety of external force fields and torques associated with conventional manufacturing operations [Gandhi and Thompson, 1985a]. Flexibility is a property of the fixturing device that makes it conform to the workpiece geometry. Flexible fixtures offer the following advantages [Thompson, 1984]: (1) reduction in lead time and effort required for designing special fixtures; (2) lower overhead cost of storing a multiplicity of fixtures required to effect a rapid changeover between different manufacturing operations; and (3) simpler programming requirements. Designs of various "resettable" fixtures for use in an FMS have been proposed [Eremin, Lysenko, and Nemytkin, 1988]. These fixtures have a common location scheme for a group of workpieces, and just resetting the clamping element is required when a new workpiece is introduced. Some special problems encountered in building fixtures for the automated factory have been outlined by Bagchi and Lewis [1986]. Zimmerman [1984] and Kuznetsov [1986a, 1986b] describe case

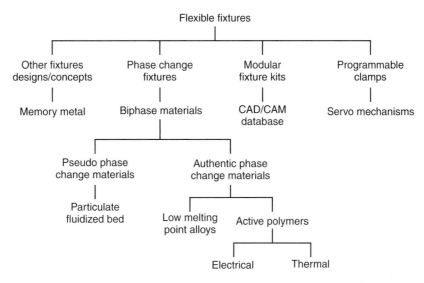

Figure 7.17   An overview of flexible fixturing methodologies [Gandhi, Thompson, and Maas, 1986].

studies on fixtures designed and built for use in FMSs. Noaker [1988] observes that, even though research and development has been ongoing in flexible fixturing systems over the last five years, the most successful systems for tough problems have cost as much as the capital equipment.

Many different approaches have been tried in flexible fixturing, as shown in Figure 7.17. A number of individuals [Thompson, 1984; Gandhi and Thompson, 1985a; Grippa, Thompson, and Gandhi, 1988; Youcef-Toumi and Buitrago, 1988] have surveyed flexible-fixturing methodologies. Broadly, there are two major groups of flexible fixtures: discrete contact and continuous contact. In the discrete-contact type, there are a finite number of contact points that can be arranged in space to give different configurations. A continuous-contact fixture is a fixture in which the number of contact points is infinite, such as a line or area contact. A point contact would completely constrain the motion in a direction normal to the workpiece surface only. Motions parallel to the workpiece surface would not be completely constrained, because of the limited friction in point contact. Surface contact would not only constrain the motion of the workpiece along the three axes, but also would constrain the applied moments. We next outline the salient features of some flexible-fixturing techniques.

7.7.1 Modular Fixtures

By incorporating a system of interchangeable and reusable components, it is possible to accommodate a wide variety of workpieces. Modular tooling systems employ a system of individual components, assembled on a base, to suit the workpiece that requires fixturing. These systems are typically used for prototype tooling, for short production runs of a limited number of parts, or as a backup work holder to replace dedicated tooling that requires repair. By using a number of standard parts, and by eliminating the use of special parts as much as possible, the time required to design a complete work holder can be significantly reduced. The assembly time is also a fraction of the time required to build a dedicated fixture. Many companies have eliminated the use of dedicated fixtures, except where they are absolutely essential, and found considerable savings in modular tooling systems [Koch, 1988a, 1988b].

Modular fixtures permit the design of a fixture parallel with part design, because most modular system components can be accessed and retrieved from a CAD database. For example, elements of the CATIC modular fixturing system have been coded and stored in a database called PALCO-FIXTURE [Ranky, 1983]. This fixture database incorporates graphics information and mating-surface descriptions for each element. Modifications in the modular fixture, due to alterations in the part geometry, are comparatively simple. Depending on the basic construction, modular fixturing is classified into three major systems: a subplate system, a T-slot system, and a dowel-pin system.

Subplate systems are comprised of a baseplate or subplate to which all other components are attached. A subplate can be a simple plate that is either drilled or tapped to accept threaded fasteners; or it can be machined with T-slots to accept nuts and bolts. The subplate may also be mounted vertically and may be single sided or multisided. Subplate systems do not typically include accessories such as locators, supports, and clamps; these must be purchased separately. Common applications of subplate systems are for large parts, irregular parts, or short runs of simple parts. Parts can be mounted directly on the subplate, or the subplate can be used as a mounting device for other work holders, such as box parallels, V-blocks, toolmaker's knees, and slotted angle plates. The Challenge System, ATCO System, Matrix Positioning System, Midstate System, and Stevens Modular Tooling Systems are the premier systems in this category [Hoffman, 1987].

T-slot modular systems are rectangular, square, or round baseplates across which T-slots are machined exactly perpendicular or exactly parallel to each other. Round baseplates usually have tangential and radial slots. T-slot systems are a complete fixturing set or system; that is, each element within the set is designed to be used to completely build or assemble an entire work holder. The advantage of T-slot systems is in the flexibility of positioning the various parts. Principal systems in this category are the Halder system [Krauskopf, 1984; Erwin Halder KG, 1987], CATIC modular fixturing system [Lewis, 1983; Xu et al., 1983], Warton Unitool system [Hoffman, 1987], Block Build Jig system 64 [Horic, 1988], and Cessloc system [Hoffman, 1987].

Dowel-pin modular tooling systems are the newest form of modular component tooling. Examples of this category are the QU-CO system [Qu-Co, 1987], Bluco Technik Modular system [Quinlan, 1984] SAFE system [Smith, 1982], and Yuasa Modula-Flex Fixturing system [Hoffman, 1987]. All these systems use some combination of precisely positioned dowel holes along with tapped holes to accurately align, locate, and secure fixturing elements. The holes are arranged in a rectangular pattern in a baseplate and larger structural elements. Like the T-slot systems, dowel-pin systems incorporate all locating and workholding elements into a single set. However, this set does not include the baseplate, which the user can buy or machine separately. Dowel-pin systems have the advantage of consistently accurate placement of each fixturing element at fixed coordinate points, so these systems are ideally suited for use with CAD, NC machining, and inspection systems and for assembly by robots.

Special T-slot overlay elements are available for dowel-pin patterns, giving both T-slot flexibility and dowel-pin accuracy [Quinter, 1988].

In addition to the baseplates or mounting plates, modular fixturing elements include locating and supporting elements, mounting blocks, and clamping elements. The locating and the supporting elements closely resemble those used in conventional jigs and fixtures. Mounting blocks are a form of locating and supporting elements that are used primarily to position locators or clamping devices at specific heights off the mounting base. The commonly used clamping elements are strap clamps and screw clamps. Generally, all fixturing elements are made of high-grade alloy steel to tolerances of ±0.005 to ±0.01 mm in flatness, parallelism, and size. These tolerances assure accurate alignment and referencing. Modular fixtures are increasingly using some type of power clamping based on air or oil pressure. Devices for power clamping are clamping cylinders, swing clamps, rotating or pivoting clamps, retracting clamps, toe clamps, and power toggle clamps. Power-operated positioning and supporting devices are also commercially available.

A cost comparison for FMSs showed that modular fixturing yielded up to 75% savings over dedicated fixturing [Lewis, 1983]. Significant reductions in lead times were also observed. Cost equations for economically justifying modular fixtures have been developed [Xu et al., 1983; Friedmann, 1984].

7.7.2 Phase-Change Fixtures

The phase-change fixture exploits the property of some materials to change from liquid to solid and back to liquid again. The workpiece is immersed to the desired depth in the material while it is still liquid. The special material is then changed to solid by altering certain conditions, and machining of the workpiece is carried out. To remove the workpiece from the fixture, the immersion material is liquified again. These materials are of two kinds: those which undergo a pseudo phase change, such as particulate fluidized beds, and those which undergo a true phase change, such as low-melting-point alloys. The evaluation of phase-change materials is based on the following properties [Gandhi and Thompson, 1984]: (1) The phase change must be rapid, reversible, and uniform in space and time; (2) there should be no adverse effect on the geometry or the surface properties of the workpiece; and (3) the power required to initiate the change of phase must be carefully quantified. Particulate fluidized-bed fixtures [Gandhi and Thompson, 1985a, 1985b; Gandhi, Thompson, and Maas, 1986; Thompson, Gandhi, and Desai, 1989] are based on the two-phase property of a particulate fluidized bed. As shown in Figure 7.18, the fixture consists of a porous bottom container that is filled with spherical particles of a material. When air is supplied at a controlled rate through the porous base, the particulate bed achieves a fluid state of dynamic equilibrium, permitting the workpiece to be immersed. On switching off the air supply, the particles compact under gravity, thereby fixing the workpiece. After machining, the air supply is switched on and the workpiece is removed. This fixture falls in the intermediate category between the discrete-contact and the continuous-contact type of flexible fixtures. The degree of performance of such a fixture can be measured by its ability to resist extraction forces in the vertically upward direction, because lateral movement of the workpiece is already constrained by the compacted particulate bed. Experiments [Gandhi, Thompson, and Maas, 1986] indicate that the extraction force F may be maximized by maximizing the depth of immersion, using a bed material of the largest specific weight, and by choosing a bed material with a

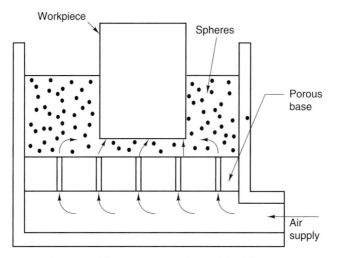

Figure 7.18   Schematic of a fluidized-bed fixture [Gandhi, Thompson, and Maas, 1986].

high coefficient of friction for the given workpiece material. Mathematical models to calculate the extraction force F and computer simulations to examine the role of different bed-material parameters and loading conditions have been devised [Gandhi and Thompson, 1985b, 1985c; Thompson, Gandhi, and Desai, 1989].

Low-melting-point alloys have been used to partially encapsulate the workpiece prior to machining. These bismuth-based alloys, which have melting points as low as 47°C, tend to expand on solidification. The technique has been applied to the manufacture of awkwardly shaped high-tolerance components, such as gas turbine and compressor blades [Nyamekye and Black, 1987], and for grinding wedges for milling cutters [Kellock, 1986]. An injected metal encapsulation machine is the main feature of this system. The workpiece, which may be an unmachined blade, is precisely positioned with locators into a cavity in the machine prior to the injection of a molten alloy. As soon as the alloy solidifies, the encapsulated blade is ejected from the injection machine. The block of alloy is thus cast in a shape that gives the reference and clamping surfaces required during machining. After machining, the capsule is mechanically cracked open and the alloy is reused. Because a different die is required for each workpiece, the use of low-melting-point alloys is generally quite expensive.

Electrically induced phase-change fixturing (EPF) is an innovative concept that employs modern polymeric materials such as polycrylonitile as the medium undergoing the phase change [Gandhi and Thompson, 1985b]. In these materials, a phase change is induced by an electric field. This class of fixtures has the advantage of a limited compliance in the solid phase, which is beneficial in robot assembly.

7.7.3 Flexible Fixturing Requirements

Among other less common fixturing schemes are the Multi-Leaf Vise and the Petal Collet [Thompson, 1984], electromagnetic chucks [Kellock, 1985], magnetic cubical fixtures [Kellock, 1986], and electrostatic fixtures [Tazetdinov, 1969]. A survey of various magnetic force applications in work holding has been conducted ["Magnetic Work Holding," 1973–1974]. Although the choice of a flexible-fixturing technique will depend on the application, it should satisfy certain requirements in order to function successfully in a flexible manufacturing cell. Some key requirements are as follows [Youcef-Toumi and Buitrago, 1988, 1989]:

1. *Reconfiguration of conformable surface.* The intermediate medium, between the rigid part of the fixture and the workpiece surface, should have the dual properties of compliance and stiffness, with a quick "phase" change between the two.

2. *Clearance from machining paths.* Flexible fixtures, by their nature, tend to occupy larger surface areas of the workpiece in certain locations. Therefore, the design of the fixturing layout is more critical than in conventional fixtures.

3. *Clearance for workpiece loading and unloading.* The workpiece must be located and referenced to fixed supports, to adapt to the workpiece surface before the conformable surfaces of the fixture approach. Thus, some initial reference surfaces should be provided to place the workpiece in the future.

4. *Ease of operation by a robot manipulator.* An example of this property can be found in modular fixtures in which special mating surfaces and locking devices guarantee positive location of the modules, thereby significantly reducing the accuracy requirements of the robot.

5. *Actuation.* The fixture should be self-contained; that is, it should contain at least one actuation element that provides the fixturing force. Actuating forces could also be provided by a robot.

7.8 ERRORS DUE TO LOCATION, CLAMPING, AND MACHINING

Tolerances are specified on a drawing, because it is not possible to manufacture a part exactly to the specified dimensions. The zone in which the outline of the finished part is to lie is provided by the tolerance specification. A prismatic part may be considered to be placed in a system of three mutually perpendicular planes called *datums*. A datum is a theoretical plane from which dimensions are specified. The surface in contact with the datum is called the *datum surface*. Tolerances control not only the dimensions, but also the geometrical properties of the part, such as flatness, perpendicularity, parallelism, straightness, runout, and surface roughness. It is the primary objective of the process planner to produce the part to the desired tolerance (accuracy); therefore, any fixture design procedure will have to take tolerances into consideration.

Location (mounting) errors have a significant effect on workpiece accuracy. These errors may be considered to be the inaccuracy of the workpiece position with respect to the pallet surface, cutting toolholder, machine tool guideways, and so on. Rigidity of the locating elements and of the workpiece material will also give rise to mounting errors. (For an analysis of rigidity effects, see Shuleshkin and Gromov [1960].) Another important source of inaccuracy is the nonsimultaneous application of clamping forces on the workpiece. Both mounting and clamping errors combine to result in tilting and turning moments and also shear forces that displace the workpiece with respect to the desired position. Bazrov [1982] has formulated criteria that show

how the overall accuracy attained in machining is influenced not only by the precision mounting of the workpiece, but also by the choice of coordinate systems employed for setting the pallet, cutting tool, fixture elements, and so on. Analyses and experiments to study the influence of the sequence of clamping forces on the accuracy of the workpiece have been conducted by various investigators. Bazrov and Sorokin [1982] established how a rational sequence of clamping forces, force magnitudes, and friction coefficients may be selected to adjust the workpiece displacement in the desired direction. Experiments by Batyrov [1984, 1986] show that adverse quality of the surfaces in contact between the fixture and the workpiece may enhance the errors caused by clamps being actuated in a particular sequence.

For a part being produced on a machining center, the operations performed most commonly are side milling, end milling, and hole making. Knowledge of the mechanics of each of these processes is vital to fixture design. Whereas a clamping force is static and fixed in both direction and magnitude, the cutting force is dynamic, with a magnitude and point of application that vary during the machining operation. These will require the development of rules or strategies to ensure that the position and orientation of clamps and supports will suffice to locate and hold the workpiece, so that it does not deflect more than a certain "critical" amount. This "critical" deflection limit will depend on workpiece material properties and machining parameters. During the metal-removal operation, the workpiece deflects and assumes a certain profile, due to the bending and torsional deflection produced in the workpiece by the machining operation. After machining, the workpiece will "spring back" and assume a profile that will differ both from its original profile and from the one assumed during machining. It is this final profile that will determine the acceptability of the workpiece in terms of the specified tolerances on its blueprint.

7.9 SUMMARY

In this chapter, the basics of tool engineering have been presented. Although several analytical models for jig and fixture design have evolved over the past decade, fixture design is still an "art" that can dictate the profitability of a manufacturing process. Tooling and fixturing also impose some of the most rigid flexibility constraints in today's flexible manufacturing environment.

7.10 KEYWORDS

3–2–1 principle	Alloys
Brinell hardness number (BHN)	Carbide
Carbon steel	CBN
Ceramic	Chip thickness
Cutting force	CVD
Datum surface	Deformation zone
Depth of cut	Feed
Fixture	Flank
Flank wear	Free-machining steel

High-speed steel, HSS

Machining parameters

Phase-change fixture

Rake angle

Speed

Taylor equation

Tool life

Jig

Modular fixture

PVD

Setup

Stainless steel

Tool failure

Tungsten carbide

7.11 REVIEW QUESTIONS

7.1 Define the following terms: *rake face, flank face, rake angle, cutting edge,* and *shear angle.*

7.2 Describe the situations in which positive and negative rake angles are used.

7.3 What are the major causes of rake wear and flank wear?

7.4 What is the purpose of coating HSS tools? What kind of materials are used in coating?

7.5 What are the most commonly used carbide substances for tools?

7.6 Why is a negative rake angle always used with ceramic tools?

7.7 What are the advantages of using ceramic tools?

7.8 What are the most commonly used criteria in measuring machinability?

7.9 How does one determine that a tool has failed?

7.10 What is free-machining steel?

7.11 What is a jig and what is a fixture? What are they used for?

7.12 List important considerations in fixture design.

7.13 What is the 3–2–1 principle?

7.14 What are the six points in 3–2–1 used for?

7.15 After the six points have been determined, how do we ensure that the workpiece will not move?

7.16 What is the relationship between locating points and datum surfaces?

7.17 What is the basic principle in selecting supporting methods?

7.18 What principle is involved in selecting clamping arrangements?

7.19 What is a modular fixture?

7.20 What are the advantages and disadvantages of using modular fixtures?

7.21 What kind of fixture is used in a flexible manufacturing system? What are the differences between this kind of fixture and conventional fixtures?

7.22 Design a fixture for the following operation: The flat surface of a disk 4 inches in diameter and 0.5 inch tall is to be milled. The disk is made of steel.

7.23 Design a fixture to hold the part shown in Figure 2.38.

7.24 Ten thousand of the parts shown in Figure 2.39 are to be machined. The true position tolerance is 0.001 inch. The operation needed is to drill the two holes. Design a jig–fixture for the process.

7.12 REVIEW PROBLEMS

7.1 Design a drilling fixture for the part shown in Figure 3.9. Assume that 5000 parts will be produced on a radial drill.

7.2 If the part shown in Figure 3.9 was to be a very low-volume part (only four to five parts were to be produced), what kind of fixture would be used?

7.3 Suppose the same part shown in Figure 3.9 was to be produced in high volume (more than a million). What kind of holding device and machine would be used?

7.13 REFERENCES

American Society for Metals. *Metals Handbook*. New York: 1985.

Applied Power, Inc. *S.A.F.E. by Enerpac,* Product Catalog. Butler, WI: API, 1988.

Bagchi, A., and R. L. Lewis. "On Fixturing Issues for the Factory of the Future," in *Proceedings of the 1986 ASME International Computers in Engineering Conference and Exhibition*. New York: American Society of Mechanical Engineers, 2 (1986):197–202.

Batyrov, U. D. "Factors Affecting Setting Errors of Pallet Fixtures," *Soviet Engineering Research,* 4(4), 1984, 67–69.

———. "More Accurate Clamping of Pallet Fixtures," *Soviet Engineering Research,* 6(9), 1986, 58–60.

Bazrov, B. M. "Selection of Support and Reference Surfaces (Locations) for Mounting the Exchangeable Elements of the MFTW System," *Soviet Engineering Research,* 2(5), 1982, 94–98.

Bazrov, B. M., and A. I. Sorokin. "The Effect of Clamping Sequence on Workpiece Mounting Accuracy," *Soviet Engineering Research,* 2(10), 1982, 92–95.

Boerma, J. R., and H. J. J. Kals. "FIXES, A System for Automatic Selection of Set-Ups and Design of Fixtures," *Annals of the CIRP,* 37(1), 1988, 443–446.

———. "Fixture Design with FIXES: The Automatic Selection of Positioning, Clamping and Support Features for Prismatic Parts," *Annals of the CIRP,* 38(1), 1989, 399–402.

Boulger, F. W., H. A. Moorhead, and T. M. Govey. "Superior Machinability of MX Steel Explained," *Iron Age,* 1678, 1951, 90–95.

Boyes, W. E., Ed. *Low-Cost Jigs, Fixtures and Gages for Limited Production*. Dearborn, MI: Society of Manufacturing Engineers, 1986.

Carr Lane. *Product Catalog*. St. Louis: Carr Lane Manufacturing Co., 1988.

Colding, B. N. "Machinability of Metals and Machining Cost," *International Journal of Machine Tool Design Research,* 1, 1961, 220–248.

Czapliki, L. "L'usinabilité et la coupe des metaux," *Res. Soc. Roy. Belge Ingeniere,* 12, 1962, 708–730.

Darvishi, A. R., and K. F. Gill. "Knowledge Representation Database for the Development of a Fixture Design Expert System," *Proceedings of the Institute of Mechanical Engineers,* Part B, 202(B1), 37–49, 1988.

De-Sta-Co. *The World of Clamping,* Product Catalog. Troy, MI: De-Sta-Co., 1988.

Drake, A. "Fixture Design: Working with Modules," *Manufacturing Engineering,* 92(1), 1984, 35–38.

Eary, D. F., and G. E. Johnson. *Process Engineering for Manufacturing*. Englewood Cliffs, NJ: Prentice Hall, 1962.

Eremin, A. V., N. V. Lysenko, and S. A. Nemytkin. "Designing Fixtures for FMSs," *Soviet Engineering Research,* 8(1), 1988, 84–86.

Erwin Halder, K. G. *Halder Modular Jig and Fixture System* 70, Parts Catalog. Berlin: Erwin Halder KG, 1987.

"Fast Clamping, More Machining—They Go Together" (1975). *Automation,* 22, 1975, 42–44.

Ferreira, P. M., B. Kochar, C. R. Liu, and V. Chandra. "AFIX: An Expert System Approach to Fixture Design," in *Computer Aided Intelligent Process Planning, ASME Winter Annual Meeting.* New York: American Society of Mechanical Engineers, 1985, 73–82.

Ferreira, P. M., and C. R. Liu. "Generation of Workpiece Orientations for Machining Using Rule-Based System," *Robotics and Computer-Integrated Manufacturing,* 4(3/4), 1988, 545–555.

Friedmann, A. "The Modular Fixturing System, a Profitable Investment," in *Proceedings of the International Conference on Advances in Manufacturing.* Dearborn, MI: Society of Manufacturing Engineers, 1984, 165–173.

Gandhi, M. V., and B. S. Thompson. "Phase-Change Fixturing for FMS," *Manufacturing Engineering,* 93(6), 1984, 79–80.

———. "Flexible Fixturing Based on the Concept of Material Phase-Change," in *Proceedings of the CAD/CAM, Robotics and Automation International Conference.* Tucson, AZ: 1985a, 471–474.

———. "The Integration of CAD and CAM in Adaptive Fixturing for Flexible Manufacturing Systems," in *Proceedings of the 1985 ASME International Computers in Engineering Conference and Exhibition.* New York: American Society of Mechanical Engineers, 1985b, 301–305.

———. "Phase Change Fixturing for Flexible Manufacturing Systems," *Journal of Manufacturing Systems,* 4(1), 1985c, 29–39.

Gandhi, M. V., B. S. Thompson, and D. J. Maas. "Adaptable Fixture Design: An Analytical and Experimental Study of Fluidized-Bed Fixturing," *Transactions of the ASME, Journal of Mechanisms, Transmissions, and Automation in Design,* 108(1), 1986, 15–21.

Gilbert, W. W. "Economics of Machining," *Machining Theory and Practice.* New York: 1950.

Gouldson, C. J. "Principles and Concepts of Fixturing for NC Machining Centers," in W. E. Boyes, ed., *Jigs and Fixtures,* 2d ed. Dearborn, MI: Society of Manufacturing Engineers, 1982, 331–349.

Grippa, P. M., B. S. Thompson, and M. V. Gandhi (1988). "A Review of Flexible Fixturing Systems for Computer Integrated Manufacturing," *International Journal of Computer Integrated Manufacturing,* 1(2), 1988, 124–135.

Hamner, J. R. "Tool Designer's Notebook," in W. E. Boyes, ed., *Jigs and Fixtures,* 2d ed. Dearborn, MI: Society of Manufacturing Engineers, 1982, 3–10.

Hatschek, R. L. "Workholding," *American Machinist,* 121(7), Special Report 697, SR-1 to SR-12, 1977.

Henkin, A., and J. Datsko. "The Influence of Physical Properties on Machinability," *Journal of Engineering for Industry,* November 1963, 321–327.

Henriksen, E. K. *Jig and Fixture Design Manual.* New York: Industrial Press, 1973.

Hoffman, E. G. *Modular Fixturing.* Lake Geneva, WI: Manufacturing Technology Press, 1987.

Horic, T. "Adaptability of a Modular Fixturing System to Factory Automation," *Bulletin of the Japan Society of Precision Engineering,* 22(1), 1988, 1–5.

Inui, M., H. Shuzuki, F. Kimura, and T. Sata. "Generation and Verification of Process Plans Using Dedicated Models of Products in Computers," in *Knowledge-Based Expert Systems for Manufacturing, ASME Winter Annual Meeting.* New York: American Society of Mechanical Engineers, 1985, 275–286.

———. "Extending Process Planning Capabilities with Dynamic Manipulation of Product Models," in *19th CIRP International Seminar on Manufacturing Systems.* University Park, PA: Pennsylvania State University, 1987, 273–280.

Janitzky, E. J. "Machinability of Plain Carbon, Alloy and Austenitic Steels and Its Relation to Yield Stress Ratios when Tensile Strengths Are Similar," *Transactions of the ASME,* 66, 1944, 649–652.

Karyakin, V. N. "Tooling and Fixtures for Flexible Manufacturing Systems," *Soviet Engineering Research,* 5(11), 1985, 45–48.

Kellock, B. "Maintaining a Grip on Changing Needs," *Machinery and Production Engineering,* 143(3681), 1985, 47–53.

———. "Never Forget You Have a Choice," *Machinery and Production Engineering,* 144(3699), 1986, 65–69.

Koch, D. H. "Modular Fixtures Build Fast, Hold Fast," *Tooling and Production,* 54(7), 1988a, 55–56.

———. *Undedicated Fixturing,* SME Technical Paper TE88-104. Dearborn, MI: Society of Manufacturing Engineers, 1988.

Krauskopf, B. "Fixtures for Small Batch Production," *Manufacturing Engineering,* 92(1), 1984, 41–43.

Kuznetsov, Y. I. "Clamping Devices for NC Machines," *Machines and Tooling,* 46(9), 1975, 45–48.

———. "Automated Fixtures for Flexible Manufacturing Systems," *Soviet Engineering Research,* 6(7), 1986a, 49–51.

———. "A Range of Resettable Fixtures," *Soviet Engineering Research,* 6(10), 1986b, 74–75.

Lewis, G. "Modular Fixturing Systems," in *Proceedings of the 2nd International Conference on Flexible Manufacturing Systems.* London: 1983, 451–461.

Machinability Data Center. *Machining Data Handbook,* 3d ed. Cincinnati: Metcut Research Associates, 1980.

"Magnetic Work Holding." *Singapore Polytechnic Engineering Society (S. P. E. S.) Annual Journal,* 11, 1973–1974, 125–130.

Niebel, B. W., A. B. Draper, and R. A. Wysk. *Modern Manufacturing Process Engineering.* New York: McGraw-Hill, 1989.

Noaker, P. M. "Workholding: Firm but Flexible," *Production,* 100(8), 1988, 50–55.

Nyamekye, K., and J. T. Black. "Rational Approach in the Design Analysis of Flexible Fixtures for an Unmanned Cell," in Society of Manufacturing Engineers, Dearborn, MI, *15th North American Manufacturing Research Conference Proceedings,* 1987, 600–607.

Olson, W. W. *Machinability Data Bases for Metal Cutting,* NIST Technical Report, ARLCB-TR-85030. Washington, DC: U.S. Department of Commerce, National Institute of Standards and Technology, 1985.

Owatonna Tool Co. *OTC Hytec Catalog No. H-8401.* Owatonna, MN: OTC, 1988.

Pirovich, L. Y. "Systematization of Fixtures for Unit-Construction-Type Machine Tools and Transfer Lines," *Machines and Tooling,* 43(5), 1972, 41–43.

Qu-Co. *Qu-Co Modular Fixturing System Catalog.* Union, OH: Qu-Co, 1987.

Quinlan, J. C. "New Ideas in Cost-Cutting, Fast-Change Fixturing," *Tooling and Production,* 50(1), 1984, 44–48.

Quinter, K. "Modular Systems Expand the Fixture Continuum," *Tooling and Production,* 54(7), 1988, 57–58.

Ranky, P. *The Design and Operation of FMS.* London: IFS (Publications), 1983.

Shatz, A. S., and A. F. Ishchuk. "Machining Workpieces Located in Composite Fixtures," *Machines and Tooling,* 42(2), 1971, 27–28.

Shuleshkin, A. V., and N. V. Gromov. "Setting-up of 'Body Type' Work-Pieces for Increased Accuracy," *Russian Engineering Journal,* 40(6), 1960, 45–49.

Smith, W. F. "A New Approach to Positioning and Holding Workpieces," in W. E. Boyes, ed., *Jigs and Fixtures,* 2d ed. Dearborn, MI: Society of Manufacturing Engineers, 1982, 169–175.

Taylor, F. W. "On the Art of Cutting Metals," *Transactions of the ASME,* 1906, 28, 31.

Tazetdinov, M. M. "Electrostatic Fixtures," *Machines and Tooling,* 40(8), 1969, 48–50.

Thompson, B. S. *Flexible Fixturing—a Current Frontier in the Evolution of Flexible Manufacturing Cells,* ASME Technical Paper 84-WA/Prod-16. New York: American Society of Mechanical Engineers, 1984.

Thompson, B. S., M. V. Gandhi, and D. J. Desai. "Workpiece-Fixture Interactions in a Compacted Fluidized-Bed Fixture under Various Loading Conditions," *International Journal of Production Research,* 27(2), 1989, 229–246.

Trent, E. M. *Metal Cutting.* London: Butterworths, 1984.

Vektek. *VEKTORFLO,* Product Catalog. Emporia, KS: Vektek, Inc., 1988.

Weck, M., and H. Bibring. *Handbook of Machine Tools,* Vol. 1. New York: John Wiley, 1984.

Wilson, F. W., and J. M. Holt, Jr., eds. *Handbook of Fixture Design.* New York: McGraw-Hill, 1962.

Xu, Y., G. Liu, Y. Tang, J. Zhang, R. Dong, and M. Wu. "A Modular Fixturing System (MFS) for Flexible Manufacturing," *FMS Magazine,* 1(5), 1983, 292–296.

Youcef-Toumi, K., and J. H. Buitrago. "Design of Robot Operated Adaptable Fixtures," in *Proceedings of Manufacturing International '88.* Atlanta: Society of Manufacturing Engineers; Dearborn, MI: 3 (1988):113–119.

———. "Design and Implementation of Robot-Operated Adaptable and Modular Fixtures," *Robotics and Computer-Integrated Manufacturing,* 5(4), 1989, 343–356.

Zimmerman, H. *Considerations for Special Machines, Processes and Fixturing as Integrated into a Flexible Manufacturing System,* SME Technical Paper MS84-929. Dearborn, MI: Society of Manufacturing Engineers, 1984.

Chapter 8

Statistically Based Process Engineering

Objective

This chapter introduces process capabilities from a stochastic point of view. The concepts of accuracy and precision are presented from a feature-based production capability. The concept of statistically based decisions is then introduced, and the cost of production as a function of product variability and defects is addressed. The chapter is presented in a case study form, and contains significant mathematical and statistical models. A fundamental discussion of these models proceeds the development sections so that readers without the proper math/stat background can still apply the concepts developed.

Outline

8.1 Background
8.2 Case Study to Illustrate Process Variability
8.3 Generalizations from the Case Study
8.4 Product Examples and Illustrations
8.5 Summary
8.6 Keywords
8.7 Review Questions
8.8 Review Problems
8.9 References

8.1 BACKGROUND

Successful factories want to manufacture products that have no defects. Unfortunately, all factories have some degree of inherent variability; although management wants to eliminate all variability, that goal is unachievable. Using techniques like Six Sigma and Taguchi Methods, defects and variability can be reduced, but they cannot be eliminated.

Chapter 6 discussed process engineering and the benefits of process tolerance charts. This chapter will show that process tolerance charts are not deterministic, and that treating them as deterministic can lead to uneconomic practices. The discussion will begin with a case study; then, it will present models to illustrate these methods.

8.2 CASE STUDY

This case study will illustrate product and process engineering. Figure 8.1 shows a simple bracket part, where all features (holes) on the part are specified with respect to a single datum system, A-B-C. These features have specifications for both dimensional tolerance (tolerance of diameter) and location tolerance (true position). The "regardless of feature size" specification indicates that the location requirement is *independent* of the actual hole size. Thus, the hole's central axis must always be centered inside of an imagined cylinder of .008″ diameter—even if the diameter of the hole is 0.508″, 0.507″, 0.506″, etc.

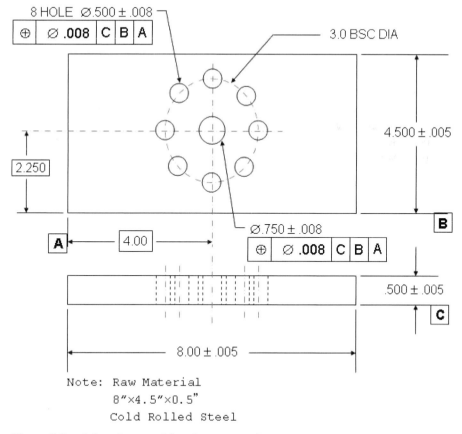

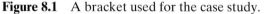

Figure 8.1   A bracket used for the case study.

TABLE 8.1   Process Variability for Hole Making

Process	Dimensional Accuracy $(\pm 3\sigma^D)$	Location Accuracy (on X or Y axis) $(\pm 3\sigma^L)$
Drill (twist)	±.008	±.003
Reaming	±.0025	±.003
Bore (semi-finish)	±.0035	±.0025
Bore (finish)	±.002	±.002

For this case, the bracket will be produced on a CNC machining center. The process tolerance characteristics can be obtained through experimentation. Table 8.1 shows the typical accuracies of the standard process variability or the precision of the machine tool. For example, the location accuracy of the twist drill reflects the precision of the movement of the drill on both the X and Y-axes. This illustration does not consider variables that cannot be predicted in advance, such as tool wear, and the variation of hole size caused by process mechanics, local differences in material hardness, and chip removal. The variation of hole size is not dependent on time or tool wear and is normally distributed. For manufacturing errors such as tool wear, the variation of produced diameters of eight small holes will most likely be correlated. This issue will not be addressed yet.

8.2.1 Notation

Several terms and symbols will be used in this case study:

C = cost of one part

C_m = manufacturing cost per part

C_d = cost of defective part (warranty cost)

Ⓜ = maximum material condition (MMC)

Ⓛ = least material condition (LMC)

$P\{E\}$ = probability of an event E

G = making a good part with all features satisfying the specifications

B = making a bad part with one or more features not satisfying their specifications

$N(\mu, \sigma)$ = Normal distribution with mean μ and standard deviation σ

n = number of features on one part

D = related to the dimensional tolerance (used as a superscript)

L = related to the location tolerance (used as a superscript)

σ_i^D = standard error of the dimensional accuracy of feature i in one machining process

σ_i^L = standard error of the location accuracy of feature i in one machining process

T_i^D = tolerance for the dimensional parameter of feature i (i.e., tolerance of a hole's diameter)

T_i^{D+} = upper tolerance limit specified for the dimensional parameter of feature i

T_i^{D-} = lower tolerance limit specified for the dimensional parameter of feature i

T_i^{L} = tolerance allowed for the location error of feature i (i.e., true position tolerance zone of a hole)

T_{nominal}^{L} = nominal tolerance allowed for the location error of feature i given RFS

T_M^{L} = true position tolerance given maximum material condition

T_L^{L} = true position tolerance given least material condition

x = actual variation of the dimension of a feature from its nominal value

y = actual distance from the location of a feature to its ideal location

z = actual distance from the location of a datum feature to its ideal location

w = actual distance from the location of a controlled feature to its ideal location

t = actual variation of the dimension of a datum feature from its nominal value

u = actual variation of the dimension of a controlled feature from its nominal value

Q_R^{H} = Probability of making a good datum feature (hole location) when RFS is specified

Q_M^{H} = Probability of making a good datum feature (hole location) when MMC/LMC is specified

Q_R^{S} = Probability of making a good datum feature (slot) when RFS is specified

Q_M^{S} = Probability of making a good datum feature (slot) when MMC/LMC is specified

In this case study, the cost function consists of two main sources: manufacturing, which is $60 per hour for the machine and operator; and producing a bad part, which is arbitrary assumed to be $100 per part. Incremental costing is used so material cost does not affect the objective. Once the manufacturing process is chosen, the total cost can be estimated using probability models, which are derived in the next section.

8.2.2 Location Modifier—Regardless of Feature Size

When the RFS call-out is used, the hole features on the bracket part are assumed to be independent of each other. If the size (diameter) is assumed to be independent of the location of the holes, then the probability of producing a good part is simple: Find the product of the probabilities from each independent feature (when the RFS condition is specified). Thus,

$$P\{G\} = \prod_{i=1}^{n} P\{G_i\} = \prod_{i} P\{G_i^D\} P\{G_i^L\} \tag{8.1}$$

where $P\{G_i^D\}$ is the probability that feature size i will be acceptable, and is calculated by

$$P\{G_i^D\} = 2\Phi\left(\frac{T_i^D}{\sigma_i^D}\right) - 1 \tag{8.2}$$

where Φ is the normal distribution cumulative density function (shaded area under the normal distribution curve), and $P\{G_i^L\}$ is the probability that the location of feature i will be acceptable.

There is a greater probability that the product will have a good location $(P\{G_i^L\})$ than a good feature size $(P\{G_i^D\})$. This occurs because geometric tolerance is defined by a three-dimensional tolerance zone. The location error data can be collected, and a distribution can be empirically fit to describe the location error performance. In practice, this is frequently used to characterize location errors.

If the axes errors are known, and if they are normal, then the following argument can be used: First, assume that a Cartesian machine tool (like a standard vertical or horizontal machining center) is used. Then assume that, when it moves in the X–Y plane, it remains perpendicular to the third datum plane. Thus, the hole location becomes a two-dimensional variable, which is determined by the tool's moving accuracy on its X and Y axes only. If the moving accuracies are independent and identical distributed normal variables, and they each have a mean zero and variance σ^2, the hole's location accuracy will be a binormal variable. Thus, it will follow bivariate normal distribution, $B(0, 0, \sigma_x^2, \sigma_y^2, \rho_{xy})$, where σ_x^2 and σ_y^2 are the variances of errors along the X and Y axes and ρ_{xy} is the correlation coefficient between the errors along the X and Y axes. Then, assume that the machine tool's movement on the X and Y axes are uncorrelated and has the same standard error. Thus, the distance from the actual location of the hole to the ideal location $(D = \sqrt{X^2 + Y^2})$ will be Rayleigh distributed, $R(\sigma^2)$ [Hahn & Shapiro, 1967], which gives the bivariate density of X and Y as follows:

$$f(X, Y) = \frac{1}{2\pi\sigma^2} \exp\left(-\frac{(X^2 + Y^2)}{2\sigma^2}\right) \tag{8.3}$$

That is, the variate $\dfrac{(X^2 + Y^2)}{\sigma^2}$ is χ^2-distributed with two degrees of freedom [Bjorke, 1978]. Transforming differential areas using $dX\, dY = Dd\, Dd\varphi$ gives the joint probability density function as

$$f(D, \varphi) = \frac{D}{2\pi\sigma^2} \exp\left(-\frac{D^2}{2\sigma^2}\right) \tag{8.4}$$

Since this is independent of phase, the random variables D and φ are statistically independent. Therefore,

$$f(D, \varphi) = f(D) \times f(\varphi) \tag{8.5}$$

Then

$$f(\varphi) = \int_0^\infty f(D, \phi)\, dD = \frac{1}{2\pi} \quad \text{for } 0 \le \varphi \le 2\pi \tag{8.6}$$

As a consequence, the density of D may be found from

$$f(D) = \int_0^{2\pi} f(D, \varphi)\, d\varphi = \frac{D}{\sigma^2} \exp\left(-\frac{D^2}{2\sigma^2}\right) \quad \text{for } D \geq 0 \quad (8.7)$$

which is the density of a Rayleigh distribution. Therefore, the probability of making a good location for a feature i can be calculated by integrating to the specified limits:

$$P\{G_i^L\} = P\left\{D \leq \frac{T_i^L}{2}\right\} = \int_0^{T_i^L/2} \frac{D}{\sigma^2} \exp\left(-\frac{D^2}{2\sigma^2}\right) dD = 1 - \exp\left[-\frac{(T_i^L/2)^2}{2(\sigma_i^L)^2}\right] \quad (8.8)$$

8.2.2.1. Quality Issues for Proscess Plan #1. Figure 8.1 and Table 8.1 demonstrate that twist drilling on a CNC machine could produce the desired part. The resultant process plan is shown in Table 8.2, which includes a typical feed (f) and cutting speed (v) for twist drilling. The calculation shows that it takes 0.21 minutes to drill the large hole, and 1.7 minutes to drill the eight small holes. Once the loading and unloading times are added, the total manufacturing time for each part is 3.41 minutes. The resultant manufacturing cost is $\$60 \times 3.41/60 = \$3.41$.

Although process plan #1 is quite efficient, some bad parts still may be produced. For instance, the dimensional tolerance for the hole size (all nine holes) corresponds exactly with the $\pm 3\sigma^D$ from Table 8.1 ($t^D/\sigma^D = 0.008/0.008/3 = 3$). Even though the twist drill size is set to the nominal size of the hole, random variation will result from the process. To determine whether a hole will be the correct dimension, note the area under the normal curve between $\pm 3\sigma^D$ or 99.73%. If the hole dimensions are independent of

TABLE 8.2 Process Plan #1 for Case Study Part

Operation	Description	Tooling	V	f	d	Time (min)
10	Load part into fixture					.5
20	Drill large hole	.750 dia	60	.010		.21
30	Drill small hole (8)	.500 dia	50	.008		1.70
40	Unload and visually inspect					1.00
Total time				3.41 min		

$$t_m = \frac{\pi D l}{12 f v}$$

Large hole: $t_m = \dfrac{\pi \times .750 \times .650}{12 \times .010 \times 60.0} = .2127$ min $= 12.76$ sec

Small hole: $t_m = \dfrac{\pi \times .5 \times .650}{12 \times 50 \times .008} = .2127$ min $= 12.76$ sec

Here, t_m = manufacturing time, minutes; V = speed, feet per minute; D = diameter, inches; f = feed, inches per revolution; l = depth of hole plus clearance height, inches. Let the clearance height be 0.15".

each other, the likelihood that all hole dimensions are good is $0.9973^9 = 0.9760$. This implies that, on average, 2.4 of every 100 parts produced $[(1 - 0.9760) \times 100]$ will have at least one hole that does not meet the dimensional specifications.

Assuming that hole size and location are independent, their effects can be examined separately and combined from a total-part-specification point of view. The hole location is specified as RFS (regardless of feature size) and is set at 0.008. As shown in Table 8.1, the $\pm 3\sigma^L$ of location accuracy is 0.003. This implies that

$$\sigma^L = \frac{0.003}{3} = 0.001$$

The likelihood that a location will not meet its specifications is then

$$P[y > (0.008/2)] = \exp\left[-\frac{(0.008/2)^2}{2 \times 0.001^2}\right] = 0.000335$$

which is very small, ~ 3.35 out of every 10,000.

The final equation shows the production cost of one part:

$$\text{cost per part} = \text{cost of machining} + \text{cost of defect}$$

$$C = C_m + C_d \times P\{B\} = C_m + C_d \times (1 - P\{G\}) = C_m + C_d \times \left(1 - \prod_i P\{G_i\}\right)$$

$$C = \$3.40 + \$100 \times (1 - 0.9973^9 \times 0.9997^9) = \$6.07 \tag{8.9}$$

8.2.2.2. Quality Issues for Process Plan #2.   In process plan #1, a significant portion of the cost is due to the production of "bad hole dimensions." Process plan #2 (Table 8.3) adds a reaming operation to improve the dimensional accuracy of the hole. With this process plan, the total processing time increases to 5.32 minutes, thus the manufacturing cost increases to $5.32 per part. The probability that one hole dimension is within the specification is

$$P\{G_i^D\} = 2\Phi\left(\frac{T_i^D}{\sigma_i^D}\right) - 1 = 2\Phi\left(\frac{0.008}{0.0025/3}\right) - 1 = 2\Phi(9.6) - 1 \approx 2 - 1 = 1$$

The likelihood that a location will not meet its specification becomes

$$P[y > (0.008/2)] = \exp\left[-\frac{(0.008/2)^2}{2 \times 0.001^2}\right] = 0.000335$$

The probability that a defect will occur is then

$$(1 - P\{G\}) = \left(1 - \prod_i P\{G_i\}\right) = (1 - 1.0^9 \times 0.9997^9) = (1 - .9973) = 0.0027$$

TABLE 8.3   Process Plan #2 for the Case Study Part

Operation	Description	Tooling	V	f	d	Time (min.)
10	Load part into fixture					0.5
20	Drill large hole	47/64 dia	60	.010		0.21
30	Ream large hole	.750 dia	300	.010	.008	
40	Drill small holes (8)	31/64 dia	50	.010		1.70
50	Ream small holes (8)	.500 dia	250	.008	.008	
40	Unload and visually inspect					1.00
Total time				5.32 min		

Thus, the total cost becomes

$$C = \$5.32 + \$100 \times 0.0027 = \$5.59$$

By using process plan #2, the total cost is reduced by \$.48 per part. For a high volume production, the cost savings is very significant.

Other process plans exist, but are unlikely to generate more savings; thus, they are not investigated here.

8.2.3 Location Modifier—Maximum Material Condition

Previous models assumed that the hole-making tool moved independently on the X and Y axes. However, this may not be true for every kind of machining mechanism. To take into account the interaction between these two contributors to location, a new model is necessary. The following generalized model of Equation (8.8) considers the correlation of this location accuracy on the X and Y axes:

$$P\{G_i^L\} = \iint_\Omega \frac{1}{2\pi\sigma_x^L\sigma_y^L\sqrt{(1 - \rho_{xy}^2)}} \exp\left\{ -\frac{1}{2(1 - \rho_{xy}^2)} \right.$$

$$\left. \times \left[\frac{x^2}{(\sigma_x^L)^2} - \frac{2\rho_{xy}xy}{\sigma_x^L\sigma_y^L} + \frac{y^2}{(\sigma_y^L)^2} \right] \right\} dx\, dy \tag{8.10}$$

Here, ρ_{xy} is the correlation coefficient and Ω is the region of the true hole position indicated by the location specification (in this case, a circular region with a diameter of .008"). To simplify the computation, the hole's location error should be a univariate random variable instead of a bivariate random vector; that is, the location accuracy should be the consolidated effect of the tool's moving accuracies on both the X and Y axes. Then, the location variation can be modeled by the half-normal distribution. This will bring a slightly higher value of $P\{G_i^L\}$, but the difference is usually negligible [Lehtihet and Gunasena, 1990].

8.2.4 Other Location Specifications

Holes are not always specified as RFS; there are good reasons to specify them using MMC or LMC. In fact, MMC is a common geometric tolerance modifier used to ensure that a product can be assembled. Table 8.4 illustrates the location tolerance requirements under MMC and LMC. Since the location tolerance zone is a function of the feature size, the location and dimensional tolerances are no longer considered to be independent variables. Therefore, the relationship between the location tolerance and actual feature size is as follows:

For the maximum material condition,

$$T^L = 0.016 + x \tag{8.11}$$

For the least material condition,

$$T^L = 0.016 - x \tag{8.12}$$

The probability of manufacturing an acceptable hole is

$$P\{G^D \cap G^L\} = P\{G^D\}P\{G^L|G^D\} \tag{8.13}$$

The conditional probability of B given A, denoted by $P(B|A)$, is defined by

$$P(B|A) = \frac{P(A \cap B)}{P(A)} \text{ if } P(A) > 0.$$

TABLE 8.4 Location Requirements for Maximum and Least Material Conditions

Small hole			Large hole		
Actual diameter (.500 + x)	Location tolerance (T^L)		Actual diameter (.750 + x)	Location tolerance (T^L)	
	MMC	LMC		MMC	LMC
.492	.008	.024	.742	.008	.024
.493	.009	.023	.743	.009	.023
.494	.010	.022	.744	.010	.022
.495	.011	.121	.745	.011	.021
.496	.012	.120	.746	.012	.020
.497	.013	.119	.747	.013	.019
.498	.014	.018	.748	.014	.018
.499	.015	.017	.749	.015	.017
.500	.016	.016	.750	.016	.016
.501	.017	.015	.751	.017	.015
.502	.018	.014	.752	.018	.014
.503	.019	.013	.753	.019	.013
.504	.020	.012	.754	.020	.012
.505	.021	.011	.755	.021	.011
.506	.022	.010	.756	.022	.010
.507	.023	.009	.757	.023	.009
.508	.024	.008	.758	.024	.008

This equation gives the probability that event B will occur when some event A has occurred. From the inspection or measurement point of view, it implies that the probability of an acceptable location of a feature is determined by the probability of an acceptable dimension, since location tolerance is affected by actual feature size, as shown in Equations (8.11) and (8.12). Thus,

$$P\{G^D \cap G^L\} = P\{G^D\}P\{G^L|G^D\} = P\{T^{D-} \le x \le T^{D+}\}$$

$$P\left\{0 \le y \le \frac{T^L}{2} \middle| T^{D-} \le x \le T^{D+}\right\}$$

$$= \int_{T^{D-}}^{T^{D+}} g(x)\,dx \times \int_{0}^{T^L/2} r(y)\,dy = \int_{T^{D-}}^{T^{D+}} g(x)\,dx \times \int_{0}^{(|T^{D+}|+|T^{D-}|+x)/2} r(y)\,dy$$

$$= \int_{T^{D-}}^{T^{D+}} \left(\int_{0}^{T^L/2} r(y)\,dy \right) g(x)\,dx \tag{8.14}$$

where $r(y)$ is the Rayleigh probability density function, given that the hole's diameter is known, and $g(x)$ is the normal (Gaussian) probability density function. This uses the conditional probability of the random variable y, given that x is not a fixed value, but $T^{D-} \le x \le T^{D+}$. Typically, the conditional distribution is used in the case of a joint probability distribution for multiple random variables, so the Rayleigh probability density function is used and the integral form of the interval is a function of a variable x ($|T^{D+}| + |T^{D-}| \pm x$), reflecting the characteristic of the conditional probability of the variables. The resultant integral value for the Rayleigh probability density function then becomes a form of the cumulative distribution of a variable x:

$$r(y) = \frac{y}{(\sigma^L)^2} \exp\left[-\frac{y^2}{2\sigma^L} \right] \qquad y \ge 0 \tag{8.15}$$

The interval in its integral form is a function of a variable x:

$$g(x) = \frac{1}{\sqrt{2\pi}\sigma^D} \exp\left[-\frac{x^2}{2(\sigma^D)^2} \right] \tag{8.16}$$

By substituting the values used in this case, Equations 8.14–8.16 are solved numerically for both MMC and LMC. Earlier, we found that their solutions are identical when using the same diameter and true position tolerance specifications. Thus, the statistical rejection rates of these holes are identical under MMC and LMC specifications, even though their manufacturing and inspection processes may be quite different. Obviously, when MMC and LMC have identical specifications, they will produce the same statistical result for a symmetric distribution even though their physical interpretations are different.

8.3 GENERAL METHODOLOGY

So far, the case study demonstrated how statistical models were formulated for the various location modifiers (such as RFS, MMC, and LMC) when all features (holes) on the part were specified with respect to a single-datum system. It also demonstrated that statistical-based process engineering can guide decision-making to compare several process plans. In addition, the models can identify and evaluate feasible design alternatives, as well as deterministic or experienced-based information (such as process capability found in machinist handbooks). Even though these statistical models can help process/product engineering decision-making, there is a need for a generalized model that also takes into account parametric tolerances, geometric tolerances (i.e., positional tolerance) under multiple-datum reference frames, datum features subject to size variations, and so on.

As discussed in Chapters 2 and 3, two types of tolerancing schemes are allowed in the specification of geometric entities: parametric and geometric. Most engineers and draftspersons learn parametric tolerancing in basic engineering drawing. The term *parametric* is used because limiting conditions, or control parameters, are defined on the basis of normal Cartesian (or polar) dimensions. Datum surfaces show critical dependencies in feature locations, especially in the case of parametric tolerancing. The general methodology uses a simple prismatic part with parametric tolerancing, as shown in Figure 8.2. Since no geometric tolerance is specified in the drawing, only the dimensional accuracy of the desired feature is considered, and the probability that the feature size Mx_i is acceptable is calculated. For this example, it is assumed that the width variations from one surface to another, x_1 and x_2, are independent of each other and, again for illustration purposes, normally distributed. It is also assumed that the part will be machined in one setup with respect to the leftmost plane in the drawing.

In general, the actual deviation of the feature width from its nominal value can follow any distribution and can be identified through experiments. To calculate the probability of making an acceptable feature dimension, one method use an integral form for any distribution or density function (i.e., normal, uniform, etc.). Figure 8.2 illustrates various ways to specify tolerances according to the datum surfaces. The tolerances stack differently depending on the tolerancing scheme in the drawing. When a dimension is specified from another dimensioned surface, the surface tolerances become additive (stack). For example, in Figure 8.2(a), the leftmost surface (labeled ⓪) becomes an implicit datum, because all measurements are made with respect to it. In Figure 8.2(b), each face is specified as a function of the previous datum face. Since face ② is specified from face ①, the tolerance of the first flat's width convolutes the second flat's width. In general, dimensions chained together in this manner accumulate as $\max d_n = \Sigma \max d_i$. This fact explains the effect of changing reference or datum faces on design interpretation in the parametric tolerancing of the drawing. However, assuming variability of the resultant feature (from the machined-part-specification point of view), the previously discussed statistical model is used to estimate or simulate the probability of producing a good feature.

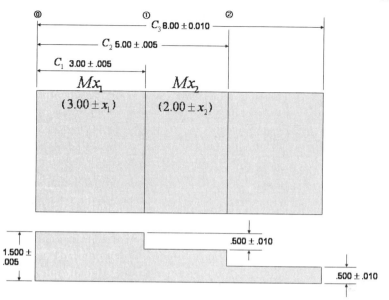

(a) Parametric tolerancing using an implicit datum (Design alternative 1)

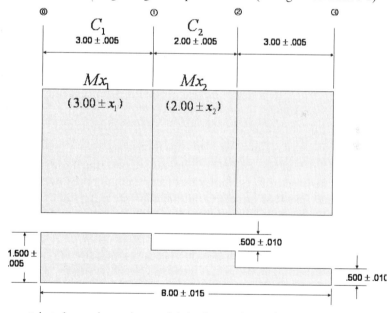

(b) Parametric tolerancing using multiple datum faces (Design alternative 2)

Figure 8.2   A simple prismatic part with parametric tolerancing.

The discussion that follows makes a number of assumptions. First, as before, assume that, x_1 and x_2 follow $N(0, \sigma_1)$ and $N(0, \sigma_2)$, respectively, if they are IID (independent and identically distributed) normal random variables. In addition, assume that σ_1 and σ_2 are identical when the same milling tool is used for two features. The milling tool's dimensional accuracy, as shown in the reference (i.e., the process tolerance charts), should be less than each dimensional tolerance or stacked tolerance in the drawing. Next, the dimensional accuracies for available milling tools must satisfy this constraint (i.e., $3\sigma_1 = 3\sigma_2 = \pm 0.005$). Hence, $P\{G_i^D\}$ is the probability that the feature size (width in this case) Mx_1 or Mx_2 will be acceptable, and it is calculated as follows:

- For the case shown in Figure 8.2(a),

$$P\{G_{Mx1}^D\} = \int_{T_{C_1}^{D-}}^{T_{C_1}^{D+}} g(x_1)\,dx_1 = \int_{-0.005}^{0.005} g(x_1)\,dx_1$$

$$= 2\Phi\left(\frac{T_i^D}{\sigma_i^D}\right) - 1 = 2\Phi\left(\frac{0.005}{0.0017}\right) - 1 \tag{8.17}$$

$$P\{G_{Mx2}^D\} = \int_{T_{C_{2-1}}^{D-}}^{T_{C_{2-1}}^{D+}} g(x_2)\,dx_2 = \int_{T_{C_1}^{D-}+T_{C_2}^{D-}}^{T_{C_1}^{D+}+T_{C_2}^{D+}} g(x_2)\,dx_2$$

$$= \int_{-0.010}^{0.010} g(x_2)\,dx_2 = 2\Phi\left(\frac{0.010}{0.0017}\right) - 1 \tag{8.18}$$

where
$$g(x_i) = \frac{1}{\sqrt{2\pi}\sigma^D}\exp\left[-\frac{x^2}{2(\sigma^D)^2}\right] \tag{8.19}$$

Φ = the standard normal distribution function

$\sigma_i^D = 0.005/3 = 0.0017$ (from the assumption)

Therefore $Mx_2 = Mx$ (manufactured dimension of a feature from a face ⓪ to a face ②) $-Mx_1$. Since all measurements are performed with respect to the datum ⓪, the dimension of Mx_2 cannot be directly obtained. However, by tolerance stacking equations that have an additive characteristic, tolerances aimed for the feature Mx_2 can be calculated using C_i, design constraints for the feature on the drawing, as

$$T(C_{2-1}) \geq T(Mx_2) = T(C_1) + T(C_2))$$

Since x_1 and x_2 are independent of each other,

$$P\{G\} = \prod_{i=1}^{n} P\{G_i\} = P\{G_{Mx1}\} \times P\{G_{Mx2}\} \tag{8.20}$$

- For the case shown in Figure 8.2(b),

$$P\{G_{Mx1}\} = \int_{T_{C_1}^{D-}}^{T_{C_1}^{D+}} g(x_1)\,dx_1 = \int_{-0.005}^{0.005} g(x_1)\,dx_1 = 2\Phi\left(\frac{0.005}{0.0017}\right) - 1 \tag{8.21}$$

$$P\{G_{Mx2}\} = P\{G_{Mx2}|G_{Mx1}\} = P\{T_{c_2}^{D-} - x_1 \le x_2 \le T_{c_2}^{D+} + x_1 | T_{c_1}^{D-} \le x_1 \le T_{c_1}^{D+}\}$$

$$= \int_{T_{c_1}^{D-}}^{T_{c_1}^{D+}} \left(\int_{T_{c_2}^{D-}-x_1}^{T_{c_2}^{D+}+x_1} g(x_2)\, dx_2 \right) g(x_1)\, dx_1 \times \frac{1}{P\{G_{Mx1}\}} \tag{8.22}$$

(The interval in its integral form for a variable x_2 is a function of a variable x_1.) The main difference between cases (a) and (b) is the tolerance chaining due to the surface relationships affecting the manufacturing (both sequence and process) and fixturing of the part. Case (a) uses a single datum face for the manufacturing process. Also, tolerance constraints for the feature Mx_2 are not directly specified on its drawing. In other words, tolerance limits for calculating the probability of an acceptable feature size Mx_2 can be obtained in only one way: by summing the tolerances for relative features in the drawing. As shown in Equation (8.18), the maximum tolerances for Mx_2 are already stacked into the interval of the integral form. Since each feature is manufactured independently with respect to the datum ⓪, there is no relationship between the probability of the acceptable feature Mx_1 and the probability of the acceptable feature Mx_2. The factor that can affect the result of Equation (8.18) is the accumulation of tolerance errors caused by tolerance limits in the integral form.

However, case (b) uses multiple datum faces ⓪ and ① for each desired feature. It implies that the resultant face ① or the feature size Mx_1 somehow affects the probability of the acceptable feature Mx_2, since Mx_2 is determined with respect to face ①. Hence, the tolerance limits for calculating the probability of an acceptable feature Mx_2 in its integral form are not the summation of tolerances for other features in the drawing, but the summation of the direct variation x_1 from the resultant feature Mx_1 and its own tolerance, which is already specified in the drawing. Thus, the effect of the accumulation of tolerance errors is reduced for case (b), as shown in Figure 8.3. For a feature Mx_2, there is a bonus tolerance, which is caused by the size variation from Mx_1; it is acceptable if face datum ② is between 4.990 and 5.010 from datum face ⓪. Therefore, the probability of the acceptable feature size Mx_2 is determined by the probability of the acceptable feature size Mx_1. Again, the normal probability density function with other variables in its integral form is used, since the interval in its integral form is a function of a variable x_1 in this case. The equation

$$\int_{T_{c_1}^{D-}}^{T_{c_1}^{D+}} \left(\int_{T_{c_2}^{D-}-x_1}^{T_{c_2}^{D+}+x_1} g(x_2)\, dx_2 \right) g(x_1)\, dx_1$$

is divided by the probability $P\{G_{Mx1}\}$, as the goal is to find the probability of an acceptable feature. We thus have

$$\int_{-0.005}^{0.005} \left(\int_{-0.005-x_1}^{0.005+x_1} g(x_2)\, dx_2 \right) g(x_1)\, dx_1 \times \frac{1}{P\{G_{Mx1}\}} \tag{8.23}$$

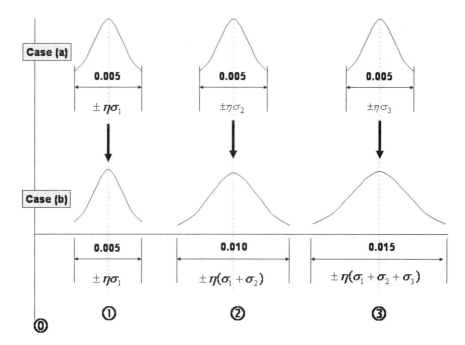

Figure 8.3   Different phases of accumulation of tolerance errors.

where

$$g(x) = \frac{1}{\sqrt{2\pi}\sigma^D}\exp\left[-\frac{x^2}{2(\sigma^D)^2}\right] \tag{8.24}$$

$$g(x_2) = \frac{1}{\sqrt{2\pi}\sigma^D}\exp\left[-\frac{x_2^2}{2(\sigma^D)^2}\right] \tag{8.25}$$

(The interval in its integral form is a function of a variable x_1.)

Therefore, the probabilities that feature size Mx_1 and Mx_2 will be acceptable are dependent on each other.

$$P\{G\} = \prod_{i=1}^{n}P\{G_i\} = \int_{T_{C_1}^{D-}}^{T_{C_2}^{D+}}\left(\int_{T_{C_2}^{D-}-x_1}^{T_{C_2}^{D+}+x_1}g(x_2)\,dx_2\right)g(x_1)\,dx_1 \tag{8.26}$$

The preceding discussion has shown that the calculation of the probability that a part will be acceptable changes according to the specification of reference features in order to eliminate possible tolerance errors using a simple prismatic part. Since datum features are frequently needed to control tolerance stacks and feature placement, a control datum may establish a datum reference frame [Meadows, 1995]. By selecting different classes of datum features in its design stage, the probability that a part will be acceptable can also change. Therefore, it is very important to develop a formal statistical model for general datum selection and geometrical characteristics of the part. Because several alternative process plans can generally exist for a part design case, the method must consider both the design alternatives and the statistical implications for producing the required features, in order to find the best process for the part in terms of manufacturing and warranty cost. Now, a generic procedure to obtain an optimal process plan can be based on the statistical model. This procedure can ease the decision-making process during the selection of appropriate processes and design specifications.

In the sections that follow, more realistic product examples are used to demonstrate how formal statistical models are formulated to find design alternatives for the various classes of datum features. If a traditional or deterministic method is used to create the previously discussed plan, then it will not produce an optimal plan. It might even produce a significant number of defective parts. As a result, a company might fail to produce parts economically. Therefore, this chapter will illustrate alternative process plans, a method for selecting a plan (based on the cost model for manufacturing, tolerance, and warranty cost), and a formal cost model (including inputs from the statistical models using example parts).

In constructing a datum reference frame, features can be specified in many ways. Generally, datum features are chosen based on part functionality and their relationships with mating features [Meadows, 1995]. Both manufacturing engineers and quality assurance personnel use these features to maximize available tolerances and enhance conformance quality so that they can improve the most critical features of the part in terms of form, fit, function, processing, tooling, and inspection [Curtis, 2002]. According to ASME Y14.5M-1994 [American Society of Mechanical Engineers], multiple datum reference frames are used during physical separation or when the functional relationship of features requires datum reference frames to be applied at specific locations on the part. Product designers can establish datum reference frames according to specific criteria based on the functional relationship and requirement. The designers can specify the appropriate features as datum features among candidates. If necessary, multiple datum reference frames can be specified in the feature-control frame related to the geometric tolerances in a part. ASME Y14.5M-1994 provides various classes for establishing datum features from the candidates. The following classes can be used as criteria to establish datum features:

- Datum features that are not subject to size variations (surfaces)
- Datum features that are subject to size variations (holes, slots, etc.)
- The specification of RFS, MMC, and LMC datum features
- Multiple datum features
- Pattern of features
- Screw threads, gears, and splines
- Partial surfaces as datum features
- Multiple datum reference frames
- Simultaneous requirements

Datum features such as diameters for shafts or holes and widths for internal or external slots differ from singular flat surfaces because they may vary in size. In other words, they are geometrically controlled and specified at MMC or LMC. If a feature uses an additional datum reference frame that includes datum features subject to size variations, statistical models should also reflect the implicit variations in those features.

Figure 8.4 shows four possible design cases that establish datum features for a hypothetical part. Any case could be selected on the basis of the functional requirements for the geometrically controlled features (i.e., a pattern of features). In Figure 8.4(a), all part features are specified with respect to a single datum reference frame, C-A-B in the feature control frame, and they have specifications for both dimensional and positional tolerances. In this case, the datum features are at RFS condition. In Figure 8.4(b), small holes, a large hole, and a slot are specified with respect to the different datum reference frames, such as C-A-B and C-D-E, respectively. These datum features are at RFS condition as well. In Figures 8.4(c) and (d), each feature uses a different datum reference frame and the small hole feature is specified with respect to the datum feature D at MMC. For cases (a) and (b), all the features are specified with respect to datum features that are not subject to size variations (flat surfaces); thus, the Equations (8.1) to (8.16) can be used to calculate the probability that a part or feature

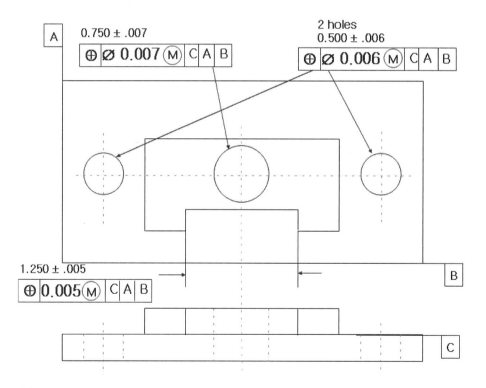

(a) A single datum reference frame; datum features not subject to size variations

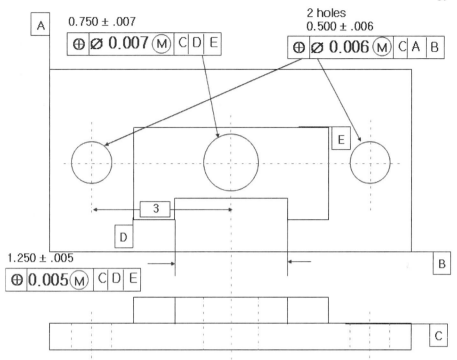

(b) Multiple datum reference frames; datum features not subject to size variations

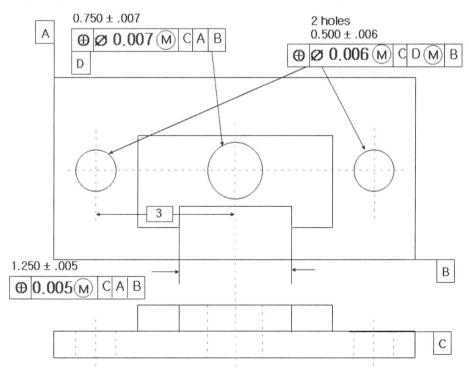

(c) Multiple datum reference frames; datum features subject to size variations (hole)

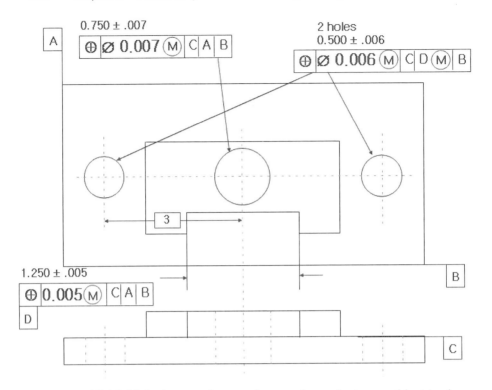

Figure 8.4   (d) Multiple datum reference frames; datum features subject to size variations (slot).

will be acceptable, regardless of the existence of multiple datum reference frames. However, cases (c) and (d) show that features (small holes) have positional tolerances and that their relationship to other features (i.e., locating datum features) is subject to size variations, such as a pilot hole (diameter) in (c) and an internal slot (width) in (d). In this case, the critical requirement become the location of the surrounding features relative to the pilot hole (i.e., the large hole in Figure 8.4(c)). The pilot hole is implied at its virtual condition and permitted orientation tolerance in its control. Hence, this datum feature should be recognized in controlled features' "pickup" in fixturing or inspection and as pertinent to the part function [Foster, 1982]. If an RFS modifier is specified to the pilot hole in the datum reference frame, this datum feature should not be subject to size variations. In other words, the probability equations for this case become the same as in cases (a) and (b).

When the MMC or LMC is specified to the surrounding holes and the pilot hole, the full advantages of assembly are realized, even though the accuracy of the features is decreased. In this case, the dimension of the pilot hole departs from its MMC in production, and a shift of the surrounding (controlled) holes is permissible. It must be relative to the pilot hole in proportion to the difference between the MMC size and the actual mating size of the datum feature. It can be stacked into the location tolerances of surrounding holes as a result of displacement. This implies that the effect of the extra allowed shift should be considered in constructing the statistical model. However, each controlled hole gains no direct extra individual location tolerance as the size of the datum feature departs from its MMC or virtual condition [Meadows, 1995]. Therefore, the statistical models in the next sections are formalized according to this implication.

8.3.1 Formal Statistical Models for the Datum Establishment

This section will use the fundamental assumptions developed in Section 8.2 to obtain formal statistical models when multiple datum reference frames are established. The variations in hole size are independent of each other and normally distributed. Even though the variation in the produced diameters of two small holes is probably correlated, it is not considered in the formulation. The machine tool's movements on the X and Y axes are uncorrelated and have the same standard errors. Tool wear is not considered (or is negligible during a short run). It is also assumed that the variation of hole size is caused by process mechanics, local differences in material hardness, chip removal, and other random variables that cannot be predicted in advance.

Furthermore, some critical mating features might have other geometric tolerances (e.g., relating to form, orientation, location, and so on), as well as dimensional tolerances in their drawing according to their functional requirements. However, this section only addresses two issues: features that have their position tolerances and other geometric tolerances (i.e., perpendicularity, concentricity, symmetry, etc.) specified in the feature control frame, and features that propose statistical models for some critical features in order to determine probability of making a good part. Sections that follow will describe cases where RFS and MMC/LMC are specified for the controlled features and there are multiple reference frames subject to size variations, as in Figure 8.4(c) and (d). Figure 8.4(a) and (b) might use equations that are similar to Equations (8.1) to (8.16).

8.3.1.1. Datum Features for the Multiple Datum Reference Frames.   When multiple reference frames appear in the part drawing, the datum features can either affect the probability of making a good controlled feature or be totally independent of each other. Each characteristic will determine whether the datum feature is subject to size variation. Hence, to determine the probability of producing interrelated features using multiple datum reference frames, it is important to identify which material side condition is used in the datum feature and datum reference frame for the controlled features. Typically, when the RFS call-out is used in the datum feature, that feature is not subject to size variation; thus, this controlled feature will be independent of the variability of the datum feature, even though multiple datum reference frames are specified in the drawing:

- For a pilot hole, as shown in Figure 8.4(c),

   when RFS is specified in the datum feature,

$$P\{G_{di}\} = P\{G_{di}^D \cap G_{di}^L\} = P\{G_{di}^D\}P\{G_{di}^L\} = Q_R^H \qquad (8.27)$$

$$P\{G_{di}^D\} = 2\Phi\left(\frac{T_{di}^D}{\sigma_{di}^D}\right) - 1 \qquad (8.28)$$

$$P\{G_{di}^L\} = 1 - \exp\left[-\frac{(T_{di}^L/2)^2}{2(\sigma_{di}^L)^2}\right] \tag{8.29}$$

When MMC or LMC is specified in the datum feature,

$$P\{G_{di}\} = P\{G_{di}^D \cap G_{di}^L\} = P\{G_{di}^D\}P\{G_{di}^L|G_{di}^D\} \tag{8.30}$$

$$P\{G_{di}^D \cap G_{di}^L\} = \int_{T_{di}^{D-}}^{T_{di}^{D+}}\left(\int_0^{T_{di}{}^L/2} r(z)\,dz\right)g(t)\,dt = Q_M^H \tag{8.31}$$

where

$$r(z) = \frac{z}{(\sigma_{di}^L)^2}\exp\left[-\frac{z^2}{2\sigma_{di}^L}\right] \qquad z \geq 0 \tag{8.32}$$

(the interval in its integral form is a function of a variable t)

$$g(t) = \frac{1}{\sqrt{2\pi}\sigma_{di}^D}\exp\left[-\frac{t^2}{2(\sigma_{di}^D)^2}\right]$$

$$T_{di}^L = |T_{di}^{D+}| + |T_{di}^{D-}| + t \quad \text{(for the datum feature di at MMC)}$$

$$|T_{di}^{D+}| + |T_{di}^{D-}| - t \quad \text{(for the datum feature di at LMC)} \tag{8.33}$$

- For an internal/external slot (open feature), as shown in Figure 8.4(d), the slot feature can be also specified as a datum feature, and it has a location tolerance for the true position; thus, a corresponding probability model is used with regard to its width. The internal or external slots are assumed to have location errors only in the X direction, where the IIDN variables have a zero mean and standard deviation σ. In other words, it follows a normal distribution, rather than Rayleigh distribution such as dimensional error. Since the location tolerance zone is still a function of feature size when MMC or LMC is specified, the following equations are obtained:

When RFS is specified in the datum feature,

$$P\{G_{di}\} = P\{G_{di}^D \cap G_{di}^L\} = P\{G_{di}^D\}P\{G_{di}^L\} = Q_R^S \tag{8.34}$$

$$P\{G_{di}^D\} = 2\Phi\left(\frac{T_{di}^D}{\sigma_{di}^D}\right) - 1 \tag{8.35}$$

$$P\{G_{di}^L\} = 2\Phi\left(\frac{T_{di}^L}{\sigma_{di}^D}\right) - 1 \tag{8.36}$$

When MMC or LMC is specified in the datum feature,

$$P\{G_{di}\} = P\{G_{di}^D \cap G_{di}^L\} = P\{G_{di}^D\}P\{G_{di}^L|G_{di}^D\} \tag{8.37}$$

$$P\{G_{di}^D \cap G_{di}^L\} = \int_{T_d^{D-}}^{T_d^{D+}} \left(\int_0^{T_{di}^{L/2}} g(z) \, dz \right) g(t) \, dt = Q_M^S \tag{8.38}$$

where

$$g(z) = \frac{1}{\sqrt{2\pi}\sigma_{di}^L} \exp\left[-\frac{z^2}{2(\sigma_{di}^L)^2} \right] \tag{8.39}$$

(the interval in its integral form is a function of a variable t)

$$g(t) = \frac{1}{\sqrt{2\pi}\sigma_{di}^D} \exp\left[-\frac{t^2}{2(\sigma_{di}^D)^2} \right]$$

$$T_{di}^L = |T_{di}^{D+}| + |T_{di}^{D-}| + t \quad \text{(for the datum feature } di \text{ at MMC)}$$

$$|T_{di}^{D+}| + |T_{di}^{D-}| - t \quad \text{(for the datum feature } di \text{ at LMC)} \tag{8.40}$$

8.3.1.2. Regardless of Feature Size. The features on the hypothetical part are assumed to be independent of each other if the RFS call-out is used. However, in some cases, the MMC or LMC is specified to the datum feature in the datum reference frame for the controlled features (i.e., small holes); thus, the location tolerance caused by the extra allowed shift is proportional to the departure of the datum feature size at MMC or LMC. Here, the independence in diameters of the controlled holes is still valid, and the probability of producing a good feature is the product of two independent probabilities for location and dimension; however, the probability equation should be changed as follows:

- For surrounding (controlled) holes at RFS,

  If the modifier for the small holes in Figure 8.4(c) and (d) is changed to RFS, the location dimension for these small holes (controlled features) should be based upon the location tolerance of the pilot hole. This is important because the location tolerances for these holes are independent of the actual hole size, and the extra allowed shift produced from the variability of the cylindrical location tolerance zone of the pilot hole's true position applies only to small holes. Furthermore, it is pointless to calculate the probability of only making a good controlled feature, since that probability is controlled by variability of the datum feature and is closely related to the probability of making a good datum feature. Hence, it is better to calculate the probability of making two good related features first and then obtain the probability of making a good dependent feature by dividing the result by Q_M^H or Q_M^S. (See Section 8.4.1.) This equation is suggested only for probability for a feature point of view and is written

$$P\{G_{ci}\} = P\{G_{ci}^D \cap G_{ci}^L\} = P\{G_{ci}^D\}P\{G_{ci}^L | G_{di}^D \cap G_{di}^L\} \tag{8.41}$$

$$P\{G_{ci}^D\} = 2\Phi\left(\frac{T_{ci}^D}{\sigma_{ci}^D}\right) - 1$$

$$P\{G_{ci}^L\} = P\{G_{ci}^L | G_{di}^D \cap G_{di}^L\} = \frac{P\{G_{ci}^L \cap G_{di}^D \cap G_{di}^L\}}{P\{G_{di}^D \cap G_{di}^L\}} = \frac{P\{G_{ci}^L \cap G_{di}^D \cap G_{di}^L\}}{Q_M^H} \qquad (8.42)$$

$$= \int_{T_{di}^{D-}}^{T_{di}^{D+}} \left[\int_0^{T_{di}^L/2} \left[\int_0^{T_{ci}^L/2} r(w)\, dw \right] r(z)\, dz \right] g(t)\, dt \cdot \frac{1}{Q_M^H} \qquad (8.43)$$

where

$$r(w) = \frac{w}{(\sigma_{ci}^L)^2} \exp\left[-\frac{w^2}{2\sigma_{ci}^L}\right] \qquad w \geq 0$$

$$T_{ci}^L: \quad z/2 + T_{no\,min\,al}^L = \frac{|T_{di}^{D+}| + |T_{di}^{D-}| + t}{2} + T_{no\,min\,al}^L \qquad (8.44)$$

(for the feature *ci* at RFS and *di* in the datum reference at MMC)

$$z/2 + T_{no\,min\,al}^L = \frac{|T_{di}^{D+}| + |T_{di}^{D-}| - t}{2} + T_{no\,min\,al}^L$$

(for the feature *ci* at RFS and *di* in the datum reference at LMC)

8.3.1.3. Maximum Material Condition and Least Material Condition. Section 8.2 indicated that solutions from both MMC and LMC were identical when using the same diameters and true position tolerance specifications from their experimentations. The section also implied that identical statistical models for MMC and LMC would be used in this chapter to produce the same statistical result.

For surrounding holes at MMC/LMC, as shown in Figure 8.4(c) and (d), the surrounding holes have a datum feature specified at MMC, so the location tolerances for these holes should be relative to the variability of the location tolerance zone of the datum feature (i.e., pilot hole), as well as to the dimensional tolerances of themselves. In this case, the location tolerances of these holes are dependent on their dimensional tolerances, and the location tolerance zones of the holes are determined by sum of the dimensional tolerance and the extra allowed pattern shift from the datum feature. Also, the true hole positions are controlled by the true datum feature position, and the displacement of the true positions is dependent on the proportionate difference between MMC/LMC size and the actual size of the datum feature. Therefore, the displacement of the controlled features as a unit radially equals $T_{di}^L/2$, and this displacement can be stacked into the location tolerances for the surrounding holes as an extra allowed pattern shift so that it can ensure part functionality.

Figures 8.5 and 8.6 further illustrate the relationship between a datum feature and the surrounding holes where the extra allowed shift resulted in tolerance stacking

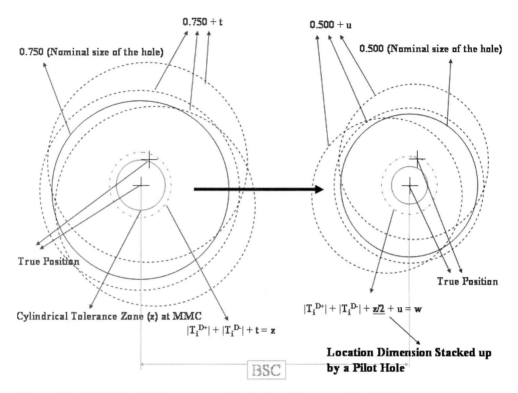

Figure 8.5   Relationship between a pilot hole and surrounding hole.

of the location diameter zone when the conditional probability of the statistical model is employed. Accordingly, the results obtained from the equations that follow are a function of variation of the datum feature dimension from its nominal value (t). The following equations consider the case of the datum feature (hole) subject to the size variations shown in Figure 8.4(c) (similar equations for the case depicted in Figure 8.4(d) can also be derived by replacing Q_M^H with Q_M^S in Equation 8.45):

$$P\{G_{ci}\} = P\{G_{ci}^D \cap G_{ci}^L\} = P\{G_{ci}^D\}P\{G_{ci}^L | G_{ci}^D \cap (G_{di}^D \cap G_{di}^L)\}$$

$$P\{G_{ci}^D \cap G_{ci}^L\} = P\{G_{ci}^D\} \frac{P\{G_{ci}^L \cap G_{ci}^L \cap (G_{di}^D \cap G_{di}^L)\}}{P\{G_{ci}^L \cap (G_{di}^D \cap G_{di}^L)\}}$$

$$= P\{G_{ci}^D\} \frac{P\{G_{ci}^L \cap G_{ci}^L \cap (G_{di}^D \cap G_{di}^L)\}}{P\{G_{ci}^L\}P\{G_{di}^D \cap G_{di}^L\}} \qquad (8.45)$$

$$= \int_{T_{di}^{D-}}^{T_{di}^{D+}} \left[\int_0^{T_{di}^{L}/2} \left[\int_{T_{ci}^{D-}}^{T_{ci}^{D+}} \left\{ \int_0^{T_{ci}^{L}/2} r(w)\,dw \right\} g(u)\,du \right] r(z)\,dz \right] g(t)\,dt \cdot \frac{1}{Q_M^H} \qquad (8.46)$$

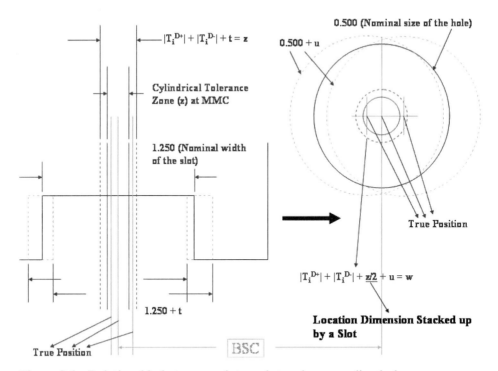

Figure 8.6   Relationship between a datum slot and surrounding hole.

In the foregoing equations,

$$r(w) = \frac{w}{(\sigma_{ci}^{L})^{2}}\exp\left[-\frac{w^{2}}{2\sigma_{ci}^{L}}\right] \qquad w \geq 0 \qquad (8.47)$$

(the interval in its integral form is a function of variables u and z)

$$T_{ci}^{L} = |T_{ci}^{D+}| + |T_{ci}^{D-}| + z/2 + u = |T_{ci}^{D+}| + |T_{ci}^{D-}| + \frac{|T_{di}^{D+}| + |T_{di}^{D-}| + t}{2} + u$$

(for the feature ci at MMC and di in the datum reference at MMC)

$$|T_{ci}^{D+}| + |T_{ci}^{D-}| + z/2 + u = |T_{ci}^{D+}| + |T_{ci}^{D-}| + \frac{|T_{di}^{D+}| + |T_{di}^{D-}| - t}{2} + u$$

(for the feature ci at MMC and di in the datum reference at LMC)

$$|T_{ci}^{D+}| + |T_{ci}^{D-}| + z/2 - u = |T_{ci}^{D+}| + |T_{ci}^{D-}| + \frac{|T_{di}^{D+}| + |T_{di}^{D-}| + t}{2} - u$$

(for the feature ci at LMC and di in the datum reference at MMC)

$$|T_{ci}^{D+}| + |T_{ci}^{D-}| + z/2 + u = |T_{ci}^{D+}| + |T_{ci}^{D-}| + \frac{|T_{di}^{D+}| + |T_{di}^{D-}| - t}{2} - u$$

(for the feature ci at LMC and di in the datum reference at LMC)

Therefore, equations for the statistical model calculate the probabilities that a part will be acceptable; these equations are formalized using $P\{G\} = \prod_{i=1}^{n} P\{G_i\}$, based on Equations 8.1 to 8.16. For the hypothetical part, it is assumed that four essential features (three holes and one slot) have critical functional requirements. The following cases show the formal models for obtaining the probability of making a good part for each of the datum reference frames or design alternatives selected:

- **Case 1:** Independent features with RFS and a single datum reference frame/multiple datum reference frames not subject to size variations (Figure 8.4(a) and (b) with RFS)

$$P\{G\} = \prod_{ii=1}^{4} P\{G_{ii}\} = \prod_{ii} P\{G_{ii}^{D}\} P\{G_{ii}^{L}\} =$$

$$\prod_{ii=1}^{2} \left[\left\{ 2\Phi\left(\frac{T_{ii}^{D}}{\sigma_i^{D}}\right) - 1 \right\} \cdot \left\{ 1 - \exp\left[-\frac{(T_{ii}^{L})^2}{8(\sigma_i^{L})^2} \right] \right\} \right] \cdot Q^S \cdot Q^H$$

$$\text{where} \quad Q^S = Q_R^S \text{ or } Q_M^S$$

$$Q^H = Q_R^H \text{ or } Q_M^H \tag{8.48}$$

- **Case 2:** Independent features with MMC/LMC and a single datum reference frame/multiple datum reference frames not subject to size variations (Figure 8.4(a) and (b))

$$P\{G\} = \prod_{ii=1}^{4} P\{G_{ii}\} = \prod_{ii=1}^{2} \left[\int_{T_{ii}^{D-}}^{T_{ii}^{D+}} \left(\int_{0}^{T_{ii}^{L}/2} r(y) \, dy \right) g(x) \, dx \right] \cdot Q^S \cdot Q^H$$

$$\text{where} \quad Q^S = Q_R^S \text{ or } Q_M^S$$

$$Q^H = Q_R^H \text{ or } Q_M^H \tag{8.49}$$

- **Case 3:** Controlled features with RFS and multiple datum reference frames subject to size variations (Figure 8.4(c) and (d) with RFS)

$$(1) \; P\{G\} = \prod_{i=1}^{4} P\{G_i\} = Q^S \cdot Q_M^H$$

$$\cdot \prod_{ci=1}^{2} \left[\left\{ \int_{0}^{T_{ci}^{L}/2} r(w) \, dw \right\} \cdot \left\{ 2\Phi\left(\frac{T_{ci}^{D}}{\sigma_i^{D}}\right) - 1 \right\} \right] \tag{8.50}$$

$$\text{where } Q^S = Q_R^S \text{ or } Q_M^S$$

$$(2)\ P\{G\} = \prod_{i=1}^{4} P\{G_i\} = Q^H \cdot Q_M^S$$

$$\cdot \prod_{ci=1}^{2}\left[\left\{\int_{0}^{T_{ci}^{L}/2} r(w)\,dw\right\} \cdot \left\{2\Phi\left(\frac{T_{ci}^{D}}{\sigma_i^D}\right) - 1\right\}\right] \tag{8.51}$$

where $Q^H = Q_R^H$ or Q_M^H

- **Case 4:** Controlled features with MMC/LMC and multiple datum reference frames subject to size variations (Figure 8.4(c) and (d))

$$(1)\ P\{G\} = \prod_{i=1}^{4} P\{G_i\}$$

$$= \prod_{ci=1}^{2}\left[\left\{\int_{T_{ci}^{D-}}^{T_{ci}^{D+}}\left(\int_{0}^{T_{ci}^{L}/2} r(w)\,dw\right)g(u)\,du\right\}\right] \cdot Q_M^H \cdot Q^S \tag{8.52}$$

where $Q^S = Q_R^S$ or Q_M^S

$$(1)\ P\{G\} = \prod_{i=1}^{4} P\{G_i\}$$

$$= \prod_{ci=1}^{2}\left[\left\{\int_{T_{ci}^{D-}}^{T_{ci}^{D+}}\left(\int_{0}^{T_{ci}^{L}/2} r(w)\,dw\right)g(u)\,du\right\}\right] \cdot Q_M^S \cdot Q^H \tag{8.53}$$

where $Q^H = Q_R^H$ or Q_M^H

8.3.2 Optimal Process Plans Using a Cost Model

Section 8.2 suggested a cost model that consists of two main cost sources in order to find the best process plans using statistical-based models among process alternatives. In this model,

$$C_p = \text{cost to produce} + \text{warranty cost}$$
$$= C_{pr} + P\{\text{defect}\}*C_d = C_{pr} + [1 - P\{G\}]*C_d$$

where

C_{pr} = production cost per a part = t_m (total manufacturing time of a part)

C_m = manufacturing cost per one part

C_d = cost of a defective part

At this point, the simple cost model is extended using the production cost model and the tolerance cost model discussed in the literature. The formal production cost per a part (C_{pr}) with some constraints (i.e., feed constraint) is presented in Section 6.4 and can be written as

$$C_{pr} = \frac{C_b}{N_b} + C_m\left[t_m + t_h + \frac{t_m}{t}\left(t_t + \frac{C_r}{C_m}\right)\right] \tag{8.54}$$

where

C_b = setup cost for a batch
C_m = total machine and operator rate (including overhead)
C_r = tool cost
N_b = batch size

In addition, process planning decisions, such as process selection, and machine/tool selection, are influenced by the type of tolerance and can also affect the cost. As discussed by [Malek and Asadathorn, 1997] and [Zhang et al, 1999], a certain amount of expenditure (which includes the cost of acquiring and operating accurate machine tools) is needed to achieve certain levels of dimensional and positional accuracy. Also, the tolerance cost is affected mainly by the tolerance parameters such as tolerance limit set by a company. Even though the tool cost is already included in the production cost, it is useful to include a tolerance-cost model in the optimization equation. Hence, the exponential function is traditionally suggested to find the cost–tolerance relationship [Zhang et al, 1999], depicted in Figure 8.7.

Therefore, the formal optimization cost model can use the information from experience-based manufacturing inputs and statistical models to evaluate whether each plan is a whole part (offline) operation or an individual (feature i) in (online) operation. A mathematical representation of the optimization model is

$$\text{Min } C_p = C_{pr} + \sum_{i=1}^{n}[g(\delta_i^D, \delta_i^L)] + \left[1 - \prod_i P\{G_i\}\right] \times C_d \tag{8.55}$$

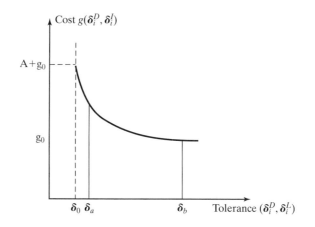

Figure 8.7   A cost model for tolerance [Zhang et al., 1999. Used by permission of Taylor & Francis, Ltd. (http://www.tandf.co.uk/)].

where

A, g_0, and δ_0 are constants used to determine the position of the cost-tolerance curve

δ_i^D = the tolerance aimed for each feature for dimension

δ_i^L = the tolerance aimed for each feature for location

$g(\delta_i^D, \delta_i^L) = A \cdot \{e^{-B(\delta_i^D - \delta_0)} + e^{-B(\delta_i^L - \delta_0)}\} + g_0$ is a cost model for tolerance

B is a constant used to control the curvature of its function

These equations are subject to the following constraints (see Section 6.4):

1. Spindle-speed constraint:

$$n_{\min} < n_w < n_{\max} \quad \text{(for a part)}$$

$$n_{t, \min} < n_t < n_{t, \max} \quad \text{(for tool)}$$

2. Feed constraint:

$$f_{\min} < f < f_{\max}$$

3. Cutting-force constraint:

$$F_c < F_{c, \max}$$

4. Power constraint:

$$P_m < P_{\max}$$

5. Surface-finish constraint:

$$R_a < R_{a, \max}$$

6. Tolerance constraint [Zhang et al., 1999]:

$$\delta_a^D < \delta_i^D < \delta_b^D$$

$$\delta_a^L < \delta_i^L < \delta_b^L$$

where

δ_a, δ_b is a region where a company determines that the tolerance is economically feasible

$\delta_0 \leq \delta_a^D$, $\delta_b^D \leq Min\{|T_i^{D-}|, |T_i^{D+}|\}$ for the dimension of a feature

$\delta_0 \leq \delta_a^L$, $\delta_b^L \leq T_i^L$ for the location of a feature

Traditionally, manufacturing features can be directly identified from a product drawing created using a feature-based CAD system; then an operation routing can be determined by comparing the manufacturing features to a set of manufacturing capabilities. Once these plans are generated, they are normally detailed for a set of generic equipment and tooling. In the next step, the alternative plans for each manufacturing task are identified. These alternative plans can be represented by an AND–OR digraph, which maintains the manufacturing precedence identified in the feature extraction step. In the next step, specific machine, tooling, and fixture combinations that are capable of creating the feature are identified for each generic alternative plan. These alternative plans are explicitly maintained by the AND–OR graph structure. By maintaining the alternatives as long as possible, specific processing decisions can be made online using the most up-to-date system status information [Wysk et al., 1995].

For our case study, an initial operation routing can be generated using experience-based criteria (such as manufacturing or process capabilities) and reference material (such as the process variability chart in Table 8.1). For example, hole-making guidelines might be obtained from a machinist handbook, as shown in Figure 8.8.

Based on these criteria, possible hole-making operations can be tentatively determined for the required accuracy. For example, the dimensional and positional tolerances of the small holes (diameter 0.5) are 0.006. Both drilling and finish-boring may be required to get the accuracy of the feature according to the guidelines in Figure 8.8. However, the reaming operation may be able to obtain the required accuracy if it uses guidelines from other factories. Therefore, it seems that the deterministic guidelines are not quite generic and cannot be considered reliable. It might be necessary to identify and evaluate alternative plans for each manufacturing task using the schematic simulation procedure instead of keeping a single sequential plan as a reference. Specific processing decisions among alternative plans maintained in the library can be made either online or offline in conjunction with results from the statistical model for a feature or a whole part in order to find an optimal process plan. This is further described in the next section. Table 8.5 shows an operation routing summary that includes a set of possible alternative plans obtained by deterministic guidelines for the hypothetical part.

```
If Diameter ≤ 0.5"

A. True position > 0.010"
        1. Tolerance > 0.010"
            Drill the hole.
        2. Tolerance ≤ 0.010"
            Drill and ream the hole.

B. True position ≤ 0.010"
        1. Tolerance ≤ 0.010"
            Drill, then finish bore the hole.
        2. Tolerance ≤ 0.002"
            Drill, semi–finish bore,
            then finish bore the hole.
```

Figure 8.8   Hole-making guidelines from the process planner's handbook.

TABLE 8.5   An Operation Routing Summary for the Hypothetical Part in Figure 8.2

Part No: hypothetical part in Figure 8.2 (a)		*Raw Material:* SAE 1040 HRS	*Machine:* Vertical Machining Center (capable of milling and drilling)	
No.	**Op. type**	**Feature**	**Tooling**	**Time (min.)**
1	Side mill side	Datum A	1.0″ end mill	0.58
2	Side mill side	Datum B	1.0″ end mill	0.58
3	Face mill		1.0″ end mill	0.85
4	Slot mill	Slot	0.25″ end mill	1.02
(5)	Re-fixture			
6	Twist drill	Large hole	0.495″ twist drill	0.13
7	Finish hole	Large hole		
7.1	Ream	Large hole	0.50″ ream	0.18
7.2	Semi-finish bore	Large hole	0.498″ bore	0.19
7.3	Finish bore	Large hole	0.50″ bore	0.17
(8)	Re-fixture			
9	Twist drill	Small holes (2)	0.745″ twist drill	0.34
10	Finish hole	Small holes (2)		
10.1	Ream	Small holes (2)	0.75″ ream	0.31
10.2	Semi-finish bore	Small holes (2)	0.748″ bore	0.33
10.3	Finish bore	Small holes	0.75″ bore	0.29
11	Inspect		-	1

$Tm = \dfrac{l}{f_v}$ for milling operation

where $f_v = f*RPM*z = 0.007*3820*2 = 53.5$ ipm

$RPM = \dfrac{12V}{\pi D} = \dfrac{12(250)}{\pi 0.25} = 3820$

$1 = (3.250 - 0.25)*2 + (1.75 - 0.25)*2 = 9$ in.

For all the cases of a part drawing, an operation routing summary with process alternatives can be established to generate an AND–OR graph of possible alternatives. However, the AND–OR graphical representation is not discussed in this chapter. Hence, 36 combinations of alternatives can be generated for the part in Figure 8.4, by changing the operation sequences and adding another fixture. For each alternative, the total manufacturing time and tolerances focusing on the specific features could also be changed. Therefore, all the alternatives can be evaluated by the optimization model using the formal statistical-based engineering approach proposed in this chapter. Decisions regarding the efficient design, datum establishment, and parameter selections might be derived using a curve-fitting approach based on experimental data. In Section 8.4.2, further illustration is provided to show how to find the best plan among alternatives. The next section presents a schematic representation of a solution-seeking procedure to find an optimal plan, as well as an efficient design for both product and process engineers.

8.3.3 Solution Procedure

The flowchart in Figure 8.9 illustrates a solution procedure using the proposed optimization cost and statistical models. It can provide a useful guideline for both product

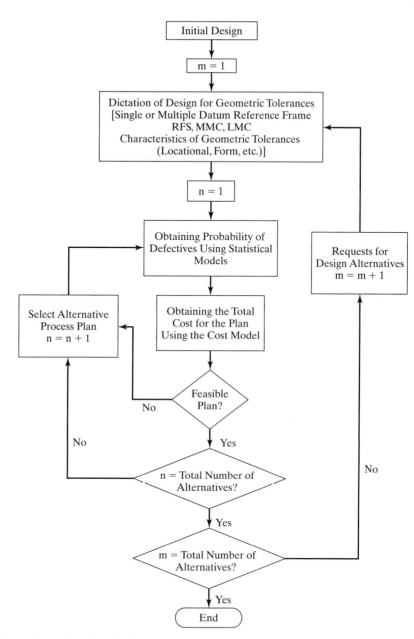

Figure 8.9   A flowchart for finding the best plan using statistical and cost-based models.

and process engineers, who can evaluate both design alternatives and alternative process plans. For the part in Figure 8.4, there are four design cases and nine process alternatives, so the flowchart's values for m and n are 4 and 36, respectively.

Product designers suggest an initial design. The drawing can dictate the design characteristics for the part, so the engineers can choose an appropriate statistical model among the general cases. For each alternative process plan, the probability of defective parts ($[1 - \prod_i P\{G_i\}]$) can be calculated using the statistical model; this calculation uses the values for dimensional and positional accuracies provided by an experimental reference in the process capability. Based on the results from the statistical model, the total cost for a plan can be obtained using the objective function and constraints. If a plan does not satisfy some constraints, other alternatives will be evaluated. After all the total costs of the design and process alternatives are compared, the optimal solutions can be determined.

In the next chapter, the experimental results are based on the hypothetical sample parts; they illustrate the effect of changing process accuracy on the acceptance rate of manufacturing features and parts.

8.4 EXPERIMENTAL RESULTS FOR PRODUCT EXAMPLES AND ILLUSTRATIONS

To illustrate the methodology to be presented, and to perform a feasible comparison among design and process alternatives, probabilities of making two good features (such as a large hole and small hole for the part in Figure 8.10) are calculated with respect to the transitions of σ_i^D and σ_i^L for each design case examined in Section 8.3.1. This calculation can be performed by a computer.

The process tolerance characteristics used for the dimensional and positional accuracies (σ_i^D and σ_i^L) can be obtained from Table 8.1. These accuracies represent the

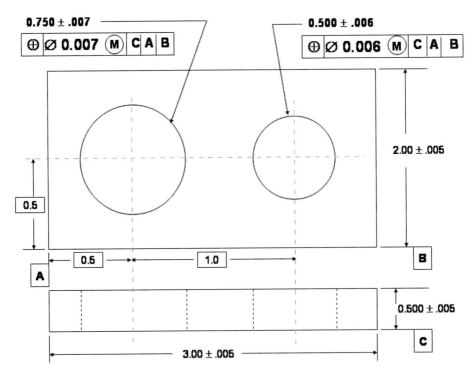

Figure 8.10 A simple part for the experiment.

standard process variability and/or the precision of the machine tool. For example, the positional accuracy of the twist drill reflects the precision of the movement of the drill on the X and Y axes. Both the dimensional and positional accuracy change from 0.0007 to 0.0034 and from 0.0007 to 0.0025, respectively, so that they can cover process variability for all hole-making processes in this example. To calculate the probabilities regarding parts that contain features requiring other than hole-making processes, other process tolerance characteristics can be used, such as those for milling operations. However, for purposes of illustration, this section refers to the information in Table 8.1.

The part in Figure 8.10 currently follows the case 2 categorized in Section 8.3.1. However, product designers can change drawings according to company requirements, part functionalities, and so on. Hence, several design alternatives can be derived from the same part. Table 8.6 shows those cases based on the categorization given in Section 8.3.1 by changing its reference frames for the features in Figure 8.8.

8.4.1 Effects of Selecting Material Side Modifier and Datum Feature on Acceptance Rate

Experimentation is performed in the mathematical simulation model using Equations 8.1 to 8.53. For example, an equation is used to calculate the probability of making

TABLE 8.6   The Cases for Changes in the Datum Reference Frames

Cases	Description	Reference Frame for Feature 1 (Large Hole)	Reference Frame for Feature 2 (Small Hole)
Case 1	Features with RFS and a single datum reference frame/multiple datum reference frames not subject to size variations	⊕ ⌀.007 C B A or ⊕ ⌀.007 C B A / D	⊕ ⌀.006 C B A or ⊕ ⌀.006 C D
Case 2	Features with MMC/LMC and a single datum reference frame/multiple datum reference frames not subject to size variations	⊕ ⌀.007 Ⓜ C B A or ⊕ ⌀.007 C B A / D	⊕ ⌀.006 Ⓜ C B A or ⊕ ⌀.006 Ⓜ C D
Case 3	Features with RFS and multiple datum reference frames subject to size variations	⊕ ⌀.007 Ⓜ C B A / D	⊕ ⌀.006 C D Ⓜ
Case 4	Features with MMC/LMC and multiple datum reference frames subject to size variations	⊕ ⌀.007 Ⓜ C B A / D	⊕ ⌀.006 Ⓜ C D Ⓜ

good holes (a large and small hole) for case 4 in Table 8.6 by changing values for σ_i^D and σ_i^L dynamically. The variables j and k are used to substitute for σ_i^D and σ_i^L, respectively. Even though there are two features ($i = 2$) in this example, it is assumed that the values of σ_i^D and σ_i^L are the same for each feature; this ensures that the equation observes the overall effect of changing process accuracies on the acceptance rate of manufacturing a good part and decreases the number of variables in the equation. It can also find the effect of changing process accuracies on the probability of making a feature and its dependent or controlled feature (i.e., a small hole with a datum feature subject to size variation in its reference frame). However, it might be quite difficult to calculate the probability for this feature only, since it has three variables in the equation and the feature is closely related to its datum feature. As discussed in Section 8.3.1, these calculations of the probabilities for making a whole part in a part point of view are as follows:

$$i(j,k): = \int_{-0.007}^{0.007} \left[\int_0^{\frac{(0.014+x)}{2}} \frac{y}{k^2} \cdot e^{\left(\frac{-y^2}{2k^2}\right)} dy \right] \cdot \left[\frac{1}{\sqrt{2 \cdot \pi \cdot j}} \cdot e^{\left(\frac{-x^2}{2j^2}\right)} \right]$$

$$\cdot \left[\int_{-0.006}^{0.006} \left[\int_0^{\frac{(0.012+0.028+2x+u)}{4}} \frac{w}{k^2} \cdot e^{\left(\frac{-w^2}{2k^2}\right)} dw \right] \frac{1}{\sqrt{2\pi j}} e^{\left(\frac{-u^2}{2j^2}\right)} du \right] dx \qquad (8.56)$$

Hence, Figures 8.11 and 8.12 illustrate the effect of changing process accuracy, which means changing a plan or tool according to alternatives, on the acceptance rate of manufacturing a part for various design alternatives. Several conclusions can be drawn based on these figures. First, for a given accuracy of a process, RFS brings a higher rejection rate due to out-of-specification features than any other two material-sided conditional geometric tolerance call-outs. Second, in the RFS call-out, in the specification of the feature's dimensional and geometric tolerances has a similar effect on the acceptance rate of this feature; however, when the material-sided conditional geometric tolerance call-out is used, the specification of the feature's dimensional tolerance tends to take a more important role than the geometric tolerance. From the perspective of quality improvement, this implies that the economical approach of increasing manufacturing capability for a feature with an MMC or LMC call-out improves the machine's dimensional accuracy instead of its location accuracy.

Generally, the RFS concept has several benefits: It preserves balance better than the MMC, it protects mating as well as the MMC, and it protects wall thickness and material preservation as well as the LMC. Even though the RFS can create tighter, more restrictive tolerances on the average part, it also risks increasing the cost of the overall product [Meadows, 1995]. Thus, there exists a trade-off between accuracy and cost in selecting material conditions for the part design, since it can produce defective parts more often and the requirement for the RFS can hardly be archived. Experimentation results also prove this fact. Figures 8.11 and 8.12 show that cases 3 and 4 in Table 8.6 have higher probabilities of making a good part than the other two cases when either dimensional or positional accuracy changes. For a given process accuracy, an RFS that does not have multiple datum reference frames subject to size variations will bring a higher rejection rate due to out-of-specification features than any other two material-sided conditional geometric tolerance call-outs. Also, this deficiency is rapidly enlarged

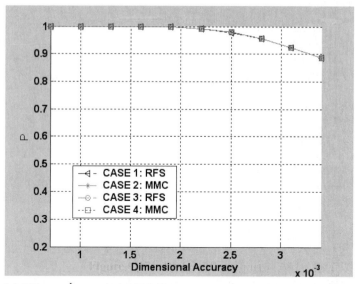

(a) When σ_i^L equals 0.0007 (lower extreme case)

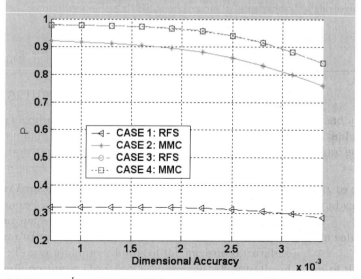

(b) When σ_i^L equals 0.0025 (upper extreme case)

Figure 8.11 Probability of making good holes (one large and one small hole) when positional accuracy is set.

when the process (or machine) location accuracy is slightly deteriorated, as shown in Figures 8.11(a) and 8.12. However, a design alternative (case 3) that has controlled features with RFS and multiple datum reference frames subject to size variations can bring a lower rejection rate than other alternatives, even when the positional accuracy

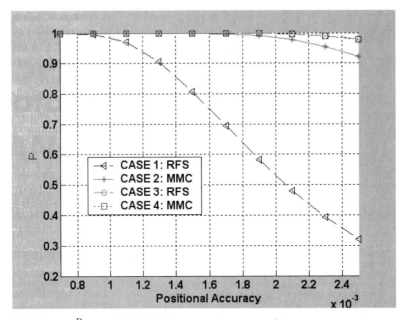

(a) When σ_i^D equals 0.0007 (lower extreme case)

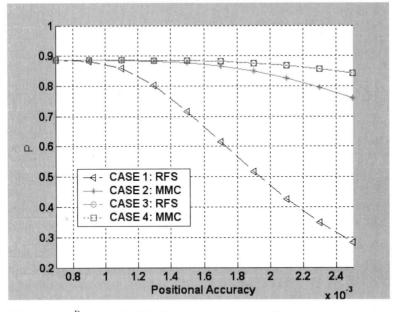

(b) When σ_i^D equals 0.0034 (upper extreme case)

Figure 8.12   Probability of making good holes (one large and one small hole) when dimensional accuracy is set.

is deteriorated. This design alternative shows almost the same result in case 4 (which has controlled features with MMC/LMC and multiple datum reference frames subject to size variations). In addition to selecting proper material condition modifiers to controlled features, this explains the importance and significant cost savings of selecting the appropriate class of datum features.

Figures 8.13–8.16 further illustrate the effect of changing process accuracy on the acceptance rate of manufacturing two holes for each design alternative. When the RFS or MMC/LMC call-out does not have multiple datum reference frames subject to size variations (case 1 and 2), the specification of the features' dimensional tolerance takes a more important role than the geometric tolerance. This occurs because the variation of probability is larger when the dimensional accuracy changes than when the positional accuracy changes. When the RFS or MMC/LMC call-out has multiple datum reference frames subject to size variations (case 3 and 4), the specification of the features' dimensional and positional tolerances has a similar effect on the acceptance rate. This occurs because the variation in probability is almost the same when the positional or dimensional accuracy changes.

A cost saving results from both the location tolerance and the extra allowed shift from the datum feature when multiple datum reference frames subject to size variations are used. The appropriate selection of material condition and datum features on the design of a part can greatly affect the quality and cost of a part.

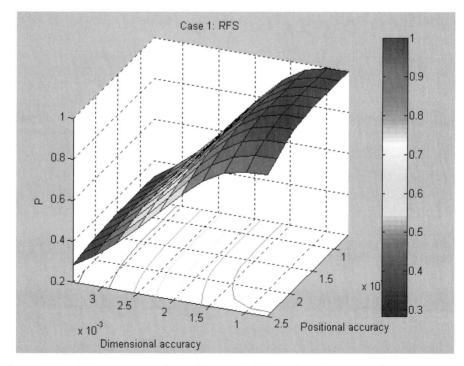

Figure 8.13   3-D representation of the probability of making a good part for case 1.

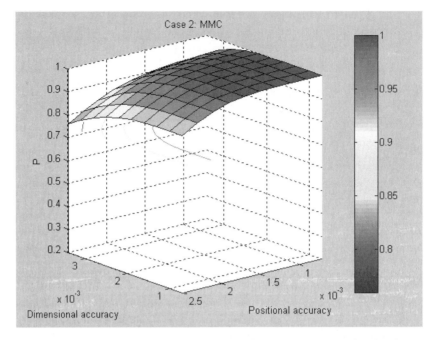

Figure 8.14   3-D representation of the probability of making good holes for case 2.

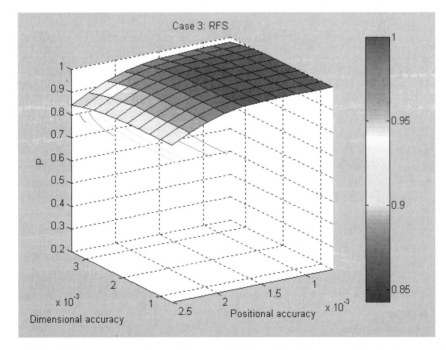

Figure 8.15   3-D representation of the probability of making good holes for case 3.

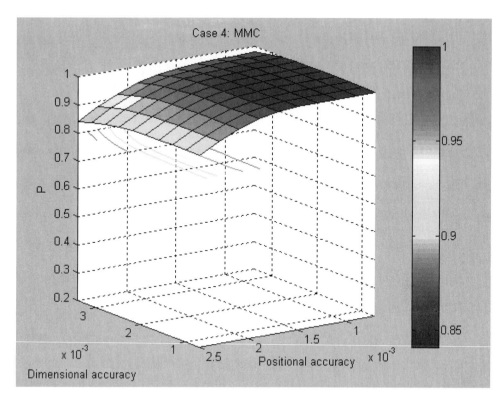

Figure 8.16   3-D representation of the probability of making good holes for case 4.

MMC is widely applied on the geometric tolerance in industry, largely because MMC ensures the interchangeability of individual parts. However, the implication of MMC on the quality/cost control is overlooked. This case study shows that significant cost savings can result by using conditional tolerance specifications and the appropriate datum feature selection. Process variability is an inherent characteristic of manufacturing, but it can prove to be very expensive if not considered properly. However, as shown in this example, applying standard statistical methods is an effective way to avoid excessive rejection costs. By separating feature size and feature location, the best process to produce a feature could be found.

This chapter has used several probability models to describe the likelihood of producing good products and the cost of these products. Probability distributions such as the truncated or skewed normal distributions can be more realistic in practice. Empirical probability distribution functions obtained directly from the company or experimentation data can also be used. Therefore, for the industrial application, an analysis of the distribution of process errors should always be performed before the process planning is done.

8.4.2 Illustration for Finding the Best Plan

At this point, the probabilities from the statistical models have been obtained when the specific operation's accuracy (i.e., boring) is known. Now, the cost optimization model can be calculated using this result and experimental information provided by a company. Therefore, product and process designers can suggest these statistical models for analysis of the distribution of process errors, as well as other reference guides, as the solution procedure suggested in Section 8.3.3. Table 8.7 shows an operation-routing summary, including a set of alternative plans for manufacturing the hypothetical part in Figure 8.10.

For the part in Figure 8.10, 32 alternative plans can be generated either by altering the combination of operation sequences or by adding another fixture mounted in the datum feature. Table 8.7 shows the total manufacturing time for the parts in those alternatives. Therefore, the optimization model, which uses the formal statistical-based engineering approach, can be used to evaluate every alternative.

For example, consider a simplified cost model that includes the total manufacturing time and manufacturing cost for a single part. In addition, assume that every alternative satisfies the manufacturing and tolerance constraints in Equation 8.55. Among four cases for the design alternatives, only case 4 is evaluated for this case study. Thus, 16 process alternatives will be considered; they are the plans determined by this design alternative in Section 8.3.3. Assuming that the other design alternatives and associated process alternatives were already evaluated, this illustration may be the final step of the evaluation procedure. Finally, all the results from the cost model can be compared

TABLE 8.7   An Operation-routing Summary for a Hypothetical Part in Figure 8.8

Operation	Description	Tooling	V	f	d	Time (min.)
10	Load part into fixture					0.5
20	Drill a large hole	47/64 dia	60	.010		0.208
(30)	Finish a large hole					
(30.1)	Ream a large hole	.750 dia	300	.008		0.053
(30.2)	Semifinish boring a large hole	.750 dia	300	.008		0.053
(30.3)	Finish boring a large hole	.750 dia	300	.008		0.053
(40)	Load a part for re-fixture					0.5
50	Drill a small hole	31/64 dia	50	.010		0.165
(60)	Finish a small hole					
(60.1)	Ream a small hole	.500 dia	250	.008		0.043
(60.2)	Semifinish boring a large hole	.500 dia	250	.008		0.043
(60.3)	Finish boring a large hole	.500 dia	250	.008		0.043
70	Unload and visually inspect					1.00

For a part with a single datum reference frame,
alternative 1: 10–20–50–70 (1.873 min.)
alternatives 2–4: 10–20–30–50–70 (1.926 min.)
alternatives 5–7: 10–20–50–60–70 (1.916 min.)
alternatives 8–16: 10–20–30–50–60–70 (1.969 min.)

For a part with multiple datum reference frames,
alternative 17: 10–20–40–50–70 (2.373 min.)
alternatives 18–20: 10–20–30–40–50–70 (2.426 min.)
alternatives 21–23: 10–20–40–50–60–70 (2.416 min.)
alternatives 24–32: 10–20–30–40–50–60–70 (2.469 min.)

in order to find the best plan among alternatives. Therefore, the cost model for this case study is as follows:

$$\min C_p = C_{pr} + \sum_{i=1}^{2} [g(\delta_i^D, \delta_i^L)] + \left[1 - \prod_i P\{G_i\}\right] \times C_d$$

$$= t_m \times C_m + \sum_{i=1}^{2} [A \times \{e^{-B(\delta_i^D - \delta_0)} + e^{-B(\delta_i^L - \delta_0)}\} + g_0] + \left[1 - \prod_i P\{G_i\}\right] \times C_d$$

where

$A = 2$

$g_0 = 2$

$\delta_0 = 0.0005$

$B = 2$

δ_i^D for the drill operation $= 0.0027$

δ_i^D for the reaming operation $= 0.00083$

δ_i^D for the semi-finish boring operation $= 0.0017$

δ_i^D for the finish boring operation $= 0.00067$

δ_i^L for the drill operation $= 0.001$

δ_i^L for the reaming operation $= 0.001$

δ_i^L for the semi-finish boring operation $= 0.00083$

δ_i^L for the finish boring operation $= 0.00067$

$C_m = \$30$ per hour for the machine and labor ($\$0.5$ per minute)

$C_d = \$50$ per part

$$P\{G\} = \prod_{i=1}^{2} P\{G_i\} = \int_{T_{di}^{D-}}^{T_{di}^{D+}} \left[\int_0^{T_{di}^L/2} \left[\int_{T_{ci}^{D-}}^{T_{ci}^{D+}} \left\{ \int_0^{T_{ci}^L/2} r(w|u, z)\, dw \right\} \right. \right.$$

$$\left. \left. g(u)\, du \right] r(z|t)\, dz \right] g(t)\, dt$$

Table 8.8 shows that alternative process plan 28 generates the lowest cost among all the plans. Its operation sequence is 10–20–30.2–40–50–60.2–70. Even though the cost difference is relatively small in this example, the opportunity for cost savings is significant when a large volume of this part is produced (i.e., 1,000,000 parts). Using the same technique, all the design cases can be concurrently evaluated with the process alternatives. In addition, although the results can differ according to the constant values used in the cost model and process accuracies as its inputs, these values can be adjusted

TABLE 8.8 Total Costs per Part for the Alternative Design Process Plans

Plan	Manufacturing Time	Operation Accuracy for Feature 1	Operation Accuracy for Feature 2	Cost per part
Alternative 17	2.373	$\delta_i^D = 0.0027$ $\delta_i^L = 0.001$	$\delta_i^D = 0.0027$ $\delta_i^L = 0.001$	\$ 14.965
Alternative 18	2.426	$\delta_i^D = 0.00083$ $\delta_i^L = 0.001$	$\delta_i^D = 0.0027$ $\delta_i^L = 0.001$	\$ 14.499
Alternative 19	2.426	$\delta_i^D = 0.0017$ $\delta_i^L = 0.00083$	$\delta_i^D = 0.0027$ $\delta_i^L = 0.001$	\$ 14.496
Alternative 20	2.426	$\delta_i^D = 0.00067$ $\delta_i^L = 0.00067$	$\delta_i^D = 0.0027$ $\delta_i^L = 0.001$	\$ 14.501
Alternative 21	2.416	$\delta_i^D = 0.0027$ $\delta_i^L = 0.001$	$\delta_i^D = 0.00083$ $\delta_i^L = 0.001$	\$ 13.694
Alternative 22	2.416	$\delta_i^D = 0.0027$ $\delta_i^L = 0.001$	$\delta_i^D = 0.0017$ $\delta_i^L = 0.00083$	\$ 13.691
Alternative 23	2.416	$\delta_i^D = 0.0027$ $\delta_i^L = 0.001$	$\delta_i^D = 0.00067$ $\delta_i^L = 0.00067$	\$ 13.696
Alternative 24	2.469	$\delta_i^D = 0.00083$ $\delta_i^L = 0.001$	$\delta_i^D = 0.00083$ $\delta_i^L = 0.001$	\$ 13.228
Alternative 25	2.469	$\delta_i^D = 0.00083$ $\delta_i^L = 0.001$	$\delta_i^D = 0.0017$ $\delta_i^L = 0.00083$	\$ 13.225
Alternative 26	2.469	$\delta_i^D = 0.00083$ $\delta_i^L = 0.001$	$\delta_i^D = 0.00067$ $\delta_i^L = 0.00067$	\$ 13.23
Alternative 27	2.469	$\delta_i^D = 0.0017$ $\delta_i^L = 0.00083$	$\delta_i^D = 0.00083$ $\delta_i^L = 0.001$	\$ 13.225
Alternative 28	**2.469**	$\boldsymbol{\delta_i^D = 0.0017}$ $\boldsymbol{\delta_i^L = 0.00083}$	$\boldsymbol{\delta_i^D = 0.0017}$ $\boldsymbol{\delta_i^L = 0.00083}$	**\$13.222***
Alternative 29	2.469	$\delta_i^D = 0.0017$ $\delta_i^L = 0.00083$	$\delta_i^D = 0.00067$ $\delta_i^L = 0.00067$	\$ 13.227
Alternative 30	2.469	$\delta_i^D = 0.00067$ $\delta_i^L = 0.00067$	$\delta_i^D = 0.00083$ $\delta_i^L = 0.001$	\$ 13.23
Alternative 31	2.469	$\delta_i^D = 0.00067$ $\delta_i^L = 0.00067$	$\delta_i^D = 0.0017$ $\delta_i^L = 0.00083$	\$ 13.227
Alternative 32	2.469	$\delta_i^D = 0.00067$ $\delta_i^L = 0.00067$	$\delta_i^D = 0.00067$ $\delta_i^L = 0.00067$	\$ 13.232

for a company's circumstances, since they are only reference data that result from a company's history or experience through experiments. They imply that the process tolerance chart is not deterministic and that it can be revised by using the simulated data from the statistical and cost model by means of reverse methods (such as the curve-fitting approach). With the proposed methods, the plans and designs for complex parts can be also evaluated by using a computer-aided statistical-based process tool. Therefore, the proposed methodology can vastly improve the decision making processes of companies or engineers, particularly when they have been dependent on only deterministic data or experienced-based guidance.

8.5 SUMMARY

This chapter introduced statistical models relating production errors that are used in conjunction with a process tolerance chart and the effect of tolerance on product cost. To formalize the statistical models, the chapter analyzed the effects of changing a product's design specifications—specifically, the selection of material condition and datum features—on the acceptance rate and total cost; it then proposed a decision tool that can obtain the optimal plan and feasible design in terms of the total manufacturing cost. The proposed methodology includes formal statistical and cost optimization models, and it can be concurrently used by both product and process designers to decide the optimal design, as well as the optimal process plan.

The tolerance setting in product specification has not received significant attention, in spite of the effect of tolerance on product cost. Most companies set limits on process capability, regardless of the type and number of features on a part. However, this chapter demonstrated the effect of these specifications on product cost and process planning. The interaction is quite pronounced and can significantly affect the product cost, as shown in the example presented.

The chapter also presented a statistical model to calculate the percentage of defective parts based on MMC and LMC specifications. For parts requiring assembly, MMC specification has become the standard. This procedure could be implemented on an engineering workstation platform to help the product engineer set limits and identify the most economic production method.

From the example presented, it is clear that the appropriate selection of material condition and datum features can largely affect the quality and the cost of a part. The example also shows that some design alternatives, which have controlled features with RFS and multiple datum reference frames subject to size variations, can bring lower rejection rates than other alternatives.

Therefore, using the results from the statistical models, the solution procedure can aid decision making to create an optimal plan and a cost-saving design. The procedure compares the total costs of several alternative plans and design alternatives. In the next step, more realistic models are developed; they include other probability distribution models for industrial application, several cases for alternative part design specifications, and a computer-aided decision-making tool using the proposed mathematical models.

8.6 KEYWORDS

Cumulative density function
Distribution density function
Feature size
Least material condition
Maximum material condition
Normal distribution
Process capability
Rayleigh distribution
Regardless of feature size

8.7 REVIEW QUESTIONS

8.1 In the hole-making process, will the hole size typically follow a normal distribution? Why or why not?

8.2 Assume that a bivariate normal distribution describes location errors in the x–y plane for the hole-making process. How can the Rayleigh distribution be derived from it? (*Hint:* $\rho_{xy} = 0$.)

8.3 Assume the hole size is uniformly distributed instead of normally distributed. How can the probability of manufacturing an acceptably large hole for the part in Figure 8.1 be calculated? Construct the equation and answer for the MMC case. (*Hint:* Use Table 8.4.)

8.4 When a material condition modifier (i.e., RFS, MMC, and LMC) is called out in the drawing's datum reference frame, what are its implications and benefits for each condition in terms of the part's quality and cost?

8.5 How can a statistical model with deterministic information benefit decision making for both process engineers and product designers?

8.6 If the MMC or LMC is specified to a feature, why are the full advantages of assembly realized when there is a mating feature?

8.7 Consider the bracket in Figure 8.1. If the datum reference frames for the large hole and small holes are changed to

and

respectively, which case can be applied? Construct the equation for the probability of a good part.

8.8 Define the following terms: controlled feature, datum feature, dependent feature, and independent feature.

8.9 What is the "extra allowed shift"? What is its implication for interrelated features, such as a datum feature, and the probability of making a good controlled feature?

8.8 REVIEW PROBLEMS

8.1 For all the cases in Table 8.6, calculate the probabilities of making a good small hole (only this feature) for the part in Figure 8.10. (*Hint:* $\sigma_i^D = 0.0027$ and $\sigma_i^L = 0.001$.)

8.2 Calculate the probabilities of making a good part for the part shown in Figure 8.10. Compare the results for four design cases. (*Hint:* $\sigma_1^D = 0.0017$, $\sigma_1^L = 0.00083$, $\sigma_2^D = 0.0027$, and $\sigma_2^L = 0.001$, where feature 1 is a large hole and feature 2 is a small hole.)

8.3 Using the input data for the cost model in Section 8.4.2, find the costs per part for all the alternative plans and design cases 1, 2, and 3. Which plan has the lowest total cost?

8.4 The part shown in Figure 8.17 is made of high-strength steel 4340. Prepare alternative process plans using the information in Table 8.1. (*Hint:* The milling tool is pre-selected as a 1-inch-diameter tool.)

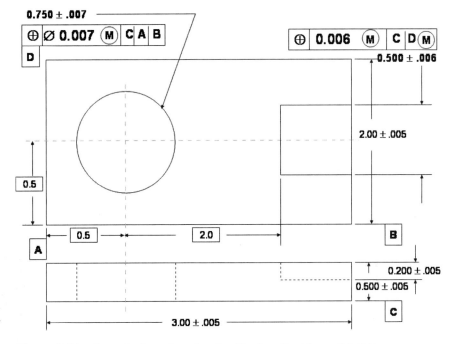

Figure 8.17   A workpiece drawing for Review Problems 8.4–8.7.

8.5 What are the other possible alternative designs for the part in Figure 8.17?

8.6 Derive the equations for the probability of making a good part for the part shown in Figure 8.17. Now calculate the probability of making a good part. (*Hint*: $\sigma_1^D = 0.0017$, $\sigma_1^L = 0.00083$, $\sigma_2^D = 0.003$, and $\sigma_2^L = 0.002$, where feature 1 is a large hole and feature 2 is a slot.)

8.7 Find the best process plan among the alternatives obtained from Review Problem 8.4. Use the same cost model as in Section 8.4.2.

8.9 REFERENCES

Abdel–Malek, L., and N. Asadathorn. "On Process Tolerance: A Brief Review and Models," *Integrated Product, Process, and Enterprise Design*, edited by Ben Wang. New York: Chapman & Hall, 1997, 90–116.

American Society of Mechanical Engineers. *Mathematical Definition of Dimensioning and Tolerancing Principles*, ASME Y14.5.1M-1994. New York, 1994.

———. *Dimensioning and Tolerancing*, ASME Y14.5M-1994. New York, 1995.

Bjorke, O. *Computer-Aided Tolerancing*. Trondheim, Norway: Tapir Publishers, 1978.

Clarke, A., and R. Disney. *Probability and Random Processes for Engineers and Scientists*. New York: John Wiley & Sons, Inc.

Deming, W. Edwards. "On Some Statistical Aids Toward Economic Production," *Interfaces*, 5(4) 1975.

Foster, L. W. *Modern Geometric Dimensioning and Tolerancing with Workbook Section*, 2d ed. Ft. Washington, MD: National Tool, Die, & Precision Machining Association, 1982.

Guttman, I., and S. Wilks. *Introductory Engineering Statistics*. New York: John Wiley & Sons, Inc., 1965.

Hahn, J., and G. Shapiro. *Statistical Methods in Engineering*. New York: John Wiley and Sons, 1967.

Harry, Mikel. *The Nature of Six Sigma Quality*. Motorola, Inc., Government Electronics Group 1988.

Hong, Y. S., and T. C. Chang. "A Comprehensive Review of Tolerancing Research," *International Journal of Production Research*, 40(11), 2002, 2425–2459.

——— "Tolerancing Algebra: A Building Block for Handling Tolerance Interactions in Design and Manufacturing (Part 2: Tolerance Interaction)," *International Journal of Production Research*, 41(1), 2003, 47–63.

Kiemele, M., S. Schmidt, and R. Berdine. *Basic Statistics: Tools for Continuous Improvement*, 4th ed. Colorado Springs, CO: Air Academic Press, 1997.

Lehtihet, E. A., and U. N. Gunasena. "Statistical Models for the Relationship between Production Errors and the Position Tolerance of a Hole," *Annals of the CIRP*, 39(1), 569–572.

Meadows, J. D. *Geometric Dimensioning and Tolerancing (Applications and Techniques for Use in Design, Manufacturing, and Inspection)*. New York: Marcel Dekker, Inc., 1995.

Pan, R., R. A. Wysk, and M. J. Chandra. *A Statistical Approach to Process Planning*, Pennsylvania State University, 1999.

Thimm, G., G. A. Britton, K. Whybrew, and S. C. Fok. "Optimal Process Plans for Manufacturing and Tolerance Charting," *Proc. Instn. Mech. Engrs.,* (215)B, 2001, 1099–1105.

Wang, B. *Integrated Product, Process, and Enterprise Design*. New York: Chapman & Hall, 1997.

Wei, C., C. Chen, and C. Tsai, "Fuzzy Assignment of Manufacturing Process Tolerance," *IEEE Transactions on Electronics Packaging Manufacturing*, (22)3, 1999, 191–194.

Wysk, R. A., B. A. Peters, and J. S. Smith. "A Formal Process Planning Schema for Shop Floor Control," *Engineering Design and Automation*, (1)1, 1995, 3–20.

Zhang, C., J. Luo, and B. Wang, 1999, "Statistical Tolerance Synthesis using Distribution Function Zones," *International Journal of Production Research*, (37)17, 1999, 3995–4006.

Chapter 9

Fundamentals of Industrial Control

Objective

This chapter is intended to provide a general background on the fundamentals of industrial control. It introduces programmable logic controllers (PLCs) and numerical control (NC), and discusses mathematical analysis, hardware definition, and hardware implementation. No background in control is assumed.

Outline

9.1 REVIEW OF CONTROL THEORY

Most computer-aided manufacturing systems can be modeled as closed-loop control systems. A closed-loop control system typically consists of a process being controlled, a controller, and a sensory feedback loop.

This chapter will use a tank-filling process as an example (Figure 9.1). The process is a liquid-filling operation, where a liquid-level sensor (like a plunger in a toilet) provides feedback. The controller receives a liquid-level command from a human operator and a current liquid-level reading from the sensor. If the current liquid level

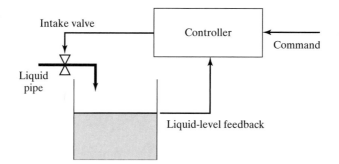

Figure 9.1   A liquid-filling process.

is lower than the command value, an intake valve is opened. When the liquid level has reached a limiting upper condition or command value, the intake valve is closed. The control system includes an actuator for turning the intake valve "on" and "off," a liquid-level sensor, and a controller.

A simple controller consists of a comparator and a switch. With the use of the controller, the liquid-level command is compared with the current liquid-level feedback. If the command value is higher, the switch turns on the intake valve. When the two values are the same, the switch turns off the intake valve. This type of controller is called an **on–off controller**; it runs the thermostat that heats houses or the mechanical system that flushes toilets.

A slightly more complex control system may control the filling speed. In this case, the intake valve can control the flow rate, and the flow rate is directly proportional to the rotational angle of the valve. The controller may turn the intake valve based on the difference between the liquid-level command value and the feedback value. This controller is called a **proportional error controller**, since the control action (turning the valve) is proportional to the difference (error) between the command and the feedback values.

9.1.1 Process Model

The liquid-filling system just described can be modeled with the formula

$$A\frac{dh(t)}{dt} = k\theta(t) \tag{9.1}$$

where

$$A = \text{cross-sectional area of the tank, in in}^2 \text{ or m}^2$$

$$h(t) = \text{liquid level at time } t, \text{ in in or m}$$

$$k = \text{valve constant, in } \frac{\text{in}^3}{\text{sec} \times \text{rad}} \text{ or } \frac{\text{m}^3}{\text{sec} \times \text{rad}}$$

$$\theta(t) = \text{valve rotational angle, in } rad$$

To simplify the problem, assume that the output of the controller is the valve rotational angle. In real life, the output would most likely be a voltage that energizes a solenoid to rotate the valve. The controller can be written as

$$\theta(t) = K_p[h_r - h(t)] = K_p e(t) \tag{9.2}$$

where

$$K_p = \text{controller gain, with no unit}$$

$$h_r = \text{liquid-level command, in in. or m}$$

$$e(t) = \text{error, the difference between the command and the current level (from the feedback device)}$$

Such a feedback control system is usually represented by a **block diagram**. (See Figure 9.2.) The input to the system is the command $R(s)$ (**set point** or **reference value**). $R(s)$ and other variables are all represented as functions of the complex variable s. $R(s)$ is obtained by means of a ***Laplace transformation*** (which yields a Laplace transform). The difference between the set point and the feedback is shown by a plus ($+$) set-point signal and a minus ($-$) feedback signal, which go into the circle. The output of the circle, $E(s)$, equals $R(s) - H(s)$. In addition, each box in the block diagram contains a **transfer function**, which is the ratio between the output and the input. For example, the transfer functions for the controller and the process are

$$D(s) = \frac{\theta(s)}{E(s)} \tag{9.3}$$

$$G(s) = \frac{H(s)}{\theta(s)} \tag{9.4}$$

To obtain the transfer function, apply the Laplace transform to the original differential equation. (Refer to Appendix C for a brief introduction to Laplace transformations.) Apply the Laplace transformation to both sides of Equation (9.1):

$$AsH(s) = k\theta(s)$$

$$G(s) = \frac{H(s)}{\theta(s)} = \frac{k}{As} = \frac{k/A}{s} \tag{9.5}$$

The Laplace transformation of Equation (9.2) is

$$\theta(s) = k_p E(s) \tag{9.6}$$

$$D(s) = \frac{\theta(s)}{E(s)} = k_p$$

Figure 9.2 captures the basic model of a feedback control system. This block diagram can be simplified and analyzed. The transfer functions and the feedback loop can

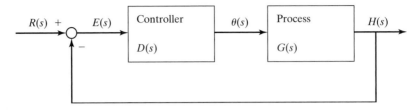

Figure 9.2 Block diagram of a feedback control system.

be combined to form an equivalent system transfer function (system output over system input) $\dfrac{H(s)}{R(s)}$:

$$H(s) = E(s)D(s)G(s)$$

$$\frac{H(s)}{R(s)} = \frac{D(s)G(s)E(s)}{R(s)} = \frac{D(s)G(s)[R(s) - H(s)]}{R(s)}$$

$$H(s) = D(s)G(s)R(s) - D(s)G(s)H(s)$$

$$[1 + D(s)G(s)]H(s) = D(s)G(s)R(s)$$

$$\frac{H(s)}{R(s)} = \frac{D(s)G(s)}{1 + D(s)G(s)} \qquad (9.7)$$

The system stability is determined by the denominator, $1 + D(s)G(s)$; the **characteristic equation** of the system is $1 + D(s)G(s) = 0$. To determine whether the system is stable, we examine the roots of the characteristic equation, which are called the **poles**. Then, write the characteristic equation as factors of the poles,

$$1 + D(s)G(s) = 1 + D(s)G(s) = \prod_{i=1}^{n}(s - p_i) = 0,\text{ where } p_i \text{ are the poles. These poles}$$

can contain real or complex numbers. If the real parts of the poles are negative, the system is stable. The Inverse Laplace transformation of the poles will turn each pole into an exponential function of time having the general form $\sum_{i=1}^{n} K_i e^{p_i t}$. When the pole is negative, the **steady-state** (when $t \to \infty$) system function will converge to a constant value. It takes only one non-negative pole to move the steady-state function value to infinity. In contrast, roots for $D(s)G(s) = 0$ are called **zeros** of the system. Zeros are not as important as poles in system stability. In the liquid-filling example, the system transfer function is

$$\frac{H(s)}{R(s)} = \frac{k_p \dfrac{k}{As}}{1 + k_p \dfrac{k}{As}} = \frac{k_p k}{As + k_p k} \qquad (9.8)$$

Therefore, the characteristic equation is $As + k_p k = 0$.

The only pole (root) is $s = -k_p k / A$. Since k_p, k, and A are all positive, the root is negative. Therefore, the system is stable.

The preceding analysis is based on the system transfer function. The system output, $h(t)$, equals the system transfer function times the input, $R(s)$. Typically, a step function is applied to the system to see whether the output will reach the step input value. The **final value theorem** can be applied to find the steady-state value:

$$\lim_{t \to \infty} h(t) = \lim_{s \to 0} s H(s)$$

Again, if the real part of all poles has a negative value, the system is stable. The steady-state value of $h(t)$ is $h(\infty)$. It can be found from $H(s)$.

Example 9.1.

The engine of a car generates a force of 500 Newtons. The viscous friction coefficient of the air is 5 Kg/s. The mass of the car is 1,000 Kg, and the initial speed is zero. The process equation can be written as

$$1{,}000 \frac{d^2 x(t)}{dt^2} + 5 \frac{dx(t)}{dt} = F_c(t)$$

where $x(t)$ is the location of the car and $F_c(t)$ is the force generated by the engine.

$$\text{Let } v(t) = \frac{dx(t)}{dt}.$$

The equation can be rewritten as

$$1{,}000 \frac{dv(t)}{dt} + 5v(t) = F_c(t)$$

The Laplace transformation of the equation (when finding the transfer function, ignore the initial conditions) is

$$1{,}000 s V(s) + 5 V(s) = F_c(s)$$

The process transfer function is

$$G(s) = \frac{V(s)}{F_c(s)} = \frac{1}{1{,}000 s + 5}$$

The cruise control has a transfer function of $D(s) = 1$.

From Equation (9.7), the system transfer function is

$$\frac{V(s)}{R(s)} = \frac{1 \cdot G(s)}{1 + 1 \cdot G(s)} = \frac{\dfrac{1}{1{,}000 s + 5}}{1 + \dfrac{1}{1{,}000 s + 5}} = \frac{1}{1{,}000 s + 6}$$

The pole of the system is located at $s = -0.006$; therefore, the system is stable.

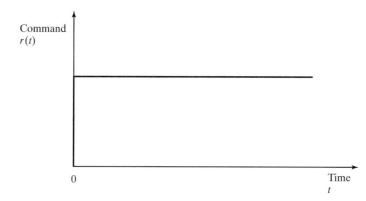

Figure 9.3 Step input.

If the input command to the system is a step function (i.e., a constant input value (see Figure 9.3) is added), then the system response can be plotted as in Figure 9.4.

The system output is plotted using the MatLab command

Step([1],[1000 6]);

When applying a step input with amplitude of 1, the system output converges to a value close to 0.17. This can be calculated using the final value theorem. The Laplace transform of a

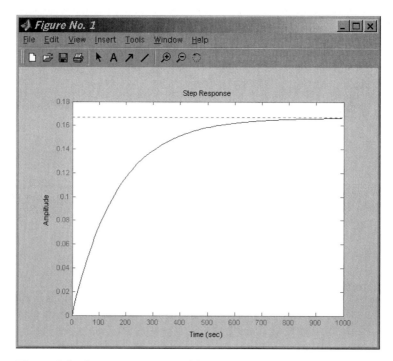

Figure 9.4 System response with a controller gain of 1.

step input is $R(s) = \dfrac{1}{s}$. Therefore,

$$V(s) = R(s) \cdot \frac{V(s)}{R(s)} = \frac{1}{s} \cdot \frac{1}{1{,}000s + 6}$$

$$v(\infty) = \lim_{s \to 0} sV(s) = \lim_{s \to 0} \frac{1}{1{,}000s + 6} = \frac{1}{6} = 0.1667$$

The result is consistent with the output graphed by MatLab. This value is much lower than the set point, which is 1. The steady-state error of the system is $1 - 0.1667 = 0.8333$. Although it is stable, the controller used in the system is not very good.

What happens if the **controller gain** is set at 100 (i.e., set $D(s) = 100$)? Then

$$\frac{V(s)}{R(s)} = \frac{100 \cdot G(s)}{1 + 100 \cdot G(s)} = \frac{\dfrac{100}{1{,}000s + 5}}{1 + \dfrac{100}{1{,}000s + 5}} = \frac{100}{1{,}000s + 105}$$

$$V(s) = \frac{1}{s} \cdot \frac{100}{1{,}000s + 105}$$

When the final value theorem is applied, the steady-state speed becomes

$$v(\infty) = \lim_{s \to 0} sV(s) = \lim_{s \to 0} \frac{100}{1{,}000s + 105} = \frac{100}{105} = 0.9524$$

It is now 95.24% of the command speed.

The new system can be plotted with the MatLab command step([100],[1000 105]). Then the system will reach the steady-state value much faster (see Figure 9.5). The **time constant** (how fast the steady-state will be reached) is the τ value in the following exponential output equation:

$$y(t) = 1 - e^{-\frac{t}{\tau}}$$

When $t = \tau$, $y(t) = 1 - e^{-1} = 0.632$. At one time constant, the output is 63.2% of the steady-state value. At three time constants ($t = 3\tau$), $y(3\tau) = 1 - e^{-3} = 0.9502$, the output reaches 95% of the steady-state value. From Figure 9.5, the time constant is about 10 seconds (corresponding to the amplitude of $0.9624 \times 0.63 = 0.6$). In Figure 9.4, the time constant is about 100 seconds (corresponding to an amplitude of $0.1667 \times 0.63 = 0.1$).

9.1.2 Controller

In the previous example, the controller algorithm was very simple:

$$U(s) = E(s) \times k_p$$

where

$U(s)$ = the output of the controller, called **manipulation**
$E(s)$ = error (input)
k_p = controller gain (constant).

The transfer function can be expressed as

$$D(s) = \frac{U(s)}{E(s)} = k_p$$

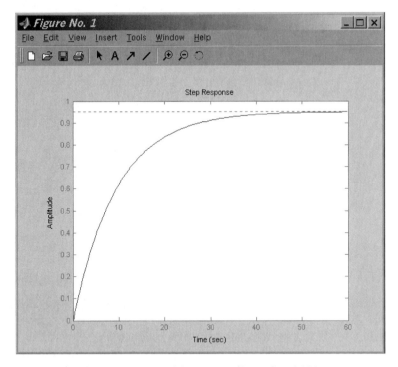

Figure 9.5   Step response with a controller gain of 100.

This controller is called a **proportional error (PE)** controller. It is part of a classical controller called a **proportional**, **integral**, and **derivative** (PID) controller. It describes the relationship between the manipulation and the error. As can be seen in the PE controller example, the manipulation is proportional to the error with a gain value of k_p. In an integral controller, the manipulation equals the integral of the error over time, multiplied by a gain k_I. A derivative controller uses the derivative of the error instead of the integral. The gain for a derivative controller is k_D. Integral and derivative controllers are never used on their own. They are always used in conjunction with a proportional controller. Equations for the controller can be written as follows:

Proportion error (PE) controller:

$$u(t) = k_p e(t)$$
$$U(s) = k_p E(s)$$
$$\frac{U(s)}{E(s)} = k_p \tag{9.9}$$

Integral (I) controller:

$$u(t) = k_I \int_0^t e(\tau) \, d\tau$$

$$U(s) = k_I \frac{1}{s} E(s)$$

$$\frac{U(s)}{E(s)} = \frac{k_I}{s} \qquad (9.10)$$

Derivative controller:

$$u(t) = k_D \frac{de(t)}{dt}$$

$$U(s) = k_D s E(s)$$

$$\frac{U(s)}{E(s)} = k_D s \qquad (9.11)$$

PID controller:

$$\frac{U(s)}{E(s)} = k_p + \frac{k_I}{s} + k_D s \qquad (9.12)$$

Example 9.2.

A proportional and derivative (PD) controller is used in the cruise controller discussed in Example 9.1. The gain values are $k_p = 100$, $k_I = 0$, $k_D = 500$. Compare the system response with that in Example 9.1.

The controller transfer function is $D(s) = 100 + 500s$.

The system transfer function is

$$\frac{V(s)}{R(s)} = \frac{(100 + 500s)\dfrac{1}{1,000s + 5}}{1 + (100 + 500s)\dfrac{1}{1,000s + 5}} = \frac{100 + 500s}{1000s + 5 + 100 + 500s} = \frac{500s + 100}{1500s + 105}$$

The steady-state speed is

$$V(s) = \frac{1}{s} \cdot \frac{500s + 100}{1500s + 105}$$

$$v(\infty) = \lim_{s \to 0} sV(s) = \lim_{s \to 0} \frac{500s + 100}{1,500s + 105} = \frac{100}{105} = 0.9524$$

The steady-state speed is the same as that of the PE controller. To plot the system response, use the MatLab command

step([500 100],[1500 105])

To compare the response with the output of the previous example, the following additional MatLab commands are used:

hold % Keep the previous plot

step([100],[1000 105]) %

This is from the PE controller example.

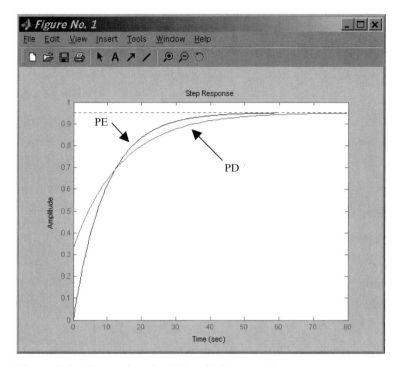

Figure 9.6   Comparing the PE and PD controller system response.

In Figure 9.6, the curve starting at (0,0) represents the one using a PE controller. The PD controller shows an initial jump in speed increase. This is due to the effect of the derivative controller. The slope of the increase in speed is slowed down, and it takes longer to reach steady state.

Example 9.3.

Try the same cruise control problem using a PI controller:

$$k_p = 100, k_I = 500, k_D = 0$$

The controller transfer function is $D(s) = 100 + \dfrac{500}{s}$.

The system transfer function is

$$\frac{V(s)}{R(s)} = \frac{\left(100 + \dfrac{500}{s}\right)\dfrac{1}{1{,}000s + 5}}{1 + \left(100 + \dfrac{500}{s}\right)\dfrac{1}{1{,}000s + 5}}$$

$$= \frac{\dfrac{100s + 500}{1000s^2 + 5s}}{\dfrac{1000s^2 + 5s + 100s + 500}{1000s^2 + 5s}} = \frac{10s + 500}{1000s^2 + 105s + 500}.$$

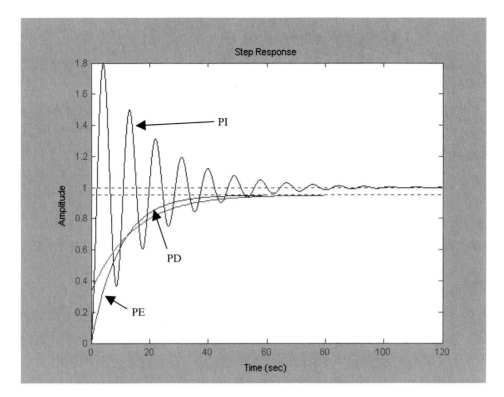

Figure 9.7   The system response including the PI controller.

The steady-state speed is

$$V(s) = \frac{1}{s} \cdot \frac{10s + 500}{1000s^2 + 105s + 500}$$

$$v(\infty) = \lim_{s \to 0} sV(s) = \lim_{s \to 0} \frac{100s + 500}{1000s^2 + 105s + 500} = \frac{500}{500} = 1.0$$

The steady-state speed is 100% of the command value. There is no steady-state error. To plot the system response, use

step([100 500],[1000 105 500])

The PI controller output (Figure 9.7) is radically different from the other two controllers. It initially oscillates widely, but eventually settles at the perfect 1.00 value. To reduce the amplitude of the oscillation, select a smaller integral gain k_I. When reducing the integral gain, the overshoot from the command value is reduced. However, it also lengthens the time it takes to reach the steady state. This is illustrated in Figure 9.8 using these commands:

```
>>  hold % Current plot held
>>  step([100 10],[1000 105 10]) % PI gain of 10
>>  step([100 1],[1000 105 1]) % PI gain of 1
```

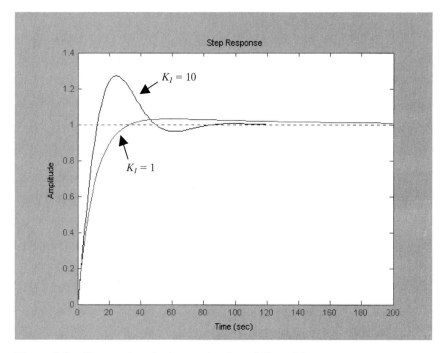

Figure 9.8 Comparing the integral gains of 10 and 1.

9.1.3 Discrete Control System

So far, the control systems presented have been in the **continuous time domain**. This means that all the functions are functions of time t. The analysis is completed using functions in the s domain, which is obtained with the Laplace transformation. Today, many control systems are implemented using digital controllers. In digital control, time is discrete. The control loop works at a fixed time interval T, which is the **sample time period**. For every T seconds, the input of the system is sampled and fed back to the controller. An analogy of this type of control system is the educational system. In this case, the sample time period is usually a semester. Student performance is evaluated at the end of each semester and the GPA is calculated only at the end of the semester. During the entire next semester, the GPA of a student does not change. Thus, the concept of an **ideal sampler** (Figure 9.9) is used. The ideal sampler samples the feedback signal at precisely T seconds. The output of the sampler is denoted as f^*.

Sample every T seconds

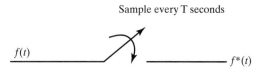

Figure 9.9 Ideal sampler.

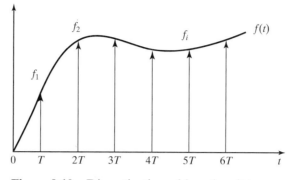

Figure 9.10 Discretization of function $f(t)$.

The input is made discrete through this ideal sampler. Figure 9.10 illustrates the discrete signal versus the original continuous signal:

$$f_1 = f(T), f_2 = f(2T), f_i = f(iT)$$

The vertical arrows show the discrete signal at each period. This signal can be represented by a **train of impulses** $f^*(t)$, where $f^*(t)$ is defined using the unit impulse $\delta(t)$, which has a duration of zero ($\Delta t \to 0$) and an amplitude of ∞. The area under the function equals 1 (Figure 9.11).

Therefore, the train of impulses $f^*(t)$ can be written as a function of the unit impulse function:

$$f^*(t) = f(0)\delta(t) + f(T)\delta(t - T) + f(2T)\delta(t - 2T) + \cdots$$

$$= \sum_{n=0}^{\infty} f(nT)\delta(t - nT)$$

$$= \sum_{n=0}^{\infty} f_n\delta(t - nT) \tag{9.13}$$

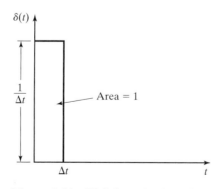

Figure 9.11 Unit impulse function.

Here, $\delta(t - nT)$ is a delayed unit impulse at time $t = nT$. The Laplace transform of the unit impulse is

$$L[\delta(t)] = 1 \tag{9.14}$$

$$L[\delta(t - nT)] = e^{-nsT} \tag{9.15}$$

The Laplace transform of the train of impulses is therefore

$$L[f^*(t)] = f_0 + f_1 e^{-sT} + f_2 e^{-2sT} + \cdots$$

$$= \sum_{n=0}^{\infty} f_n e^{-nsT} \tag{9.16}$$

A new transformation, the Z-transform, can be defined on the basis of this result. Let $z^{-n} = e^{-nsT}$, where

$$Z[f^*(t)] = f_0 + f_1 z^{-1} + f_2 z^{-2} + f_3 z^{-3} + \cdots$$

$$= \sum_{n=0}^{\infty} f_n z^{-n} \tag{9.17}$$

The Z-transform is a transformation method for discrete systems.

Example 9.4.

The Z-transform of a unit step is shown in Figure 9.12 and is obtained as follows:

$$u^*(t) = \delta(t) + \delta(t - T) + \delta(t - 2T) + \cdots$$

$$L[u^*(t)] = 1 + e^{-sT} + e^{-2sT} + \cdots$$

$$= \frac{1}{1 - e^{-sT}}$$

$$Z[u^*(t)] = \frac{1}{1 - z^{-1}}$$

$$= \frac{z}{z - 1}$$

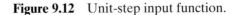

1

0   T $2T$ $3T$ $4T$ ...

Figure 9.12   Unit-step input function.

A discrete function can be expressed with a difference equation, such as

$$y_n = y_{n-1} + f_n \tag{9.18}$$

Let $y_i = 0, i < 0, f_n = 0$ if $n < 0$, and $f_n = 1$ if $n \geq 0$. The difference equation represents a discrete function:

$$
\begin{aligned}
n = 0 \quad & y_0 = y_{-1} + f_0 = 0 + 1 = 1 \\
n = 1 \quad & y_1 = y_0 + f_1 = 1 + 1 = 2 \\
n = 2 \quad & y_2 = y_1 + f_2 = 2 + 1 = 3 \\
\cdots \quad & \cdots
\end{aligned}
$$

y_n can be found iteratively.

The Z-transform can be directly applied to the difference equation (see Appendix D):

$$Z[f_n] = F(z) = \sum_{n=0}^{\infty} f_n z^{-n} \tag{9.19}$$

$$Z[f_{n-i}] = z^{-i} F(z) \tag{9.20}$$

Applying Equations (9.19) and (9.20) yields the Z-transform of Equation (9.18):

$$Y(z) = z^{-1} Y(z) + F(z)$$

Similarly, a **discrete transfer function** $G(z)$ can be found:

$$G(z) = \frac{Y(z)}{F(z)} = \frac{1}{1 - z^{-1}}$$

This result can also be used to find y_n. The method is called **long division**. The result of long division is a train of impulses. For example, the input function f is a step function. Therefore, $F(z) = \dfrac{1}{1 - z^{-1}}$ (see Example 9.4), and

$$
\begin{aligned}
Y(z) &= \frac{1}{1 - z^{-1}} \cdot \frac{1}{1 - z^{-1}} \\
&= \frac{1}{1 - 2z^{-1} + z^{-2}}
\end{aligned}
$$

In long division form, the last expression looks like this:

$$
\require{enclose}
\begin{array}{r}
1 + 2z^{-1} + 3z^{-2} + 4z^{-3} + \cdots \\[2pt]
1 - 2z^{-1} + z^{-2} \enclose{longdiv}{1 } \\
\underline{1 - 2z^{-1} + z^{-2}} \\
2z^{-1} - z^{-2} \\
\underline{2z^{-1} - 4z^{-2} + 2z^{-3}} \\
3z^{-2} - 2z^{-3} \\
\underline{3z^{-2} - 6z^{-3} + 3z^{-4}} \\
4z^{-3} - 3z^{-4} \\
\underline{4z^{-3} - 8z^{-4} + 4z^{-5}} \\
\cdots
\end{array}
$$

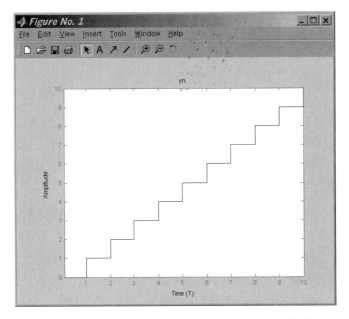

Figure 9.13   System response y_n over 10 time periods.

Thus, $Y(z) = 1 + 2z^{-1} + 3z^{-2} + 4z^{-3} + \cdots$, and y_n could be plotted directly from $Y(z)$ by using the coefficients of the function (Figure 9.13).

The analysis of a discrete system is similar to that for a continuous system, but is beyond the scope of this text.

9.2 LOGIC CONTROL

Another class of control is logic control. The process is turned on or off based on whether several conditions are met. For example, a machine is turned on if the workpiece is on the machine table and the correct tool has been mounted. The control algorithm can be written as a logical statement. To model and analyze such systems, **Boolean algebra** is used.

9.2.1 Boolean Algebra

In Boolean algebra, ON is represented using a logical TRUE, or the binary number 1. OFF is represented by FALSE or 0.

A Boolean variable can take one of the two states shown in Table 9.1. These variables coexist with a set of operators and associated laws. Operators can best be shown using **truth tables**. A truth table shows the outcome of an operator for all possible combinations of input states. The Boolean operators can also be implemented in electronics

TABLE 9.1	Equivalent States	
Switch	Logic	Binary
ON	TRUE	1
OFF	FALSE	0

with **logic gates**. The following are truth tables and the corresponding logic gates for one set of Boolean operators:

(a) Complement, or NOT

X	X′
0	1
1	0

(b) AND

$C = A \cdot B$ Often the symbol $\cdot$ is omitted. The expression can be written as $C = AB$.

A	B	C
0	0	0
0	1	0
1	0	0
1	1	1

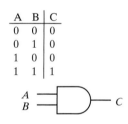

(c) OR

$C = A + B$

A	B	C
0	0	0
0	1	1
1	0	1
1	1	1

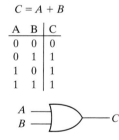

(d) XOR

$C = A \oplus B$

A	B	C
0	0	0
0	1	1
1	0	1
1	1	0

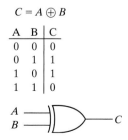

Boolean algebra laws include the following:

Commutative law:

$$XY = YX$$
$$X + Y = Y + X$$

Associative law:

$$(XY)Z = X(YZ) = XYZ$$

Distributive law:

$$X(Y + Z) = XY + XZ$$
$$X + YZ = (X + Y)(X + Z)$$
$$(X + Y)' = X'Y'$$
$$(XY)' = X' + Y'$$

These laws can be proven with the use of truth tables. Also, the following set of simplification theorems can be established:

$$XY + XY' = X$$
$$(X + Y)(X + Y') = X$$
$$X + XY = X$$
$$X(X + Y) = X$$
$$(X + Y')Y = XY$$
$$XY' + Y = X + Y$$

Example 9.5.

Simplify the logic statement $Z = A'BC + A'$:

$$\text{Let } X = A', Y = BC$$
$$Z = X + XY$$

Apply the third rule, $Z = X = A'$

A ——▷○—— Z

Example 9.6.

Simplify $Z = (A + B'C + D + EF)[A + B'C + (D + EF)']$

Let $X = A + B'C, Y = D + EF$

$$Z = (X + Y)(X + Y') = X = A + B'C$$

The corresponding logic gate circuit is shown in Figure 9.14.

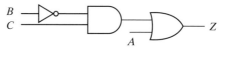

Figure 9.14

9.2.2 Logic Control Modeling

There are several ways that control logic can be modeled, such as a **state table**, **state transition diagram**, etc. This section will discuss only the state-table approach. A state table consists of a set of Boolean variables and the combination of all possible input states. The truth table for the operator AND can be used to illustrate a simple state table. In the truth table, there are three variables: A, B, and C. The two input variables are A and B. Since both A and B can be either 0 or 1, there are four possible combinations of states. The output C is true only if both A and B are 1. A state table may have multiple inputs and outputs. The state table shows when an output variable can be true.

In the following state table, A, B, and C are input variables, while D and E are output variables:

A	B	C	D	E
0	0	1	1	0
0	1	0	0	0
...			...	

Note that the table shows only partial states. A complete table should show all possible combinations of A, B, and C states. In the partially completed table, D is true only when C is true and both A and B are false. This can also be written in a logic statement as $D = A'B'C$. The '0' in the input is written as the complement of the input variable. All the variables on the same row use AND together. When there are multiple rows with $D = 1$, D is the sum of all the rows with $D = 1$.

Example 9.7.

Convert the following state table into a logic statement and simplify it:

A	B	C	D	
0	0	0	0	
0	0	1	0	
0	1	0	0	
0	1	1	1	$A'BC$
1	0	0	1	$AB'C'$
1	0	1	1	$AB'C$
1	1	0	1	ABC'
1	1	1	1	ABC

$$D = A'BC + AB'C' + AB'C + ABC' + ABC$$

$$= A'BC + AB' + AB$$

$$= A'BC + A$$

$$= BC + A$$

This logic statement can be implemented in a programmable logic controller, in a control computer, or in logic gates. The equivalent logic gate circuit is shown in Figure 9.15.

Figure 9.15   Logic gate circuit for $D = BC + A$.

Example 9.8.

Design a one-digit adder to add two one-bit binary numbers A and B. The result is a two-bit sum.

The adder can be represented by a state table:

		Sum	
A	B	X	Y
0	0	0	0
0	1	0	1
1	0	0	1
1	1	1	0

$$X = AB$$

$$Y = A'B + AB'$$

The logic gate circuit is shown in Figure 9.16.

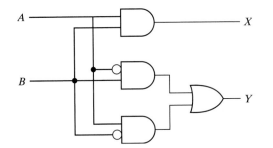

Figure 9.16   Logic gate circuit for $X = AB$ & $Y = A'B + AB'$.

Example 9.9.

A process is started by pushing a start pushbutton switch. The process continues even if the start switch is released. It can be turned off only if the stop pushbutton switch is pushed. Let PB1 be the start pushbutton switch and PB2 be the stop pushbutton switch. R is the

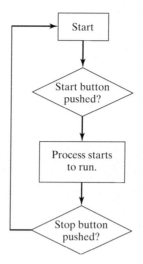

Figure 9.17   Process flowchart.

process. In this case, the logic cannot be directly represented in a logic table. The PB1 state has to be captured (memorized). The process can be represented by the logic statement

$$R = (PB1 + R) \cdot PB2'$$

The actual controller will be presented in the next chapter.

9.3 SENSORS AND ACTUATORS

In a feedback control system, sensors are used to provide feedback and actuators are used to manipulate the process. There are numerous types of sensors and actuators. In this section, only the most commonly used ones will be discussed. Before going further, several terms need to be defined.

Sensor/Transducers: A device that changes one physical quantity to another.
Accuracy: Agreement between the actual value and the measured values.
Resolution: Smallest measurable increment.
Repeatability: Variation over repeated measurements of a given value.
Range: Lower and upper limits of the measured variable.
Dynamic response: How fast a device can adapt to the changes of the measured variable—for example, the measuring speed (tachometer) of car going through a bumpy road. It can also be defined by increasing the input frequency until the output signal strength looses 3 db. Each sensor is characterized by these measurements.

9.3.1 Sensors

a. Position Sensors. Position sensors can be either linear (distance) or rotary (angle). They can also be used to calculate speed (change in position over time) or other quantities of interest (such as fuel quantity). Many kinds of position sensors can be found in feedback systems. This section lists several common position sensors.

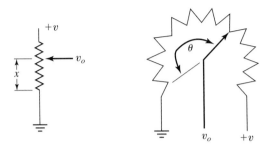

Figure 9.18 Linear and rotary potentiometers.

Potentiometer: A potentiometer is a variable resistor (Figure 9.18). A typical potentiometer is the volume control on the radio. When the knob is turned, the output DC voltage v_o is a fraction of the input voltage v. It can also be arranged in a linear fashion. The displacement is directly proportional to the output voltage.

Linear: $v_o = kvx$, x is the linear displacement.
Rotary: $v_o = kv\theta$, θ is the angular displacement.

The advantages of potentiometers are their low cost and ease of use. However, since they rely on mechanical contacts, potentiometers are subject to wear.

Linear variable differential transformer (LVDT). LVDT uses the principle of **induction**. A tube is covered by two copper coil windings (Figure 9.19). A movable ferromagnetic core is inserted into the tube. The secondary winding is separated into two pieces and placed on either side of the primary winding. An AC **exciting signal** (voltage) is applied to the primary winding. When the core is centered with respect to the two windings, the output of the secondary winding is zero. As the core moves away from the center, the induction to the two parts of the secondary winding becomes

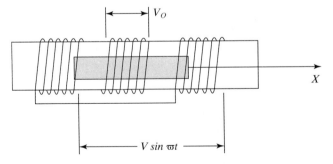

Figure 9.19 An LVDT.

unbalanced, thus producing an output AC voltage. The output voltage is proportional to the displacement of the core and is given by

$$v_o = kVx \sin \varpi t \qquad (9.21)$$

where k is LVDT constant, $V \sin \varpi t$ is the input signal, and x is the displacement. LVDTs have a range of motion of 1–50 mm and a resolution of 1 micron or better. A phase-sensitive rectifier can remove $\sin \varpi t$.

Rotary resolver.   The rotary resolver uses phase difference to detect rotational angle. Again, it is based on the principle of induction. The primary winding is placed on the stator and it creates an electromagnetic field when excited with an AC signal (Figure 9.20). The secondary winding is then mounted on the rotor. Since there is no charge on the core, the output voltage is fixed. However, as the rotor turns, the phase of the AC output will change. The phase difference between the input signal and the output signal is equal to the rotational angle. The resolution of a rotary resolver is ± 10 minutes of an arc.
Input to the two stator windings is

$$v_s = V \sin \varpi t \qquad (9.22)$$

The output signal due to one input winding is $v = v_s \cos \theta \sin \varpi t$.
When the two input windings are combined, the final output is

$$v_o = v_{s1} \cos \theta - v_{s2} \sin \theta \qquad (9.23)$$

where θ is the rotation angle.

Linear resolver.   A linear resolver uses the same principle as a rotary resolver, except that the windings are printed linearly on a flat surface. The two sets of windings have relative linear motion with an accuracy of 10 microns or better. The range is limited only by the size of the sensor.

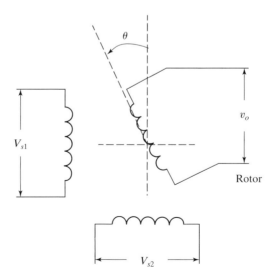

Figure 9.20   Two-phase rotary resolver.

Optical-angle encoders.   Optical-angle encoders use light sensors that generate electric pulses. The simplest encoder has an opaque disk with a cut slot. A light source is placed on one side of the disk and a light detector on the other side. Whenever the slot is in between the light source and the light detector, the light detector generates an electric output. Each rotation of the disk generates one electric pulse. By increasing the number of the slots on the disk, the resolution of the encoder increases. A more common way of designing an encoder is to print a checkerboard pattern on a glass disk. A much finer resolution can be achieved this way. There are two types of encoders: absolute and incremental. Absolute encoders print binary position codes around the encoder disk. An array of light detectors reads the position code and outputs it. An incremental encoder only outputs electric pulses as the disk spins; there is no direct position information in the output. An encoder can have a resolution of 4096 pulses per revolution or better.

Encoders can also be arranged in a linear configuration. They use enclosed or exposed glass scales. Grating lines for the encoder may be printed on glass or metal. The spacing between grating lines is in the micron scale. An absolute linear encoder uses several rows of grating to identify the location. A read head uses several photo sensors, travels over the linear scale, and outputs the location information.

b. Velocity Sensors.   The most commonly used velocity sensor is a tachometer, which measures the rotational speed of a shaft. The linear speed can then be inferred from the rotational speed. A DC tachometer converts the shaft rotational speed into a DC voltage. It is actually similar to a DC motor, but works in reverse:

$v_a = K\varpi$, where v_a is the output voltage, K is a constant, and ϖ is the rotational speed in rpm.

When measuring the speed of a gas or liquid, the ramp pressure or the propeller's rotational speed is used.

c. Acceleration Sensors.   Acceleration can be measured indirectly. Given **Newton's law** $F = ma$, if force and mass are known, acceleration can be calculated. One type of device that can measure force effectively is a **piezoelectric** sensor. Piezo crystals are either quartz or ceramic crystals. When such a crystal is subject to an external force, it deforms. The deformation produces a voltage that is proportional to the amount of deformation. Acceleration can be measured by attaching a Piezo crystal to a known free mass. Since the piezo effect is reversible, piezo crystals can also be used as actuators. Good examples include inkjet heads and solid-state buzzers.

d. Temperature Sensors.   Temperature can be measured with resistance thermometers, thermocouples, or semiconductor thermistors. Resistance thermometers and thermocouples are based on the change in metal electrical conductivity due to temperature. Resistance thermometers use a conductor such as copper, nickel, or platinum. For a given wire, the temperature is directly measured based on its resistance value. Thermocouples use a pair of wires made of different metal alloys. One wire is placed at the reference temperature and another at the measuring temperature. The

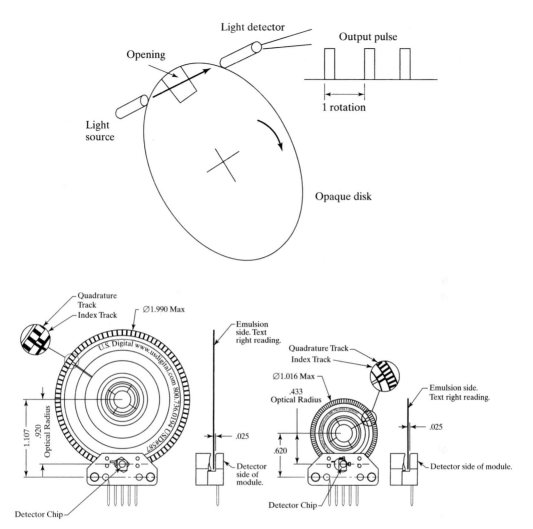

Figure 9.21   An encoder.

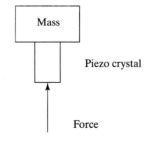

Figure 9.22   Acceleration sensor.

resistance difference between the wires is converted into a temperature difference. Thermistors are made of semiconductor materials. The temperature is correlated to the resistance value.

9.3.2 Motors and Actuators

The controller uses actuators to manipulate the process under control. For example, in a machine tool control, electric motors are used to move the tool and the machine table. The actuators are electric motors. In this section, only a few of the most commonly used motors and actuators are briefly introduced.

(a) **DC servo motor.** The **servo** in the term DC servo motor means slave. A servo motor is one that is controlled through a feedback loop. It consists of a DC motor, a tachometer, and/or a position sensor. The DC motor is usually a **permanent magnet** (PM) motor. The stator consists of two permanent magnets, which create a permanent magnetic field. The rotor has a Ferro metal core and three separate copper wire windings oriented 60° (Figure 9.23) apart. Two carbon brushes are used to conduct DC current to the rotor windings through commutation bars (see Figure 9.24). Each of the six commutation bars are connected to the ends of the three copper windings. At any given time, only the winding at a particular position (position A in Figure 9.23) is energized. The magnetic field produced by this winding attracts it toward either the N or the S pole. This force produces a torque on the rotor. As soon as the rotor rotates to a new position, the energized winding is de-energized and the next winding that rotates to position A will be energized. The motion continues, and the motor turns. To rotate counterclockwise, change the commutator bar polarity and the pulling force will become a repelling force, changing the direction of rotation.

The control loop of a DC servo motor is shown in Figure 9.25. The command voltage determines the speed as well duration of the rotation.

To analyze a DC motor, the following model is used. The torque generated by a motor is $T_m = K_m i_a$, where K_m is the motor constant and i_a is the **armature** (rotor) current. The torque required to drive the load is $T = J\ddot{\theta} + b\dot{\theta}$. J is the rotor and load **inertia**, b is the **viscous friction coefficient**, and θ is the angular speed. The motor torque is equal to the load torque; therefore,

$$J\ddot{\theta} + b\dot{\theta} = K_m i_a \qquad (9.24)$$

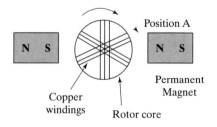

Figure 9.23   DC motor structure.

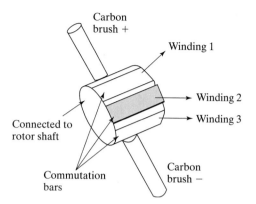

Figure 9.24   A commutator.

As the motor rotates, the winding generates a DC voltage called **back emf** that counters the input voltage. Therefore, the electric equation is

$$Ri_a + L_a\dot{i}_a = v_a - K_e\dot{\theta} \tag{9.25}$$

(b) **Stepper motor**. A stepper motor is controlled digitally. Each control signal pulse sent to the motor causes it to rotate one step. The step angle θ of a motor depends on the motor design. It is usually from $0.9°$ to $3.6°$, which is 400 or 100 steps per revolution, respectively. Since the rotation angle is controlled directly by the input pulses, it is not necessary to use an additional position sensor. The speed of a stepper motor is thus controlled by the pulse rate. Let N be the number of steps per revolution and f be the pulse rate in Hz. $N = 360/\theta$. The rotational speed in rpm is $N \cdot f \cdot 60$. The number of rotations made by a stepper motor, Δ, can be calculated by the number of pulses received, n: $\Delta = n/N$.

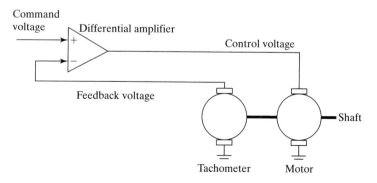

Figure 9.25   A DC servo motor control loop.

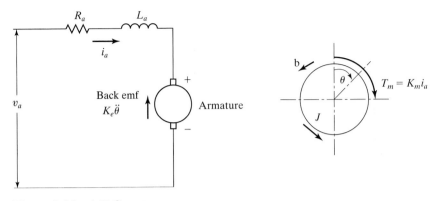

Figure 9.26 A DC motor.

Unlike DC motors, stepper motors have a permanent magnet rotor and four windings on the stator (see Figure 9.27). There are a total of six leads; the two center leads are v_{cc} (positive voltage) leads. The other four leads control each of the four electromagnets. A logic 0 is represented by 0 volts and a logic 1 is represented by v_{cc}. In Figure 9.27, the input signal is 0101 and the rotor is locked in the upright position. When the input signal changes to 1010, the rotor makes a 90° turn. In this simplified model, the rotor has only one permanent magnet. The step angle is thus 90°. When increasing the number of permanent magnets on the rotor, the step angle reduces.

Since the positioning and speed of stepper motors can be controlled without additional sensors, stepper motors are often used in open-loop control systems. Due to the differences in voltage and current between stepper motors and control computers, a stepper-motor drive circuit normally serves as the interface between the two. A stepper-motor drive also converts the input pulse train into the necessary signals for the four signal leads.

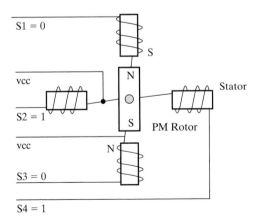

Figure 9.27 Stepper motor.

The torque generated by a stepper motor is related to the stepping speed. Therefore, in designing an application, it is important that the torque curve supplied by the manufacturer be consulted; the motor is allowed to operate only within a certain speed range.

(c) **AC servo motor.** There are several types of AC motors: induction motors, switched reluctance motors, and synchronous motors. They all operate on different principles. Here, the discussion is limited to induction motors. An induction motor uses the field generated by the stator electromagnets (poles) to induce an AC current in the rotor. In turn, the current in the rotor produces a magnetic field slightly behind the phase of the stator field. The attraction force between the two fields pulls the rotor to catch up with the stator field. Since the AC field rotates around these poles, the rotor follows. The motor speed is thus proportional to the input AC frequency and the number of stator poles. The motor speed is

$$\text{rpm} = 120 \times \frac{f_{AC}}{n}$$

where f_{AC} is the AC frequency and n is the number of stator poles.

The analysis of AC motors is more complex. Usually, performance data are collected empirically. Curve fitting is used to find coefficients of equations similar to those for DC motors.

(d) **DC brushless motor**. In DC motors, carbon brushes are designed for commutation, which is what makes the motor rotate. The use of carbon brushes generates sparks and causes wear, so they are not desirable for many applications. Since the commutation is mechanical, it imposes a speed limitation. A brushless DC motor uses electronic devices to switch the current (commutator). The control is more complex than that for the brush motor.

There are two types of brushless motors:

1. The split-phase permanent-magnet (PM) motor uses a PM rotor. The stator windings change field directions using oscillator circuits to generate commutation actions.

2. The Hall effect motor also uses a PM rotor. It uses built-in Hall effect sensors in the stator to generate proper signals to drive the stator windings. The position of the PM rotor is picked up by the Hall effect sensors.

(e) Solenoids. Solenoids are simple devices that contain an electric magnet and a ferromagnetic push rod. There are two types of solenoids: linear and rotary solenoids. When a current is applied, the push rod moves linearly in a linear solenoid or rotates in a rotary solenoid. When the current is removed, the push rod returns to its original position. The other important characteristic of the solenoid is the **duty cycle**, which tells how many power on–off cycles per second the solenoid can perform without losing the rated output force.

9.4 SUMMARY

In this chapter, an overview of industrial control systems was presented. The chapter was divided into three sections: control theory, logic control, and sensors and actuators. The overview of control theory included all fundamental concepts of classical control theory and a brief introduction of discrete control systems. It also provided a minimum knowledge to understand the theory behind modern control systems. The section on logic control introduced the concepts of Boolean logic, truth tables, and logic gates, providing an essential background for the understanding of programmable logic controllers, presented in the next chapter. Finally, the section on sensors and actuators links the theoretical knowledge with practical devices. Sensors and actuators are the inputs and outputs of controllers in a closed-loop control system. This chapter touches upon only the most basic knowledge in control systems. For further reading, there are hundreds of books on these topics. Specific device-related information can be found on the Web. Readers are encouraged to explore the vast information available.

9.5 KEYWORDS

Armature
Back emf
Block diagram
Boolean algebra
Characteristic equation
Commutation bars
Continuous time domain
Controller gain
Discrete transfer function
Duty cycle
Exciting signal
Final value theorem
Ideal sampler
Induction
Inertia
Laplace transformation
Logic gates
Long division
Manipulation

Newton's law
On–off controller
Permanent magnet (PM)
PID controller
Piezoelectric
Poles
Proportional error controller
Proportional, Integral, and Derivative
Reference value
Sample time period
Set point
State table
State transition diagram
Time constant
Train of impulses
Transfer function
Truth table
Viscous friction coefficient
Zeros

9.6 REVIEW QUESTIONS

9.1 What is a transfer function? How can a transfer function be obtained from the process defining differential equation?

9.2 What is the meaning of *steady state*?

9.3 Under what circumstances can a system be considered unstable?

9.4 How does the time constant affect system performance? How is the 0-to-60-acceleration time related to the time constant of a car's performance?

9.5 Using house temperature control as an example, what are set point, error, and feedback? Can you guess what kind of control algorithms are used in traditional and programmable thermostats?

9.6 In a discrete control system, how often is the input sampled?

9.7 What kind of controllers is used in discrete control systems?

9.8 What kind of control model is best suited for home security systems?

9.9 Answer the following questions regarding velocity sensors:

(a) What velocity sensors are used in automobiles? Check the Internet and see if you can find the answer. What would you suggest?

(b) What velocity sensors are used in an airplane? What is the most commonly used speed sensor? What is the operating principle of this sensor?

(c) Can boats use the same sensors for boat speed? If not, what could be used?

9.10 What are piezoelectric sensors and actuators? What applications can they do?

9.11 Discuss the functions and limitations of the resistance thermometer, thermocouple, and thermistor. What temperature sensors are used at home and in your car engine?

9.12 What kind of sensor is used in an automobile fuel gauge?

9.13 Where are DC brushless motors used? Discuss the advantages and disadvantages of a DC brushless motor compared with a DC brush motor and an AC servo motor.

9.14 What kind of control algorithm is suitable for solenoid control?

9.7 REVIEW PROBLEMS

9.1 Given the block diagram in Figure 9.28, prove that the system transfer function is

$$\frac{H(s)}{R(s)} = \frac{A(s)}{1 + A(s)B(s)}$$

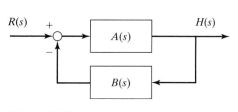

Figure 9.28

9.2 Given the block diagram in Figure 9.29, what is $H(s)/D(s)$?

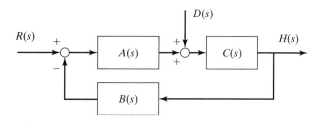

Figure 9.29

9.3 As shown in Figure 9.30, a car weighing 2000 lb is pulled by an engine that generates $F(t)$ lb of force. The viscous friction coefficient is 0.1 lb second/ft. The process can be written in a differential equation: $F(t) - 0.1\dot{x}(t) = 2000\ddot{x}(t)$, where x is the position of the car. What is the transfer function $G(s) = X(s)/F(s)$?

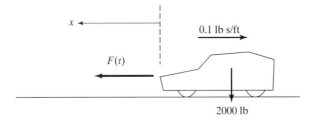

Figure 9.30

9.4 In problem 3, replace x with v (speed) so that $v = \dfrac{dx}{dt} = \dot{x}$. What is the transfer function

$$G(s) = \frac{V(s)}{F(s)}?$$

9.5 As shown in Figure 9.31, a satellite requires attitude control so that the antenna can point to the ground station. The rotational angle θ is the control variable, and a gas jet is used to provide the force (F) to rotate the satellite. The process can be modeled as $F(t)d = I\ddot{\theta}(t)$, where d is the distance between the jet and the center of rotation. What is the transfer function

$$G(s) = \frac{\theta(s)}{F(s)}?$$

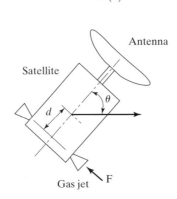

Figure 9.31

9.6 Find the transfer function for the angular speed $\varpi = \dot{\theta}$ of the satellite in problem 5, where

$$G(s) = \varpi(s)/F(s).$$

9.7 A DC servo motor can be modeled by the torque relationship and the electric circuit. The torque generated is $K_m i_a$. K_m is the motor constant, which depends on the motor design; i_a is the armature (rotor) winding current. The combined moment of inertia for the rotor and the load is J, the viscous friction is b, the armature resistance is R_a, the armature reluctance is L_a, and the input voltage is v_a. The back emf constant is K_e. The equations are

$$J\ddot{\theta} + b\dot{\theta} = K_m i_a$$

$$L_a \frac{di_a}{dt} + R_a i_a = v_a - K_e \dot{\theta}$$

What is the transfer function of the motor $G(s) = \theta(s)/v_a(s)$, where $\varpi(t) = \dot{\theta}(t)$?

What is the transfer function of the motor speed $G(s) = \varpi(s)/v_a(s)$?

(*Hint*: Combine two equations before applying the Laplace transformation.)

9.8 A constant force $F(t) = F_0$ is applied to the car in problem 3. Show that the speed of the car is stable. What will be the steady-state speed of the car? (*Hint:* Find the $V(s)$ function. The stability is determined by the characteristic equation (denominator). The steady-state speed is found using the final value theorem.)

9.9 In problem 5, the inertia of the satellite is 10 slug·ft^2, for $d = 2$ ft. In this problem, $F(t)$ is a step function and equals 20 lb. What is the steady-state angular speed (rad/sec)? In the satellite attitude equation $F(t)d = I\ddot{\theta}(t)$, the angular acceleration $\ddot{\theta}$ is in rad/sec^2. Note that $slug = (lb \cdot sec^2)/(ft \cdot rad)$

9.10 Ten volts of power (step function) is applied to a DC motor (see problem 9.7). What is the steady-state motor speed, $\varpi(\infty) = \dot{\theta}(\infty)$?

9.11 A PID controller is used to control the car speed (see problems 9.3 and 9.4). The control equation is $F(s) = \left(1 + \dfrac{0.1}{s} + 0.5s\right) E(s)$, where $E(s) = R(s) - V(s)$. In this problem, R is the speed set by the driver; let it be a step function with a value of 65 mph (95 ft/s). What is the steady-state car speed? What is the steady-state error (difference between the set speed and the actual speed)?

9.12 Repeat problem 11 using the controller $D(s) = 1 + 0.5s$.

9.13 A PE controller is used to control the DC motor's rotational angle. Let the controller gain $K_p = 10$. The rest of the motor parameters are as follows (to simplify the problem all units are ignored, we assume that they are balance):

$J = 100$, $b = 1$, $K_t = 5$, $L_a = 10$, $R_a = 10$, and $K_e = 1$. Note that θ is in rad. Let the input set angle equal 3 rad. What is the steady-state θ? If you have access to MATLAB, plot the result. Use the result from problem 7 for this problem.

9.14 Change the controller in problem 13 to a PD controller. The derivative gain, $K_d = 5$. What is the steady-state θ? If you have access to MATLAB, plot the result.

9.15 The discrete transfer function for the car control problem is

$$G(z) = \frac{X(z)}{F(z)} = \frac{0.00025(z^{-1} + z^{-2})}{1 - 2z^{-1} + z^{-2}}$$

What is the difference equation of x_n as a function of F_n?

9.16 Assume that $x_n - 2x_{n-1} + x_{n-2} = 2F_{n-1} - 3F_{n-2}$. Let F_n be a step function such that $F_n = 0$ if $n < 0$ and $F_n = 1000$ if $n \geq 0$. Also, $x_n = 0$ for $n < 0$. What is x_4? Try to plot x_n for $n = 0$ to 20, using Excel.

9.17 The discrete satellite attitude-control-process equation can be written as the difference equation $\theta_n - 2\theta_{n-1} + \theta_{n-2} = \dfrac{T}{2}\dfrac{d}{I}(F_{n-1} + F_{n-2})$, where T is the sample period. To simplify the problem, let $d/I = 1$ and $T = 2$. What is the discrete transfer function of the process?

9.18 The Z-transform of a step function is $1/1 - z^{-1}$. Let the force function in the car control problem (problem 9.15) be a step function. Using long division, plot x_n if

(a) $\dfrac{X(z)}{F(z)} = \dfrac{0.00025(z^{-1} + z^{-2})}{1 - 2z^{-1} + z^{-2}}$

(b) $X(z) = \dfrac{0.00025(z^{-1} + z^{-2})}{1 - 2z^{-1} + z^{-2}} \cdot \dfrac{1}{1 - z^{-1}}$

9.19 The Z-tranform for an impulse function is 1. Use long division to plot θ_n of the satellite control problem in problem 17.

9.20 A DC servo motor is used in a closed-loop control system. Figure 9.32 is a system block diagram. The input to the system is a train of pulses represented by a frequency f_r. The first circle and the block represent an up–down counter. Its output equals 1/s times $(f_r - f_e)$. Note that the output of the up–down counter is a binary number. However, the inputs are two pulse trains. The unit conversion in the block diagram is handled by the constants used in the transfer functions. This control system uses a PE controller, DAC acts as the controller, and the output of the DAC is a voltage v_a. Again, unit conversion has been handled. The output of the motor is angular speed, a reduction gear is used to increase the torque, and this angular speed is converted to a linear speed by a leadscrew. The leadscrew gain, k_l, equals the pitch/(2π). Note that $\varpi = \dot\theta$ from Equation 9.25.

(a)   What is the function $v(s)$? Let f_r be a unit-step function.
(b)   What is the steady-state velocity $v(\infty)$?

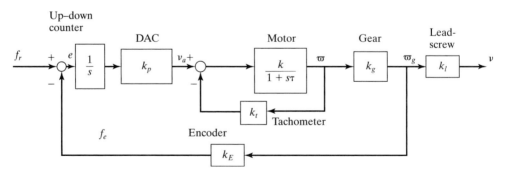

Figure 9.32

9.21 A one-bit full adder adds two binary numbers A, B, and the carry from the previous bit, C_{in} together. The result is placed in S, and the carry is placed in C_{out}. Design a full adder using logic gates. The logic table of a full adder is as follows:

A	B	C_{in}	S	C_{out}
0	0	0	0	0
0	0	1	1	0
0	1	0	1	0
0	1	1	0	1
1	0	0	1	0
1	0	1	0	1
1	1	0	0	1
1	1	1	1	1

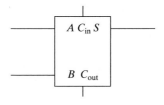

Figure 9.33

9.22 Express the following logic statements in the simplest forms:

(a) $ABC + ADB$

(b) $(A + B')BBA + ABCD$

(c) $((AB)' + (AB')')'$

9.23 Let positive voltage represent logic 1 and ground represent logic 0. A switch can be used to input logic by opening or closing the circuit to connect to the positive voltage. A light bulb is used to show the output. Construct the wiring diagram for the logic statement shown in problem 22. Use the electric element symbols in Figure 9.34 to draw your circuits.

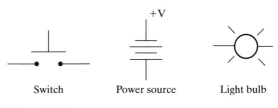

Switch   Power source   Light bulb

Figure 9.34

9.24 Design a logic gate circuit based on the following logic table, where A and B are inputs and C and D are outputs:

A	B	C	D
0	0	1	0
0	1	1	1
1	0	0	0
1	1	0	0

9.8 REFERENCES

Bishop, R. H. *The Mechatronics Handbook*. Boca Raton, FL: CRC Press, 2002.

Bollinger, J. G., and N. A. Duffie. *Computer Control of Machines and Processes*. Reading, MA: Addison Wesley, 1988.

Boylestad, R. and L. Nashelsky. *Electronic Devices and Circuit Theory*, 8th ed. Upper Saddle River, NJ: Prentice Hall, 2002.

Dorf, R. C., and R. H. Bishop. *Modern Control Systems*, 9th ed. Upper Saddle River, NJ: Prentice Hall, 2001.

Franklin, G. F., J. D. Powell, and A. Emami-Naeini. *Feedback Control of Dynamic Systems*, 4th ed. Upper Saddle River, NJ: Prentice Hall, 2002.

Ogata, K. *Modern Control Engineering*, 4th ed. Upper Saddle River, NJ: Prentice Hall, 2002.

Chapter 10

Programmable Logic Controllers

Objective

Chapter 10 provides an overview of programmable logic controllers (PLC) and their applications in manufacturing. It will explore the basic features of a PLC and its programming, as well as the roles of a PLC in automated manufacturing. Understanding this chapter requires a knowledge of industrial control fundamentals.

Outline

10.1 INTRODUCTION

A manufacturing system consists of a group of machines along with material-handling, storage, and control devices. To automate the system, two factors must be considered: the control of the equipment and the flow of information.

Today, the word *automation* usually implies a system controlled by computers. However, this is not the only form of automation used in modern industry. Along with sophisticated computer controls, there are conventional control devices, such as mechanical controllers with cams and linkages, relay panels, NC controllers, and programmable logic controllers. This chapter will focus on programmable logic controllers (PLCs)—their capabilities, programming, and applications.

10.1.1 Functions of Controllers

The functions of controllers used in a manufacturing system can be classified as

1. on–off control
2. sequential control
3. feedback control
4. motion control

The most simple form of control is the on–off control. The controller switches a device on or off, based on the state of a sensor. For example, a temperature controller uses a thermostat to turn a furnace on or off, and a power switch turns a conveyor on or off. In many instances, a simple switch and basic wiring are the only necessary controller components. When the control logic becomes more complex or the operating voltage and current are higher than safety regulations allow, relays are used.

Sequential controls are used to control a fixed sequence of events. Many manufacturing operations go through a fixed sequence of events. Each event takes a fixed time period to complete or may be terminated by a trigger from a sensor. For example, a drum sequencer is a control device that was once used widely. It is very similar to a music box, in that many spikes on a drum represent the events. However, a drum sequencer is used for discrete event control; it basically turns a series of switches on and off in a certain pattern. For smoother and analog control, mechanical cams can be used.

For processes that require more precise control or that are subject to uncertainty, a feedback device is required. For example, if a furnace must remain at a stable temperature, it is necessary to use the feedback of a temperature transducer to determine the amount of fuel injected into the burner. This control action must be done continuously or the temperature will vary over time. Thus, a thermostat is not sufficient for this application. The controller must also consider the characteristic of the burner. Either a specially designed analog device or a computer-based controller is necessary.

Motion control is another type of control that is often used in factory processes. Quite often, factories need to maintain the speed of a machine at a precise rate. For example, to obtain a good quality part, an NC machine must run at a predefined feed and speed. Another example of motion control is robotic control. The details of NC machines and robotic control are discussed in Chapters 12 and 15. Usually, a machine drive system consists of a motor and a transmission. Given any load change, the speed can be represented by a second-order differential equation. A motion controller is needed to move the machine to the desired position quickly and to maintain that speed.

10.1.2 Control Devices

In industry, several types of control devices are used to satisfy the previously mentioned control needs:

1. mechanical control
2. pneumatic control
3. electromechanical control
4. electronic control
5. computer control

Mechanical controls include cams and governors. Although they have been used to control very complex machines, today they are employed primarily for simple and fixed-cycle task control. Some automated machines, such as screw machines, still use cam-based control. But mechanical control can be difficult and expensive to manufacture and is subject to wear.

Pneumatic control is still very popular for certain applications. Compressed air, valves, and switches can be used to construct a simple controller. Because standard components are used to construct the logic, it is easier to build than a mechanical controller. Pneumatic control is generally used for fixed automation, because reprogramming is not easy, and it requires rewiring air ducts. Pneumatic control parts are subject to wear.

Like a mechanical controller, electromechanical controls use switches, relays, timers, counters, and so on, to construct control logic. They are similar to pneumatic controllers, except electric current is used instead of compressed air. They also share the same limitation as pneumatic controllers. However, because electric current is used, it is faster and more flexible. The controllers built using electromechanical control are called relay devices.

Electronic control is similar to electromechanical control, except that the moving mechanical components in an electromechanical-control device are replaced by electronic switches, which work faster and are more reliable. The inherent problems of electromechanical control can also apply to an electronic controller.

Computer control is the most versatile control system. The logic of the control is programmed into the computer memory using software. It can be used for machine and manufacturing-system control, as well as for data communication. Very complex control strategies with extensive computations can be programmed. However, speed suffers when complex logic is handled by the computer. Even with modern super-microcomputers, some fast feedback-control applications may still need a special controller coupled with a computer.

The difficulty of using a computer as a controller stems from two facts: its interface and its software. First, the computer uses a low voltage (5 to 12 volts) and a low current (several milliamps). However, machinery requires much higher voltages (24, 110, or 220 volts) and currents (measured in amps). Thus, an interface to convert the voltage difference and filter out the electric noise usually found in the shop must be developed. This interface is normally custom built for each application. Second, software development can be more difficult and more costly than hardware development. Knowledge of a lower-level language is usually needed for device-level programming. However, it is difficult to write the code and even more difficult to debug the software; therefore, software development is a long and costly process.

To take advantage of all these controllers and eliminate many of the difficulties, the programmable logic controller (PLC) was invented. A PLC is a computer-based device that has standard interface modules. Initially, it replaced relay devices, which are programmed with the use of a ladder diagram (a standard electric wiring diagram). As PLCs became more flexible, both high-level and low-level languages became available to PLC programmers. PLCs have the flexibility of computers, as well as a standard interface with processes and other devices. They are widely accepted in industry for controlling a single device to a complex manufacturing facility.

10.1.3 Programmable Logic Controllers

Programmable logic controllers (PLCs) were first introduced in 1968 as a substitute for hardwired relay panels. The original intent was to replace mechanical switching devices (relay modules). Bedford Associates (now the Modicon Division of Gould, Inc.) first coined the term and patented the invention. However, since 1968, the capabilities of the PLC have been enhanced significantly. Modern PLCs have many more functions. Their use extends from simple process control to manufacturing system control and monitoring. They are used for high-speed digital processing, high-speed digital communication, high-level computer-language support, and, of course, for basic process control.

The National Electrical Manufacturing Association (NEMA) defines the programmable controller as "a digitally operating electronic apparatus which uses a programmable memory for the internal storage of instructions by implementing specific functions such as logic sequencing, timing, counting, and arithmetic to control, through digital or analog input/output modules, various types of machines or processes. The digital computer which is used to perform the functions of a programmable controller is considered to be within this scope. Excluded are drum and other similar mechanical sequencing controllers." This definition implies that a PLC is an electronic interface device used to perform logic operations on input signals in order to generate a set of desired output signals or responses. Input for a basic PLC typically comes from discrete signal devices (such as pushbuttons, microswitches, photocells, limit switches, and proximity switches) or analog devices (such as thermocouples, voltmeters, and potentiometers). Output from a basic PLC is normally directed to switching closures for motors, valves, motor starters, and so on. More sophisticated PLCs may also include a mathematical processor, a color graphics display, serial communication ports, and a local-area-network interface.

PLCs vary in size and power, and are constructed with modules. Each module has a specific function (such as the processor module, I/O module, or interface module), and all the modules are connected together to form the PLC. A large PLC can have thousands of I/O points and support all the functions discussed earlier. There are also expansion slots to accommodate PCs and other communication devices. For many applications, a small PLC is sufficient, particularly since the speed of PLCs is constantly improving. Even the low-end PLCs perform at high speeds.

Figure 10.1 shows a high-end PLC system. Figure 10.2 shows a small PLC (Allen Bradley MicroLogix 1000). It measures only 4.72″ × 3.15″ × 1.57″, and has 32 I/O points and a standard RS-232 serial communication port. The major differences between

Figure 10.1   A PLC system: CPU module (left) and an I/O rack (right) (Allen Bradley PLC-5) (Courtesy of Allen-Bradley).

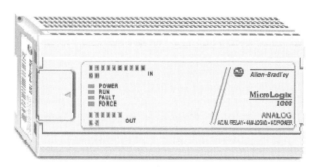

Figure 10.2   A small PLC (Allen-Bradley MicroLogix 1000; 4.72" × 3.15" × 1.57") (Courtesy of Allen-Bradley).

high-end and small PLCs are the number of I/Os allowed, the number of control functions supported, and the availability of different functional modules.

10.2 RELAY-DEVICE COMPONENTS

PLCs were primarily intended to replace relay devices, so it is important to become familiar with the components used in relay devices. A relay device consists of a front display panel with switches, relays, timers, and counters. Each of these is discussed briefly in the following sections.

10.2.1 Switches (Contact)

A switch is a device that either opens or closes a circuit. Although there are numerous types and styles of switches, they can be classified into the following categories (Figure 10.3):

1. locking and nonlocking
2. normally open and normally closed

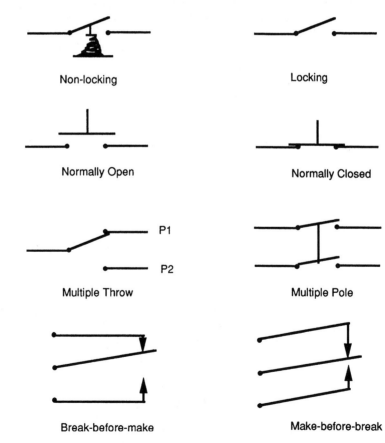

Figure 10.3 Types of switches.

3. single throw and multiple throw
4. single pole and multiple pole
5. break-before-make (interrupt transfer) and make-before-break (continuity transfer)

The first category of switches is easy to understand. A nonlocking switch simply returns to its initial state.

In the second category, a normally open switch contact is made by physically depressing the switch ("make" contact). Normally closed switches operate in the opposite manner: Contact is actively interrupted (broken).

In the third category, a single-throw switch has two states: on and off. There are some switches that have three states: a release and two operating positions. In this case, it can select either a neutral circuit or connect to one of two circuits. This kind of switch is called a double-throw, or multiple-throw switch.

In the fourth category, a multiple-throw switch has several states. These switches all have a single pole (moving part), and, subsequently, are called single-pole switches. In order to close (or break) two or more contacts at the same time, multiple-pole switches are necessary. The most widely used multiple-pole switch is the double-pole switch.

The last category is more complex. For some circuits, contact can be made or broken several times in succession. There are two types of transfer contacts in which "makes" and "breaks" can be combined. A "break-before-make," or interrupt transfer, contact does as the name suggests—it breaks one contact before another is made. When the switch is operated, there is a certain amount of time when the common spring is in contact with neither contact. Thus, a break-before-make results. A "make-before-break," or continuity transfer, contact provides the same function as a transfer contact. However, continuity always exists for one or the other contact.

Typical switches used in the control circuits include

selector switches
pushbutton switches
photoelectric switches
limit switches
proximity switches
level switches
thumbwheel switches
slide switches

Among them, photoelectric switches and proximity switches are noncontact switches. A proximity switch changes its state when an object is moved in close proximity to the sensing face of the switch. Photoelectric switches are activated by a light beam. Usually, the switch includes a light source and a reflector. The light beam is reflected back to the switch sensor located beside the light source.

Switches are also rated by their currents and voltages. For industrial applications, standard switch ratings are

24 volts AC/DC
48 volts AC/DC

120 volts AC/DC

230 volts AC/DC

TTL level (transistor-to-transistor, ±5 V)

10.2.2 Relays

A switch whose operation is activated by an electromagnet is called a relay (Figure 10.4). The contact and symbology for relays is usually the same as for switches. A small current passes through the magnet, causing the pole to switch. Usually, the magnet is rated between 3 to 100 volts and a few hundred milliamps. Therefore, it is operated at very low power (current and voltage). A circuit carrying a much heavier rating can be switched using a relay, however, the two circuits are totally separated.

When a relay operates, the contacts do not all open or close instantaneously. There may be a delay of several milliseconds between the operation of two contacts of the same relay. In the design of a relay circuit, this delay must always be taken into account.

On the basis of the preceding discussion, it is clear that a relay is really a magnet-operated contact switch. The contact switch inside a relay also can be classified by the number of poles and throws. Although most relays are single throw, it is very common to have multiple-pole relays.

10.2.3 Counters

On the basis of their structure, counters can be classified as mechanical or digital. Mechanical counters, such as an odometer, usually give readings as their output. Because mechanical counters are generally not used in a relay panel circuit, they are not discussed here. Digital counters output in the form of a relay contact when a preassigned count value is reached. A digital counter consists of a count register, an accumulator, and a relay contact (Figure 10.5). The count register holds the preassigned count value. The accumulator is used to either increment or decrement a count each time an input pulse is received. When the accumulator value equals the register value, the relay contact is activated.

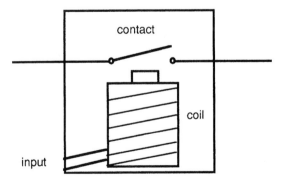

Figure 10.4 A relay.

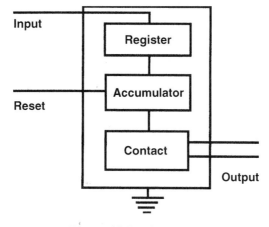

Figure 10.5 A counter.

The operation of a counter can be best shown by a timing diagram (Figure 10.6). The preassigned count register value is 5. There are up counters and down counters. An up counter counts starting from zero and increments the value when there is an input. A down counter, on the other hand, counts down from an initial value. They both serve the same purpose, to count a certain number of inputs and then output to a relay contact. A typical counter is characterized by the number of counting digits, the input electric rating, the reset system, the output contact rating, and the power source. Counters can be cascaded to form a larger counter. For example, an eight-digit counter can be made by cascading two four-digit counters. Input to a counter is normally activated from a contact. To initialize the counting, a reset input is used. The output contact is rated the same as a regular contact switch.

10.2.4 Timers

A timer, as its name implies, is used for some timing purpose. It consists of an internal clock, a count-value register, and an accumulator (Figure 10.7). In process control, a significant number of operations must be timed. For example, in a chemical

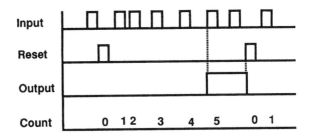

Figure 10.6 Counter timing diagram (the count value is 5).

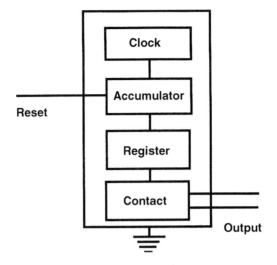

Figure 10.7 A timer.

process, the curing of certain products, the mixing of chemicals, and so on, all require a certain period of time to complete. In process control, synchronization of operations is also essential. There are two ways to synchronize operations, namely, event-triggered synch and time-controlled synch. Event-triggered synch can be achieved by using sensors and switches to detect the event. For time-controlled synch, each operation is given a fixed time period to finish; therefore, a clock or timer is necessary.

A timer starts timing after receiving a start signal. When a preassigned timing value is reached, it outputs a signal. The operation of a timer can be shown by a timing diagram (Figure 10.8). It is very similar to that of a counter, except that the counting pulses are generated internally in the timer. In the diagram, the preassigned timing value is 5 s. Each clock pulse represents 1 s.

Depending on the application, the timing diagram for other timers may have some small deviation. However, the basic principle is the same. Usually, timers are

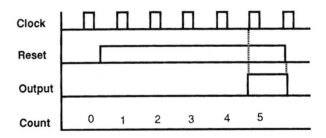

Figure 10.8 Timer timing diagram (the timing value is 5).

characterized by the following factors: features, such as size, mounting, and display type; time range; rated operating voltage; accuracy; contact rating; and output contact classification.

10.2.5 An Example of Relay Logic

For a control process, it is desired to have the process start (by turning on a motor) 5 s after a part touches a limit switch. The process is terminated automatically when the finished part touches a second limit switch. An emergency switch stops the process any time when it is pushed.

The circuit design for the process control is shown in Figure 10.9. In the diagram, LS_1 is the first limit switch. It is a normally open type. PB_1 is a pushbutton switch; it is normally closed. LS_2 is the second limit switch; it is normally closed. R_1 is a relay with a double-pole contact. R_2 is a relay connected to a motor. (The ladder diagram is discussed in greater detail in Section 10.4.1.) A wiring diagram for this circuit is shown in Figure 10.10.

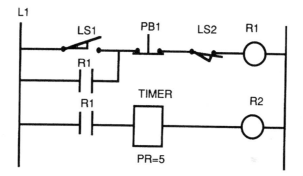

Figure 10.9   Ladder diagram for the circuit.

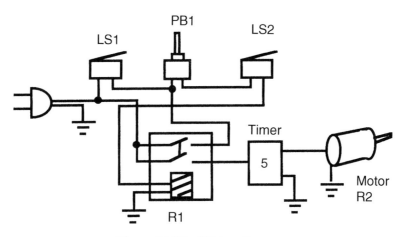

Figure 10.10   Wiring diagram.

10.3 PROGRAMMABLE CONTROLLER ARCHITECTURE

In the preceding sections, the basic components in a relay panel circuit and the design of a relay ladder diagram have been discussed. In the implementation of such a circuit, not only does the logic problem have to be analyzed, but the electrical compatibility and wiring layout also have to be considered. Breadboarding a relay panel circuit is a tedious task; it requires a lot of careful planning and work. The debugging and changing of a circuit is even more difficult. Programmable logic controllers replace most of this wiring by software programming. Therefore, the task is made much easier. Because the wires and the moving mechanical components (relay contacts) are mostly replaced by software, the system is much more reliable. In this section, the basic architecture of programmable logic controllers is discussed.

Like a general-purpose computer, a programmable controller consists of five major parts: the CPU (processor), memory, input/output (I/O), power supply, and peripherals (Figure 10.11).

10.3.1 The Processor

Although early PLCs used special-purpose logic circuits, most current PLCs are microprocessor-based systems. The processor (central processing unit, or CPU, as it is often called) scans the status of the input peripherals, examines the control logic to see what action to take, and then executes the appropriate output responses.

The microprocessor-based PLC has significantly increased the logical and control capacities of programmable logic controllers. High-end PLC systems allow the user to perform arithmetic and logic operations, move memory blocks, and interface with computers, a local-area network, functions, and so on.

10.3.2 Memory

The memory of the PLC is important because the control program and the peripheral status are stored there. Memory size in a PLC is measured in either bits, bytes, or

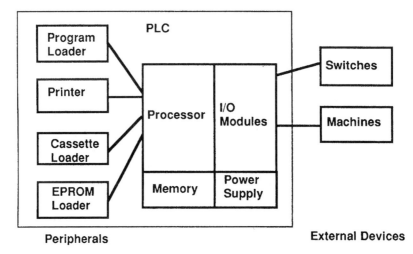

Figure 10.11   Programmable logic controller system structure.

words. Because many words of memory are required, it is usually measured in "k" increments (where 1 k = 1024).

Although several types of memory are used in modern PLCs, memory can be classified into two basic categories: volatile and nonvolatile. Volatile memory loses state (the stored information) when power is removed. This may seem perfectly appropriate. However, you must remember that the program is stored in memory, and if the power fails (the plug is pulled, and so on), the program must be rekeyed or reread into memory, a potentially time-consuming activity. Nonvolatile memory, on the other hand, maintains the information in memory even if the power is interrupted. Some types of memory used in a PLC include

1. ROM (read-only memory)
2. RAM (random-access memory)
3. PROM (programmable read-only memory)
4. EPROM (erasable programmable read-only memory)
5. EAPROM (electronically alterable programmable read-only memory)
6. bubble memory

Of these memory types, the only volatile memory is RAM. (Oddly enough, it is probably the most commonly used memory.) The other memory types maintain their status even after power is lost. Many RAM-based memory systems use battery backups to preserve the contents of memory in case of power failure. These systems are built using CMOS technology. CMOS devices consume minimal amounts of power.

Memory can be further classified as read-only or read/write. Of the memory types listed, RAM and bubble memory are read/write. They can be easily changed by the processor. The other types of memory require additional hardware to alter or program.

PLC memories are usually expandable. Memory modules can be added. User memory ranges from less than 1k byte to several megawords.

10.3.3 Input and Output

The input and output (I/O) for a PLC is normally a set of modular plug-in peripherals (notice the difference between this definition and the one used in computer I/Os). (See Figure 10.12.) The I/O modules allow the PLC to accept signals from a variety of external devices, for example, limit switches, optical sensors, and proximity switches. The signals (two state signals for the devices mentioned, open or closed) are converted from an external voltage (115 VAC, 230 VAC, 24 VDC) to a TTL signal of ±5 VDC. The PLC processor then uses these signals to determine the appropriate output response. A 5 VDC signal is transmitted to the appropriate output module, which converts the signal to the appropriate response domain (115 VAC, 230 VAC, 24 VDC).

Normally, a peripheral-interface adapter (PIA) is used to transfer the status of the input peripherals to some prespecified memory location. The user defines the location of the peripheral on the I/O housing in the program. Each I/O location is assigned to a specific memory location. This makes accessing input by the CPU a task of loading the contents of a specific memory into a storage register. Output changes are equally easy for the CPU to perform. The contents of a particular memory location are then altered. Due to the electrical differences between the CPU and external I/O peripheral,

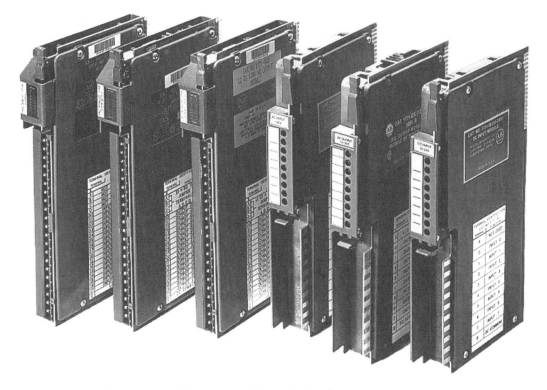

Figure 10.12   I/O modules (Courtesy of Allen-Bradley).

the I/O points and internal memory are actually electrically isolated. In a more advanced design, a separate I/O processor is used to bring the external I/O status to an internal memory location.

I/O modules are typically housed in a rack separate from the PLC. Light indicators are usually included in the I/O module to provide the current state (ideal for troubleshooting). In addition, each module is normally fused and isolated from the processor. Some typical I/O modules include

1. AC voltage input and output
2. DC voltage input and output
3. numerical input and output
4. special-purpose modules, for example, high-speed timers and stepping motor controllers

For discrete I/Os, either DC or AC are most commonly used. A discrete input is usually connected to a switch. Internally, the input module always electrically isolates the external signal from the internal circuit. An LED (light emitting diode) indicator on each of the input points of the input module shows the logic state of the switch. A discrete output point is turned on or off by the controller. It is also electrically isolated from the internal circuit.

TTL inputs are used to connect to TTL-compatible devices, including solid-state controls and sensing instruments, and some photoelectric sensors. TTL outputs are connected to devices such as LED displays and other 5-VDC devices.

Analog inputs are usually used for sensor interface. Analog outputs are used to drive analog devices such as actuators, motors, and so on. Following is a list of both analog input and output devices:

ANALOG INPUTS

Flow sensors

Humidity sensors

Potentiometers

Pressure sensors

Temperature sensors

ANALOG OUTPUTS

Analog meters

Analog valves and actuators

DC and AC motor drives

Numerical I/O exchange multiple-bit data with the outside devices. Typical inputs include thumbwheel switches, bar-code readers, and encoders. Typical outputs include seven-segment displays, intelligent displays, and so on. For seven-segment displays, BCD code (Appendix E) is used.

There is a wide range of special-purpose I/O modules. Not every PLC provides the full range of special-purpose I/O modules. Usually, they are supported by midsize to large-size PLCs. A few of those special-purpose I/O modules follow:

- *Thermocouple input:* Low-level analog signal, filtered, amplified, and digitized before sending to the processor through the I/O bus.
- *Fast input:* 50- to 100-microsecond pulse-signal detection.
- *ASCII I/O:* Communicates with ASCII devices.
- *Stepper-motor output:* Provides direct control of a stepper motor.
- *Pulse width modulation (PWM)*: Output for DC device control.
- *Servo interface:* Controls DC servo motor for point-to-point control and axis positioning. May use either resolver or encoder interface. (For more details, see Chapter 9).
- *PID control:* The proportional integral derivative is used for closed-loop process control. It is applied to any process that requires continuous closed-loop control. For an example, see Section 10.4.3.
- *Network module:* Provides LAN (Chapter 11) capability to the controller. It supports vendor-specific protocol, ethernet, DevicNet (Chapter 12), or ControlNet (Chapter 12.)

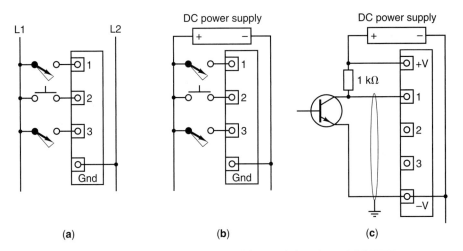

Figure 10.13 Power input connections: (a) AC, (b) DC, and (c) TTL.

10.3.4 Power Supply

The power supply operates on AC power to provide the DC power required for the controller's internal operation. It is designed to take either 115 or 220 VAC. Some power supplies can take either voltage with a jumper switch for selection. The operation of I/O modules is also supported by the PLCs' power supply. However, separate power sources are required in order to close the circuits of switches, motors, and external devices. Figure 10.13 shows typical AC, DC, and TTL input connections.

In the rest of the text, whenever an input or output module is shown, the power connection will not be shown. However, it is implied that proper wiring is necessary. The same is true for output (Figure 10.14).

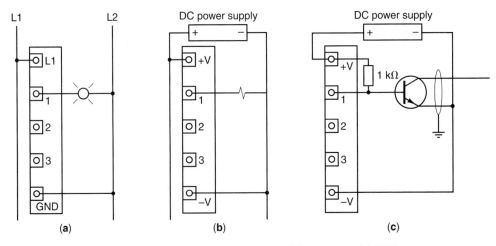

Figure 10.14 Power output connections: (a) AC, (b) DC, and (c) TTL.

10.3.5 Peripherals

A number of peripheral devices are available. They are used to program the PLC, prepare the program listing, back up the program, and display the system status. Old PLCs may still have handheld programmers and CRT programmers; today they have been replaced by a PC-based software programming environment. Following is a partial list of peripherals:

> Operator console (or human interface terminal)
> Printer
> Simulator
> EPROM loader (or EEPROM backup device)
> Network communication interface
> PC-based programming software

Figure 10.15 shows a human interface terminal. It can display both text and graphics. The human operator can also enter data or control commands though this terminal, which includes the human operator in the control loop. Monitoring and control of the process can be done by the human operator, but the PLC program cannot be changed through the terminal.

Each peripheral device has its own function. A printer is used to print PLC programs and operation messages. A simulator is a board usually consisting of some lights and switches. It is used to debug a program and can be connected to the PLC I/O module. Then, the PLC program logic can be tested by flipping switches and observing the lights. Due to the use of PCs for PLC programming, the hardware simulator has often been replaced by the software simulator. In the latter case, program logic debugging is done through the PC's simulation software.

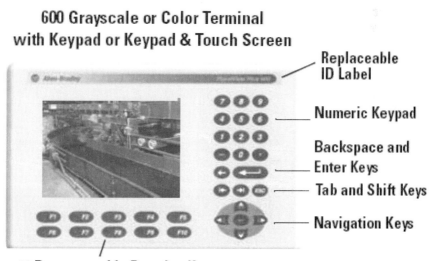

Figure 10.15   Human interface terminal (Courtesy of Allen-Bradley).

EPROM or EEPROM loaders can load a program from an EPROM/EEPROM to a PLC's memory. They can also be used to backup the program from a PLC to the EPROM/EEPROM.

The network communication module allows the PLCs to be integrated into a network of controllers and computers. In an integrated manufacturing system, all machines and material handling devices are coordinated. Thus, PLCs and other control devices can exchange data and commands. A LAN (local area network) makes such data communication fast and reliable. Traditionally, sensors and actuators used in an automated manufacturing system are wired directly to its controller. This type of configuration requires a large number of wires to be laid between the process (machine) and the controller. However, smart sensors and actuators contain their network interface card, so they can be directly tied into a LAN. This arrangement greatly reduces the wiring requirement.

The PC-based programming software became the norm for programming PLCs. A simple programming environment, such as the one shown in Figure 10.16, utilizes

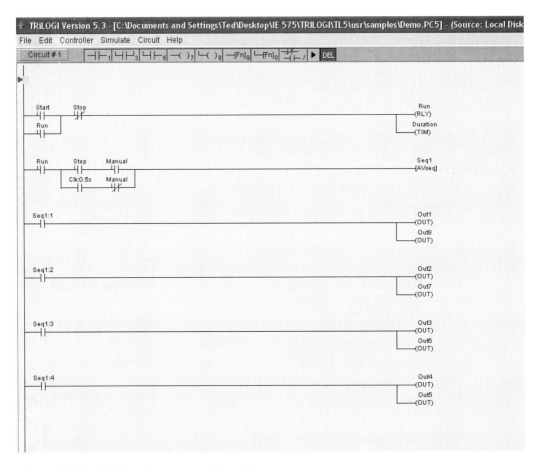

Figure 10.16   PC-based programming software.

the graphic user interface on a PC to enter and edit PLC programs. Some software enables the user to link several different software programs together to form a control program. See the next section for different programming languages.

10.4 PROGRAMMING A PROGRAMMABLE LOGIC CONTROLLER

Programmable logic controllers were initially developed to replace relay devices. The programming language used was similar to that used by electrical technicians to design electric circuits—the ladder diagram. However, as PLCs grew more powerful and flexible, the limitations of the ladder diagram soon became apparent. Not only does the ladder diagram have no easy way to represent data manipulation, but it is also extremely difficult to write and debug a large and complex ladder diagram. In recent years, many high-end PLCs began to introduce high-level languages. Some of them are English-like, some use BASIC language or BASIC-like language, and some others developed PASCAL-like structured language. Often, codes written in such high-level languages can coexist with the ladder-diagram program. An international effort to standardize the PLC programming language has resulted in an IEC 1131-3 standard that was published in 1993 [Lewis, 1995]. Under IEC 1131-3, the ladder diagram, structured text, function block diagram, instruction list, and sequential function chart are all included in the standard. Before the standard was fully adopted, each PLC vendor had developed its own language. In this section, we will try to follow the IEC standard.

What language to use for an application depends on several factors. First, most current PLCs use a vendor-specific language. For a chosen PLC, the choice becomes limited. Often there are only two choices: a high-level language and the ladder diagram. Ladder diagrams were designed to solve logic problems involving simple timing, counting, and sequencing. For these applications, a ladder diagram can be easily written. However, if math functions or communications are required, ladder-diagram programming is very difficult, if not impossible.

Because the ladder diagram is still the most basic, this section begins with a brief introduction to ladder-diagram programming. A few programming examples are also given. High-level language programming using the IEC 1131-3 standard will follow. Finally, some discussion on advanced PLC functions and their programming will be discussed. One such function is PID control, which is widely used in the process industry.

10.4.1 Ladder Diagram

A ladder diagram (also called contact symbology) is a means of graphically representing the logic required in a relay logic system. Ladder diagrams have long preceded the PLC and still represent the basic logic required by a relay device or PLC. The fundamental ladder diagram consists of a series of inputs, timers, and counters. Most simply, the ladder diagram represents the actions required (relay closure) as a function of a series of inputs that are either on or off. Each ladder-diagram element is represented using some standard symbols; some commonly used ones are shown in Figure 10.17.

A ladder diagram consists of two rails of the ladder and various control circuits, or rungs (Figure 10.18). Each rung starts from the left rail and ends at the right rail. We can consider that the left rail is the power wire and the right rail is the ground wire.

Limit switch	Normally open		LS 
	Normally closed		LS 
	Held open		
	Held closed		
Proximity switch	Open		
	Closed		
Toggle switch			
Rotary selector	Nonbridging contacts		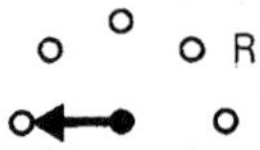 RSS
	Bridging contacts		 RSS
Push button	Single circuit	Normally open	 PB
		Normally closed	 PB
	Double circuit		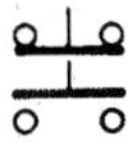 PB
Contacts	Relay	Normally open	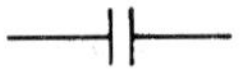
		Normally closed	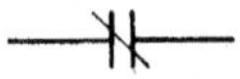
Coils	Relays		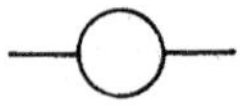 CR
	Solenoids		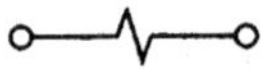 SOL
Motor	DC armature		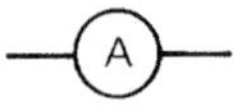 A — MTR
Pilot lights			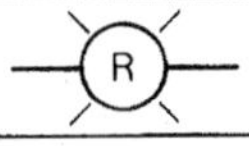 R — LT

Figure 10.17 Some relay diagram symbols.

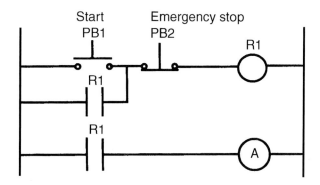

Figure 10.18   A ladder diagram.

Power flows from the left rail to the right rail, and each rung must have an output to prevent a short. The output is connected to physical devices, such as motors, lights, and solenoids. To control the output, some switches are used on the rung to form the AND and OR logic. Different rungs are not connected except through the rails. Each rung can contain only one output.

Functionally, the components in a ladder diagram consist of those used internally to construct the logic, such as some relays, timers, and counters, and those used to connect to the physical devices, such as switches and motors. The internal components are the ones replaced by a programmable logic controller (Figure 10.19).

Because of the operating-voltage difference and the logic-circuit requirement, the output is usually not connected to the motor or other devices directly. Instead, a relay is normally used. In Figure 10.18, PB_1 and PB_2 are the input pushbutton switches and R_1 *is the internal relay.* Motor-starter relay A is connected to a DC motor.

In the circuit of Figure 10.18, when pushbutton switch PB_1 is pushed, relay R_1 energizes and turns on the output circuit. In the circuit, the relay has a double-pole

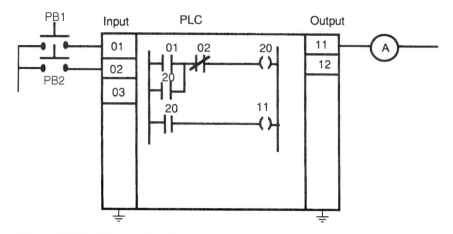

Figure 10.19   PLC wiring diagram.

Figure 10.20   PLC scan.

contact. The first contact forms a parallel circuit with PB_1. Even when PB_1 is released, current can still flow through rung 1. The second contact of R_1 is connected to A, turning A on or off.

In a hardwired relay circuit, current flows through all rungs simultaneously. However, in a programmable logic controller, the relay circuit is implemented in software and executed sequentially. Each ladder diagram element is assigned a working memory location (register). A PLC resolves the logic of a ladder diagram (program) rung by rung, from the top to the bottom. Usually, all the outputs are updated based on the status of the internal registers. Then the input states are checked and the corresponding input registers are updated. Only after the I/Os have been resolved is the program then executed. This process is run in an endless cycle. The time it takes to finish one cycle is called the *scan time*. The shorter the scan time, the faster the sample rate, and, thus, the faster the response to the event.

A scan cycle is illustrated in Figure 10.20. The reason that I/Os are done at the beginning of the cycle is that the program scan (execution) will not be interrupted by I/O changes. Input always affects the program from the first rung, and the output is always the result of the entire program execution from the first to the last rung. If I/Os are read and written during the program scan, depending on the timing, the same set of input states may take effect at a different rung, thus creating different results. On a PLC, scan length depends on program length. In order to obtain a fixed scan time, some PLCs insert an idle state after the program execution. The idle time makes up the difference between the preset scan time and the I/O and program scan times. This gives a predictable scan time.

For some time critical applications, immediate I/O might be desirable. In this case, a special program code is used to force immediate input or output. The scan cycle is used regardless of the programming method used.

As shown in Figure 10.19, a PLC uses ladder logic programming. The program is very similar to a standard ladder diagram. The following sections show how such programming is done. The term *ladder diagram* is used to denote the program normally input to a PLC. There are basically seven types of PLC instructions:

1. relay
2. timer and counter
3. program control
4. arithmetic

5. data manipulation

6. data transfer

7. others, such as sequencers

The first types of instructions, relay and timer and counter instructions, are the most fundamental because they correspond to the ladder diagram, and are available on all PLCs. The other five types of instructions are available on more advanced PLCs. This discussion begins with the logic of the PLC before it discusses PLC instructions.

10.4.2 Logic

By using serial and parallel connections, various types of logic can be represented in a ladder diagram. The logic states of a component are either on (true, contact closure, energize) or off (false, contact open, deenergize). The ladder diagram takes the input state from the input module and outputs results to the output module. In Figure 10.18, on rung 1, PB_1 and R_1 are in parallel and are connected with PB_2 and the R_1 coil in series. Complex control logic can be represented using this simple graphical logic. In this section, the fundamentals of logic construction are discussed. Ladder diagrams and equivalent logic representations are presented. Additional information on logic and truth tables can be found in Chapter 9.

10.4.2.1. Basic Logic.   The ladder diagram in Figure 10.21 depicts the most simple circuit logic. The output is solely determined by the input. Rung 1 uses a normally open contact, where the state of the output is the same as the state of the input. On rung 2, output R_2 has the opposite state of input PB_2. In the control diagram, when switch PB_2 is pushed, output R_2 is turned off.

NOT represents negation. AND and OR are logic operators. These symbols are used throughout the text.

AND Logic.   As mentioned before, AND logic is achieved by connecting two components in series. The two rungs shown in Figure 10.22 consist of AND logic. AND logic is the most commonly used logic and is required in most applications. For example, in a punch-press control, the operator has a foot pedal to control the stroke of the punch. A safety guarding device, such as a photodiode detector, is also used to prevent the accidental triggering of the punch while part of the operator's body is under the

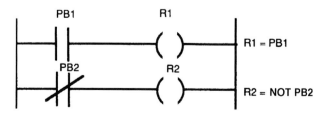

Figure 10.21   Basic logic.

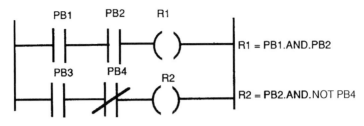

Figure 10.22   And logic.

punch. The circuit on rung 2 can be used for this process. Let PB_3 be the pedal switch, PB_4 be the photodiode detector, and R_2 be the control of the clutch.

OR Logic.   OR is used when either one of two switches (or one or more switches out of several) is pushed and the logic output needs to be true. In Figure 10.23, either switch PB_1 or PB_2 can turn on output R_1.

Combined AND and OR logic.   A combination of AND and OR logic also can be included in one rung. For example, to turn on R_1 when either (1) PB_1 is pushed or (2) PB_2 and PB_3 are pushed, the logic can be represented by an equation:

$$R_1 = PB_1 \quad \text{OR} \quad (PB_2 \quad \text{AND} \quad PB_3)$$

The corresponding rung is shown in Figure 10.24.

10.4.2.2. Relays.   A relay consists of two parts, the coil and the contact(s). The following are instructions (symbols) related to different contacts:

CONTACTS
(a) normally open: -| |-
(b) normally closed: -|/|-
(c) positive transition—sensing: -|P|-
(d) negative transition—sensing: -|N|-

A transitional contact output is a brief pulse. When the reference is triggered, it outputs only in one scan and functions like a one-shot. A positive transitional contact output is a pulse resulting when the reference makes a transition. A negative transitional contact occurs when the transition is from on to off.

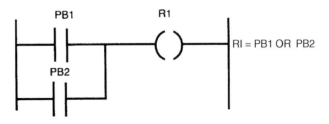

Figure 10.23   OR logic.

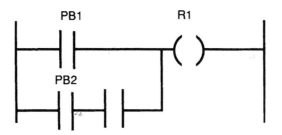

Figure 10.24   Combined AND and OR logic.

The coil is shown in the previous examples. The following instructions are related to different coils:

COILS

(a) coil: -( )-

(b) negative coil: -(/)-

(c) set coil: -(S)-

(d) reset coil: -(R)-

(e) retentive memory coil: -(M)-

(f) set retentive-memory coil: -(SM)-

(g) reset retentive-memory coil: -(RM)-

(h) positive transition-sensing coil: -(P)-

(i) negative transition-sensing coil: -(N)-

When the rung condition is true, the coil is energized (which, in turn, closes its normally open internal contacts). A negative coil works in the opposite way. When the rung condition is false, the coil is energized. The set coil latches the state of output. As soon as the rung condition turns true, the coil keeps the energized state, even though the rung condition becomes false. The reset coil is the only one that can deenergize the set coil. A retentive coil behaves as the normal coil, except that the state of the coil is retained on PLC power failure. A positive transition-sensing coil is set on for one scan when the power flow on the left-hand link changes from off to on. On the other hand, a negative transition-sensing coil is set on when the opposite happens.

10.4.2.3. Timers and Counters.   The retentive-on-delay timer starts counting when the run condition is true. The retentive-off-delay timer starts counting when the rung condition is false. The timing value is specified on the timer symbol. The timing diagram of Figure 10.8 is for a retentive-on-delay timer. The following instructions apply to timers and contacts:

TIMERS

(a) retentive on delay: -(RTO)-

(b) retentive off delay: -(RTF)-

(c) reset: -(RST)-

COUNTERS

(a) counter up: -(CTU)-

(b) counter down: -(CTD)-

(c) counter reset: -(CTR)-

An up counter increments the accumulator value each time there is an input. The down counter decrements the accumulator value. The counter reset is used to reset the counter value.

10.4.2.4. Programming Example. Given the background knowledge presented about PLC programming, the example that follows shows its applications. In order to keep the example brief, the problem has been simplified.

Example 10.1. Robotic Material-Handling Control System

A robot is used to load/unload parts to a machine from a conveyor. The layout of the system is shown in Figure 10.25. The process is shown in Figure 10.25.

A part comes along the conveyor. When it touches a microswitch, it is scanned by a bar-code reader to identify it. If the part is the desired one, a stopper is activated to stop it. A robot picks up the part and loads it onto the machine if it is idle. Otherwise, the robot waits to unload the machine. The following is the assignment of control components:

ID	Description	State	Explanation
MS_1	Microswitch	1	Part arrives
R_1	Output to barcode reader	1	Scan the part
C_1	Input from barcode reader	1	Right part
R_2	Output robot	1	Loading cycle
R_3	Output robot	1	Unloading cycle
C_2	Input from robot	1	Robot busy
R_4	Output to stopper	1	Stopper up
C_3	Input from machine	1	Machine busy
C_4	Input from machine	1	Task complete

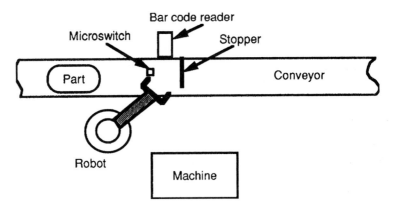

Figure 10.25 Cell layout.

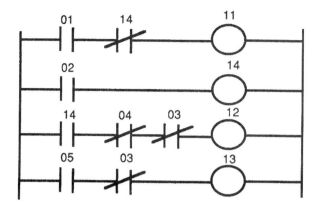

Figure 10.26   Program.

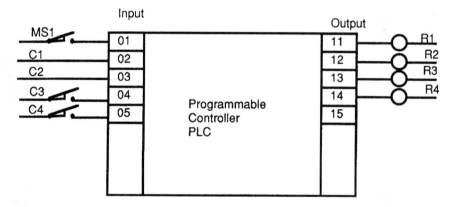

Figure 10.27   Wiring diagram.

The program and wiring diagrams are shown in Figures 10.26 and 10.27, respectively.

EXPLANATION OF THE PROGRAM

Rung 1. If a part arrives and no part is stopped, trigger the barcode reader.

Rung 2. If it is a right part, activate the stopper.

Rung 3. If the stopper is up, the machine is not busy and the robot is not busy; load the part onto the machine.

Rung 4. If the task is completed and the robot is not busy, unload the machine.

10.4.3 Structured Text Programming

Structured text is a high-level language that can be used to express the behavior of functions, function blocks, and programs. In the IEC 1131-1 standard, structured text has a syntax very similar to PASCAL. In this section, a brief introduction to structured text programming is presented.

Structured text is a strongly typed language. That means that all variables used in the program have to be declared before they can be used. The language also provides the following functionalities:

assignments

expressions

statements

operators

function calls

flow control, such as conditional statements and iteration statements.

10.4.3.1. Data Types.   IEC defines a wide range of data types. Before a variable can be used, it must be declared as one of the data types. Data types include the following:

SINT	short integer	1 byte
INT	integer	2 bytes
DINT	double integer	4 bytes
LINT	long integer	8 bytes
USINT	unsigned short integer	1 byte
UINT	unsigned integer	2 bytes
UDINT	unsigned double integer	4 bytes
ULINT	unsigned long integer	8 bytes
REAL	real	4 bytes
LREAL	long real	8 bytes
TIME	time duration	
DATE	calendar date	
TOD	time of day	
DT	date and time of day	
STRING	character strings	
BOOL	boolean	1 bit
BYTE	byte	1 byte
WORD	16-bit bit string	16 bits
DWORD	32-bit bit string	32 bits
LWORD	64-bit bit string	64 bits

User data type also can be derived from the built-in data types. When declared, they are placed inside a TYPE define statement.

```
TYPE    (*user-defined data types, this is a comment*)
    pressure    :    REAL;
    temp        :    REAL;
    part_count  :    INT;
END_TYPE;
```

The data type of structure is also allowed. For example, the following code declares a data structure called "data_packet," which contains four data elements: input, t, out, and count.

```
TYPE data_packet:
    STRUCT
        input :     BOOL;
        t     :     TIME;
        out   :     BOOL;
        count :     INT;
    END_STRUCT:
END_TYPE;
```

10.4.3.2. Variable Declaration.   Variables are declared before they are used. There are global variables, local variables, input variables, output variables, input–output variables, external variables, and directly represented variables. Global variables are declared outside a program organization unit (POU), such as a function. Local variables are declared in a POU. Input variables are used as input parameters to a POU, and output variables are used as output parameters.

To declare a variable, use the following code:

```
a, b, c :    REAL;
input   :    INT;
```

To declare special types of variables, a keyword corresponding to the variable type is used. VAR, VAR_INPUT, VAR_OUTPUT, VAR_IN_OUT, VAR_GLOBAL, and VAR_EXTERNAL are used for local, input, output, input–output, global, and external variables, respectively. The declaration ends with an END_VAR keyword. For example, to declare a set of local variables, the following statements are used:

```
VAR
    I,j,k  :INT;
    v      :REAL;
END_VAR
```

10.4.3.3. Assignment Statements.   An assignment statement has the following format:

```
A: = 1.234;
X: = Y;
```

10.4.3.4. Operators.   Operators include the following:

( )	parenthesized expression
function()	function
**	exponentiation
—	negation

NOT	Boolean complement
+−*/	math operators
MOD	modulus operation
<><=>=	comparison operators
=	equal
<>	not equal
AND, &	Boolean AND
XOR	Boolean XOR
OR	Boolean OR

10.4.3.5. Expressions and Statements.   An expression contains one or more constants, variables, and/or functions linked by operators. An expression always produces a value of a particular data type. In a statement, there may be several expressions. A statement ends with a semicolon. The following are a few example statements:

```
y := a AND b;
v := (v1 + v2 + v3)/3
output := (light = open) OR (door = shut);
```

10.4.3.6. Conditional Statements.   There are two types of conditional statements:

```
IF ... THEN ... ELSE ... END_IF;
```

and

```
CASE ... OF ... ELSE ... END_CASE;
```

The first conditional statement selects the statement to execute based on the result of the Boolean expression after IF. The second conditional statement switches the execution statement based on an integer value given after the CASE keyword. The following examples show how these two statements are used:

```
IF a > 100 THEN
        redlight := on;
ELSEIF a > 50 THEN
        yellowlight := on;
ELSE
        greenlight := on;
END_IF;
CASE dial_setting OF
    1 :    x := 10;
    2 :    x := 15;
    3 :    x := 18;
    4,5:   x := 20; (*either 4 or 5*)
ELSE
    x := 30;
END_CASE
```

10.4.3.7. Iteration Statements.   Iteration statements are used to create looped execution. There are three types of iteration statements: FOR … DO, WHILE … DO, and REPEAT … UNTIL. The FOR … DO statement repeats a set of statements depending on the iteration variable after FOR. For example,

```
FOR I:= 0 to 100 BY 1 DO
     light[I] := ON;
END_FOR
```

The WHILE … DO checks the Boolean expression after WHILE to decide when to exit the loop. For example,

```
I:= 0;
WHILE I < 100 DO
    I := I + 1;
    light[I]:= on;
END_WHILE
```

The REPEAT … UNTIL is similar to the WHILE … DO except the Boolean statement is placed after UNTIL:

```
I:= 0;
REPEAT
    I := I + 1;
    light[I] := on;
UNTIL I > 100;
END_REPEAT
```

10.4.3.8. Functions.   A function is a program unit that takes a set of inputs, performs some operations, and then returns a value to the calling statement. A function is typed based on the data type of the return value. The following is an example of a function that adds two real numbers and returns the result:

```
FUNCTION add_num    :REAL
    VAR_INPUT
         I,J   :     REAL
    END_VAR
    add_num:= I + J;
END_FUNCTION
```

In the preceding function, the VAR_INPUT section defines the input variables (argument list), and the output is returned through the function name.

To call a function, a function name is referenced with a set of input variables. For example, to call function "add_num", use

```
x := add_num(1.2, 5.6);
```

This statement will return a value 6.8 and assign it to the variable x.

Built-in functions include math functions (ABS, SQRT, LN, LOG, EXP, SIN, COS, TAN, ASIN, ACOS, ATAN, ADD, MUL, SUB, DIV, MOD, EXPT, MOVE), logic functions (AND, OR, XOR, NOT), bit-string functions [SHL (shift bit string left), SHR (shift bit string right), ROL (rotate bit string left), ROR (rotate bit string left), and so on.]

10.4.3.9. Programs.   A program begins with the keyword PROGRAM and ends with END_PROGRAM. The structure of a program is similar to that of a function. However, the program body can be described using one of the IEC languages, including a ladder diagram. The following is a sample program for the application in Example 10.1:

```
PROGRAM Example7.1
    VAR_INPUT
            MSI :       BOOL;
            C1  :       BOOL;
            C2  :       BOOL;
            C3  :       BOOL;
            C4  :       BOOL;
    END_VAR
    VAR_OUTPUT
            R1  :       BOOL : FALSE;
            R2  :       BOOL : FALSE;
            R3  :       BOOL : FALSE;
            R4  :       BOOL : FALSE;
    END_VAR
    R1:=MS1 AND (NOT R4);
    R2:=R4 AND (NOT C3) AND (NOT C2);
    R3:=C4 AND (NOT C3);
    R4:=C1;
END_PROGRAM
```

10.4.4 Functional Block Programming

In the IEC 1131-3 standard, a functional block (FB) is a well-packaged element of software that can be reused in different parts of an application or even in different projects. Functional blocks are the basic building blocks of a control system and can have algorithms written in any of the IEC languages. A function block type contains two parts: (1) data declarations, and (2) an algorithm expressed using a structured text, a function block diagram, a ladder diagram, an instruction list, or a sequential function chart. A functional block also can be used directly in a ladder diagram.

A functional block for an upcounter is defined as shown in Figure 10.28.

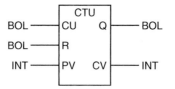

Figure 10.28   An up-counter function block.

The algorithm in structured text is as follows:

```
FUNCTION BLOCK CTU
    VAR_INPUT
        CU:BOOL R_TRIG;
        R:BOOL;
        PV:INT;
    END_VAR
    VAR_OUTPUT
        Q:BOOL;
        CV:INT;
    END_VAR
    IF R THEN
        CV := 0;
    ELSIF CU
            AND (CV<PV) THEN
            CV:= CV+1;
    END_IF;
    Q:= (CV>=PV);
END_FUNCTION_BLOCK
```

The CTU block counts the number of input CUs. In the block, R is the reset, PV is a preset value, Q is the contact output, and CV is the counter value. When a signal from CU is detected, the CV value increments by one. When the CV value reaches the PV value, Q is set to true. When the R signal is detected, the CV value is reset to zero and Q is set to false.

The IEC 1131-3 standard defines a small number of basic function blocks. Counters, timers, real-time clocks, edge detectors, and bistable are all predefined. More complex function such as the PID control block also can be defined. An example PID block is shown in Figure 10.29. (The PID Control concept is discussed in Section 9.1.2 of Chapter 9.)

The variables can be found in the control block diagram of Figure 10.30. The control algorithm is represented by the following equation:

$$V_{\text{out}} = K_p E + T_r \int E \, dt + T_d \frac{dE}{dt}$$

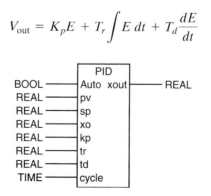

Figure 10.29 A PID-control function block.

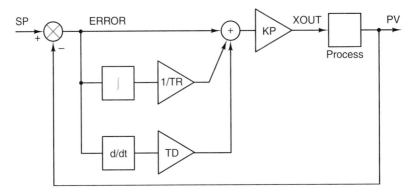

Figure 10.30   Block diagram of a PID controller.

A PID controller consists of three components: proportional control, integral control, and derivative control. It is commonly used in the process industry for closed-loop continuous-process control. For example, in a chemical-reaction control, SP may be the set temperature, XOUT may be the current going into a heating element, and PV may represent the temperature-sensor reading. It also can be used in the control of servo motors in a machine like a robot arm. In this case, SP will be the target rotational angle, XOUT will be the voltage signal sent to the servo motor, and PV will be the position-transducer feedback. KP, TR, and TD are gains for the proportional error, the integral, and the derivative control, respectively. The user must set these values. If AUTO is true, the function block calculates the value for XOUT. The "cycle" defines the time between function-block execution. XO is for manual output adjustment.

10.4.5 Instruction List

The instruction list (IL) is a low-level language that has a structure similar to an assembly language. Because it is simple, it is easy to learn and ideal for small handheld programming devices. The instruction list has a simple syntax. Each line of code can be divided into four fields: label, operator, operand, and comment. Label and comment fields are optional. Basic operators include the following:

OPERATOR	MODIFIERS	DESCRIPTION
LD	N	Load operand into register
ST	N	Store register value into operand
S		Set operand true
R		Reset operand false
AND	N, (	Boolean AND
&	N, (	Boolean AND
OR	N, (	Boolean OR
XOR	N, (	Boolean XOR
ADD	(	Addition
SUB	(	Subtraction

MUL	(	Multiplication
DIV	(	Division
GT	(	Greater than
GE	(	Greater than and equal to
EQ	(	Equal
NE	(	Not equal
LE	(	Less than and equal to
LT	(	Less than
JMP	C, N	Jump to label
CAL	C, N	Call function block
RET	C, N	Return from function or function block
)		Execute last deferred operator

Modifier "N" means negate, "(" defers the operator, and "C" is a condition modifier; the operation is executed if the register value is true.

By using the instruction list, the example in Figure 10.25 can be written as follows:

```
PROGRAM example 7.1
      VAR_INPUT
            MSI :     BOOL;
            C1  :     BOOL;
            C2  :     BOOL;
            C3  :     BOOL;
            C4  :     BOOL;
      END_VAR
      VAR_OUTPUT
            R1  :     BOOL:FALSE;
            R2  :     BOOL:FALSE;
            R3  :     BOOL:FALSE;
            R4  :     BOOL:FALSE;
      END_VAR
      LD      MS1
      ANDN    R4
      ST      R1
      LD      R4
      ANDN    C3
      ANDN    C2
      ST      R2
      LD      C4
      ANDN    C3
      ST      R3
      LD      C1
      ST      R4
END_PROGRAM
```

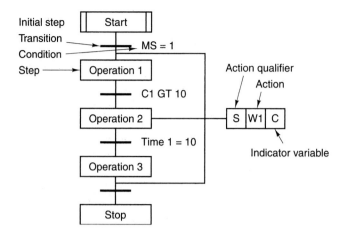

Figure 10.31   A sequential function chart.

10.4.6 Sequential Function Chart

The sequential function chart (SFC) is a graphics language used for depicting sequential behavior. The IEC standard grew out of the French standard, Grafcet, which in turn is based on Petri-net. An SFC is depicted as a series of steps shown as rectangular boxes connected by vertical lines (Figure 10.31). Each step represents a state of the system being controlled. A horizontal bar indicates a condition; it can be a switch state, a timer, and so on. A condition statement is associated with each condition bar. Each

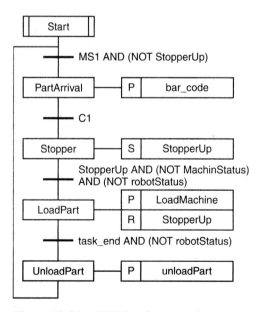

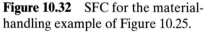

Figure 10.32   SFC for the material-handling example of Figure 10.25.

step also can have a set of actions. The action qualifier causes the action to behave in certain ways. The indicator variable is optional; it is for annotation purposes.

The action can be described as part of the SFC, or on another diagram or page. The action qualifiers are as follows:

N	nonstored; executes while the step is active
R	resets a store action
S	sets an action active
L	time-limited action; terminates after a given period
D	time-delayed action
P	a pulse action; executes once in a step
SD	stored and time-delayed
DS	time-delayed and stored
SL	stored and time-limited

The example problem in Figure 10.25 can be written in a sequential function chart, as shown in Figure 10.32.

10.5 TOOLS FOR PLC LOGIC DESIGN

In this section, two analytical tools for PLC logic design are introduced. The PLC logic design problem takes the description of a control problem and converts it into a PLC program. However, the solution is not always obvious. For very simple problems, the problem description can be translated directly into a ladder diagram or other PLC programs. When problems are more complex, this translation is either very difficult or produces inefficient program. The two tools introduced in this section can help organize the problem description and convert the description into logic statements. (Logic statements can be further simplified using the methods discussed in Chapter 9.) Since there is a one-to-one correspondence between logic statements and ladder diagrams, PLC programs can be written easily.

10.5.1 Design Using a Truth Table

A truth table was introduced in Chapter 9, Section 9.2.2. These tables are effective for resolving problems without involving an explicit sequence of events. In fact, a logic control problem can be modeled using a truth table. It represents the logical relationship between a set of input variables and an output variable (Figure 10.33). In the figure, A, B and C are input variables and f is the output variable.

Input variables represent events; for example, an input variable can indicate whether a switch is on, a door is closed, a timer has been tripped, etc. The output variable

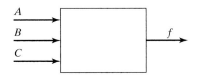

Figure 10.33   System I/O.

A	B	C	f	
0	0	0	0	
0	0	1	0	
0	1	0	0	
0	1	1	1	$A'BC$
1	0	0	1	$AB'C'$
1	0	1	1	$AB'C$
1	1	0	1	ABC'
1	1	1	1	ABC

Figure 10.34   Truth table.

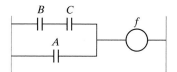

Figure 10.35   Ladder diagram
for a three-switch system.

represents an action; for example, it can specify a particular instruction, such as turn the
light on, turn the motor on, close the door, etc. Each variable can either be TRUE (ON
or 1) or FALSE (OFF or 0). A truth table, such as the one shown in Figure 10.34, can
represent the desired control function.

As discussed in Chapter 9, the expression for f can be written and simplified:

$$f = A'BC + \underbrace{AB'C' + AB'C}_{} + \underbrace{ABC' + ABC}_{}$$

$$= A'BC + \underbrace{AB'}_{} + \overbrace{AB}^{}$$

$$= A'BC + \underbrace{\qquad A \qquad}_{}$$

$$= BC + A$$

This expression can be converted into a ladder diagram as shown in Figure 10.35.

Example 10.2.

Design an adder to add two one-bit binary numbers, A and B. The sum should be a two-bit
number.

The problem can be represented using a truth table. When only A or B is "1," the
sum is "01." When both A and B are "1," the sum is "10."

Input		Sum	
A	B	X	Y
0	0	0	0
0	1	0	1
1	0	0	1
1	1	1	0

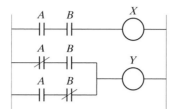

Figure 10.36 Ladder diagram for Example 10.2.

The logic statements for X and Y are

$$X = AB$$
$$Y = A'B + AB'$$

No simplification is needed. The ladder diagram is shown in Figure 10.36.

10.5.2 Control Using a State Diagram

A truth table is not able to capture the sequencing information, such as two actions that follow each other. It is difficult to represent this relationship in a truth table. Thus, state diagrams are used when the problem involves sequencing and timing. A state diagram consists of a set of nodes and arcs, where the nodes represent states and the arcs represent state transitions. A state can transit to another state only if a condition is fulfilled. Conditions, therefore, are attached to the arcs. Let s_i represent states and x_i represent conditions on the arcs. A state diagram is shown in Figure 10.37.

The state diagram can be easily interpreted: In state s_0, when x_0 is true, stay in the state. When x_2 is true, change to state s_1. The same can be applied to state s_1. Outputs can be attached to the states if necessary; but the input is always part of the condition statement. Note that a state has only two values: "0" and "1". Only one state can be true at any given time. In addition, the following conditions must hold:

$$s_i = \{0, 1\}$$
$$x_i = \{0, 1\}$$

$$\sum s_i = 1$$

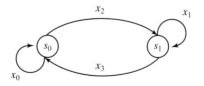

Figure 10.37 State diagram for a two-state and transition system.

To convert a state diagram into a set of logic statements, the following rule is applied: "A state is entered if and only if one of the arcs pointing to the state is true and all the arcs emanating from the state are NOT true." This condition can also be stated as "A state is true if it is entered and not left." To make an arc true, the emanating state must be true and the condition on the arc must also be true:

$$ s_i = \left(\sum x_j s_j \right) \left(\prod x_k s_k \right)' $$

In the preceding equation, $x_j s_j$ represents an entering arc and $x_k s_k$ represents an exiting arc. The state diagram in Figure 10.37 can be converted into the following logic statements:

$$ s_0 = (s_0 x_0 + s_1 x_3)(s_1 x_2)' \tag{10.1} $$

$$ s_1 = (s_1 x_1 + s_0 x_2)(s_0 x_3)' \tag{10.2} $$

In Equation 10.1, $s_0 x_0$ represents staying in the current state and $s_1 x_3$ represents a transition from state s_1 to state s_0 through an arc with the condition x_3. There are two entering paths to this state. To make s_0 true, it should not transit to other states. There is only one path to transit to state s_1, and it should not be true. This ensures that only one state is true at all times. To write a ladder diagram based on Equations 10.1 and 10.2, the "NOT" operator must be removed from outside the parentheses in the equations. The equations can then be rewritten as

$$ s_0 = (s_0 x_0 + s_1 x_3)(s_1' + x_2') \tag{10.3} $$

$$ s_1 = (s_1 x_1 + s_0 x_2)(s_0' + x_3') \tag{10.4} $$

In reviewing the two equations, if all variables are initialized to zero, neither state will ever be entered. Therefore, there must be a one-time external event to initialize one of the states to "1". This can be done simply by adding the equation

$$ s_0 = a \tag{10.5} $$

where a is an input from an external switch.

Note that the purpose of x_0, x_1 is to keep the state active if there is no transition to other states. However, Equations 10.3 and 10.4 may be rewritten as

$$ s_0 = (s_0 + s_1 x_3)(s_1' + x_2') \tag{10.6} $$

$$ s_1 = (s_1 + s_0 x_2)(s_0' + x_3') \tag{10.7} $$

This formulation has a "latch" effect: As soon as a state is entered, the program will stay in the state even if the triggering event is no longer true. Example 10.3 illustrates this latch effect.

Example 10.3. Latch circuit

A machine tool is controlled by two pushbutton switches: START and STOP. As soon as the START button is pushed, the machine starts. Even when the button is released, the machine will continue to run. It can be stopped only by pushing a STOP button.

The typical ladder diagram for this application is shown in Figure 10.38.

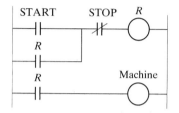

Figure 10.38   Ladder diagram for Example 10.3.

R is a DPST (double-pole, single-throw) relay. The logic statements for this ladder diagram are

$$R = (R + START)STOP'$$

$$Machine = R$$

The equation for R is exactly the same as that in Figures 10.6 and 10.7. After the START is pushed, state R is entered. It will stay in state R until it departs through STOP. The momentary event of START being true is latched.

Finally, examine Figure 10.39, which shows the ladder diagram for the state diagram in Figure 10.37. Contact a is connected to a normally open switch. To initialize the controller, the external switch is pushed once. States are represented by relays. There is a one-to-one correspondence between the logic statements and the rungs in a ladder diagram.

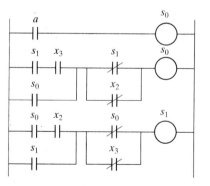

Figure 10.39   Ladder diagram for the state diagram in Figure 10.27.

Example 10.4. Single entering and departing path

An application calls for two sequential operations. After the process is initialized, switch A turns a warning light on. After the warning light is on, switch B turns the machine on. (At this moment, the light is turned off.) The machine can be turned off by switch C, at which point the process reinitializes.

On the basis of the problem description, the machine has three states:

s_0: begin the process.
s_1: warning light is on.
s_2: machine is on.

The state diagram is shown in Figure 10.40.

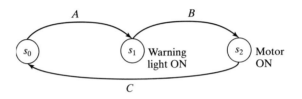

Figure 10.40   State diagram for Example 10.4.

For this example, the outputs are attached to the states. There is only one output for each state. It is appropriate to attach any number of outputs to a state. In this example, it is assumed that when the output is not explicitly defined as ON, then it must be OFF. Of course, an output may be turned either ON or OFF when a state is entered. The logic statements from the preceding state diagram are

$$s_0 = (s_0 + s_2 C)(s_1 A)'$$
$$= (s_0 + s_2 C)(s_1' + A')$$
$$s_1 = (s_1 + s_0 A)(s_2 B)'$$
$$= (s_1 + s_0 A)(s_2' + B')$$
$$s_2 = (s_2 + s_1 A)(s_0 C)'$$
$$= (s_2 + s_1 A)(s_0' + C')$$

$$s_0 = start$$
$$light = s_1$$
$$machine = s_2$$

where *start* is the input SPST (single-pole, single-throw) pushbutton switch and *light* and *machine* are outputs on the output module. Since the ladder diagram is rather straightforward, its construction is left as an exercise to the reader.

Example 10.5. Multiple entering and departing paths

The state diagram in Figure 10.41 shows multiple entering and departing paths.

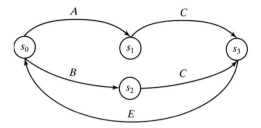

Figure 10.41   State diagram for Example 10.5.

The logic statements are as follows:

$$s_0 = (s_0 + s_3E)(s_1A + s_2B)'$$
$$s_1 = (s_1 + s_0A)(s_3C)'$$
$$s_2 = (s_2 + s_0B)(s_3C)'$$
$$s_3 = (s_3 + s_1C + s_2C)(s_0E)'$$
$$s_0 = start$$

Note the difference between this example and the preceding one. Before the preceding statements can be translated into a ladder diagram, the departing terms must be simplified.

Example 10.6. Timing and sequencing

A machine produces two different liquid products, each of which is identified by the barcode on the container. When the container arrives at the machine, a quick scan is made. If the container is for product 1, an ingredient is added for 5 seconds. On the other hand, if it is for product 2, the same ingredient is added for 6 seconds. After the operation is completed, the machine reinitializes.

The T1 and T2 represent timers for 5 and 6 seconds (Figure 10.42). Timers are initialized as soon as the states they enter a particular state. When the time value is reached,

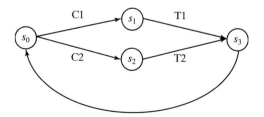

Figure 10.42   State diagram for Example 10.6.

it outputs an ON to the controller. The timer and other conditions should not be located on the same arc. The logic statements are as follows:

$$s_0 = (s_0 + s_3)(s_1 C1 + s_2 C2)'$$
$$s_1 = (s_1 + s_0 C1)(s_3 T1)'$$
$$s_2 = (s_2 + s_0 C2)(s_3 T2)'$$
$$s_3 = (s_3 + s_1 T1 + s_2 T2)s_0'$$
$$T1 = s_1$$
$$T2 = s_2$$
$$s_0 = start$$

10.6 SUMMARY

This chapter introduced the basic functions of a programmable logic controller. Because the expanded functions of a PLC vary in symbology and capability from vendor to vendor, they were not directly addressed. Instead, a hypothetical PLC symbology was presented that should simplify the process of using any PLC and should expand PLC command libraries.

The PLC is an inexpensive flexible control device that is quickly becoming a standard control device for individual processes as well as for process integration; however, it is worthwhile to note that the PLC is not the only means of control. The conventional relay and computer control are two other alternatives. Although the PLC is both attractive and economical for most control applications, relay logic and/or computer control may be more appropriate in some applications. This is especially true for controlling simple applications such as simple machines and toys where a single-board (or single-chip) computer-based controller is most economical. No feasible tool should be overlooked.

The major advantages of the PLC over other devices are the ease of its programming interface and robustness. When an application requires only 10 or fewer relays, hardwired relay logic is probably more economical. However, when an application requires more complex control, then the PLC becomes a more attractive alternative.

When compared with computer control, the PLC is not as flexible. However, because the PLC is microcomputer based and specifically designed for industrial control, it is much easier to use in the harsh shop environment, where a computer will not survive as long (except for industrial-hardened computers). A PLC can operate without any problems; yet, if the application requires some function that a PLC does not support or a sampling rate that is greater than the shortest scan time can provide, then a computer is more desirable. Readers who have experience with computer interfacing and low-level programming will appreciate how easy it is to use a PLC for control.

Actually, a PLC can be seen as a computer dedicated to process control. It has some standardized interface modules and runs a special interpretive control language. Following the advance of microcomputer technology and the development of interface and language, the PLC has become more like a computer. More and more mathematical and database functions have been added. Its color-graphics display, network capability, and so on have totally changed the image of a PLC as a relay-panel substitute. Future users will see a greater use of PLCs for all kinds of manufacturing-systems control and device control. The failure to recognize and utilize these devices will undoubtedly handicap manufacturing, electrical, and mechanical engineers.

10.7 KEYWORDS

Coil	Contact
Counter	EEPROM
Functional block programming	Human operator terminal
I/O module	Instruction list programming
Ladder diagram	Ladder-diagram programming
Logic	PLC scan
Pole and throw	Relay
Relay device	Sequential-function-chart programming
State diagram	Structured-text programming
Timer	Truth table

10.8 REVIEW QUESTIONS

10.1 Why is scan time critical to some applications?

10.2 What are the advantages of using a PLC over a relay-panel circuit?

10.3 What are the advantages of using a PLC over a microcomputer?

10.4 Why does the machine start and stop control usually use a latch circuit?

10.5 What are the advantages and disadvantages of using ladder-diagram programming over other methods of programming?

10.6 Although the Basic language is not an official IEC standard language for PLCs, some vendors use it for their PLCs. Find some examples of PLCs based on the Basic programming language.

10.7 List some manufacturing system control functions that might be controlled by PLCs.

10.8 Most NC machine tools have built-in PLC functions. Find the purpose of having those PLC functions.

10.9 Find the smallest-size (physical dimension) PLC on the market. What features are available on it? How much does it cost? What programming language is supported?

10.10 In many applications, PLCs are not alone. They must communicate with each other and with other computers. How are such communications done?

10.9 REVIEW PROBLEMS

10.1 Design a ladder diagram that will latch the state of a switch LS_1. It is unlatched only when switch LS_2 is triggered. The state of switch LS_1 is shown by a pilot light, PL_1.

10.2 Design a ladder diagram to function XOR logic.

10.3 Design a PLC ladder diagram that will turn on light PL_1 5 s after pushbutton PB_1 is pressed. The light is turned off when the switch is released.

10.4 A two-axis modular robot is controlled by a PLC. Each axis is driven by a pneumatic cylinder. An output signal to the pneumatic cylinder causes it to extend (on state). When there is no signal, the pneumatic cylinder retracts (off state). The gripper is also driven by a pneumatic cylinder. The gripper open position is an off state and the closed position is an on state. In order to perform a task, the following sequence of operations must be performed:

Sequence No.	Axis 1	Axis 2	Gripper
0	off	off	off
1	on	off	off
2	on	off	on
3	on	on	on
4	off	on	on
5	off	off	on
6	off	off	off

The process begins when the start pushbutton is on and a part triggers a limit switch. For safety, an emergency switch is used to shut off the entire system. Design a circuit and a PLC ladder diagram for this application.

10.5 For the application described in Review Problem 4, prepare a list of system components needed. Also prepare a programmable logic controller specification for the application. Assume that all the pneumatic cylinders are controlled by a 24-V signal.

10.6 Call a local PLC vendor and find out the cost of the cheapest PLC, relays, timers, and counters. What is the break-even point of switching from relays to a PLC (in terms of numbers of internal relays)?

10.7 Design a ladder diagram that has a light indicator connected to output 10 that flashes on six times for 1 s with a 2-s delay between flashes. A switch connected to 02 initiates and resets the program.

10.8 The ABC Company is in the toy business. It produces two types of toy cars, wooden cars and metal cars. A conveyor belt carries the product through the production floor (see Figure 10.43). Wooden cars are sensed by a photoelectric switch and metal cars are sensed by a proximity switch. Cars are pushed to diverters by means of solenoids. The conveyor belt is activated by a push button. However, in case of an emergency, the whole system is shut down by means of an emergency pushbutton. Assume that wooden cars are diverted to diverter A by means of solenoid 1 and that metal cars are diverted to diverter C by means of solenoid 3. Write a ladder diagram for the control of this system. See Review Problem 10.9 for a hint.

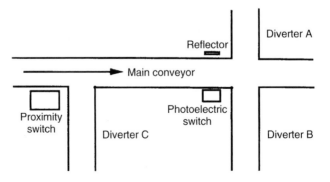

Figure 10.43 System layout.

10.9 From Review Problem 10.8, the company increases the production of wooden toy cars. The production increase requires two diverters for the wooden cars. Both diverters have a capacity of four cars. Diverters *A* and *B* are controlled by solenoids 1 and 2, respectively. Assuming that the parts are removed when the diverters are full, draw a ladder diagram and test it on a simulation board, using the following information:

Addresses for Input Devices

Switch	Address
Motor start	1
Emergency stop	2
Photoelectric	3
Proximity	4

Addresses for Output Devices

Output	Address
Solenoid 1	29
Solenoid 2	30
Solenoid 3	31
Motor	32

Addresses for Counters

Output	Address
Counter 1	901
Counter 2	902

Hint for Review Problems 8 and 9. There are two pieces representing the cars, a wooden piece and a metal piece. When the wooden piece is passed through the photoelectric switch output, 29 or 30 is the address in the output module. When a metal piece is passed through the proximity switch, output 31 is turned on. However, notice that a metal piece is sensed by both the proximity switch and the photoelectric switch. That is not the case for wooden parts, which are only activated when the beam of light is broken.

The motor that controls the conveyor is represented by address 32. This means that whenever the motor start button is pressed, light 32 in the output module of the programmable controller is turned on. However, when the emergency stop is pressed, light 32 changes its status to off.

10.10 Figure 10.44 shows a PLC ladder diagram and a wiring diagram. Outputs 10, 11, and 12 are connected to lights. Fill in the state of the lights in the table.

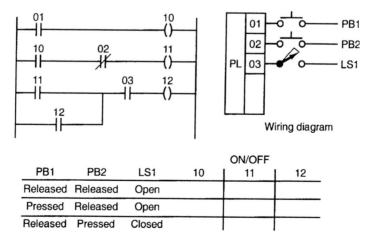

				ON/OFF	
PB1	PB2	LS1	10	11	12
Released	Released	Open			
Pressed	Released	Open			
Released	Pressed	Closed			

Figure 10.44

10.11 Design a ladder-diagram program to control a two-speed motor. The motor can be started only at a low speed. The motor can be switched to high speed only after 10 seconds of operation. The motor cannot be switched from high to low speed.

10.12 In a PE (proportional error) controller, the output is proportional to the error between the set point (SP) and the sensor feedback (PV):

$$\text{xout} = K(SP - PV)$$

For example, the goal is to maintain a certain temperature. The set point is the desired temperature and PV is the thermometer reading. However, to regulate the temperature, the controller has to provide more electrical current to the heating element. The Xout is the current.

Write a program in structured text for a functional block called PE control. The inputs are RUN, SP, PV. The output is Xout.

Note: The ladder-diagram programming problems in this chapter may be solved by using structured text, functional blocks, instruction lists, or sequential function charts.

10.13 Convert the following sentences into a Boolean equation: (define your own variables)

(a) The process should be turned on if the part has arrived, the time is between 8 a.m. and 5 p.m., and it is not a holiday.

(b) The spindle motor should be running if and only if
   i.   The workpiece is on the machine table.
   ii. An end-of-program signal is not present, and
   iii. The machine-program start button has been pressed and the spindle on code is read, or both the manual run button and the spindle run button have been pressed.

(c) The sound system will squeal if the microphone is turned on, and the microphone is too close to the speaker or the volume control is set too high.

10.14 Convert the following Ladder diagram into logic statements:

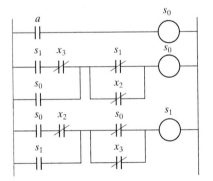

Figure 10.45

10.15 Convert the following logic statements into ladder diagrams:

$$R1 = (A + B')(CD + AB)'$$

$$R2 = R1 + AD + B'C$$

10.16   Simplify the following truth table and design a ladder diagram to provide the output *F*.

A	B	C	F
0	0	0	0
0	0	1	0
0	1	0	1
0	1	1	1
1	0	0	0
1	0	1	0
1	1	0	1
1	1	1	0

10.17   Write a ladder diagram for the following problem (use a state diagram to solve the problem):
- A process is controlled by four buttons: start, choice1, choice2, and end.
- Three seconds after the start button is pressed, the process may begin.
- If the choice1 button is pressed, motor one is turned on for 5 seconds.
- If the choice2 button is pressed, motor one is turned on for 10 seconds.
- After the motor is stopped, the end button will reset the process.

10.10 REFERENCES

Bertrand, R. M. *Programmable Controller Circuits.* Albany, New York: Delmar Publishers, 1996.

Bolton, W. *Programmable Logic Controllers*, 3d ed., Newnes, Elsevier, 2003.

Crispin, A. J. *Programmable Logic Controllers and their Engineering Applications.* New York: McGraw-Hill, 1990.

Friedman, S. B. *Logical Design of Automation System.* Englewood Cliffs, NJ: Prentice Hall, 1990.

Gilles, M. *Programmable Logic Controllers.* New York: John Wiley, 1990.

Lewis, R. W. *Programming Industrial Control Systems Using IEC 1131-3.* London: Institution of Electrical Engineers, 1995.

Otter, J. D. *Programmable Logic Controllers.* Englewood Cliffs, NJ: Prentice Hall, 1988.

Simpson, C. D. *Programmable Logic Controllers.* Englewood Cliffs, NJ: Prentice Hall, 1994.

Stenerson, J. *Fundamentals of Programmable Logic Controllers, Sensors, and Communications.* Englewood Cliffs, NJ: Prentice Hall, 1993.

Chapter 11

Data Communication and LANs in Manufacturing

Objective

Chapter 11 explains the importance of data communication for planning, monitoring, and control within a factory or between production sites. Since communication is a rapidly growing technology, the chapter explores the fundamentals of data communication, as well as the specifics of local-area networks.

Outline

In an automated manufacturing facility, there are typically several different types of machines and controllers. Machines are controlled by controllers, which are commonly computer based. In order to maintain a smooth operation, all of these devices must be coordinated. Communication between devices thus becomes essential.

In a shop where devices cannot talk to each other, communication is normally carried out by human operators. A human operator typically goes to a machine, reads the display from its controller, and then goes to another machine and enters the data to its controller. This human-assisted communication is slow and can be prone to error.

Fortunately, most modern controllers are computer based, and thus have a built-in communication capability. They are able to send data and read data in one of several ways. At the lowest level, such communication may be conducted through discrete input and output points. That is, the machines communicate through an on–off signal. Because limited information (on–off) can be carried, this type of communication is normally used only for linking simple devices.

A more popular method of communication is to link two computers via their serial communication ports. A byte (8 bits) of data is serialized and transmitted through a pair of wires (actually, at least three wires, one for transmit, one for receive, and one for ground). This method allows encoded data, such as text, numbers, and so on, to be transmitted from one device to another.

Another similar approach is parallel communication, in which a wire is used for each data bit. A byte of data (rather than a bit) is normally sent via parallel communication. These methods of data communication are inexpensive to implement, but are limited by their communication speed and one-to-one communication topology. In recent years, local-area-network (LAN) techniques have become more popular in the shop floor and in office automation. A LAN is able to perform much faster than the previously mentioned (point-to-point) communication methods. It also allows many-to-many communication on the same network through the same cable. As the price of implementing LANs becomes relatively low, many factories are installing some type(s) of LAN.

At the beginning, there was little standardization in the way LANs were implemented. Devices manufactured by different vendors typically could not communicate with each other. A separate computer had to be dedicated to translate data between two networks from two different vendors, and sometimes even from the same vendor. According to a study [General Motors, 1984], 50% of the costs spent on shop-floor computers was spent on networks. This cost was far too high to be justifiable. In the early 1980s, GM spearheaded an effort to establish a common standard—Manufacturing Automation Protocol (MAP)—that defined the physical and logical communication standard for manufacturing facilities. MAP has since been adopted as the standard for the shop-floor intervendor device data communication. The International Standards Organization (ISO) is adopting MAP as the international standard, although many other communication protocols are still being used.

This chapter will discuss the fundamentals of computer data communication, LANs, and the MAP standard. The purpose is to provide a comprehensive introduction to various data-communication methods employed on the shop floor today.

11.1 FUNDAMENTALS OF DATA COMMUNICATION

11.1.1 Basic Concepts in Communication

Data communication means passing data from one computer device to another. Internally, a digital computer-based device stores data in registers or RAM memory. (See Figure 11.1.) Data are exchanged between memory locations or registers through a data bus. Depending on the computer architecture, the data bus can have a width of 8, 16, 32, or 64 bits. Usually, the bus size determines the nomenclature of a CPU—that is, whether it is 8 bits, 16 bits, 32 bits, or 64 bits. When there is a need to communicate with external devices, input/output (I/O) hardware is used. This communication can be performed either through isolated I/O or through memory-mapped I/O. Isolated I/O uses

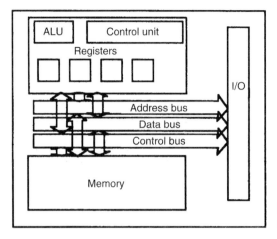

Figure 11.1   A simplified computer architecture.

specially assigned ports and locations. A limited number of locations are available. Special instructions that allow the input and output of data through those ports are available in the CPU instruction set.

In the case of memory-mapped I/O, a digital device (IC chip) is assigned to a memory address and treated like any other computer memory location. To address a memory location, the address is placed on the address bus. The decoder decodes the address and enables the device (or the memory byte) to obtain the data. Only the device that is enabled by the decoder replies to the control signal, which came from a control bus. The control signals can be either *read* or *write*. Note that the I/O port previously mentioned also needs one of these digital I/O devices.

A digital I/O device can be a buffer, a latch, a bidirectional driver, a parallel interface adapter, or a serial interface adapter (i.e., UART, universal asynchronous receiver/transmitter, or USRT, universal synchronous receiver transmitter). Whenever a read or write access is addressed to that memory location, the I/O device responds to the instruction. To the CPU, the operation is the same as that of an ordinary memory read or write. However, some synchronization with the external device is needed. The simplest I/O device can be a TTL buffer. Whenever the buffer is accessed, the state of the data bus is stored and made available to the external device (Figure 11.2). Other I/O devices work similarly to a buffer. Two of the most frequently used data-communication methods are serial data communication and parallel data communication. Before these methods are discussed, data-coding methods must be addressed.

11.1.2 Data Coding

Computers store data in a binary format, so everything in a digital computer has to be represented in binary. Unfortunately, only integer numbers have equivalent binary numbers. All other data, such as real numbers, text (alphanumeric), and graphics, do not have natural binary equivalents. A data-coding scheme is used to represent those data. To exchange data between two computers, the data must be in a form that is recognizable to both. Most computers store alphanumeric data in ASCII (American Standard Code for Information Interchange) form. Some manufacturing devices also use

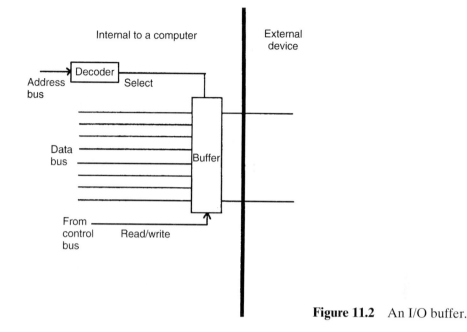

Figure 11.2   An I/O buffer.

the EIA (Electronic Industries Association) form. Another code which only applies to IBM mainframes, is the EBCDIC (Expanded Binary-Coded Decimal Interchange Code.) These codes include all letters, digits, punctuation marks, and control characters. Control characters are those nonprintable characters used to control printer devices, CRT displays, coordinate data communication, and to mark line breaks and the end of a file. For example, line feed (LF), carriage return (CR), backspace (BS), form feed (FF), escape (ESC), and delete (DEL) are some control characters. Each control character has a commonly accepted meaning. Numerical data can be stored in ASCII or (as in most programming languages) in a different format. However, device vendors can use control characters differently.

These control characters are listed in the first two columns of the ASCII table presented in Table 11.1. The table shows the characters represented and their codes. The first row contains the high bits—that is, the first three bits of the seven-bit code. The first column shows the low bits. For example, to find the ASCII code for LF, locate LF in column 2 of the table. The high bits for column 2 are 000. The low bits for the row in which LF is located are 1010. Therefore, the code for LF is 0001010_2. Uppercase and lowercase letters are distinguished by different codes. The differences are the high bits. For example, S is 1010011_2 and s is 1110011_2. To convert an uppercase letter to lowercase, 0100000_2 (32_{10}) is added to the code.

In Table 11.1, a character is stored in one byte. (Seven bits are for the basic code and one bit for parity.) The parity bit is used for error checking. A code may be set to be at even parity; in this case, the parity bit is set to make the total number of ones in the binary representation even. Parity can be even parity, odd parity, no parity, space parity, or mark parity. Even parity means that the total of the data bits plus the parity

TABLE 11.1 ASCII Code Chart

					High Bits				
	low	000	001	010	011	100	101	110	111
	0000	NUL	DLE	SP	0	@	P	`	p
	0001	SOH	DC1	!	1	A	Q	a	q
	0010	STX	DC2	"	2	B	R	b	r
	0011	ETX	DC3	#	3	C	S	c	s
	0100	EOT	DC4	$	4	D	T	d	t
	0101	ENQ	NAK	%	5	E	U	e	u
	0110	ACK	SYN	&	6	F	V	f	v
Low Bits	0111	BEL	ETB	'	7	G	W	g	w
	1000	BS	CHN	(	8	H	X	h	x
	1001	HT	EM	)	9	I	Y	i	y
	1010	LF	SUB	*	:	J	Z	j	z
	1011	VT	ESC	+	;	K	[	k	{
	1100	FF	FS	,	<	L		l	|
	1101	CR	GS	–	=	M	]	m	}
	1110	SO	RS	.	>	N	^	n	~
	1111	SI	US	/	?	O	_	o	DEL

bit yields an even number. Odd parity means that the total of the data bits plus the parity bit yields an odd number. When no parity is selected, the parity bit is not generated by the transmitting device and not checked by the receiving device. Space, or zero, parity means that the parity bit is always set to zero. Mark parity means that the parity bit is always set to one. For example, the ASCII code for character S is 1010011_2, which has four ones. In even parity, it is 01010011_2; in odd parity, it is 11010011_2. In space parity, the code is 01010011_2; in mark parity, it is 11010011_2. The parity bit can be inserted either by the hardware or software. The receiving end checks parity when it is used. An odd number of bit-transmission errors can be detected by this method. However, if there are an even number of errors, the transmission error will not be detected.

When a data file consists of only numbers, the numbers can be stored in binary format, which is the natural format of numbers. For example, the integer number 20 in binary format is 10100. Depending on the application, an integer number can be stored in two or four bytes. When representing a number in ASCII code, the length of the code can depend on the size of the number. For the previous case, two bytes are needed. The first byte is for 2, which is 0110010_2 in ASCII, and the other for 0, which is 0110000_2 in ASCII (Table 11.1). In ASCII, eight bits are usually used to represent one character. Seven of the eight are used to represent the data, and one bit is used to represent parity. With even parity, 2 is 10110010_2; with odd parity, it is 00110010_2. Although it does not make any difference in the memory requirement whether a single-digit integer number is stored in binary format or in ASCII, it does make a difference for a multiple-digit integer number. For a number like 32767, two bytes are needed in binary format and 5 bytes are needed in ASCII. Also, before the number can be used, the computer must convert the ASCII-coded number into binary.

For data communication, either format may be used. However, formats normally do not mix. ASCII is still the most popular way of transferring text files, including NC part programs (the new BCL format is binary; see Chapter 12) and machine-control instructions. Most terminals, except some IBM terminals, use ASCII code. A keyboard transmits the corresponding ASCII code of the key typed to the computer, and the computer, in turn, transmits the ASCII code of the data to be displayed on the CRT screen.

Graphics encoding is different from text encoding. A picture can be represented by a bit map. Depending on the number of colors used, each pixel (picture element) is represented by a certain number of bits. This data can be either transferred in binary or encoded into ASCII using a conversion standard before transfer. Some applications have their own representation format. For example, the PostScript, GIF, JPEG, and TIFF formats are used for various communication requirements. For CAD applications, drawings and geometric models can be represented in IGES, PDES/STEP, and DXF formats. All these formats are encoded in ASCII. So far, graphics representation is still very application-specific.

11.1.3 Serial Data Communication

Serial data communication, as indicated by its name, is a way to communicate data in a serial fashion. A data byte is serialized and sent out on a line one bit at a time. Serial communication is the most widely used data-communication method. A serial communication port usually is built into many computers, machine controllers, telephone modems, and terminals. It is inexpensive and easy to program. The cable connection is also simple. In serial data communication, both communication devices need a serial port and a cable with at least three wires connecting them (Figure 11.3). Of the three required wires, one is for transmitting data (TX), another is for receiving data (RV), and the third is for ground (GND). In the figure, the transmit and receive wires are crossed, so the receive pin of device 1 is connected to the transmit pin of device 2. The UARTs shown in Figure 11.3 convert output data from parallel to serial and input data from serial to parallel. Coordination with the UARTs is required to ensure data are transmitted or received properly. Several error-checking schemes are built into the UARTs.

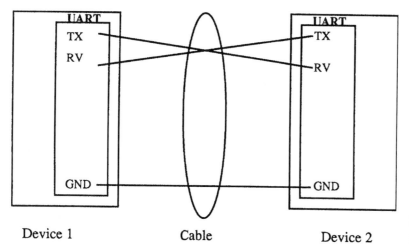

Figure 11.3   Serial communication.

Device 1   Cable   Device 2

On a serial communication line, data are transmitted as square waves (or very close to square waves). The state "zero" is represented by a positive voltage, and the state "one" is represented by a negative voltage. Positive voltages are between +5 and +15 volts for output and between +3 and +15 volts for input. Negative voltages are between −5 and −15 volts for output and between −3 and −15 volts for input. The most common signal is ±5 V. To coordinate both UARTs, the same signal frequency must be used. UARTs are asynchronous, which means clocks are set individually at the transmission and the receiving sides. A signal is resynchronized at the receiving side. The clock rate is also called the baud rate (bit per second) in most cases, however, some modern modems may pack more than one bit in a band. For example, 110 baud means 110 bits are transmitted/received per second. Most serial communications are set at 110, 300, 1200, 2400, 4800, 9600, 14.4, 19.2, or 28.8 baud.

Data are sent one byte at a time. Such a byte is sent in a *frame*. In this frame, start and stop bits are used to signal the beginning and end of a data byte. An agreed-upon word length (five to eight bits), parity type (even, odd, none), baud rate, and start and stop bits (one or two) must be set at both ends identically. For every byte of data sent to a UART, parity is generated and the data bits are shifted into serial form. Start and stop bits are inserted, and the entire signal is sent out through the transmit wire. A complete frame for the letter S (ASCII 1010011_2) is shown in Figure 11.4. Note that when transmitting a byte, the least significant bit is sent first, whereas in writing, the most significant bit is always presented first. The rate of this transmission is set by the baud rate clock. On the receiving side, the start bit is detected by sampling the input line at the clock rate. After the start bit is detected, the parity is checked and the data bits are sent to a shift register to convert them back into a parallel byte format. The converted byte is held in an output register, ready to be sent to the data bus. When a byte of data is ready in the output register, it also sends out a control message. To the UART, the type of code is immaterial. For the purpose of explanation, we use only ASCII code in this chapter.

The transmission length of a byte varies, depending on the number of start bits, stop bits, and data bits used. As a rule of thumb, it generally takes 10 bits to transmit a character; therefore, a simple formula to estimate the number of characters transmitted in a second (n) is

$$n = \frac{\text{baud rate}}{10}$$

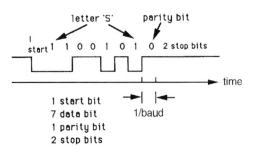

Figure 11.4   Transmitting the letter S.

In this formula, the delay between two consecutive bytes is not taken into account. The problem is that the gap between two bytes varies. It depends on how fast the transmitting device can provide the next byte to the UART. The actual n is thus smaller.

Often, it is necessary to let the transmitting device know that the receiving device is ready to receive the next byte. This need occurs when the receiving device cannot respond as fast as the transmission. For example, a printer cannot print as fast as the computer can send the data. Under these circumstances, information must be sent back from the receiving device to the transmitting device to signal when it is ready. This is called flow control, or handshaking, which may be done either by hardware or software. Hardware handshaking is done through a dedicated wire in the serial port (discussed in the next section). Software handshaking is done by sending a control character, such as DC1 and DC2. DC1 signals that the device is ready and DC2 signals that the device is busy.

11.1.3.1 Serial Communication Standards.   There are standards defining the physical (electrical and mechanical) and logical specifications of the serial connection. The two most commonly used standards in industry are RS-422 and RS-232-C (also CCITT V.24, its international equivalent; CCITT stands for International Consultative Committee for Telephony and Telegraphy, and V stands for voice). RS-232-C, published in 1969 by the Electronic Industries Association (EIA), is still the most widely used serial communication standard. It was developed for modem communication (transmitting data through voice signal); therefore, many idiosyncrasies of modem communication are in the specification. Devices are classified as data-terminal equipment (DTE) and data-communication equipment (DCE). The transmit and receive signal lines in DTE and DCE are swapped. Pin 2 of the DTE is for transmit and pin 3 is for receive. For DCE, pin 2 is for receive and pin 3 is for transmit. The DTE's port normally has a male connector, whereas the DCE device typically uses a female connector. Typical connectors have 9 or 25 pins, and the most common connectors are the DB9 and DB25 connectors. Not all manufacturers comply with these pin and connector arrays, and it is not always obvious whether a given device is DTE or DCE. As a result, the communication of two devices is not necessarily plug compatible. Also, there remain the problems of gender change (when two connectors of the same gender are to be connected) and pin swapping (e.g., when both pin 2's are to receive). Often, the user must experiment with the connectors by swapping wires. (One method uses a specially designed break-out box that allows the user to swap the wires easily.) When connecting two devices for the first time, if there is no communication at all, the first action is always to swap pins 2 and 3. If only garbage (unreadable) data are received, the causes may be baud rate, parity, start, stop bits, and data-bit length.

The definition of RS-232-C signals are shown in Table 11.2. "Data Set Ready" (DCE) and "Data Terminal Ready" (DTE) signals are used to indicate whether the equipment is ready to send or receive data. Many RS-232 interfaces assume the equipment is always ready and thus omit these signals. The DTE sends a "Request to Send" signal when it is ready to transmit data. When the DTE is ready to receive data, a "Data Terminal Ready" signal is sent. The DCE responds with a "Clear to Send" signal when it is ready to transmit the data. When a DCE is ready to receive data, a "Data Set

TABLE 11.2 EIA RS-232-C Standard

PIN	Name	< To DTE	To DCE >	Function	EIA	CCITT
1	FG			Frame ground	AA	101
2	TD		>	Transmitted data	BA	103
3	RD	<		Receive data	BB	104
4	RTS		>	Request to send	CA	105
5	CTS	<		Clear to send	CB	106
6	DSR	<		Data set ready	CC	107
7	SG			Signal ground	AB	102
8	CD	<		Carrier detect	CF	109
9	—			Reserved	—	—
10	—			Reserved	—	—
11	—			Unassigned	—	—
12	(S)CD	<		Sec. carrier detect	SCF	122
13	(S)CTS	<		Sec. clear to send	SCB	121
14	(S)TD		>	Sec. transmitted data	SBA	118
15	TC	<		Transmitter clock	DB	114
16	(S)RD	<		Sec. received data	SBB	119
17	RC	<		Receiver clock	DD	115
18	—			Unassigned	—	—
19	(S)RTS			Sec. request to	SCA	120
20	DTR		>	terminal ready	CD	108.2
21	SO	<		Signal quality detector	CG	110
22	RI	<		Ring indicator	CE	125
23			>	Data rate selector	CH	111
				Data rate selector	CI	112
24	(E)TC		>	Ext. transmitter clock	DA	113
25	—			Unassigned	—	—

Ready" signal is sent. The RS-232 is usually for low-to-medium data rates. Equipment using the RS-232 interface is limited to 19.2 kilobaud. Cable length for RS-232-C communication is limited to 50 feet. Although a cable of even several thousand feet may work, the longer the cable, the lower the baud rate that can be used. This problem is due to distortion and attenuation of the signal.

RS-422 is a newer standard intended for a faster communication link; rates up to 10 megabaud are possible. RS-422 is for balanced circuits; higher speeds use two wires. The ones and zeros are signaled by changes in the polarity of the two wires with reference to each other instead of a single wire changing in polarity with reference to a signal ground. By using twisted-pair wires, RS-422 devices can communicate at 100 kbps (kilobits per second) with a 120-m cable length and at 10 Mbps (megabits per second) with a 12-m cable length. RS-423-A defines an unbalanced circuit. The speed for RS-423-A specification is 3 kbps at 1000 m and 300 kbps at 10 m. It is possible to interconnect RS-423-A and RS-232-C devices. Theoretically, RS-422 is not compatible with RS-232-C. However, RS-422 ports of some devices are compatible with RS-232-C because manufacturers have made some modifications to the standard. These devices are theoretically not RS-422 devices.

When two devices communicate, they may transmit and receive data simultaneously or alternatively. When they transmit and receive simultaneously, they are called full duplex. When devices can only transmit or receive alternatively, they are called half duplex. In the case of a terminal, when a half-duplex device is used, the transmitted data are echoed back to the terminal.

The type of communication discussed thus far takes place at the "physical layer," the most fundamental layer of communication. At the higher levels, there are several data-communication protocols. A data-communication protocol defines a logical convention for data transfer. These protocols enable data to be transferred easily and without error. They are used to transfer data files from one device to another. The file type, file name, file length, and so on, are transferred in a header data block. The file is divided into small chunks called blocks or packets. Each block has a fixed size with the header information added. An error-checking procedure is built into the protocol, so that transfer errors can be detected. When such an error is detected, the block is retransmitted. A handshaking method is used to synchronize the devices. For example, XON and XOFF (control characters) may be used to signal when the receiving device is respectively ready or not ready for the next block of data. Many popular data-communication software packages support all popular protocols.

11.1.4 Parallel Data Communication

In parallel data communication, an entire byte is transmitted together. It is a popular way to transfer data between the computer and outside devices such as sensors, actuators, and peripherals. However, it is less frequently used between computers transferring data. Typically, a programmable LSI (large-scale integrated) device is used as the parallel I/O interface. Such a device usually contains two or more eight-bit ports that can be programmed to be either input or output ports. A data-direction register is used to define the direction in which the data are to be transferred by the corresponding bits of the port. It is possible to program individual bits of a port to be either input or output. For example, a "1" in the data-direction register means output, and a "0" means input. By setting 11110000 in the data-direction register, the first four bits of the port are set to be output and the last four bits are set to be input. When the data arrive, they are latched and saved in the buffer. Some other general features of a parallel port include status and control for handshaking, other control and timing signals for peripherals, and direct interface with the processor address, data, and control buses. A parallel-interface adapter block diagram is shown in Figure 11.5.

A parallel interface is normally addressed as a set of memory locations. Each port contains a direction register, a data register, a control register, and two control lines. The control register determines the active logic connections in the device and also contains the data-ready or peripheral-ready bits.

One of the parallel interface standards is IEEE 488, the standard digital interface for programmable instrumentation. It is also known as the Hewlett-Packard Interface or the general-purpose interface bus (GPIB). It is used in networks of instruments. Most Hewlett-Packard instruments have built-in IEEE 488 interfaces.

The IEEE 488 is a bus of 24 lines. Of the 24 lines, 8 are for ground, 8 for data, and 8 for control. As mentioned, IEEE 488 is designed for network instruments. Up to 15 devices can be connected to the bus, and the control indicates which devices can talk or

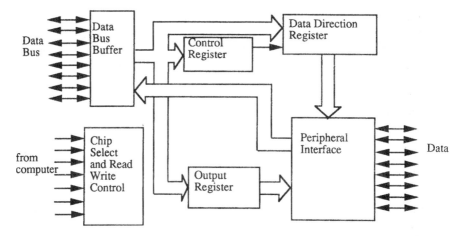

Figure 11.5 A parallel interface adapter.

listen. All devices are connected to the 8 data lines, on which, bits are transmitted in parallel and bytes are transmitted in serial. The 8 control lines are divided into 3 data-transfer lines and 5 bus-management lines. Data-transfer lines consist of DAV (data valid), NRFD (not ready for data), and NDAC (no data accepted). These lines help synchronize the flow of information between the talking (transmitting) and the listening (receiving) devices. Each line has a specific task. The bus-management lines coordinate the flow of information on the bus. The IFC line places the system in a known state. The ATN line indicates the nature of the information on the data lines. REN commands instruments to select remote operation. The SRQ line services requests. The EOI line indicates the end of a multiple-byte transfer sequence, or, in conjunction with ATN, executes a polling sequence.

In a network, one device is designated as the controller. This device controls the bus. In a typical communication cycle, the ATN line is set low so all devices monitor the lines. The commands Untalk and Unlisten are sent to eliminate all previous connections. Then the commands Listen Address and Talk Address are placed on the data bus to configure the bus. The device addresses are built into the devices and assigned by the hardware manufacturers. At any given time, only one device is assigned as the talker, whereas several can be listeners. ATN lines are then set high so that data transmission can begin. After the transmission is completed, the talker returns control of the network to the controller.

The maximum data rate in an IEEE 488 network is 1 megabit per second. The maximum cable length is 2 meters times the number of devices or 20 meters, whichever is less.

11.1.5 Data Transfer Modes

Data transfer can be carried out in two modes: simplex and duplex. Simplex mode can be viewed as a "one-way street," since data flow only in one direction. In other words, a device can be a receiver or a transmitter exclusively. However, a simplex device is not a transceiver. A good example of simplex communication mode is an FM radio station and your car radio. Information flows only in one direction; the radio station is the transmitter, and your car radio is the receiver. It is obvious that simplex is not often used in computer communications, because there is no way to verify whether data have been received.

Duplex communications overcome the limits of simplex communications by allowing the devices to act as transceivers. With duplex communication, data flows in both directions thereby allowing verification and control of data reception/transmission. Exactly when data flow bidirectionally further defines duplex communications. Duplex mode is divided into full-duplex and half-duplex modes. Full-duplex devices can transmit and receive data at the same time. For example, a cell phone is a full-duplex radio. Half-duplex devices allow both transmission and receiving, but not at the same time. Essentially only one device can transmit at a time while all other half-duplex devices receive. A walkie-talkie is an example of a half-duplex radio. Physically, a full-duplex device needs to use two communication channels (frequencies bands), and a half-duplex device uses only one channel.

11.1.6 Data Transfer Techniques

The actual data transfer between the computer and external devices uses one of three methods:

(a) polling

(b) interrupts

(c) direct-memory access

The first two methods are used widely for data communication between two computers. Direct-memory access is used mostly for computer and peripheral device communication, especially when large amounts of data need to be transferred, such as computer-to-hard-disk-drive communication.

(a) Polling. Polling means checking the status of the I/O ports. When there is only one I/O port being polled, a communication program checks the status of the port repetitively. When the output port is ready, it sends one byte to the port. When the input port is ready, the byte in the input port buffer is retrieved. The polling loop runs continuously and the status of the port(s) is checked at fixed intervals. Polling is simple and no additional hardware circuitry is required. Figure 11.6 shows a polling loop. However, polling also requires constant checking of the status, which is time-consuming (CPU time). In reality, most of the CPU time is spent checking I/O status. When there are multiple I/O ports with different data-transfer rates, the polling may not be fast enough to service the highest-data-rate I/O port. Some data may get lost while the CPU checks other I/O ports. This polling loop also shares CPU time with other applications (the polling loop can run in background and may be active only during certain time intervals, which are set by a timer). It may slow down other applications.

(b) Interrupts. To reduce the overhead in polling, interrupts can be used. The idea is to eliminate the time spent by polling devices that do not need service. Every CPU has a built-in hardware interrupt. When an interrupt signal is sent to the CPU interrupt line, the system saves the current register status and jumps to a memory location defined

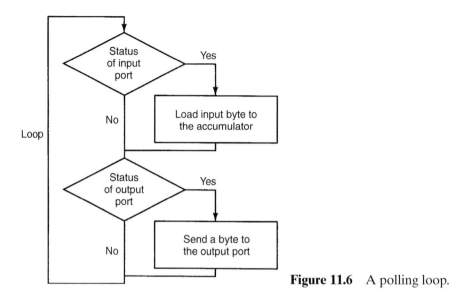

Figure 11.6 A polling loop.

by either the hardware or the user software. A routine at that location is executed. The routine is called an interrupt-service routine. I/O routines can be written as interrupt-service routines. By using interrupts, the I/O routine is called to read only when data are available to the buffer. Therefore, CPU time is not used to check the status unnecessarily. If interrupt I/O is used, hardware-interrupt circuitry must be constructed to send the interrupt signal to the CPU. An interrupt-service routine is written and stored in an appropriate memory location.

Priorities can also be assigned to different devices. The priorities are numbered as level 0, level 1, and so on. The lower the level, the higher the priority (Figure 11.7). For critical devices, a higher priority is assigned. The higher-priority interrupt can interrupt the lower-priority one, but not the other way around. Because I/O is normally slow, when interrupt I/O is used, the computer can perform other tasks. Polling and interrupts can be applied to either serial I/O or parallel I/O.

11.2 LOCAL-AREA NETWORKS

In order for a computer-aided manufacturing (CAM) facility to run smoothly, many devices in the facility must be linked. Not only do data files, such as NC part programs,

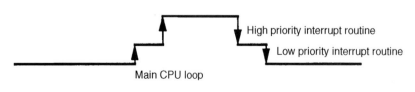

Figure 11.7 Interrupt.

design data files, and so on, have to be exchanged between devices and computers, but control commands and device-status information also have to be exchanged. The system operator needs the ability to directly log into each system device to maintain the system, check the status, alter the control, and upload or download the data files. The traditional serial or parallel I/Os are limited in this application because of the following aspects:

1. point-to-point communication requires $C_2^n = n!/(n-2)!\,2$ connections; for example, for 50 devices, $C_2^{50} = 1225$
2. slow communication speed (up to 20 kb per second, which is higher when data compression is used)
3. limited transmission distance (about 10 m)

The development of computer-networking techniques has enabled a large number of computers to be connected. Computer networks can be classified as wide-area networks (WANs), local-area networks (LANs), and high-speed local networks (HSLNs). A WAN serves a geometric area of more than 10 km. Such a network includes the Internet. Thousands of computers are connected on the Internet, enabling users to transfer files, mail electronic messages, and remotely log on to computers. HSLNs are designed for computer rooms, mainly for mainframe-to-mainframe communication and mainframe-to-disk-drive communication. They are confined to very short distances and extremely high speeds.

A LAN is confined to a 10-km distance (in practice, much shorter) in a building, a university campus, or a company. Devices in the network can perform the same type of data communication as a WAN. In a LAN, the communication speed can be as high as 300 megabits per second. However, Ethernet, which became the most popular LAN in the 1980s, runs at 10 Mbps (or 100 Mbps fast Ethernet). It is significantly faster than an RS-232 serial communication connection (19.2 kbps). A comparison of a few network technologies is shown in Table 11.3. Note that, in this table, CBX utilizes telephone service for data transmission. Several of the terms used in the table are explained in the following section.

A factory network fits the definition of a LAN, so a LAN is the network best suited for CAM. An ideal LAN has the following characteristics:

- high speed: greater than 10 Mbps
- low cost: easily affordable on a microcomputer or a machine controller
- high reliability/integrity: low error rates, fault tolerant, reliable
- expandability: easily expandable to install new nodes
- installation flexibility: easily installed in an existing environment
- interface standard: standard interface across a range of computers and controllers

In a local-area network, computers and controllers can communicate with each other. Terminals also can access any computer on the network without a physical hardwire. It is beneficial because the same terminal can now access all devices on the shop floor.

TABLE 11.3   Comparison of a Few Communication Technologies

	LAN	High Speed Local Network (HSLN)	Computerized Branch Exchange (CBX)
Transmission medium	Twisted pair Coaxial cable Optical fiber	CATV coax	Twisted pair
Topology	Bus, tree, ring	Bus	Star
Speed	1–100 Mbps	50 Mbps	9.6–64 kbps
Maximum distance	25 km	1 km	1 km
Switching technique	Packet	Packet	Circuit (no delay)
Number of devices supported	100's–1000's	10's	100's–1000's
Attachment cost	Low	High	Very low
Applications	Computers Terminals	Main frame to disk drive	Voice Terminal-to-terminal Terminal-to-host

11.2.1 Network Technology

A LAN consists of software that controls data handling and error recovery, hardware that generates and receives signals, and media that carry the signal. The software and hardware design are governed by a set of rules called a protocol. This protocol defines the logical, electrical, and physical specifications of the network. In order for devices in a network to communicate with each other, the same protocol must be followed. For a device to send a message across the network, the sender and the receiver must be uniquely identified. The message must be properly sent and error checking must be performed in order to ensure the correctness of the information. Similar to a parcel prepared for the Postal Service, the software has to package and identify the data with appropriate labels. The label contains the address of the destination and information concerning the contents of the message. The package is then converted into the appropriate electrical signals and waits to be transmitted. To transmit the signals, a conducting medium carries the electrical signals to the destination. Several techniques are used to accomplish this task. This section discusses the packaging of the data (packet), the signal itself (bandwidth), the transmission media, and the network configuration.

11.2.1.1 Packet.   Data can be sent either byte by byte, in small units, or in one package. In the case of byte-by-byte transfer, overhead is extremely high, because each byte must include the sender's and receiver's address, as well as some control information. When sending all the data in one package, there is the possibility of network congestion. Each message may occupy the network for a long time (the message can be a several-hundred-kilobyte data file), and an error causes the retransmission of the entire message. In order to utilize the network efficiently, packet switching splits data into smaller units called packets. Packets are routed and carried separately through

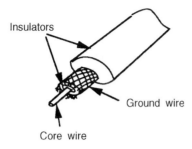

Insulators

Ground wire

Core wire

Figure 11.8   Coaxial cable.

the network and later reassembled in sequence at the receiving end. To the user of the data, the entire message comes in sequence. For example, on the Ethernet, each packet may contain between 46 and 1500 bytes of data.

11.2.1.2 Transmission Media.   Coaxial cable and twisted-pair wires are used in LANs. A coaxial cable is commonly seen in household cable-TV connections. It is a cable consisting of a solid copper-core wire, wrapped by a mesh of thin copper wires (Figure 11.8) (or, in inexpensive TV cable, aluminum foil). The wrapping wire is connected to the ground and serves as the noise insulator. There is an electric insulator between the center core and the wrapping wire. The exterior of the cable is a plastic insulator. A coaxial cable allows high-frequency analog signals to travel. A twisted-pair wire, as the name indicates, is a pair of twisted wires. The frequency of the signal carried by the twisted-pair wire is much lower than that of coaxial cable; thus, the transmission speed is slower. There are 50 Ω and 75 Ω coaxial cables.

11.2.1.3 Bandwidth.   Signals are transmitted on the cable, and these signals are modulated at certain frequencies. The range between the lowest and the highest frequency that can be carried on the communication line is called the bandwidth. The higher the bandwidth, the more information it can carry. There are two methods used in LAN: baseband and broadband. Baseband is a method of data communication in which different voltages representing binary 0 and 1 are directly applied to the communication line. The speed is limited and only one signal can be carried at a time. Broadband utilizes analog (radio) signals. A digital signal is modulated on one of the analog signals. The carrier-signal frequency is from 5 to 300 MHz, and higher frequencies are also possible. Each analog signal takes one frequency band (channel). Two channels can be separated easily by filtering the desired frequency. Because there are many channels, broadband technology allows the same cable to be used simultaneously by several LAN subnetworks, TV signals, and so on. Cable TV coaxial cable is normally used as the transmission media. More than 100 channels can be opened at the same time without interfering with each other. Each channel is able to transmit at least 5 Mbps. Needless to say, the total information transmitted is extremely high. Currently, gigabits per second is available.

11.2.1.4 Network Topologies.   A network can be arranged in several ways. It may be arranged as a ring, a star, or a bus. Such arrangements can be made either physically or

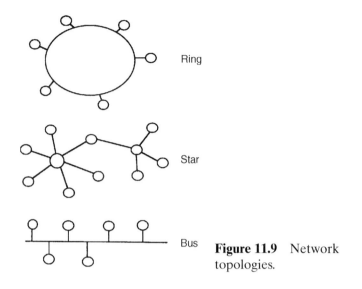

Figure 11.9   Network topologies.

logically. This chapter discusses only the physical arrangements. These topologies can be seen in Figure 11.9. The ring topology is simple; however, if the ring breaks, the entire network is down. Also, the cable must be routed back to its beginning. The star topology is easy to grow, however, the network relies on a server at the center of the star. All communication between nodes must go through the center. The bus topology is like the ring, except that it does not require the cable to be routed back to its beginning. Nodes can easily be added into the network. However, the control is conceptually more difficult. In reality, the control of the bus network is no more difficult than the others. For this reason, the use of bus topology in LANs is popular.

11.2.1.5 Medium-Access Control.   LANs use distributed control. This enables each device on the network to have equal access to the network and prevents catastrophic failure due to the failure of the master. The distributed control scheme is called the medium-access control. The most commonly used medium-access control methods are CSMA/CD (Ethernet bus), and token passing (token bus or token ring).

(a) CSMA/CD and Ethernet.   Carrier sense multiple access with collision detection (CSMA/CD) was developed by Xerox in the 1970s and implemented in Ethernet. Ethernet has since become the industry standard, used by most engineering workstations and minicomputers. Many of the industrial controller vendors also support their version of Ethernet. A bus topology is used in Ethernet (Figure 11.10). In the figure, four computers are connected to two separate buses. Each bus is a 50-ohm coaxial cable with a terminator at each end. (There is a small resistor inside the terminator. It absorbs the signal to prevent it from bouncing back and causing interference.) This cable is specially made for Ethernet, and due to its relatively small demand, it is much more expensive than TV coaxial cable. A tap connects the computer to the bus. The tap must be placed at a multiple of a fixed interval (2 ft) on the cable. An Ethernet controller board in the computer translates the data to be sent or received. The data in digital form are then sent

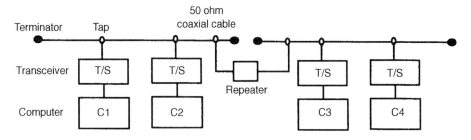

Figure 11.10   Ethernet.

to the transceiver. The transceiver makes the physical and electrical connections onto the cable and transmits and receives digital signals to and from the cable. A repeater links two buses together to form a single bus. Whenever there is no signal on the bus, any device can start transmitting. Because it takes time for a signal to travel from one end of the network to the other, it is possible that one device will begin transmitting while there is a signal already on the bus. The CSMA/CD protocol resolves this problem.

In each device, there is a built-in circuit that detects the collision of the signals on the bus. The procedure to resolve the collision is as follows:

(a) When a collision is detected by the transmitter, stop transmitting. Send a jam signal to ensure that everyone knows that there has been a collision.

(b) Each transmitter waits for a random amount of time and then transmits again.

Because the waiting time is random, the chance of having both devices transmitting at the same time is low. This process is shown in Figure 11.11. The transmission time for a message from A is 1. At t_0, A begins to transmit. It takes a amount of time for the signal to reach B. At e, the time before the signal reaches B, B begins to transmit. B detects the collision of two signals only at a time $t_0 + a$. B stops transmission for a random amount of time β. The signal from B continues to travel until A detects it at a time $t_0 + 2a - e$. At this time, A stops transmission for a random amount of time α before it retransmits.

Ethernet is defined by the IEEE 802.3 standard. There are five types of Ethernet connections: standard Ethernet (10BASE5), ThinNet Ethernet (10BASE2), twisted-pair Ethernet (10BASE-T), broadband-based Ethernet (10BASE-36) and optical-fiber-based Ethernet (10BASE-F). The coding 10BASE5 means 10 Mbps using a 500-m segment length of coaxial cable. The same Ethernet also comes in a 100-Mbps version. For the standard Ethernet, the 100-Mbps network would be coded 100BASE5. ThinNet has a segment length of 185m and rounded to 2. The segment-length limitation is due to the use of a coaxial cable with a much smaller diameter (higher impedance). 10BASE-T Ethernet uses telephone type unshielded twisted-pair wire cable. Unlike other Ethernet connections, BASE-T type Ethernet requires the use of an electronics box called a *hub*. A telephone-style connector (RJ45) is used on both the computer side and the hub side. A 10BASE-T cable uses

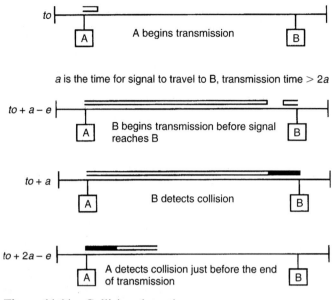

Figure 11.11 Collision detection.

four conductors, a 100BASE-T eight. In comparison, telephone uses only two conductors. Broadband Ethernet uses CATV type coaxial cable. Multiple channels can be transmitted on the same cable. Finally, the optical-fiber-based Ethernet uses optical fiber instead of electric cable. The computer is connected to an optical fiber transceiver (FOMAU) through an electric cable (15-pin AUI cable). Two strands of optical fiber connect the MAU to a repeater hub. Both broadband-based Ethernet and optical-fiber-based Ethernet are not very common. Due to the cost, most of the office and home networks use a twisted-pair network. The comparison of these five types of connections is shown in Table 11.4.

(b) Token Ring. A token-ring network uses a token for access control. A token is a unique bit pattern that is passed from one device to another. The device that has the token is allowed to transmit; all others receive only. When no device is transmitting, the token circulates continuously in the ring. (The ring is formed by linking the repeaters with ring segments.) The token and the data packet travel in one direction (Figure 11.12). The repeater always receives data from one side and sends the identical data out to the other side. To transmit a message, the computer must wait for the token to arrive at its repeater. When the token arrives, it changes one of the bits to "1". This changes the free token to a busy token, so that it is no longer available to other computers. The computer sends out the message with header information indicating the address of the receiving computer. The message goes through the repeaters in the ring in sequence. Each computer has a chance to read the message; however, only the ones addressed will copy the message to their memory. The message eventually travels back to the sender. At this time, the repeater of the sender is actually separated into two parts: one is a transmitter of the message and other is the sink of the message. When the message comes back to the sender, it knows that the transmission has been successful.

TABLE 11.4   Comparison of Different Ethernets

	BASE5 Standard	BASE2 ThinNet	BASE-T Twisted-pair	BASE-36 Broadband	BASE-F Fiber
Cable	Coaxial	Coaxial	Twisted-pair	Coaxial	Optical fiber
Segment length	≤500 m	≤185 m	≤100 m	≤3,600 m	≤2,000 m
Cable length	≤2.5 km	≤1 km	—	—	—
Transceiver cable length	≤50 m	—	—	—	—
Between transceivers or T-connectors	≤2.5 m	≤0.5 m	—	—	—
Number of transceivers per segment	≤100	≤30			
Terminators	50 ohm	50 ohm	None	75 ohm	None
Comments	PC to transceiver to cable	Direct connect to PCs	Connect to hub		

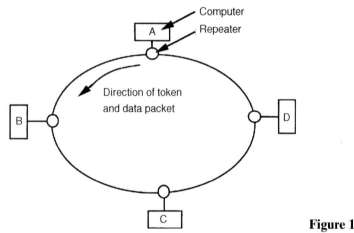

Figure 11.12   A token ring.

The message length may be variable. When the message transmission is complete, the token is reinstated onto the ring by the transmitting computer. Because only the computer with the token can talk, there are no collision problems as in an Ethernet connection. The speed of a token ring is typically 1 or 4 Mbps.

(c) Token Bus. A token-bus network is a logical token ring implemented on a physical bus (Figure 11.13). Thus, it has the advantages of both the token ring and the CSMA/CD. Like Ethernet, there is no need for a repeater, and the signal is available to all devices at the same time. The cable does not need to loop back to the initial device. A new device can be added to the network easily by tapping into the existing cable. For medium-access control, the token eliminates the problem of signal collision, which is a problem with Ethernet. It is thus appropriate for real-time applications, such as applications on the shop floor.

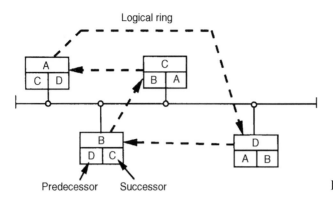

Figure 11.13   A token bus.

In a token bus, each device is assigned a unique address. A token is passed from one device to another according to a logical-ring sequence. The next device in the logical ring does not have to be physically adjacent. The ring is formed by assigning preceding and succeeding addresses to each address. A token is a packet with a zero-data byte. In the packet, the source and destination addresses are given. When the token is broadcast through the bus, only the destination device accepts the token. If transmission is desired, the device with the token begins transmitting its message in packets. After the transmission is completed, the device sends the token to its successor. In Figure 11.13, the ring is formed by assigning each device an address: a predecessor address and a successor address. The ring has a sequence of A–D–B–C. By changing the address table, the sequence of the devices in the ring can be changed.

It is the responsibility of the device that has a free token to make sure the token is properly received by the next device. In case there is no response from the next device, a new successor is selected and the failed device's address is removed from the logic ring. If one of the devices has failed, the token bus will still operate.

A token bus may use either a baseband or a broadband technology. Either kind of technology may have up to a 10-Mbps data rate. The communication medium used is 75-ohm coaxial cable. The manufacturing automation protocol (MAP) uses a token bus.

11.2.2 The ISO/OSI Model

Several LAN protocols have been developed. As mentioned before, a protocol is a set of rules that governs the operation of functional units to achieve communication. However, to ensure that data are transmitted properly, there are several tasks that must be performed. Assume the cell controller needs to check the status of NC machine 1. The application program in the cell controller issues an instruction such as "Machine 1 status" to machine 1. This message must be converted to a proper format that the network can carry. The final data must be modulated into electric signals, either digital or RF. Then the signals are properly sent through the network to the destination machine. The access control, network layout, media type, and so on, all play an important role in transmitting this data. On the receiving side, the data are decoded and sent to the machine-control executive. The message format may not be compatible with the NC machine-controller, command-language syntax. In order to establish

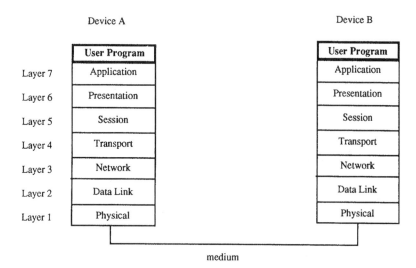

Figure 11.14   The seven-layer ISO/OSI model.

painless communication, this incoming message must be translated into a compatible format. This simple example proves that many tedious tasks must be performed before the network is usable. A set of protocols must be defined in order for the devices in the network to follow.

In the late 1970s, the International Standards Organization (ISO) began developing a model for LANs. The model is called the Open System Interconnect (OSI) model and it splits the communication process into seven layers (Figure 11.14). On each layer, standards can be set to satisfy the need. When a packet of data is sent from device *A* to device *B,* it goes down the layers. On each layer, a control message is appended to the data. The complete packet is then transmitted through the medium to device *B*. On each of the device *B* layers, the control message is stripped and proper actions are taken to convert the data into the proper format. This layered model allows the system designer to modularize the implementation task. Each layer can be developed independently and replaced without affecting the other layers. However, it is also very complicated, thus requiring significant software overhead. For a simple application, this may be overkill. However, for data communication in a complex manufacturing facility, such an approach definitely has more benefits than costs.

The next several sections explain the seven layers of the OSI model.

11.2.2.1 Physical Layer.   This layer deals with the electrical and mechanical means of transmitting data. The physical layer includes the cable, connector, voltage level, speed, bandwidth, and data-encoding method. For example, the RS-232 communication standard belongs to the physical layer. In the case of a LAN, the baseband and broadband technologies are on this layer.

11.2.2.2 Data-link Layer.   This layer transfers frames across a single LAN. Its functions include resolution of contention for use of the shared transmission medium, delimitation and selection of frames addressed to this node, detection of noise via a frame check sequence, and any error correction or retries performed within the LAN. It is used to improve the error frequency for messages moved between adjacent nodes. The data integrity transmitted between two devices in the same network is ensured on this layer. The ISO high-level data-link control (HDLC) standard is used on this layer. In HDLC, a frame begins and ends with an 8-bit flag. Between the flags are addresses of the destination, control information, variable-length data, and frame length. In a LAN, this layer is divided into two operations: The resolution of contention is the medium-access control, and there is also a logical-link control.

11.2.2.3 Network Layer.   The network layer provides the transparent transfer of data between transport entities. It is responsible for establishing, maintaining, and terminating connections. Between networks, this layer has the responsibility for inter-network routing. A global unique node address is used by the network layer.

11.2.2.4 Transport Layer.   This layer ensures that data units are delivered error-free, in sequence, with no losses or duplications. It should provide a network-independent service to the upper layer.

11.2.2.5 Session Layer.   The session layer controls the dialogue between applications during a communication session. The type of dialogue, whether it is two-way (simultaneous, alternate) or one-way dialogue, is defined. It provides a checkpoint and a resynchronizing capability. In case the dialogue breaks during the session, it provides a means to recover from the failure.

11.2.2.6 Presentation Layer.   The presentation layer takes care of the syntax of the data exchanged between applications. For example, a device may use the EBCDIC code. A message sent from it to an ASCII device is translated by this layer. Other data syntax includes teletext, videotex, encryption, and virtual terminal.

11.2.2.7 Application Layer.   This is the most difficult layer. It ensures that data transferred between any two applications are understood. Given the number of applications that exist, it is extremely difficult to define this layer. A standard for each application domain needs to be defined.

11.2.3 TCP/IP Protocols

Although MAP and TOP have been promoted as the network protocols for manufacturing and office automation, the most widely used network is probably the commercial Ethernet. With the installation of engineering workstations such as Sun, DEC, HP, Silicon Graphics, and RS6000, a common network has been established in most engineering offices. Most current engineering workstations run UNIX or an implementation of the UNIX operating system. In fact, Ethernet and Internet protocols are built into the UNIX operating system. A UNIX system can communicate locally with other UNIX machines on the network, as well as connect through a wide-area network (like the Internet) to other machines. When a gateway is established between the local network and an external network (such as ARPAnet or NSFnet), nationwide and world-wide network service is provided.

A set of Internet protocols have been defined to enable machines to communicate with each other. The TCP/IP protocols, a product of the U.S. Department of Defense, are the most widely accepted protocols on the Internet. TCP/IP can trace its roots to the early 1970s to the ARPAnet project. The Internet Protocol (IP) is a simple datagram protocol that uses globally assigned addresses to transmit data between two hosts. It allows data to be broken into small packets. When data are transferred over a wide network, the datagram may go through any number of gateways. Internet Protocol is a lower-level protocol. TCP (Transmission Control Protocol) is a connection-oriented protocol. It provides reliable flow-controlled data transfer between two nodes. TCP is built on top of the IP protocol. When a message is sent to a remote machine, a datagram is sent. The IP protocol is appropriate for this application. However, if a remote login terminal session is desired, a connection-oriented protocol, TCP is needed. Two of the popular communication applications, FTP and TELNET, are both built on top of the TCP protocol. FTP is a file-transfer application and TELNET is a remote-terminal session application.

TCP/IP corresponds to the ISO model on the network and the transport layer. Figure 11.15 shows the Internet protocol layering.

The physical layer can be the Ethernet, just as it is in all of the UNIX engineering workstation installations. For a wide-area network, the physical layer required to transmit the data from one host to another may be several different networks. The user-diagram protocol (UDP) shown in the figure is a simple protocol that provides an additional data checksum for a datagram. It does little beyond what IP does. It is used in the Internet domain for datagram communication.

To establish communication between two hosts, a socket is established. A socket is an end point of communication. It is created by a system call with a specification of the destination. There is an associated data structure that stores all the necessary information for the communication and a data buffer. Data are broken into small pockets (Figure 11.16) and sent to the socket with a specification of protocols. The TCP, IP, and Ethernet headers are added to the packet before the packet is transferred to the Ethernet cable. The TCP header includes source port, destination port, sequence number, acknowledgment number, checksum, urgent pointer, and additional flags. The IP header includes type of service, total message length, ID, fragment flags, header checksum, protocol, source and destination addresses, and so on. When the local network is not Ethernet, an appropriate protocol is used to replace the Ethernet header.

Transport Layer	TCP	UDP
Network Layer	IP	
Data-link Layer	Network interface	

Figure 11.15   Internet layering.

Ether	IP	TCP	Data

Figure 11.16   A packet.

When a message is received by the destination host, the header information is stripped off the packet and the proper action is taken to ensure the correct data transfer. Out-of-sequence packets are kept until properly sequenced packets have arrived. When using an application such as FTP or TELNET, the user only need specify the destination host name (e.g., mars.ecn.purdue.edu). The protocols are transparent. When writing programs to access the network, a set of system function calls takes care of the communication.

For example, to initialize a TELNET session with the Science and Technology Information System (STIS) at the National Science Foundation, type

```
telnet stis.nsf.gov or telnet 128.150.195.40
```

At the login prompt, log in as user *public*. The user then responds as if he or she is logged into a local terminal. The preceding *stis* is the name of the host where STIS is located. The domain name where *stis* is located is *nsf*. A government installation is indicated by *gov*. The Telnet application uses a host table on the local system to locate the Internet address. If the address is not found, a user may type the address directly, such as 128.150.195.40.

11.2.4 Other Networks

In this section, a few other manufacturing networks are introduced. Their adoption in industry is driven by either cost or functionality.

11.2.4.1 RS-485 Networks.   Multidrop means that multiple devices may be connected to the same cable. RS-485 is a multidrop LAN (Figure 11.17) 10BASE2 and 10BASE5 are other examples of multidrop networks. The cabling of RS-485 is extremely simple. A single pair of twisted wires (two wires) is used. The data transmission mode is similar to that used in RS-422 (Section 11.1.3). The signal differential between two wires determines the data. It eliminates noise and allows a high data rate on two wires. The RS-485 cable can be up to 4000 ft long and as fast as 1200K baud. Devices are connected directly to the wires.

RS-485 meets the requirements for a truly multipoint communications network, since the standard specifies up to 32 drivers and 32 receivers on a single (two-wire) bus. With the introduction of automatic repeaters and high-impedance drivers/receivers, this "limitation" can be extended to hundreds (or even thousands) of nodes on a network. To solve the data-collision problem that is often present in multidrop networks, hardware units (converters, repeaters, microprocessor controls) can be constructed to remain in a receive mode until they are ready to transmit data. Single master systems (many other communications schemes are available) offer a straightforward and simple means of avoiding data collisions in a typical two-wire, half-duplex, multidrop system.

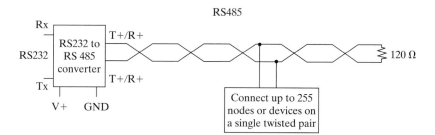

Figure 11.17   RS-485 network.

The master initiates a communications request to a slave node by addressing that unit. The hardware detects the start bit of the transmission and automatically enables the RS-485 transmitter. Once a character is sent, the hardware reverts back into a receive mode in about 1 to 2 microseconds.

Any number of characters can be sent, and the transmitter will automatically re-trigger with each new character. (In many cases, a bit-oriented timing scheme is used in conjunction with network biasing for fully automatic operation, including any Baud rate or any communications specification, e.g., 9600,N,8,1.) Once a slave unit is addressed, it is able to respond immediately because of the fast transmitter turn-off time of the automatic device.

The major advantage of using RS-485 is the low cost.

11.2.4.2 Fieldbus.   Foundation™ Fieldbus is developed by Fieldbus Foundation, a not-for-profit international consortium of leading controls and instrumentation suppliers. It is also published as ANSI/ISA 50.02 Fieldbus Standard for Use in Industrial Control Systems. Fieldbus is a digital, two-way, multidrop communication link among intelligent measurement and control devices. It is designed to serve as a LAN for advanced process control, remote input/output, and high-speed factory automation applications. In the control architecture, there are two different kinds of networks: H1, the Fieldbus, and HSE, the high-speed Ethernet. The Fieldbus runs at 31.25 kbps and the HSE runs at 100 mbps. The network hierarchy is shown in Figure 11.18. On the factory floor, the H1 network provides a linkage between "field" devices, such as sensors, actuators, and I/O devices. The HSE provides a high-speed linkup between H1 subnetworks, PLCs, and control computers. Unlike the traditional control devices that point-to-point interconnect, Fieldbus reduces the wiring tremendously. (This is the same argument as used in Section 11.2.) It also standardizes the interface between devices and thus simplifies the interconnection. The application-level protocol provides commands to transfer different kinds of data seamlessly.

Fieldbus transfers data in "blocks." A communication block can carry one of the following pieces of information: analog input, analog output, bias/gain, control selector, discrete input, discrete output, manual loader, proportional/derivative, proportional/integral/derivative, ratio, device control, output splitter, signal characterizer, lead lag, deadtime, integrator, setpoint ramp generator, input selector, arithmetic

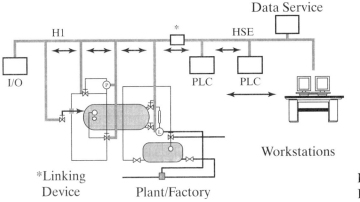

Data Service

H1

*

HSE

I/O

PLC   PLC

Workstations

*Linking
Device   Plant/Factory

Figure 11.18
Fieldbus networks.

timer, and analog alarm. Each of these types of information is assigned a code to reduce the data transmission rate.

Fieldbus HSE is a standard Ethernet; thus, it needs no further definition. H1 is based on the ISO/OSI communication model (Figure 11.19). However, it skips layers 3 to 6. On the top is the Fieldbus Message Specification protocol. Messages can say

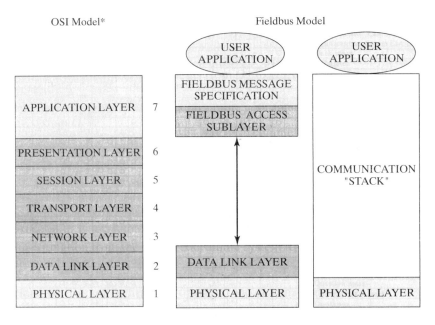

OSI Model*

Fieldbus Model

USER
APPLICATION

USER
APPLICATION

FIELDBUS MESSAGE
SPECIFICATION

FIELDBUS ACCESS
SUBLAYER

APPLICATION LAYER   7

PRESENTATION LAYER   6

COMMUNICATION
"STACK"

SESSION LAYER   5

TRANSPORT LAYER   4

NETWORK LAYER   3

DATA LINK LAYER   2

DATA LINK LAYER

PHYSICAL LAYER   1

PHYSICAL LAYER

PHYSICAL LAYER

* The user application is not defined by the OSI Model

Figure 11.19   The Fieldbus model.

"establish communications," "release communications," "reject improper service," "read a device status," "send unsolicited status," and "read vendor type and version." The Fieldbus Access Sublayer defines three major types of communication situations: client/server, report distribution, and publisher/subscriber. Those are all the communication needs in a factory-floor control environment. The Data Link Layer uses a token-passing protocol to control the media access. The Physical Layer defines a 4–20mA current-loop wires and a single-pair shielded instrument cable. It allows up to 32 devices on 1900 meters of cable.

11.2.4.3 Wireless Network (WLAN).   It is reported that the Ethernet was modeled after AlohaNet, which was installed by the University of Hawaii in the 1970s. Unlike the Ethernet, which uses cable, AlohaNet was based on packet radio. Radio-signal computers located on different islands were linked on a network and could communicate with each other. Therefore, AlohaNet could be said to be the predecessor of both the Ethernet and wireless networks. Now, the wireless network (WLAN) replaces the cable-based physical layer with radio waves. A popular term, "Wi-Fi" (or Wi-fi, WiFi, Wifi, or wifi), short for "wireless fidelity," is a set of standards for WLAN. It is based on the Institute of Electrical and Electronics Engineers (IEEE) standards.

In 1997, IEEE created the first WLAN standard. It was the beginning of the IEEE 802.11 standard (Figure 11.20). IEEE 802.11 supported a maximum bandwidth of only 2 Mbps, too slow for most applications. By 1999, the 802.11 standard had expanded to 802.11b, which supports bandwidths up to 11 Mbps. This is comparable to traditional Ethernet (10Based). Both 802.11 and 802.11b use the unregulated 2.4-GHz radio-signaling frequency; thus they are free of frequency usage charges. Unfortunately, they interfere with microwave ovens, cordless phones, and some other appliances. To avoid interference, 802.11b devices must be kept some distance away from the interference sources.

At the same time that 802.11b was created, IEEE created 802.11a, which uses a 5-GHz signaling frequency and supports bandwidths up to 54 Mbps. The frequency domain is regulated; thus, it does not have the same interference problem as 802.11b. Yet,

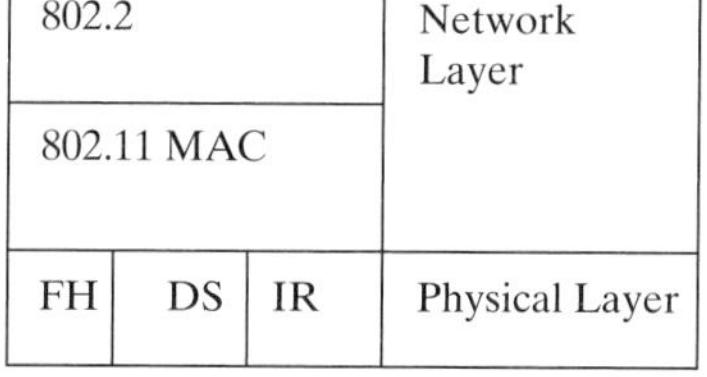

802.2	Network Layer
802.11 MAC	
FH | DS | IR	Physical Layer

MAC: Medium Access Control
FH: Frequency Hopping Spread Spectrum
DS: Direct Sequence Spread Spectrum
IR: Infrared

Figure 11.20   IEEE 802.11.

the higher frequency means that the signal is less able to penetrate walls and structures. In addition, the range of the signal is shorter. Because two standards use different frequencies, they are incompatible with each other. Some vendors offer hybrid 802.11a/b network devices that implement the two standards side by side.

In 2002 and 2003, a new standard became available. This standard, 802.11g, uses 2.4 GHz and supports bandwidths up to 54 Mbps. More improvement over the years is expected.

The physical and data link layers of the ISO/OSI model are covered by 802.11. Similar to Ethernet, it uses CSMA/CD (see Section 11.2.1.5) to access the medium. The physical layer can use either spread-spectrum radio signals or infrared signals. The network is divided into cells, called Basic Service Sets (BSSs). Each BSS has an Access Point (AP), which is also called a wireless router in commercial products. Several APs can be connected by a backbone, usually Ethernet. When a device (station) wants to access the network, it either actively sends a Probe Request Frame (a set of data) to an AP or passively waits to receive a Beacon Frame (which is periodically sent by AP and contains synchronization information). When the communication is established, each side sends a password for authentication. After a station is authenticated, an association process adds the station to the BSS. Only after the station is added to the BSS does it begin to transmit and receive data packets. Roaming from one cell to another cell is allowed.

Due to signal interference, a wireless network is expected to have higher data transmission errors than a cable-based network. The throughput for a WLAN is lower than the throughput of a LAN using the same data rate.

11.3 SUMMARY

This chapter has discussed the different methods for data communication. Each of these methods has been implemented on the shop floor. Among them, RS-232-C serial communication is the oldest, yet it is still used in many applications. Since the 1990s, LANs became increasingly important in manufacturing applications. In addition to the obvious advantages of LANs over point-to-point communication, the drastic cost reduction attained by LAN devices and the plug-and-play standardization make LAN affordable and user friendly. In the 1980s, the use of broadband Manufacturing Automation Protocol was proposed as the manufacturing shop-floor LAN standard. Due to its complexity and cost, however, it was dropped by industry by the late 1990s.

Today, Ethernet-based LANs are the backbone of many shop-floor communication networks. Wireless networks (WLANs) became popular in early 2000, and they see more applications in manufacturing as well. As LAN and WLAN functions are implemented in single IC chip devices (including sensors), many such devices have onboard LAN interfaces. Now, it is easy to exchange data among shop-floor devices. However, it is critical to ensure that the data exchanged are correct and the transmission is secure. More importantly, data collected must be used properly. This is an issue outside the scope of this text. It is essential for students of computer-aided manufacturing to understand the fundamentals of data communication, its current practice, and its future development. Since great innovation is still being made in this field, this chapter should be used only to build background understanding. Readers need to use the Internet to learn about the most current developments.

11.4 KEYWORDS

10BASE-2, 10BASE-5, 10BASE-T, 10BASE-F
Application layer
ASCII code
Bandwidth
Baseband
Broadband
Bus
Bytes and bits
Carrier
CATV
Coaxial cable
Collision detection
CSMA/CD
CPU
Data-link layer
Duplex
Ethernet
Fieldbus
Frame
I/O
IEEE 802.3
IEEE 802.11
IEEE-488
Interrupt
ISO/OSI

Local-area network (LAN)
MAP
Medium-access control
Memory address
Network layer
Network topology
Packet
Parallel communication
Physical layer
Polling
Presentation layer
RAM
RS-232
RS-422
RS-485
Serial communication
Session layer
Simplex
TCP/IP
Token bus
Token ring
Transport layer
TTL
WLAN

11.5 REVIEW QUESTIONS

11.1 Describe the different techniques in data transfer. Describe the pros and cons of each.

11.2 In connecting a PC to a cell-control computer, only garbage (unreadable) text was displayed on the screen. What are the possible causes? Both computers use RS-232 ports. What kind of action can be taken to correct the error? Assume that both the PC and the cell-control computer are in good operating condition.

11.3 What kind of communication ports is available on the CNC controller in your shop? What are the purposes of those ports? Are they being used now? If so, what kind of devices are connected to them?

11.4 Find the setting of the serial port (COM 1 and COM 2 on PCs) on your computer. What are the baud rate, numbers of start and stop bits, parity, and duplex?

11.5 Describe the incentive of installing a LAN on a shop floor. What kind of LAN products are available?

11.6 Define the following terms: baseband, broadband, medium-access control, and packet.

11.7 What are the disadvantages of using CSMA/CD for manufacturing LAN?

11.8 What is the OSI/ISO model?

11.9 What is the function of a bridge and a gateway in a multiple-network communication system?

11.10 What are the advantages of using a LAN (such as Fieldbus) to connect all the sensors on the shop floor?

11.11 What are the disadvantages of using a LAN to connect all the sensors on the shop floor?

11.12 Discuss the differences between Ethernet and RS-485 multidrop LANs. What are their network topologies, transmission media, protocols, cable length limitations, and number of nodes per segment?

11.13 Most cable-TV companies offer high-speed Internet (broadband) to their customers. A cable modem is connected to the cable-TV coaxial cable. How does this high-speed Internet service work? Check on the Internet for "broadband cable Internet."

11.14 Compare the advantages and disadvantages of using LAN and WLAN on the shop floor.

11.6 REVIEW PROBLEMS

11.1 Convert the following messages into ASCII code:

(a) MACHINE 1 OFF
(b) This is a test.

11.2 Convert the following messages into ASCII code with (a) even parity, (b) odd parity, (c) space parity, and (d) mark parity.

(a) ON
(b) OFF
(c) Idle
(d) M/C 3

11.3 A parts program is 8000 characters in length. It is to be transferred to a machine through a serial port running at 300 baud. What is the approximate time in seconds needed to complete the transfer? Assume that there is a 10% delay overhead.

11.4 A manufacturing shop has many machine tools and PLCs. The management would like to modernize the shop and link all the machines into one or more computer networks. As a CAM expert, you are asked to provide a preliminary design for this project. Assume that the shop is located in an industry park. No other company will allow you to tie into their network. In your report, specify the following:

(a) The layout of the network, including the type of each network and wiring
(b) Any additional hardware necessary to complete the network
(c) A brief justification of your design

Shop equipment and communication needs:
- 10 CNC machines with PC-based controllers
  - Primarily need parts program download/upload

- 5 PLC-based controllers for material handling and production machines
  - Coordinate the controllers, remote monitoring of the process
  - Some sensors and actuators are Fieldbus (such as the Rockwell DeviceNet) ready
- 4 PCs used for parts programming, email, accounting, etc.
  - Need to get on the Internet, download parts programs, and monitor processes

Use the Internet to find commercial products for the assignment.

11.5 Translate the following ASCII codes into text:

```
1001001
0100000
1101100
1101111
1110110
1100101
0100000
1000001
1010011
1000011
0101110
0100000
1110111
1101000
1100001
1110100
0111111
0000111
0001101
0001010
```

11.6 As a manufacturing engineer, you need to carry your laptop computer with you while solving problems on the shop floor. You also need to access data from your desktop and the Internet. The shop is 300′ by 200′. Ethernet access is available only on the side of the shop adjacent to the offices (200′ wall). How many access points do you need? Lay out the WLAN, and specify the equipment and cable needed to build the BSS. Use the Internet to find available products and their capabilities.

11.7 REFERENCES

Daconta, M. C. *Java for C/C++ Programmers*. New York: John Wiley, 1996.

Electronic Industries Association. *EIA Standard RS-232-C: Interface Between Data Terminal Equipment and Data Communications Equipment Employing Serial Binary Data Interchange.* Washington, DC: EIA, 1969.

Fieldbus Foundation. *Foundation Fieldbus Technical Overview*, FD043, Revision 3. Austin, TX: Fieldbus, 2003. http://www.fieldbus.org/

Forouzan, B. A., and B. Forouzan. *Data Communications and Networking*. New York: McGraw-Hill, 2003.

Graham, I. S. *The HTML Sourcebook: A Complete Guide to HTML 3.0*. New York: John Wiley, 1996.

Halsall, F. *Data Communications, Computer Networks, and Open Systems*, 4th ed. Reading, MA: Addison-Wesley, 1996.

Kamen, E. W. *Industrial Controls and Manufacturing*. London: Academic Press, 1999.

Roshan, P., and J. Leary. *Wireless Local-Area Network Fundamentals*. Pearson Education, 2003.

Schatt, S. *Understanding ATM*. New York: McGraw-Hill, 1996.

Thomas, R. M. *Introduction to Local-Area Networks*. San Francisco: Network Press, 1996.

APPENDIX A: A GLOSSARY OF SELECTED COMMUNICATION TERMS

ADSL (Asynchronous Digital Subscription Line): ADSL is a digital phone service.

ATM (Asynchronous Transfer Mode): ATM is a protocol for broadband ISDN. It uses large fixed-length packets. The data rate for ATM is 100 Mbps.

Asynchronous: A form of communication in which data is sent using start and stop bits, without regard for the time needed for transmission. Compare to synchronous transmission.

Backbone: A LAN or WAN that interconnects intermediate systems (bridges and/or routers).

Bandwidth: Frequency range (Hz) used by the communication system.

Baseband: Digital communication using a voltage difference to denote zeros and ones. One channel on the medium.

Baud: One audio-signal transition per second.

Bit: The units (0 and 1) used in the binary numbering system.

Bridge: A device that interconnects local or remote networks. It interprets messages from different networks using different protocols.

Broadband: Uses RF signals to transmit data through the medium (e.g., cable). Multiple channels exist on the same medium. Data are modulated on a carrier frequency. The carrier frequency is the channel center frequency. Cable TV uses broadband. It has a 5-Mbps channel bandwidth.

Carrier: A continuous frequency capable of being modulated or impressed with a second (information) signal. The carrier is stripped off by the receiver to extract the information.

CCITT V.XX Standard (Consultative Committee for International Telephone and Telegraph): V.XX is a modem standard developed by CCITT. Over the years, it progresses from V.22, V.32, to V.35.

Circuit switching: Data transfer relies on direct connection from end point to end point. Traditional analog phone networks use circuit switching.

Client/Server: A distributed system model of computing where clients access resources from servers.

Enterprise network: Private large network connecting most major points in a company.

Frame Relay: Used on an ISDN line; it assumes less transmission errors and can use larger packets.

FDDI (Fiber-Optic Distributed Data Interface): A LAN based on a fiber-optic medium.

Firewall: A router or workstation with multiple network interfaces that controls and limits specific protocols, types of traffic within each protocol, types of services, and the direction of the flow of information.

HTTP (Hypertext Transfer Protocol): Protocol used on the Internet for page display and data transfer.

ISDN (Integrated Service Digital Network): ISDN is a digital phone service that uses several channels. There are two basic types of channels: B channel and D channel. The B channel, or basic user channel, has a bandwidth of 64 kbps. It is used to transmit data. The D channel is used for control and setup, such as linking multiple channels together to increase the data rate and sending the dialing phone number. The D channel has a bandwidth of either 16 or 64 kbps. There are two levels of ISDN services: basic and primary. The basic service consists of two B channels and one D channel, a total of 144 kbps. The primary service or T-1 service consists of 23 B channels and 1 D channel. It transmits at 11.544 Mbps. It is possible to get more B channels to increase the total bandwidth.

ISO/OSI model (International Standard Organization/Open System Interconnect): A seven-layer communication model.

Medium-Access Control: Controls which device on the network gets to send data to the medium (such as cable, radio wave, optical fiber, etc.).

Modem encoding: In order to pack more data into the limited bandwidth of POTS, the signals are modulated. There are several popular modulation methods. Amplitude shift keying (ASK), the amplitude of the sinusoidal carrier is varied in accordance with the data signal. No audio signal is represented by 0; otherwise it is 1 (1 baud = 1 bps). Frequency-shift keying (FSK), the frequency of the sinusoidal carrier, is varied in accordance with the data signal, 1750 Hz for binary 1 and 1080 Hz for binary 0 (1 baud = 1 bps). Phase-shift keying (PSK), the variation of the phase of the sinusoidal carrier according to the data signal, has 0 and 1 represented by the alteration of the carrier's phase. The bit is coded at a fixed phase shift (e.g., 0, 90, 180, and 270 degrees) and has four bits per baud. Quadrature-amplitude modulation (QAM) combines both phase and frequency. At 1700 Hz or 1800 Hz of 2400 baud, each phase shift is keyed with six bps data per baud.

Modem (Modulator and Demodulator): An electronic device that converts the digital data into voice and voice back into digital data in order to transmit it through the analog telephone system.

NFS (Network File System): File-system sharing over network, remote disk mounting.

Packet: Small chunk of data.

Packet switching: Messages are split into packets and reassembled at the receiving end of the communication link. The data communication occupies a communication link only during the time of actual data communication.

POTS (Plain Old Telephone System): The analog telephone system at a channel width of 2400 Hz.

Protocol: A set of rules that governs the operation of functional units to achieve communication.

Router: Protocol-dependent device that connects subnetworks together.

Server: A computer on the network that acts as a central repository for data.

T1: Communications circuit provided by long-distance carriers for voice and data transmission at 1.544 Mbps.

T3: Digital-communication circuit standard created by AT&T that operates at 44 Mbps.

TCP/IP: Transport protocols concurrent with existing Ethernet.

USB (Universal Serial Bus): A device communication standard for replacing serial and parallel ports in computers and other computer-based devices. USB provides 12 Mbps bandwidth, 5 M per segment, and 127 devices on the same bus.

X.25: X.25 is a packet-switching protocol. Control and data are on the same channel. It uses variable-length packets.

WAN (Wide Area Network): Public or private computer network serving a wide geographic area.

WIFI: Wireless Fidelity or just wireless network.

WLAN: Wireless Local Area Network.

WWW (World Wide Web): A collection of hypertext pages located on computers around the world and logically linked together by the Internet.

Chapter 12

Fundamentals of Numerical Control

Objective

This chapter introduces the fundamentals of numeric control (NC) in product geometries. Since machine tools in a computer-aided manufacturing system are mostly NC machine tools, this subject is extremely important. Thus, the chapter will focus on the motion control of the machine tool, as well as introduce other features of modern NC machine tools.

Outline

12.1 INTRODUCTION

It has been more than half a century since the first numerical-control (NC) machine tool was demonstrated. The invention of this tool was a major technological revolution in manufacturing. It has completely changed the technology and image of the discrete product manufacturing industry. In fact, NC is the foundation for many modern manufacturing technologies, such as robotics, flexible manufacturing cells (FMC), flexible manufacturing systems (FMS), CAD/CAM, and computer-integrated manufacturing (CIM). This chapter will discuss the history and the hardware aspects of NC. In Chapter 13, the programming of NC machines will be examined.

12.2 HISTORICAL DEVELOPMENT

In tracing the history of machine-tool development, there is evidence of basic machining (turning) as early as 700 B.C. [Rolt, 1967]. However, it was not until the 15th century that people began machining metal. Eighteenth-century industrialization ushered in the demand of production-type machine tools, so machine shops were established to create more machines. But there was little change in the principles and mechanisms of machine tools and cutting tools. Early in the 20th century, F. W. Taylor invented a new tool metal—high-speed steel—and published his work on tool-life studies. Up to this point, machines could only reinforce human strength, cut hard materials, and provide a means to measure the dimensions of the parts produced.

In the first two decades of 20th century, attempts were made to mass-fabricate products and automating production equipment. Screw machines, transfer lines, and assembly lines are but a few successful examples of the era. These automated machines were controlled by mechanical devices using cams and preset stops. The cam had a motion sequence encoded in its contour. When it turns at a constant speed, a follower moves up and down the surface of the cam creating the desired displacement and velocity. In turn, the follower controls the motion of the machine table or tool carrier. These types of automation work well as long as the product is produced in large quantities and is not too complex in shape. Parts with complicated geometry or small demand still require the full attention of a skilled machinist. For most parts, current technology allows machines to replace skilled humans.

The past millennium has witnessed the increasing use of electromechanical devices. Machine tools that were driven by a common power source (e.g., a steam-engine-powered shop shaft) were fitted with their own motors. Certain operation sequences can now be controlled by mechanical, electromechanical, or pneumatic controllers. Then, the 1930s and 1940s constituted an era of fixed automation. Toward the end of that period (the eve of World War II), there was a race of arms production. More and more sophisticated military equipment was developed, and a large number of sophisticated products were needed. The quantity prompted the demand of automation, and the sophistication prompted the demand of a more skilled labor force.

During the first half of the 1940s, World War II was fought and the Cold War began. The arms race did not stop there. More sophisticated aircraft were developed for both military and civilian use. There was a technology boom, and the U.S. Air Force found that sophisticated part contours were required to make airplanes fly higher and faster. It was (and in some cases still is) not uncommon to remove up to 95% of the metal from a workpiece in order to produce a lightweight and structurally sound aircraft part. These parts must also be extremely accurate. Thus, reducing the manufacturing cost of these sophisticated parts became a major goal in aircraft development. The concept of numerical control was born as part of this effort. The intent was to replace human skill and sophistication with a programmable machine.

Generally, the concept of numerical control is credited to John Parsons. In 1947, Parsons developed a jig bore that was coupled with a computer. Using punched cards, Parsons was able to control the machine's position. In 1949, the U.S. Air Force, encouraged by Parson's success, commissioned MIT to develop a prototype of a "programmable" milling machine. The history of NC development can be found in a book written by one of its project leaders, T. F. Reintjes [1991]. In 1952, a modified three-axis Cincinnati Hydrotel milling machine was demonstrated, and the term *numerical control* (NC) was coined. The machine was controlled by a hardwired electromechanical controller, and the program was punched on cards and fed to the controller. Since then, many improvements have been made in the machine and in the control. Almost all machine tools are now equipped with NC controllers. A new class of machines called *machining centers* and *turning centers* that can perform multiple machining processes was developed. One of these machines, a mill-turn center, combines lathe features with a second active spindle and can turn as well as mill a part.

Modern NC controllers are computer based (called CNCs), and many have an interactive computer graphics interface and easy "on-machine" programming. Tool changers, pallet changers, and network communications are also available on many machines. Many can run unattended after the workpieces have been loaded. Some machines also have high-speed spindles ($>20,000$ rpm), high-feed-rate drives (>600 ipm), and high precision (<0.0001-in. accuracy). In the 1990s hexapod machines were introduced. These new machines control six legs simultaneously to position and orient the cutting tool.

CNC controllers have been applied to nearly every kind of machine tool: lathes, milling machines, drill presses, grinders, and so on. Modern NC machines are accurate, fast, strong, versatile, and easy to use. This is a result of improvements in both machine hardware and the controllers that run the machines. Now, there is a trend that places more intelligence on the machine tool (such as STEP-NC development), and NC machines of the future will include some kind of planning and adaptive control capabilities.

Figure 12.1 shows a modern machining center for large parts. In the front is a pallet changer. Four pallets have been mounted and are ready for machining. On the right is a loading and unloading station, where an operator is mounting a workpiece onto a pallet. As soon as the workpiece is mounted, the pallet can be loaded onto the index table. Then, any one of the six pallets can be placed onto the machine table, ready for machining. This machine is located at the far side of the pallet index table. In the figure, a pallet carrying a large disklike workpiece is being machined. On the left-hand side of the machine is a carousel-tool magazine on which many tools have been mounted. This type of system is capable of running unattended for a long period of time until all workpieces in queue have been completed. By using the tool changer, the machine is able to use one of the many tools stored in the tool magazine. The pallet changer serves as a queue for the workpiece. Very complex parts can be produced automatically.

Figure 12.1   An NC system for large parts (*Courtesy of Cincinnati Lamb*).

12.3 PRINCIPLES OF NUMERICAL CONTROL

Controlling a machine tool using variable input, such as a punched tape or a stored program, is known as numerical control (NC). NC has been defined by the Electronic Industries Association (EIA) as "a system in which actions are controlled by direct insertion of numerical data at some point. The system must automatically interpret at least some portion of this data." A *parts program* is the numerical data required to produce a part.

An NC machine-tool system contains a machine-control unit (MCU) and the machine tool itself (see Figure 12.2). The MCU is further divided into two elements: the data-processing unit (DPU) and the control-loops unit (CLU) [Koren, 1983]. The DPU processes the coded data that are read from the tape or some other medium. Then, it tells the CPU specific information: the position of each axis, its direction of motion feed, and its auxiliary-function control signals. The CLU operates the drive mechanisms of

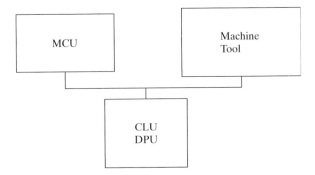

Figure 12.2   An NC machine system.

the machine, receives feedback signals about the actual position and velocity of each of the axes, and announces when an operation has been completed. The DPU reads the data sequentially when each line has completed execution.

A DPU consists of the following parts:

Data-input devices, such as a RS-232-C port

Data-reading circuits and parity-checking logic

Decoding circuits for describing data among the controller axes

An editor and graphical user interface.

A CLU, on the other hand, consists of the following:

An interpolator that supplies machine-motion commands between data points for tool motion

Position-control-loop hardware for all the axes of motion where each axis has a separate control loop

Velocity-control loops, where feed control is required

Acceleration, deceleration, and backlash take-up circuits

Auxiliary function control, such as a coolant on–off control, gear changes, and spindle on–off control.

As previously mentioned the motion control of NC machine tools is completed by translating NC codes into machine commands. The NC codes can be broadly classified into two groups, as shown in Figure 12.3: (1) commands for controlling individual machine

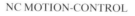
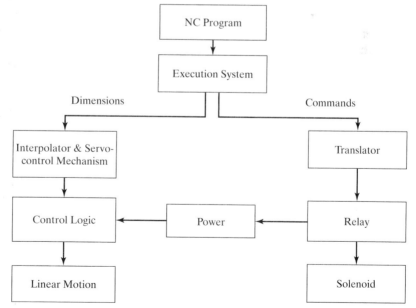

Figure 12.3 NC motion-control commands.

components, such as motor on–off control, selection of spindle speed, tool change, and coolant on–off control (these tasks are accomplished by sending electric pulses to the relay system or logic-control network); and (2) commands for controlling the relative movement of the workpiece and cutting tools. These commands consist of such information as the axis and distance to be moved at each specific time unit. They are translated into machine-executable, motion-control commands that are then carried out by the electromechanical control system.

12.4 CLASSIFICATION OF NUMERICAL CONTROL

NC machines can be classified as follows:

1. Motion control: point to point (PTP) and continuous path (contouring)
2. Control loops: open loop and closed loop
3. Power drives: hydraulic, electric, or pneumatic
4. Positioning systems: incremental and absolute positioning
5. Hardwired NC and softwired computer-numerical control (CNC)

12.4.1 Motion Control: Point to Point and Continuous Path

There are two types of motion-control systems for NC machines: point-to-point and continuous-path control. The function of a PTP motion-control system is to move the machine table or spindle to a specified position so that machining operations may be performed at that point. The path taken to reach the specific point is not defined by the programmer in this system (Figure 12.4). Because this movement is nonmachining, it is made as rapidly as possible, usually at a rate of more than 100 ipm. Figure 12.5 illustrates some paths that may be taken between the two points P and Q. If the job is to drill the same-sized holes at points P and Q, the programmer does not need to specify the path used to reach the points. In most cases, path B is taken in a PTP-NC machine. A PTP-NC machine is able to perform simple milling operations if the machine is equipped with a feed-control mechanism. With most PTP machines, the only directions that are accurately controlled are straight lines parallel to the machine axes (i.e., right

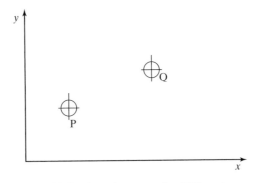

Figure 12.4   A point-to-point NC system.

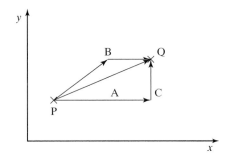

Figure 12.5   Possible paths between two locations in a PTP NC system.

and left, forward and backward). "Picture frames" can be easily done in such a manner on a PTP machine.

In a continuous (contouring) control system, the machine controls two or more axes simultaneously. The machine controls not only the destinations, but also the paths through which the tool reaches these destinations. In the process of machining, the tool contacts the workpiece and the desired shapes are made, as shown in Figure 12.6. Suppose that a slot is to be cut from left to right at 30°, as shown in Figure 12.7. When the table (or spindle) moves 4.330 in. to the right, it must move up 2.500 in. and travel 5.000 in. on the diagonal. This type of cut requires the two driving motors to run simultaneously at two different speeds. This contouring control system is more complex than PTP equipment.

Control of the travel rate in two (or more) directions, which is proportional to the distance moved, is called linear interpolation. If V_f is the desired velocity along the line of motion, the velocities along the two axes, V_x and V_y, are

$$V_x = \frac{\Delta x}{\sqrt{\Delta x^2 + \Delta y^2}} V_f \tag{12.1}$$

$$V_y = \frac{\Delta y}{\sqrt{\Delta x^2 + \Delta y^2}} V_f \tag{12.2}$$

where Δx and Δy are displacements along the X and Y axes, respectively.

Figure 12.6   Continuous-path control.

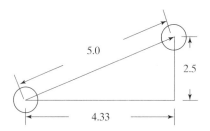

Figure 12.7 Continuous-path control using linear interpolation.

12.4.2 Control Loops

When there is position feedback, the NC system is a closed-loop system; otherwise, it is an open-loop system. Obviously, the advantage of a closed-loop system is its positioning accuracy. Because high precision is required of most NC machines, the majority of NC machines use a closed-loop control system. Most of these systems control servo motors (either DC or AC), wherein a position transducer returns the current table position. When the table reaches the programmed position, the motor stops running.

Open-loop control can be found in some light-application NC machines. Open-loop NC machines normally use stepping-motor drives, where the stepping motor is controlled through a digital signal. Each pulse of the signal turns the motor a small fixed angle (step angle). This movement is also translated to the table to move a distance of one BLU (basic length unit).

12.4.3 Power Drives

Most modern NC machines use a DC or AC servo motor to drive the axes and the spindle. Their small size, ease of control, and low cost are a few of the advantages of servo motors. In some large machines, hydraulic drives are used. A hydraulic motor has a much larger power/size ratio. For the same size as a DC motor, it can drive a much larger load at a greater acceleration. Difficult maintenance, increased noise, and a bulky off-the-machine power supply have limited the use of hydraulic drives. Another option, pneumatic drives, is rarely used in NC positioning systems, due to the difficulty of maintaining a precise continuous motion and position. They can be used to drive the auxiliary devices attached to the machine.

Although most motors are rotational motors, some are linear motors. The rotational motors use a leadscrew and nut mechanism to convert the rotational motion into linear motion. Their construction is simple. Yet, the conversion limits the precision. Linear motors eliminate this problem, but the technology is still expensive. Only special machines may use linear motors. Most NC machines still use rotational motors.

12.4.4 Positioning Systems

Whether a machine uses all incremental or absolute positioning systems depends on the position transducers used. When the transducer reports the absolute position of the machine table, the machine is an absolute positioning machine. Modern NC machines also allow the user to choose the types of positioning systems through software. Regardless of the hardware, users may program the machine in an incremental, absolute, or mixed positioning system. The coordinates are tracked by the software. In an absolute-positioning-system NC machine, the coordinate origin also can be reset. Machines that allow their coordinate origins to be reset are called floating-zero machines; others are called fixed-zero machines. Floating-zero machines have the advantage of allowing users to program a part using the part coordinates and set the machine origin at the part coordinate origin.

12.4.5 NC and CNC

The difference between NC and CNC lies in the controller technology. At the time when NC was invented, the computer was a rarity. The NC system was built using a dedicated vacuum-tube circuit, a transistor, and relay technology. These early controllers could perform only simple binary addition and subtraction. There was little system software. Every NC function had to be designed and implemented in hardware circuits. It was not until the late 1960s, when minicomputers became available on the shop floor (it was proposed in 1969), that a minicomputer was used to replace the hardwired NC controller. NC functions then could be implemented in software. Microcomputers were still more expensive than machine tools. In the late 1970s and 1980s, microcomputers were introduced to NC. Microcomputers are small in size, low in cost, and high in performance. Many of today's CNC controllers are based on existing microcomputers. They use an industrial-grade, general-purpose computer as the hardware platform. The computer operating system, the development environment, and the existing software can all be used in an NC controller. In fact many CNC controllers have data communication, color graphics, file management, editing, and so on. Some controllers even allow users to include customized functions. CNC has also made it possible to integrate machine tools with the rest of the manufacturing system. Many of them are actually MS Windows based. Since the 1980s, no hardwired NCs have been produced. Today, when the term NC is mentioned, it normally means CNC.

12.5 NUMERICAL-CONTROL SYSTEMS

The fundamental components that comprise a numerical-control installation and their functions are described in this section. Figure 12.8 shows a cutaway view of a horizontal NC machining center. The machine controller and the servo controls are located in a cabinet on the right side of the figure. The spindle, machine table, tool changer, tool magazine, and the three-axis leadscrew ways and motors can be seen in the figure. The major components of a numerical-control system are also shown in a block diagram in Figure 12.9.

   The NC system controller consists of both a mechanism that automatically reads the tape and the electronic hardware and software that converts the coded tape information into machine-tool instructions. This controller is the heart of the NC operation. The size of the machine has little effect on the size or complexity of the controller.

   Between the controller and the machine, there are several nonelectronic parts that are not normally found in conventional machine tools. The operator's console contains

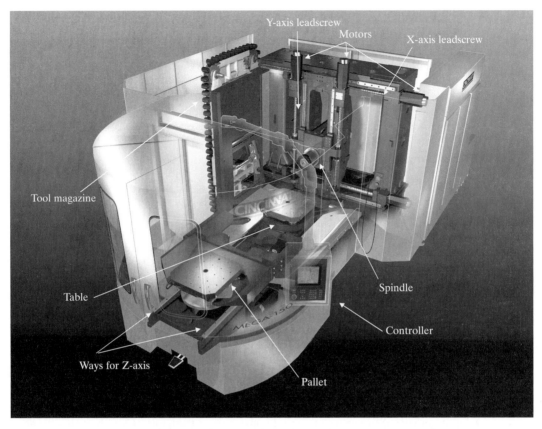

Figure 12.8 The structure of an NC machine (*Courtesy of Cincinnati Lamb*).

the necessary controls for operating the machine manually, thus offering the facility for "setting up" workpieces. The console may be on a pedestal, in a rolling cabinet, or in an overhanging pendant. The main distinction between this console and the operator panel on a conventional machine tool is in the increased number of command switches and buttons, some of which may be duplicates of those on the panel of the controller.

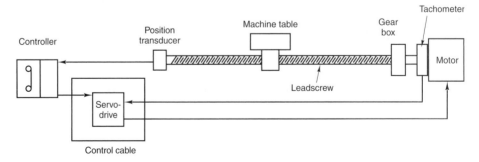

Figure 12.9 Major components comprising an NC machine tool.

Another major item is the magnetic-control cabinet. This unit contains the magnetic relays and starting switches that control the flow of electrical power to the hydraulic and coolant pumps, spindle motor, and other electric devices. These devices within the magnetic box are controlled by signals from the controller.

However, the prime mover of an NC system is the motor. Although hydraulic motors are used in some machine tools, the majority of machine tools use electric motors. The electric motor is typically connected to a leadscrew (Figure 12.10, the center screw) through a gear box or coupling, with the machine table riding on two ways (Figure 12.10). As the leadscrew turns, the nut attached to the table and constrained from rotating moves the table along the ways. A tachometer is also attached to the motor spindle to send feedback to the servo drive. Based on the location feedback received from a position transducer and a speed command, the controller sends an analog signal to the servo drive, which in turn controls the motor rotation. A transducer is a device that transforms one physical phenomenon to another, for example, from table position to a digital signal.

As the leadscrew rotates, the feedback device (which may be attached to either end of the leadscrew or may reside entirely independent of the leadscrew, e.g., on a rotating table or motor shaft) records the movement and sends a signal back to the controller.

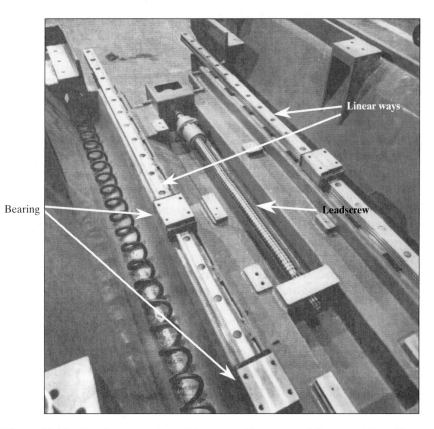

Figure 12.10   Leadscrew and machine ways (*Courtesy of Cincinnati Lamb*).

The controller compares the signal with the input instructions as described by the tape and continues to send signals to the servo drive until a balanced condition exists between the command signals (as read from the tape) and the feedback signals (as generated by movement of the machine components).

12.5.1 NC Accuracy and Repeatability

An NC machine is usually rated by the following factors:

- accuracy
- repeatability
- spindle and axis-motor horsepower
- number of controlled axes
- dimension of the workspace
- and features of the machine and the controller

The NC machine's accuracy results from a combination of the control instrumentation resolution and hardware accuracy. The control resolution is the minimum length distinguishable by the control unit. It is called the basic length unit (BLU) and is determined primarily by the axis transducer and leadscrew that are used. By contrast, hardware inaccuracies are caused by physical machine errors. There are many sources of error in a machine, all caused by component tolerances. Such errors include inaccuracies in the machine elements, machine tool assembly errors, spindle runout, and leadscrew backlash.

Another type of error is related to the machine's operation. Tool deflection, whose magnitude is a function of the cutting force, produces dimensional error and chatter marks on the finished part. Thermal error, which can be more critical than the deflection error, is caused by the thermal expansion of machine elements. The heat sources during machine operation include heat generated by the motor operation, cutting process, friction on the ways and bearings, and so on. Because the temperature on various parts of the machine differs, the error caused by the thermal effect is not uniform. Because thermal error is normally the greatest source of machine error, many methods are used to remove heat from a machine. Using cutting fluids, locating drive motors away from the center of a machine, as well as reducing friction from the ways and bearings, are a few conventional techniques used. In precision machining, a strictly controlled environment is used to ensure a constant temperature. Machine accuracy is an absolute measurement. Normal accuracy is specified as the machine error plus one-half of the control resolution. The manufacturer-rated machine accuracy does not include machine-operation-related errors. The actual accuracy can vary significantly from the rated accuracy.

Thermal error can also be compensated through software. The work initiated at the National Institute of Standards and Technology (NIST) and Purdue [Donmez et al., 1986] was later implemented in several industrial machine tools. The idea is to use regression analysis to find an error model for a specific machine tool. When a machine tool is running, thermal sensors mounted on the machine elements collect temperature data and the position error is measured. After the error profile is developed, a

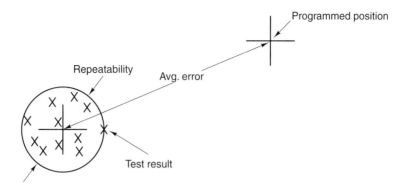

Figure 12.11   Accuracy and repeatability.

predicative model uses real-time temperature data from the machine elements and the current tool position to predict the error. This error data is fed back to the CNC machine control loop to compensate for the position error.

Repeatability is another measure of machine accuracy. It measures how closely a machine repeats a given position command. If a machine always goes to a fixed position, then it is said to be highly repeatable, even if the position is very far from the command position. Repeatability is measured as the diameter of the circle enclosing a target area produced by many repeated experiments (Figure 12.11).

Spindle power and axis power are important in the operation of a machine. The type of processed material and the material removal rate can be constrained by these factors. The number of axes of a machine determines the geometric complexity of the parts that the machine is able to produce. For most parts, a three-axis machine is sufficient; however, many aerospace and aircraft parts require four-, five-, and even six-axis machines. The additional axes make the construction of the machine and the control and the programming of the part more difficult.

12.6 NC CONTROLLERS

The NC controller is the brain of the NC system. It controls all functions of a machine tool. The controlled functions may be classified as motion control and auxiliary control. Motion control deals with the control of the tool position, orientation, and speed. Auxiliary control includes spindle rpm, tool change, coolant, pallet change, fixture clamping, etc. The auxiliary control functions are relatively simple compared with the motion-control function. Often, it can be done using sequential on–off control or proportional control. These control functions are similar to those provided by Programmable Logical Controllers (PLC). More details of PLCs can be found in a later chapter. In this section, only motion control function will be discussed.

The motion control begins with the position, orientation, and speed commands from a part program. The controller must translate these commands into appropriate individual-axis motor-control signals. Figure 12.12 illustrates the hierarchical structure of the control functions.

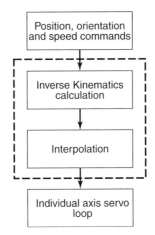

Figure 12.12   Motion control for NC.

Inverse kinematics is used to convert the position and orientation commands into the machine axes commands. Its calculation is necessary whenever the machine is not a two- or three-axis Cartesian structure. A Cartesian machine has its axis aligned with the Cartesian X, Y, and Z axes. The position command (where the spindle/tool orientation is fixed and aligned with the Z axis) can be mapped directly to the Cartesian coordinates. A traditional five-axis machine has three translational axes aligned with the Cartesian X, Y, and Z axes and two rotational axes. The position and orientation commands cannot be mapped directly into these axes controls. The conversion from the axis control to the Cartesian coordinates is called forward kinematics; the reverse is inverse kinematics. Equations for inverse kinematics are derived for the specific machine design. It is done by the machine tool builder. The equations are programmed into the controller.

The function of interpolation is to coordinate multiple axes to move the tool on a desired trajectory. For example, to cut along a line on a 2-D Cartesian machine, the two motors driving the two axes must run at different speeds. Otherwise, the cut will always be along an X, Y, or XY (45-degree) direction. The interpolator sends signals to individual axis-servo controllers. The output of the interpolator is mostly electric pulses. Each pulse represents one BLU of movement. The frequency of the output pulses depends on the BLU and the desired speed. For example, if the BLU $= 0.001''$ and the desired speed is 20 ipm, the pulse frequency would be

$$f_r = \frac{V/60}{BLU} = \frac{20/60}{0.001} = 333 \text{ Hz}$$

where 1 Hz $= 1$ pulse/second. When the steps of tool movement are small, the interpolator has to compute very fast in order to keep up with the pulse output.

At the bottom of the controller is the single-axis servo control. A typical single-axis servo control includes a servo controller, motor, leadscrew, and transducers (Figure 12.13). The leadscrew is not needed if the motor is a linear motor. A typical

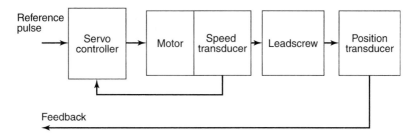

Figure 12.13   Single-axis-servo control.

machine uses rotational motors. The transducer can be a rotational transducer or a linear transducer (see Chapter 9). The transducer can be analog or digital. The servo controller makes sure that each input pulse is translated into one BLU of axis movement at the desired speed (frequency). The speed transducer provides the feedback to the servo controller to maintain the desired speed. The position transducer usually sends the feedback to the NC controller to decide when to terminate the motion.

12.6.1 Control Loops

There are open- and closed-loop NC systems. Open-loop systems are normally used only on small and low-power machines. However, the majority of NC systems use closed-loop control. Figure 12.14 illustrates the concept of open-loop control. In this scheme, the controller is used as an online processor of programs and data; it generates specific commands for the manipulation of machine and process actuators. Open-loop principles are also employed in the use of computers to manipulate stepping motors for machine-tool control. In an open-loop control system, the controller first converts time, speed, and displacement commands into pulse rate and pulse counts. Then, a timer is used to generate the desired pulse rate. The output of the timer is sent to the stepping motor and, at the same time, is sent to a down counter. The counter is loaded with the pulse count. When the count value reaches zero, the pulse is stopped from reaching the motor. With this simple control scheme, the motor will rotate at the desired speed for a desired number of rotations. The following example illustrates the operation of an open-loop system.

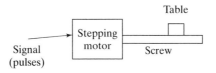

Figure 12.14   Open-loop control mechanism.

Example 12.1.

Consider a system similar to the one shown in Figure 12.14. In this system, the BLU = 0.001 in. The designers wish to move the table 5 in. at a speed (feed rate) of 6 ipm. The pulse rate and pulse count can be calculated as follows:

$$\text{pulse rate} = \text{speed/BLU} = 6 \text{ ipm/0.001 ipp} = 6000 \text{ pulses/min}$$
$$= 6000/60 = 100 \text{ pulses/s}$$
$$\text{pulse count} = \text{distance/BLU} = 5/0.001 = 5000 \text{ pulses}$$

A timer is programmed to generate 100 pulses/s, and the down counter is loaded with 5000 pulses. To completely automate a machine tool, it is necessary to develop control systems that have the capability of comparing a desired set of results with the actual results and of taking corrective action. This is the essence of closed-loop feedback control.

The system shown in Figure 12.15 is an incremental system that uses an encoder as the feedback transducer and a proportional-error-control controller (the output-control signal is proportional to the desired position and the feedback signal). In the figure, the DC motor, tachometer, gear box, encoder, and leadscrew are physically connected. The arrows show the direction of the control information flow, and the input to the system is the reference signal that was used in the stepping-motor drive. A certain number of pulses are sent to the up–down counter at a certain pulse rate. When a pulse is received, the counter increments its value. The counter value (a binary number) is converted into an analog signal by a digital-to-analog converter (DAC). The magnitude of this signal determines the motor speed. Thus, the higher the reference pulse rate, the higher the counter value can be.

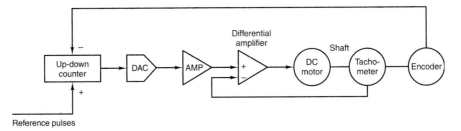

Figure 12.15   Closed-loop control mechanism.

A comparator-amplifier motor/tachometer circuit makes sure that the motor accelerates to the desired speed (see Section 12.4.2). The motor shaft is connected to an encoder through a gear box. When the motor turns, the encoder returns pulses to the up–down counter. The counter value is decremented by the feedback encoder pulses. The higher the reference pulse rate, the higher the counter value, thus, the greater the motor acceleration. When the desired speed has been reached, the motor stops accelerating, but maintains its speed. When the motor turns, the counter value is decremented. When the destination is reached, the counter value should be zero and the motor stops. The motor accelerates at the beginning of the motion, reaches a steady-state speed, and then decelerates before stopping at the desired position.

12.7 KINEMATICS OF NC MACHINES

Kinematics is defined as the "geometry of motion." The kinematics of NC machines describes how the cutting tool is controlled to move with respect to the workpiece.

As an example, a drill press has the simplest kinematical structure. A position table moves the tool on the X and Y 2-D plane. The drill depth, Z, is controlled only at the end of the X and the Y motion. The motion of the tool consists of many X–Y motions moving up and down at a depth of Z at the end of each X–Y motion. For a five-axis machine, the tool spindle can be rotated to different orientations. (a three-axis machine, by contrast, has a fixed tool-spindle orientation). The tool-tip position and the tool-spindle orientation are determined by the combination of three translational axes and two rotational axes (the numbers of translational and rotational axes depend on the machine's configuration). The motion of the tool can be very complex. To study the kinematics of an NC machine, the machine's axes configuration must be analyzed first.

Machine tools can be broadly classified as Cartesian machines (Figure 12.16) and articulated machines (Figure 12.17). Cartesian machines consist primarily of three translational axes, for X, Y, and Z motions corresponding to Cartesian coordinates. Additional axes are rotational, such as those in five-axes machines. The control of the motion is relatively simple, since the tool path is defined in Cartesian space and there is a one-to-one correspondence between the coordinates and the axis position. In contrast, an articulated machine, such as a hexapod machine, uses multiple legs to form linkages. The tool position is the result of extending and retracting multiple legs. There is no one-to-one correspondence between the X–Y–Z coordinates to the leg lengths.

12.7.1 Two- and Three-axis Cartesian Machines

Two-axis Cartesian machines are typified by a basic lathe. A simple lathe has one spindle and one tool post. The tool can move on the X and the Z axes (Figure 12.18). Since the actual cutting is done with the tool, the tool coordinate system appears on the tool tip O_t. Another coordinate system may be set on the tool post O_p. The relationship between O_t and O_p is the tool length l_t. The relationship between O_p and the machine coordinate system O_m is defined by an built-in offset to the tool post, $(\Delta x_{pm}, \Delta z_{pm})$, (not shown in the figure), and the controlled machine axes motion (x, z). The X and Z values from the parts program are used directly to control the X and the Z motors. Often, the machine coordinate system is the same as the workpiece coordinate system. At most, the workpiece

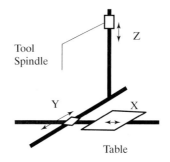

Figure 12.16   A Cartesian machine.

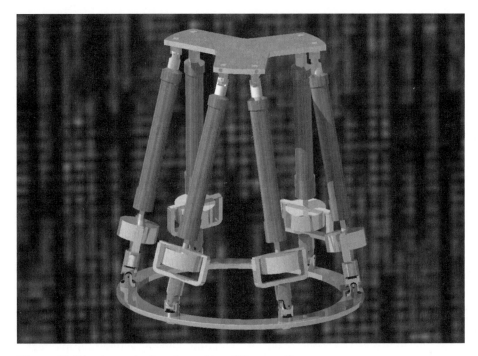

Figure 12.17   An articulated machine (Hexapod).

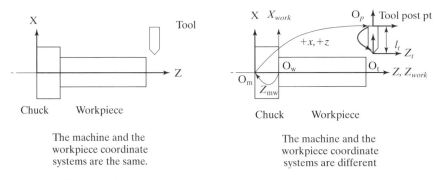

Figure 12.18   A lathe.

coordinate system is translated along the Z direction from the machine coordinate system. Let the displacement between the two origins be Δz_{mw} (displacement from O_m to O_w). T^v_{mw} is a matrix that transforms coordinates in O_m to O_w. The superscript v indicates that the transformation matrix is a "view" transformation matrix:

$$T^v_{mw} = \begin{bmatrix} 1 & 0 & 0 & 0 \\ 0 & 1 & 0 & 0 \\ 0 & 0 & 1 & -\Delta z_{mw} \\ 0 & 0 & 0 & 1 \end{bmatrix} \quad (12.3)$$

View transformation matrices are used in this chapter and can be found in Chapter 5 as well. Notice that Δz_{mw} is a positive value. It is the Z coordinate of O_w in O_m (Figure 12.18).

The distance from the tool post coordinate system to the machine coordinate system can be defined by two translations: a fixed displacement at the home position, and a controlled displacement due to X and Z movements. Since the X and Z motions are measured from the machine coordinate system, the sign in the transformation matrix should be positive. The transformation matrix is shown below:

$$T^v_{pm} = \begin{bmatrix} 1 & 0 & 0 & -\Delta x_{pm} \\ 0 & 1 & 0 & 0 \\ 0 & 0 & 1 & -\Delta z_{pm} \\ 0 & 0 & 0 & 1 \end{bmatrix} \begin{bmatrix} 1 & 0 & 0 & x \\ 0 & 1 & 0 & 0 \\ 0 & 0 & 1 & z \\ 0 & 0 & 0 & 1 \end{bmatrix} = \begin{bmatrix} 1 & 0 & 0 & -\Delta x_{pm} + x \\ 0 & 1 & 0 & 0 \\ 0 & 0 & 1 & -\Delta z_{pm} + z \\ 0 & 0 & 0 & 1 \end{bmatrix} \qquad (12.4)$$

The distance from O_t to O_p is a translation of l_t (the tool length):

$$T^v_{tp} = \begin{bmatrix} 1 & 0 & 0 & 0 \\ 0 & 1 & 0 & 0 \\ 0 & 0 & 1 & -l_t \\ 0 & 0 & 0 & 1 \end{bmatrix} \qquad (12.5)$$

Combining these three equations yields the tool-tip coordinates in the workpiece coordinate system:

$$T^v_{tw} = T^v_{tp} T^v_{pm} T^v_{mw}$$

$$T^v_{tw} = \begin{bmatrix} 1 & 0 & 0 & -\Delta x_{pm} + x \\ 0 & 1 & 0 & 0 \\ 0 & 0 & 1 & -\Delta z_{pm} + z - \Delta z_{mw} - l_t \\ 0 & 0 & 0 & 1 \end{bmatrix} \qquad (12.6)$$

For example, let $p = \begin{bmatrix} 1 & 0 & 2 & 1 \end{bmatrix}^T$ be a point on the workpiece (in O_w). To control the tool tip $\begin{bmatrix} 0 & 0 & 0 & 1 \end{bmatrix}^T$ to move to this point, the controller issues a command in the machine coordinate system:

$$T^v_{tw} \begin{bmatrix} 0 & 0 & 0 & 1 \end{bmatrix}^T = p$$

$$\begin{bmatrix} 1 & 0 & 0 & -\Delta x_{pm} + x \\ 0 & 1 & 0 & 0 \\ 0 & 0 & 1 & -\Delta z_{pm} + z - \Delta z_{mw} - l_t \\ 0 & 0 & 0 & 1 \end{bmatrix} \begin{bmatrix} 0 \\ 0 \\ 0 \\ 1 \end{bmatrix} = \begin{bmatrix} 1 \\ 0 \\ 2 \\ 1 \end{bmatrix}$$

$$\begin{bmatrix} -\Delta x_{pm} + x \\ 0 \\ -\Delta z_{pm} + z - \Delta z_{mw} - l_t \\ 1 \end{bmatrix} = \begin{bmatrix} 1 \\ 0 \\ 2 \\ 1 \end{bmatrix}$$

$$x = 1 + \Delta x_{pm}$$

$$z = 2 + \Delta z_{pm} + \Delta z_{mw} + l_t$$

Since l_t, Δx_{pm}, Δz_{pm}, Δz_{mw} are constants for the setup, the tool-tip position is controlled by x, z. Most machines can move the tool to touch the workpiece coordinate system origin and set that point as the zero. When this is done, l_t, Δx_{pm}, Δz_{pm}, Δz_{mw} are read by the controller and are automatically compensated for in the subsequent control motion.

Most milling machines have three controlled axes. A typical three-axis Cartesian machine has the structure shown in Figure 12.19. The workpiece coordinate system is aligned with the machine coordinate system, except that the origin is translated. This translation depends on the setup of the workpiece. It can easily compensated for by the controller. The machine (on the table)-to-workpiece relationship is

$$T_{mw}^v = \begin{bmatrix} 1 & 0 & 0 & -\Delta x_{mw} \\ 0 & 1 & 0 & -\Delta y_{mw} \\ 0 & 0 & 1 & -\Delta z_{mw} \\ 0 & 0 & 0 & 1 \end{bmatrix} \tag{12.7}$$

where Δx_{mw}, Δy_{mw}, Δz_{mw} are displacements from the machine origin to the workpiece origin.

The tool tip is related to the gauge reference point on the spindle by a tool length displacement l_t:

$$T_{tg}^v = \begin{bmatrix} 1 & 0 & 0 & 0 \\ 0 & 1 & 0 & 0 \\ 0 & 0 & 1 & -l_t \\ 0 & 0 & 0 & 1 \end{bmatrix} \tag{12.8}$$

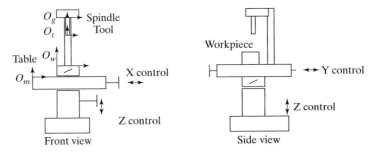

Figure 12.19   A three-axis vertical milling machine.

From the gauge reference point to the machine coordinate system, there is a fixed displacement of $\Delta x_{tm}, \Delta y_{tm}, \Delta z_{tm}$, and the machine axes motions are x, y, z. Again, since the x, y, z motions are defined from the gauge reference coordinate system to the machine coordinate system, they stay positive:

$$
T_{gm}^v =
\begin{bmatrix}
1 & 0 & 0 & -\Delta x_{gm} \\
0 & 1 & 0 & -\Delta y_{gm} \\
0 & 0 & 1 & -\Delta z_{gm} \\
0 & 0 & 0 & 1
\end{bmatrix}
\begin{bmatrix}
1 & 0 & 0 & x \\
0 & 1 & 0 & y \\
0 & 0 & 1 & z \\
0 & 0 & 0 & 1
\end{bmatrix}
=
\begin{bmatrix}
1 & 0 & 0 & -\Delta x_{gm} + x \\
0 & 1 & 0 & -\Delta y_{gm} + y \\
0 & 0 & 1 & -\Delta z_{gm} + z \\
0 & 0 & 0 & 1
\end{bmatrix}
\tag{12.9}
$$

Therefore,

$$
T_{tw}^v = T_{tg}^v T_{gm}^v T_{mw}^v
$$

$$
=
\begin{bmatrix}
1 & 0 & 0 & 0 \\
0 & 1 & 0 & 0 \\
0 & 0 & 1 & -l_t \\
0 & 0 & 0 & 1
\end{bmatrix}
\begin{bmatrix}
1 & 0 & 0 & -\Delta x_{gm} + x \\
0 & 1 & 0 & -\Delta y_{gm} + y \\
0 & 0 & 1 & -\Delta z_{gm} + z \\
0 & 0 & 0 & 1
\end{bmatrix}
\begin{bmatrix}
1 & 0 & 0 & -\Delta x_{mw} \\
0 & 1 & 0 & -\Delta y_{mw} \\
0 & 0 & 1 & -\Delta z_{mw} \\
0 & 0 & 0 & 1
\end{bmatrix}
$$

$$
=
\begin{bmatrix}
1 & 0 & 0 & -\Delta x_{gm} + x - \Delta x_{mw} \\
0 & 1 & 0 & -\Delta y_{gm} + y - \Delta y_{mw} \\
0 & 0 & 1 & l_t - \Delta z_{gm} + z - \Delta z_{mw} \\
0 & 0 & 0 & 1
\end{bmatrix}
\tag{12.10}
$$

Like two-axis machines, the controller only needs to make a simple translation from the workpiece coordinates to the machine coordinates. By setting the machine origin at the workpiece origin (floating zero), the controller is doing this translation internally. This allows the programming to be done under the workpiece coordinates. The exact location of the setup is not needed during the programming. If the machine has a fixed-zero controller (as do some older controllers), the programming must be done in the machine coordinate system. This translation is done offline by the programmer.

12.7.2 Five-Axis Machines

While three-axis machines have a fixed spindle orientation (aligned with the z-axis), five-axis machines allow the spindle to be rotated to a different orientation. There are several possible five-axis machine configurations (Figure 12.20). Most of them use three translational axes and two rotational axes. The simplest five-axis machine is a three-axis machine with a two-axis rotary attachment to the machine table (Figure 12.21). This rotary table rotates about the X-axis (or the Z-axis, depending on whether the home position of the rotary table is aligned with the X- or the Z-axis) and the Y-axis. The new rotational axes are denoted as the A and the B axes, respectively.

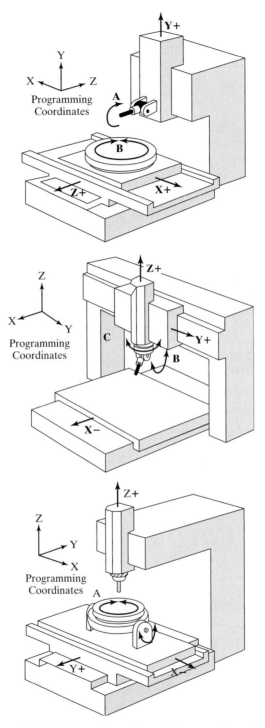

Figure 12.20   Three five-axis machine configurations.

Figure 12.21   Rotary table attachment (Courtesy of Fadal Machine).

The workpiece is now mounted on the rotary table, which is the end of the A and the B axes. The relationship between the machine coordinate system and the workpiece coordinate system is more complex. Figure 12.22 illustrates the relationships of multiple coordinate systems.

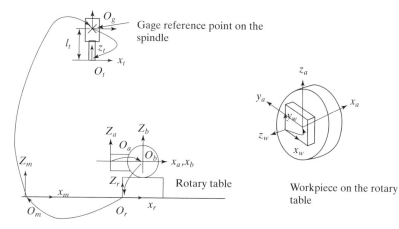

Figure 12.22   Five-axis machine coordinate systems.

The following offsets need to be considered. The transformation from the tool tip O_t to the gauge reference point, O_g, is

$$T_{tg}^v = \begin{bmatrix} 1 & 0 & 0 & 0 \\ 0 & 1 & 0 & 0 \\ 0 & 0 & 1 & -l_t \\ 0 & 0 & 0 & 1 \end{bmatrix} \tag{12.11}$$

The transformation from the gauge reference point O_g to the machine-coordinate system O_m (similar to that of a three-axis machine) is

$$T_{gm}^v = \begin{bmatrix} 1 & 0 & 0 & x - \Delta x_{gm} \\ 0 & 1 & 0 & y - \Delta y_{gm} \\ 0 & 0 & 1 & z - \Delta z_{gm} \\ 0 & 0 & 0 & 1 \end{bmatrix} \tag{12.12}$$

The transformation from the rotary table coordinate system O_r to the machine coordinate system O_m is

$$T_{mr}^v = \begin{bmatrix} 1 & 0 & 0 & -\Delta x_{mr} \\ 0 & 1 & 0 & -\Delta y_{mr} \\ 0 & 0 & 1 & 0 \\ 0 & 0 & 0 & 1 \end{bmatrix} \tag{12.13}$$

where Δx_{mr}, Δy_{mr} are the displacements from O_m to O_r. Since the Z-coordinates are on the same level, $\Delta z_{mr} = 0$.

The transformation from the rotary table coordinate system O_r to the B-axis coordinate system O_b is

$$T_{rb}^v = \begin{bmatrix} 1 & 0 & 0 & -\Delta x_{rb} \\ 0 & 1 & 0 & -\Delta y_{rb} \\ 0 & 0 & 1 & -\Delta z_{rb} \\ 0 & 0 & 0 & 1 \end{bmatrix} \tag{12.14}$$

The transformation from the B-axis coordinate system to the A-axis coordinate system O_a is

$$T_{ba}^v = \begin{bmatrix} 1 & 0 & 0 & -\Delta x_{ba} \\ 0 & 1 & 0 & 0 \\ 0 & 0 & 1 & 0 \\ 0 & 0 & 0 & 1 \end{bmatrix} \begin{bmatrix} \cos \beta & 0 & -\sin \beta & 0 \\ 0 & 1 & 0 & 0 \\ \sin \beta & 0 & \cos \beta & 0 \\ 0 & 0 & 0 & 1 \end{bmatrix}$$

$$= \begin{bmatrix} \cos\beta & 0 & -\sin\beta & -\Delta x_{ba} \\ 0 & 1 & 0 & 0 \\ \sin\beta & 0 & \cos\beta & 0 \\ 0 & 0 & 0 & 1 \end{bmatrix} \tag{12.15}$$

where Δx_{ba} is the X-coordinate of the O_a origin in O_b (a negative value) and β is the rotation angle of the B-axis.

It takes two rotations to align O_w with O_a. First, there is a $-90°$ rotation about z_a; then there is a $90°$ rotation about x'. The coordinate system is also translated by $(\Delta x_{aw}, \Delta y_{aw}, \Delta z_{aw})$, where x_a is the A-axis. Therefore, a controlled angle α is rotated about x_a. In the example in Figure 12.22, Δx_{aw} is negative, Δy_{aw} is positive, and Δz_{aw} is negative. The distance from the A-axis–coordinate system to the workpiece coordinate system, O_w, is

$$T^v_{aw} = \begin{bmatrix} 1 & 0 & 0 & 0 \\ 0 & \cos\alpha & \sin\alpha & 0 \\ 0 & -\sin\alpha & \cos\alpha & 0 \\ 0 & 0 & 0 & 1 \end{bmatrix} \begin{bmatrix} 1 & 0 & 0 & 0 \\ 0 & \cos 90 & \sin 90 & 0 \\ 0 & -\sin 90 & \cos 90 & 0 \\ 0 & 0 & 0 & 1 \end{bmatrix}$$

$$\times \begin{bmatrix} \cos 90 & \sin(-90) & 0 & -\Delta x_{aw} \\ -\sin(-90) & \cos 90 & 0 & -\Delta y_{aw} \\ 0 & 0 & 1 & -\Delta z_{aw} \\ 0 & 0 & 0 & 1 \end{bmatrix}$$

$$= \begin{bmatrix} 1 & 0 & 0 & 0 \\ 0 & \cos\alpha & \sin\alpha & 0 \\ 0 & -\sin\alpha & \cos\alpha & 0 \\ 0 & 0 & 0 & 1 \end{bmatrix} \begin{bmatrix} 1 & 0 & 0 & 0 \\ 0 & 0 & 1 & 0 \\ 0 & -1 & 0 & 0 \\ 0 & 0 & 0 & 1 \end{bmatrix} \begin{bmatrix} 0 & -1 & 0 & -\Delta x_{aw} \\ 1 & 0 & 0 & -\Delta y_{aw} \\ 0 & 0 & 1 & -\Delta z_{aw} \\ 0 & 0 & 0 & 1 \end{bmatrix}$$

$$= \begin{bmatrix} 1 & 0 & 0 & 0 \\ 0 & \cos\alpha & \sin\alpha & 0 \\ 0 & -\sin\alpha & \cos\alpha & 0 \\ 0 & 0 & 0 & 1 \end{bmatrix} \begin{bmatrix} 0 & -1 & 0 & -\Delta x_{aw} \\ 0 & 0 & 1 & -\Delta z_{aw} \\ -1 & 0 & 0 & \Delta y_{aw} \\ 0 & 0 & 0 & 1 \end{bmatrix}$$

$$= \begin{bmatrix} 0 & -\cos\alpha & -\sin\alpha & -\Delta x_{aw} \\ 0 & -\sin\alpha & \cos\alpha & -\Delta z_{aw} \\ -1 & 0 & 0 & \Delta y_{aw} \\ 0 & 0 & 0 & 1 \end{bmatrix} \tag{12.16}$$

Therefore, the transformation from the tool tip to the workpiece is

$$T^v_{tw} = T^v_{tg} T^v_{gm} T^v_{mr} T^v_{rb} T^v_{ba} T^v_{aw} \tag{12.17}$$

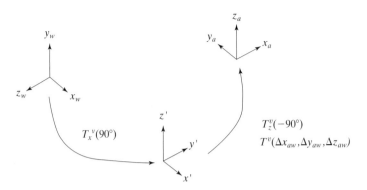

Figure 12.23 Relating the workpiece coordinate system with the A-axis coordinate system.

where

$$T_{tw}^v = T_{tg}^v T_{gm}^v T_{mr}^v T_{rb}^v T_{ba}^v T_{aw}^v$$

$$= \begin{bmatrix} 1 & 0 & 0 & 0 \\ 0 & 1 & 0 & 0 \\ 0 & 0 & 1 & -l_t \\ 0 & 0 & 0 & 1 \end{bmatrix} \begin{bmatrix} 1 & 0 & 0 & x - \Delta x_{gm} \\ 0 & 1 & 0 & y - \Delta y_{gm} \\ 0 & 0 & 1 & z - \Delta z_{gm} \\ 0 & 0 & 0 & 1 \end{bmatrix} T_{mr}^v T_{rb}^v T_{ba}^v T_{aw}^v$$

$$= \begin{bmatrix} 1 & 0 & 0 & x - \Delta x_{gm} \\ 0 & 1 & 0 & y - \Delta y_{gm} \\ 0 & 0 & 1 & z - \Delta z_{gm} - l_t \\ 0 & 0 & 0 & 1 \end{bmatrix} \begin{bmatrix} 1 & 0 & 0 & -\Delta x_{mr} \\ 0 & 1 & 0 & -\Delta y_{mr} \\ 0 & 0 & 1 & 0 \\ 0 & 0 & 0 & 1 \end{bmatrix} T_{rb}^v T_{ba}^v T_{aw}^v$$

$$= \begin{bmatrix} 1 & 0 & 0 & x - \Delta x_{gm} - \Delta x_{mr} \\ 0 & 1 & 0 & y - \Delta y_{gm} - \Delta y_{mr} \\ 0 & 0 & 1 & z - \Delta z_{gm} - l_t \\ 0 & 0 & 0 & 1 \end{bmatrix} \begin{bmatrix} 1 & 0 & 0 & -\Delta x_{rb} \\ 0 & 1 & 0 & -\Delta y_{rb} \\ 0 & 0 & 1 & -\Delta z_{rb} \\ 0 & 0 & 0 & 1 \end{bmatrix} T_{ba}^v T_{aw}^v$$

$$= \begin{bmatrix} 1 & 0 & 0 & x - \Delta x_{gm} - \Delta x_{mr} - \Delta x_{rb} \\ 0 & 1 & 0 & y - \Delta y_{gm} - \Delta y_{mr} - \Delta y_{rb} \\ 0 & 0 & 1 & z - \Delta z_{gm} - l_t - \Delta z_{rb} \\ 0 & 0 & 0 & 1 \end{bmatrix} \begin{bmatrix} c\beta & 0 & -s\beta & -\Delta x_{ba} \\ 0 & 1 & 0 & 0 \\ s\beta & 0 & c\beta & 0 \\ 0 & 0 & 0 & 1 \end{bmatrix} T_{aw}^v$$

$$= \begin{bmatrix} c\beta & 0 & -s\beta & x - \Delta x_{gm} - \Delta x_{mr} - \Delta x_{rb} - \Delta x_{ba} \\ 0 & 1 & 0 & y - \Delta y_{gm} - \Delta y_{mr} - \Delta y_{rb} \\ s\beta & 0 & c\beta & z - \Delta z_{gm} - l_t - \Delta z_{rb} \\ 0 & 0 & 0 & 1 \end{bmatrix}$$

$$
\times
\begin{bmatrix}
0 & -c\alpha & -s\alpha & -\Delta x_{aw} \\
0 & -s\alpha & c\alpha & -\Delta z_{aw} \\
-1 & 0 & 0 & \Delta y_{aw} \\
0 & 0 & 0 & 1
\end{bmatrix}
$$

$$
=
\begin{bmatrix}
s\beta & -c\beta c\alpha & -c\beta s\alpha & -c\beta\Delta x_{aw} - s\beta\Delta y_{aw} + x - \Delta x_{gm} \\
 & & & \quad - \Delta x_{mr} - \Delta x_{rb} - \Delta x_{ba} \\
0 & -s\alpha & c\alpha & -\Delta z_{aw} + y - \Delta y_{gm} - \Delta y_{mr} - \Delta y_{rb} \\
-c\beta & -s\beta c\alpha & -s\beta s\alpha & -s\beta\Delta x_{aw} + c\beta\Delta y_{aw} + z - \Delta z_{gm} \\
 & & & \quad - l_t - \Delta z_{rb} \\
0 & 0 & 0 & 1
\end{bmatrix}
\qquad (12.18)
$$

In the preceding equation, $s\alpha = \sin \alpha$, $c\alpha = \cos \alpha$, $s\beta = \sin \beta$, and $c\beta = \cos \beta$. The following MATLAB commands can be used to find the solutions:

```
syms a b dxw dyw dzw dxb dyb dzb dxr dyr dzr dxm dym dzm dxt dyt dzt x y z lt;
ttg=[1 0 0 0; 0 1 0 0; 0 0 1 -lt;0 0 0 1];
tgm=[1 0 0 x-dxg; 0 1 0 y-dyg; 0 0 1 z-dzg; 0 0 0 1];
tmr=[1 0 0 -dxm; 0 1 0 -dym; 0 0 1 0;0 0 0 1];
trb=[1 0 0 -dxr;0 1 0 -dyr; 0 0 1 -dzr; 0 0 0 1];
tba=[cos(b) 0 -sin(b) -dxb;0 1 0 0;sin(b) 0 cos(b) 0;0 0 0 1];
taw=[0 -cos(a) -sin(a) -dxw;0 -sin(a) cos(a) -dyw;-1 0 0 dzw;0 0 0 1];
ttw=ttg*ttm*tmr*trb*tba*taw
```

The tool position in the workpiece coordinate system can be found with the use of Equation 12.18. The position is determined by the fixed displacements and the axes commands: x, y, z, α, β. The tool orientation (the vector $\mathbf{V}_T V_T$) can be found from the difference between the origin of the gauge and the tool-tip coordinate systems: $O_t - O_g$. O_t is at $T^v_{tw}[0 \quad 0 \quad 0 \quad 1]^T$ and O_g at $T^v_{tw}[0 \quad 0 \quad l_t \quad 1]^T$. Thus,

$$
V_t = T^v_{tw}\left(
\begin{bmatrix} 0 \\ 0 \\ l_t \\ 1 \end{bmatrix}
-
\begin{bmatrix} 0 \\ 0 \\ 0 \\ 1 \end{bmatrix}
\right)
$$

$$
= T^v_{tw}
\begin{bmatrix} 0 \\ 0 \\ l_t \\ 1 \end{bmatrix}
\qquad (12.19)
$$

$$
=
\begin{bmatrix}
-c\beta s\alpha l_t - c\beta\Delta x_{aw} - s\beta\Delta y_{aw} + x - \Delta x_{gm} - \Delta x_{mr} - \Delta x_{rb} - \Delta x_{ba} \\
c\alpha l_t - \Delta z_{aw} + y - \Delta y_{gm} - \Delta y_{mr} - \Delta y_{rb} \\
-s\beta s\alpha l_t - s\beta\Delta x_{aw} + c\beta\Delta y_{aw} + z - \Delta z_{gm} - l_t - \Delta z_{rb} \\
1
\end{bmatrix}
$$

Note that in subtracting one homogeneous-coordinates vector from another, the last element ("1") is not subtracted.

12.7.3 Hexapod Machines

Hexapod machines are also called nonorthogonal axial configuration machines. Unlike the traditional machines, which have an orthogonal axial arrangement hexapod machines have X, Y, and Z axes that are perpendicular to each other. Hexapod machines have six nonorthogonal axes (legs or struts). The extending and retracting of each axes moves the tool and changes the tool orientation. This machine configuration is based on the Stewart platform [1965]. The advantages of such a configuration are high speed, rapid acceleration, and high stiffness. High speed and acceleration reduce the machining time. The high stiffness is the result of the six legs; it yields high precision (less distortion). The machine is best suited for freeform surface machining. A typical hexapod machine is shown in Figure 12.24. It consists of a base, a platform, six extendable struts, six universal joints, six ball and socket joints, and a tool.

A hexapod machine is controlled by controlling the length of each strut. The strut can be driven by a mechanical leadscrew, a hydraulic cylinder, or a pneumatic cylinder. The programmed tool tip position and tool orientation is converted into six strut lengths. This conversion is called *inverse kinematics* (from Cartesian coordinates to strut coordinates). Inverse kinematics for a hexapod machine is relatively simple. First, let the base coordinate system be set at the center of the base. The positions of the six universal joints are denoted as a_i, $i = 1, 2, \ldots, 6$. The length of the corresponding strut is l_i. The struts link universal joints to ball-and-socket joints, b_i. At the center of the platform is the spindle reference point, $P = [x_p \quad y_p \quad z_p]^T$. The tool tip is located at $P_t = [x_{pt} \quad y_{pt} \quad z_{pt}]^T$. Let $a_i = [x_{ai} \quad y_{ai} \quad z_{ai}]^T$ be the coordinates of universal joint i and $b_i = [x_{bi} \quad y_{bi} \quad z_{bi}]^T$ be the coordinates of ball-and-socket joint i. Since the machine coordinate frame (O) is set at the center of the base and joints a_i are on the same X–Y plane as the coordinate system origin, it follows that $z_{ai} = 0$, $i = 1, 2, \ldots, 6$. At the home position, point P is right below O, and b_i are on the same X–Y plane. Let P be the center of the platform coordinate frame and

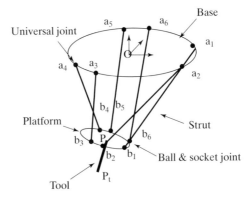

Figure 12.24   Hexapod machine.

$b_i^p = [x_{bi}^p \quad y_{bi}^p \quad z_{bi}^p]^T$ be the coordinates of ball-and-socket joints in the platform co-ordinate system. Thus, $z_{bi}^p = \text{const}, i = 1, 2, \ldots 6$, where const is the distance between the platform and the base at the home position and a_i, b_i^p are fixed for a given machine configuration.

To control a machine, the desired tool tip position P_t and tool orientation $P_t - P$ must be given. The goal of inverse kinematics is to find the length

$$l_i = \sqrt{(x_{bi} - x_{ai})^2 + (y_{bi} - y_{ai})^2 + z_{bi}^2} \tag{12.20}$$

where

$$b_i = Rb_i^p + P \tag{12.21}$$

and R is a transformation matrix. This length can be calculated from the tool orientation vector $P_t - P$. In Figure 12.25, the coordinate system X–Y–Z represents the platform coordinate system. Let Z' represent the tool orientation vector. The home orientation of the tool is $[0 \quad 0 \quad -1]^T$. The new orientation is obtained by rotating the tool about the X and the Y axes. This can be shown in the equation

$$R \begin{bmatrix} 0 \\ 0 \\ -1 \end{bmatrix} = \frac{P_t - P}{|P_t - P|} \tag{12.22}$$

$$R = T_x^v T_y^v$$

$$= \begin{bmatrix} 1 & 0 & 0 \\ 0 & Cx & -Sx \\ 0 & Sx & Cx \end{bmatrix} \begin{bmatrix} Cy & 0 & Sy \\ 0 & 1 & 0 \\ -Sy & 0 & Cy \end{bmatrix}$$

$$= \begin{bmatrix} Cy & 0 & Sy \\ SxSy & Cx & -SxCy \\ -CxSy & Sx & CxCy \end{bmatrix} \tag{12.23}$$

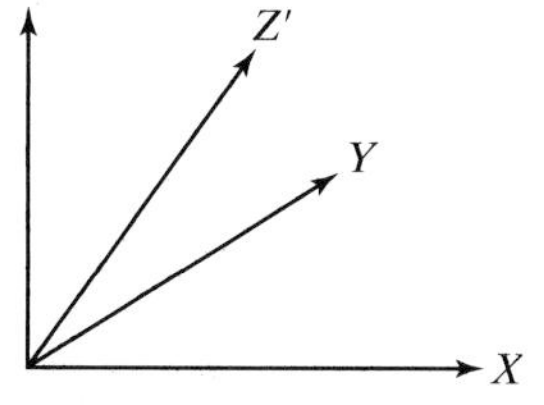

Figure 12.25 Platform coordinate system and tool orientation.

where $Cx = \cos\theta_x$, $Cy = \cos\theta_y$, $Sx = \sin\theta_x$, and $Sy = \sin\theta_y$. Combining the two equations yields

$$\begin{bmatrix} -Sy \\ SxCy \\ -CxCy \end{bmatrix} = \frac{P_t - P}{|P_t - P|} \qquad (12.24)$$

θ_x, θ_y can be solved from Equation 12.22 and then plugged back into Equation 12.21 to find R. After R is found, it can be used in Equations 12.19 and 12.18 to find l_i.

Example 12.2.

The base of a hexapod machine has a diameter of 120″, and the platform has a diameter of 60″. At the home position, the platform is 40″ below the base. The layout of the joints is shown in Figure 12.26. What are the strut lengths for $P_t = \begin{bmatrix} 1 & 2 & -50 \end{bmatrix}^T$ and $P = \begin{bmatrix} 1 & 1 & -49 \end{bmatrix}^T$? The tool orientation vector is

$$P_t - P = \begin{bmatrix} 0 \\ 1 \\ -1 \end{bmatrix}$$

Hence,

$$\frac{P_t - P}{|P_t - P|} = \begin{bmatrix} 0 \\ 1/\sqrt{2} \\ -1/\sqrt{2} \end{bmatrix} = \begin{bmatrix} 0 \\ 0.707 \\ -0.707 \end{bmatrix}$$

$$-\sin\theta_y = 0, \quad \theta_y = 0$$

$$\sin\theta_x \cos\theta_y = 0.707, \theta_x = 45°$$

$$-\cos\theta_x \cos\theta_y = -0.707, \theta_x = 45°$$

$$R = \begin{bmatrix} 1 & 0 & 0 \\ 0 & 0.707 & -0.707 \\ 0 & 0.707 & 0.707 \end{bmatrix}$$

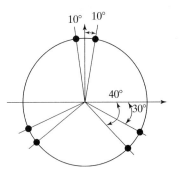

Figure 12.26   Joints on both the base and the platform (these are the designer's choice).

Let $A = [a_1 \quad a_2 \quad a_3 \quad a_4 \quad a_5 \quad a_6]$, $B^p = [b_1{}^p \quad b_2{}^p \quad b_3{}^p \quad b_4{}^p \quad b_5{}^p \quad b_6{}^p]$

$$A = \begin{bmatrix} 51.962 & 45.963 & -45.963 & -51.962 & -10.419 & 10.419 \\ -30 & -38.567 & -38.567 & -30 & 59.088 & 59.088 \\ 0 & 0 & 0 & 0 & 0 & 0 \end{bmatrix}$$

$$B^p = \begin{bmatrix} 25.981 & 22.982 & -22.982 & -25.981 & -5.210 & 5.210 \\ -15 & -19.284 & -19.284 & -15 & 29.544 & 29.544 \\ 0 & 0 & 0 & 0 & 0 & 0 \end{bmatrix}$$

Rewrite Equation 12.19 in matrix form:

$$B = RB^p + [P \quad P \quad P \quad P \quad P \quad P]$$

$$B = \begin{bmatrix} 1 & 0 & 0 \\ 0 & 0.707 & -0.707 \\ 0 & 0.707 & 0.707 \end{bmatrix}$$

$$\times \begin{bmatrix} 25.981 & 22.982 & -22.982 & -25.981 & -5.210 & 5.210 \\ -15 & -19.284 & -19.284 & -15 & 29.544 & 29.544 \\ 0 & 0 & 0 & 0 & 0 & 0 \end{bmatrix} +$$

$$\begin{bmatrix} 1 & 1 & 1 & 1 & 1 & 1 \\ 1 & 1 & 1 & 1 & 1 & 1 \\ -49 & -49 & -49 & -49 & -49 & -49 \end{bmatrix}$$

$$B = \begin{bmatrix} 25.981 & 22.982 & -22.982 & -25.981 & -5.210 & 5.210 \\ -10.605 & -13.6338 & -13.6338 & -10.605 & 20.8876 & 20.8876 \\ -10.605 & -13.6338 & -13.6338 & -10.605 & 20.8876 & 20.8876 \end{bmatrix} +$$

$$\begin{bmatrix} 1 & 1 & 1 & 1 & 1 & 1 \\ 1 & 1 & 1 & 1 & 1 & 1 \\ -49 & -49 & -49 & -49 & -49 & -49 \end{bmatrix}$$

$$= \begin{bmatrix} 26.981 & 23.982 & -21.982 & -24.981 & -4.21 & 6.21 \\ -9.605 & -12.6338 & -12.6338 & -9.605 & 21.8876 & 21.8876 \\ -59.605 & -62.6338 & -62.6338 & -59.605 & -28.1124 & -28.1124 \end{bmatrix}$$

$$l_1 = \sqrt{(51.962 - 26.981)^2 + (-30 + 9.605)^2 + (-59.605)^2}$$

$$= 67.7699$$

Similarly, the rest of the strut length can be calculated by using the values in A and B. In MATLAB, the problem can be solved as follows:

```
>> p=[1;1;-49];
pt=[1;2;-50];
R=[1 0 0;0 0.707 -0.707;0 0.707 0.707]
```

```
A=[51.962 45.963 -45.963 -51.962 -10.419 10.419;
-30 -38.567 -38.567 -30 59.088 59.088;
0 0 0 0 0 0];
Bp=[25.981 22.982 -22.982 -25.981 -5.21 5.21;
-15 -19.284 -19.284 -15 29.544 29.544;
0 0 0 0 0 0];
B=R*Bp + [p p p p p p];
for i=1:6
    l(i)=sqrt((A(1,i)-B(1,i))^2+(A(2,i)-B(2,i))^2+B(3,i)^2);
end;
>>l =
    67.7699    71.2649    71.9070    68.5324    47.0396    46.8176
```

The lengths of the six legs are output in the last line.

As can be seen, the inverse kinematics for the hexapod machine is relatively straightforward. However, forward kinematics (finding the tool-tip coordinates for the given strut lengths) is difficult. Since passive joint values are not known, there is no closed-form solution. An iterative procedure is used to solve the problem. The algorithm is left for the reader to investigate.

12.8 INTERPOLATION

Normally, at least two axes are needed to control a machine like a lathe or a drill press (for fixed-Z drilling applications). The majority of the applications require three-axis control. For a two-axis point-to-point machine, two individual single-axis controllers can simply be placed together. As long as the tool reaches the desired position, the exact path is not important. For example, in Figure 12.27(a), the tool moves from (3,2) to (10,5). Both the X- and Y-axes move at the same speed, and the resultant tool path is shown in the figure. The tool path is acceptable for drilling operations, but not for milling or turning. To control the tool to move in a straight line, it is necessary to control the speed of each axis, so that both axes reach the destination at the same time (Figure 12.27(b)).

Using the preceding example, if the desired feed rate is 6 ipm, the speed components on the X- and Y-axes are (from Equations (12.1) and (12.2))

$$V_x = 6\frac{10 - 3}{\sqrt{(10 - 3)^2 + (5 - 2)^2}} = 6\frac{7}{\sqrt{49 + 9}} = 5.5149 \text{ in./min}$$

$$V_x = 6\frac{5 - 2}{\sqrt{(10 - 3)^2 + (5 - 2)^2}} = 6\frac{3}{\sqrt{49 + 9}} = 2.3635 \text{ in./min}$$

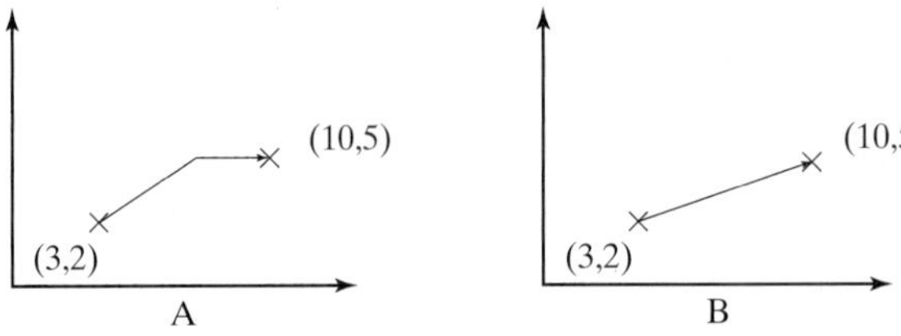

Figure 12.27   Linear path.

If BLU equals 0.001 in., the pulse rate in X is 5515 pulses/min and in Y is 2364 pulses/min. (Each pulse equals 1 BLU; see Section 12.5 for more details.) Although this method will probably work, there is no guarantee that the result is a perfect line. Even though the velocity components are set correctly, there is no synchronization beyond the start of the motion. To get as close to the desired line as possible, both axes must be coordinated closely. This is called linear interpolation. Linear interpolation is available on nearly all NC machines.

Circles are also frequently used in many parts. Circular arcs can be used to approximate high-order curves. Although a circle can be approximated by line segments, it is more desirable to control the machine to interpolate a circle directly with minimum error. The function of a circular interpolator is to move the tool automatically on a circular arc without external approximation. There are also parabolic or other higher-order interpolators available on some specialized machines. Virtually all CNC machines provide both linear and circular interpolation capabilities.

Interpolation is designed to generate a series of fixed-size steps (1 BLU) in order to approximate a geometric feature that is not directly attainable. The maximum deviation of the tool path is kept within one step (BLU). Because the interpolation is right above the servo level, speed is critical; the process must not involve excessive computation. For higher-order curves that cannot be implemented easily in an online interpolator, offline approximation algorithms are applied. Such algorithms break down the curve into line segments or circular arcs before feeding them into an NC machine. One example of higher-order curves is a free-form curve or free-form surface.

12.8.1 Linear Interpolator

A linear function of time can be represented as

$$x = Vt + x_0 \tag{12.25}$$

where x is position, t is time, V is speed, and x_0 is the initial position of the object in question.

In a digital system, the time is divided into small increments called clock cycles (which are the same as pulses), Δt. The linear function can be written in a difference form:

$$x_i = V \sum_{j=1}^{i} \Delta t + x_0 = x_{i-1} + V\Delta t \tag{12.26}$$

A simple device called a digital differential analyzer (DDA) (Figure 12.28), is built to carry out the preceding computation. This simple device can generate pulses at intervals of duration Δt. The input to the DDA is clock pulses. Each time a pulse (with frequency f) is received, the value of the register (a value) is added to the accumulator. When the accumulator overflows, the overflow bit is output to the motor control up–down counter (x value). Δt is determined by the accumulator word width.

$$\Delta t = \frac{1}{2^N} \tag{12.27}$$

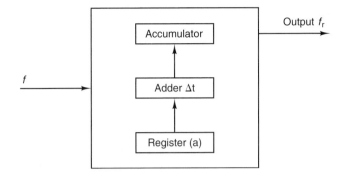

Figure 12.28   A DDA.

where N is the width of the accumulator (bits). DDA functions like a frequency divider. It can reduce the input frequency based on a binary register a value.

For example, an eight-bit accumulator has an output of $\Delta t = 1/256 = 0.00390625$.

The relationship of the output frequency and the input frequency can be expressed as

$$f_r = \frac{af}{2^N} \tag{12.28}$$

where

f_r = output frequency, Hz

a = register value

N = accumulator width, bit

f = input clock frequency, Hz

To control two axes, two DDAs are used (Figure 12.29) with a single-input clock signal driving the two DDAs. Each DDA is loaded with appropriate register values. The output pulses are sent to the axis-control up–down counters. Because a DDA acts like a pulse divider, to the feed rate can be controlled by simply adding another DDA to the input (Figure 12.29).

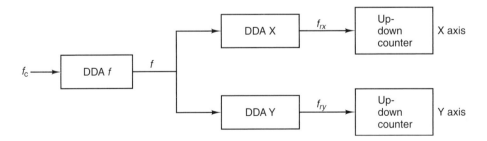

Figure 12.29   Two-axis control.

The feed-rate control DDA can be developed as follows:

$$f = \frac{a_f f_c}{2^{N_f}} \tag{12.29}$$

where

a_f = DDA f register value
N_f = DDA f width
f_c = controller clock pulse

The output to the axis control is

$$f_r = \frac{a_f f_c}{2^{N_f}} \frac{a}{2^N} = \frac{a a_f}{2^{N_f+N}} f_c \tag{12.30}$$

A conventional NC controller is not equipped to compute the speed components, V_x and V_y, from the feed rate. A scheme can be implemented to take care of this deficiency. The a value for each axis is loaded with the BLU value of each incremental motion. By using the earlier example to move the tool from (3,2) to (10,5) at the 6-ipm feed rate, the following values are loaded in the DDAs:

$$a_x = (10 - 3) \text{ in.}/0.001 \text{ in./step} = 7000 \text{ steps}$$

$$a_y = (5 - 2) \text{ in.}/0.001 \text{ in./step} = 3000 \text{ steps}$$

From the earlier calculation, f_{rx} must be 5515 and f_{ry} must be 2364. They are computed as follows:

$$f_{rx} = V_f \frac{\Delta x}{\sqrt{\Delta x^2 + \Delta y^2}} \tag{12.31}$$

(All lengths are in BLUs.) Equation 12.28 is a general equation for both f_{rx} and f_{ry}. Substitute f_{rx} for f_r and a_f for a_x into Equation 12.28, and combine the result with Equation 12.29:

$$\frac{a_x a_f}{2^{N_f+N}} f_c = V_f \frac{\Delta x}{\sqrt{\Delta x^2 + \Delta y^2}} \tag{12.32}$$

Because a_x is set to Δx, both terms can be eliminated from the preceding equation:

$$\frac{a_f}{2^{N_f+N}} f_c = \frac{V_f}{\sqrt{\Delta x^2 + \Delta y^2}}$$

Solving for a_f yields

$$a_f = \frac{V_f}{\sqrt{\Delta x^2 + \Delta y^2}} \frac{2^{N_f+N}}{f_c} \tag{12.33}$$

Notice that $2^{N_f+N}/f_c$ is a constant based on hardware design. If this value is set to A, then

$$a_f = \frac{AV_f}{\sqrt{\Delta x^2 + \Delta y^2}} \qquad (12.34)$$

If both the feed-rate-control DDA (the a_f value) and each axis DDA are loaded with the incremental displacement value, then the resultant motion will be at the desired feed rate. No additional internal computation is needed to compute the speed components. Actually, Equation 12.16 provides the definition of the inverse time code used in conventional NC feed-rate programming. Of course, in CNC machines, this is not necessary because feed-rate conversion can be done easily via computer.

The following example illustrates the operation of a simplified DDA interpolator:

$$N = 3$$

$$\Delta x = 4 \text{ BLUs}$$

$$\Delta y = 3 \text{ BLUs}$$

Figure 12.30 illustrates the operation of the DDAs. First, the Δx value, 4 (binary 100), and the Δy value, 3 (binary 011), are loaded into the corresponding DDA registers. When the clock sends a pulse to the DDAs, they each add their register value to the accumulator. In the table, the X and the Y columns show the corresponding accumulator value in binary. At clock cycle 2, there is an overflow of DDA x, and the X counter value is 1. At clock cycle 3, the Y counter also receives an overflow. The operation continues until the destination is reached at clock cycle 8. The diagram in the figure shows the tool motion. Note that each motion is within 1 BLU, the best precision that any machine can achieve.

Clock	X	X Counter	Y	Y Counter
1	100	0	011	0
2	000	1	110	0
3	100	1	001	1
4	000	2	100	1
5	100	2	111	1
6	000	3	010	2
7	100	3	101	2
8	000	4	000	3

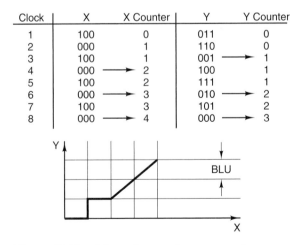

Figure 12.30   Linear interpolator example.

Clock	X		X counter	Y		Y counter
1	100		0	011		0
2	000	$\rightarrow$	1	110	$\rightarrow$	0
3	100		1	001	$\rightarrow$	1
4	000	$\rightarrow$	2	100		1
5	100		2	111		1
6	000	$\rightarrow$	3	010	$\rightarrow$	2
7	100		3	101		2
8	000	$\rightarrow$	4	000	$\rightarrow$	3

DDAs can also be simulated through software. The software can either run in a loop (not recommended) or be driven by an interrupt. (See Chapter 8.) Following is pseudocode for a single software DDA (the program must be executed in a fixed-cycle loop with a cycle time equal to Δt):

```
Max = 2^N;
C=0;
X=0;
While X < X_count Do
    C = C + A;
    If C >= Max then
        X = X + 1;
        C = 0;
        Output a pulse;
    End;
End;
```

A DDA can also be replaced by a programmable timer. The input to the timer is a clock of frequency f. The register value is a, which is loaded by the CPU. The word length of the timer is N. The output frequency of the timer is

$$f_r = \frac{af}{2^N} \tag{12.35}$$

$$a = \frac{f_r 2^N}{f} \tag{12.36}$$

Example 12.3.

Given the following data, calculate the register value:

$BLU = 0.01 \; in.$

$f = 1 \times 10^3 \; Hz$

$\Delta x = 2 \; in.$

$\Delta y = 3 \; in.$

$v_f = 6 \; in./\text{min}$

$N = 10$

Solution:

$$v_x = \frac{6 \times 2}{\sqrt{2^2 + 3^2}} \frac{1}{0.001 \times 60}$$

$$= 55.5 \text{ BLU/s}$$

$$v_y = \frac{6 \times 3}{\sqrt{2^2 + 3^2}} \frac{1}{0.001 \times 60}$$

$$= 83.3 \text{ BLU/s}$$

$$f_{rx} = 55.5 \text{ Hz}$$

$$f_{ry} = 83.3 \text{ Hz}$$

$$p_x = \frac{55.5 \times 2^{10}}{1 \times 10^3}$$

$$= 57$$

$$p_y = \frac{83.3 \times 2^{10}}{1 \times 10^3}$$

$$= 85$$

12.8.2 Circular Interpolator

Circular interpolation can also be implemented in DDAs. The DDAs used in circular interpolation is modified from the ones used in linear interpolation. A circular arc is defined by a center p_0 and two endpoints, p_1 and p_2 (Figure 12.31). The angular position of the tool is ϕ, and it is measured with respect to p_0. The feed rate is therefore

$$V_f = R\frac{d\phi}{dt} \tag{12.37}$$

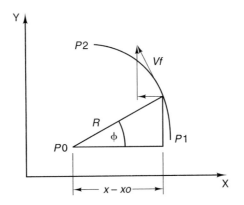

Figure 12.31 A circular arc.

where R is the radius of the circular arc. Thus,

$$x = R \cos \phi + x_0 \tag{12.38}$$

$$y = R \sin \phi + y_0 \tag{12.39}$$

$$R \cos \phi = x - x_0 \tag{12.40}$$

$$R \sin \phi = y - y_0 \tag{12.41}$$

and the X and Y components of the feed speed are

$$\frac{dx}{dt} = -R \sin \phi \frac{d\phi}{dt}$$

$$= -(y - y_0)\frac{d\phi}{dt} \tag{12.42}$$

$$\frac{dy}{dt} = R \cos \phi \frac{d\phi}{dt}$$

$$= (x - x_0)\frac{d\phi}{dt} \tag{12.43}$$

The dx/dt and dy/dt are values loaded into the DDAs. Because DDA registers are unsigned, absolute values $|y - y_0|$ and $|x - x_0|$ are used. From Equation 12.37, $d\phi/dt = V_f/R$. This provides the feed-rate control. To move on a circular arc, the speed also changes where the rate of change of dx/dt and dy/dt can be found:

$$\frac{d^2x}{dt^2} = -R \cos \phi \frac{d\phi}{dt}\frac{d\phi}{dt}$$

$$= -\frac{dy}{dt}\frac{d\phi}{dt} \tag{12.44}$$

$$\frac{d^2y}{dt^2} = -R \sin \phi \frac{d\phi}{dt}\frac{d\phi}{dt}$$

$$= \frac{dx}{dt}\frac{d\phi}{dt} \tag{12.45}$$

The change over the arc is therefore defined by Equations 12.44 and 12.45. A DDA circuit can be designed using two modified DDAs (Figure 12.32). The DDA X is loaded with $|y_1 - y_0|$, DDA Y with $|x_1 - x_0|$. The sign is added after the output is produced by changing the direction of the motor. Because the desired rate of change to DDA X equals $-dy/dt$, that is the output of DDA Y. When there is an output from DDA Y, the DDA X register is incremented. Note that both dx/dt and d^2x/dt^2 have the same sign. Similarly, the DDA Y register value is affected by the output of DDA X. This time, the

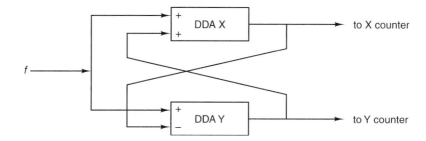

Figure 12.32 A circular interpolator.

register value is decremented (the negative sign in dx/dt is carried out through the decrement operation).

The arc shown in Figure 12.31 is in quadrant I. In quadrant II, the value of $(x - x_0)$ is negative. Because the register is an unsigned binary, the sign must be handled separately. Therefore, this circular interpolator does not work across two quadrants. When an arc extends across two quadrants, it must be broken into two segments. This hardware limitation is the reason that many old NC machines must be programmed in separate quadrants.

The feed-rate control for circular interpolation is similar to that used in linear interpolation, except that the value is

$$a_f = \frac{V_f 2^{N_f + N}}{R f_c}$$

$$= \frac{10 V_f}{R} \tag{12.46}$$

A numerical example is not given here, but it can be constructed easily from the preceding discussion. The following data may be used: A circular arc is centered at (0,0); cut it from (10,0) to (0,10); the DDA register width is 4 bits. Initially, both a_x and a_y should be loaded with a value 10.

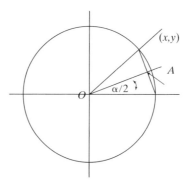

Figure 12.33 Find the step size.

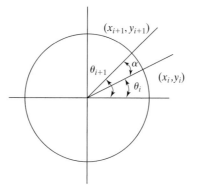

Figure 12.34   Interpolate a circle.

In CNC machines, full-circle circular interpolation can be achieved through software interpolation. Let R be the radius of the circle. The origin is at (0,0) and the step angle is α. All quantities are in BLUs. The step angle is chosen to make the interpolation error less than 1 BLU. In Figure 12.34, a step is taken from $(R, 0)$ to (x, y). The interpolation error is

$$R - \overline{OA} = 1 \tag{12.47}$$

where

$$\overline{OA} = R \cos \frac{\alpha}{2} \tag{12.48}$$

Combining Equations 12.47 and 12.48 yields

$$R - 1 = R \cos \frac{\alpha}{2} \tag{12.49}$$

$$\alpha = 2 \cos^{-1}\left(\frac{R - 1}{R}\right) \tag{12.50}$$

The position of the tool at step $i + 1$ is (x_{i+1}, y_{i+1}), with a step size of α from step i. Thus,

$$x_{i+1} = R \cos(\theta_i + \alpha) = R[\cos \alpha \cos \theta_i - \sin \alpha \sin \theta_i]$$

$$y_{i+1} = R \sin(\theta_i + \alpha) = R[\cos \alpha \sin \theta_i + \sin \alpha \cos \theta_i] \tag{12.51}$$

The speed for each axis can also be calculated:

$$V_{xi} = V \sin \theta_i$$

$$V_{yi} = V \cos \theta_i \tag{12.52}$$

Since trigonometric functions in computers are approximated with the Taylor expansion, the computation is very time consuming. To keep up with the servo loop, a simplified approximation can be used:

$$\cos\theta = \frac{1 - (\theta/2)^2}{1 + (\theta/2)^2}$$

$$\sin\theta = \frac{\theta}{1 + (\theta/2)^2} \tag{12.53}$$

12.9 AXIS SERVO CONTROL

A single-axis servo control system is shown in Figure 12.15. The operation of the servo control has been illustrated in Section 12.5.1. In this section, several system components will be discussed in more detail and a control system model will be presented.

12.9.1 Leadscrews

An important component of an NC machine is the leadscrew (Figure 12.35). The rotational motion of the motors is converted to a linear motion by a leadscrew, which is coupled with the machine table through a nut. The machine table is confined to a linear motion by two slides (ways). When the leadscrew turns, the machine table is forced to move along the slides. As previously discussed, heat generated by the friction must be reduced. A precision-ground ball-bearing leadscrew (Figure 12.35) is normally used in NC machines. To reduce backlash, the ball bearings are preloaded. Backlash is caused by the free play of the screw and the nut. When there is a gap between the screw thread and the nut thread and the rotation is reversed, the threads are not immediately engaged. This causes the nut to hesitate for a brief moment before it reverses the direction of motion. When one of the axes briefly stops and the other axes are still moving, a geometric error is produced. Preloading the ball bearing ensures the proper contact between the leadscrew thread, the ball bearing, and the nut screw thread. Some controllers also use software to further compensate the backlash error.

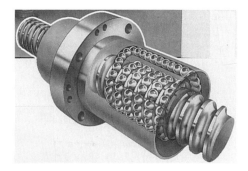

Figure 12.35   Preloaded ball nut and leadscrew.

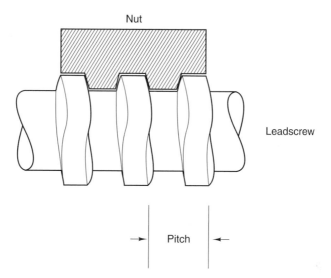

Figure 12.36   A typical screw thread.

The pitch of the leadscrew determines the resolution that a machine can achieve. The pitch (Δp) is defined as the distance between adjacent screw threads (Figure 12.36) and is related to the number of teeth per inch, given by

$$n = \frac{1}{\Delta p} \qquad (12.54)$$

For each screw rotation, the nut advances a distance equal to the pitch.

12.9.2 Motor

As discussed in Section 12.5.1, DC servo motors are the most common motors used in NC machines. In fact they are still used in spindle drives. More recently, however, AC servo motors have been replacing DC servo motors for axis drives.

The difference between the two types of motor is simple. DC servo motors are controlled by current (or DC voltage). AC servo motors are driven by the input power frequency. It requires much simpler electronics to change the voltage than to change the frequency, especially for high-power applications. Therefore, DC motors were popular right from the beginning of the NC revolution.

Unlike a stepping motor, which can be driven by a digital signal, servo motors are driven by analog signals. DC motor speed control is achieved by the input voltage and an attached tachometer. The input voltage develops a torque, which is used to accelerate the rotor. The tachometer acts like a DC power generator: The higher the speed, the higher is the output voltage. Its output is used to cancel the input voltage. As the speed of the DC motor gradually reaches the desired speed (a speed that corresponds to the input voltage), the actual input to the motor reduces. This reduces the acceleration of the motor rotor until the desired speed is reached.

An AC motor is controlled by the input power frequency. The rotor chases the changing magnetic fields in the stator of the motor. The motor speed can be calculated by

$$\text{rotor rpm} = (\text{AC frequency})/(\# \text{ of poles in the stator}) \times 120 \qquad (12.55)$$

The number of poles in the stator is a physical motor design parameter.

Neither DC nor AC servo motors have a built-in position control like stepping motors. Since these motors are rotational motors, the position control is defined as the control of the number of rotations of the rotor and the rotation angle within one revolution. Thus, a position transducer must be used in conjunction with the motor. The transducer may be attached directly to the motor shaft (rotor), to the leadscrew, or to the machine table. When the desired position is reached, the controller turns the motor off and keeps it in place. Section 10.3.1 discusses several position transducers.

The machine BLU is determined by the leadscrew pitch and the position-transducer resolution. The rotational position-transducer resolution coupled to the leadscrew has N pulses per revolution resolution. Thus, the BLU of the machine is

$$\text{BLU} = \frac{\Delta p}{N} \qquad (12.56)$$

For example, an NC machine uses a 0.1-in. pitch leadscrew and a 100-pulse/rev encoder. The BLU of this machine is

$$\text{BLU} = \Delta p/N = 0.1 \text{ in./rev} / 100 \text{ pulses/rev} = 0.001 \text{ in.}$$

The same formula can be used in stepping-motor control. For example, if the same leadscrew is used in a stepping-motor system, and the stepping motor has 100 steps/rev, then the BLU of this machine is the same as that of the preceding one:

$$\text{BLU} = 0.1/100 = 0.001 \text{ in.}$$

To size the motor (determine the horsepower), the axial force must be greater than the feed force. The axial force can be calculated by

$$F = \frac{2\pi}{16} T \times \Delta p \times eff \qquad (12.57)$$

where

T = torque

F = axial force (lb)

Δp = the leadscrew pitch

eff = the screw efficiency: ball screw: 85–95%, ACME screw: 35–45%.

The leadscrew inertia can be calculated by

$$I_{screw} = D^4 \times \text{length} \times 0.028 \tag{12.58}$$

12.9.3 Control System

Figure 12.9 illustrates one axis of an NC system. In the figure, a gear box is used between the motor and the position transducer. The gear box does not affect the resolution of the machine, because its influence is outside the position transducer and leadscrew loop.

A typical DC motor configuration is shown in Figures 9.25 and 9.26 in Chapter 9. The defining Equations, 9.24 and 9.25, can be found in Section 9.3.2 and are

$$J\ddot{\theta} + b\dot{\theta} = K_m i_a$$

$$Ri_a + L_a \dot{i}_a = v_a - K\dot{\theta}$$

Figure 9.32 shows a block diagram for the entire single-axis control based on the DC servo motor and encoder. This block diagram corresponds to the closed-loop control system shown in Figure 12.15. In the block diagram, the output of the motor $\varpi = \dot{\theta}$. The relevant equations are

$$J\dot{\varpi} + b\varpi = K_t i_a \tag{12.59}$$

$$L_a \dot{i}_a + R_a i_a = v_a - K_e \varpi \tag{12.60}$$

From Equation 12.59,

$$i_a = \frac{J}{K_t}\dot{\varpi} + \frac{b}{K_t}\varpi \tag{12.61}$$

and

$$\dot{i}_a = \frac{J}{K_t}\ddot{\varpi} + \frac{b}{k_t}\dot{\varpi} \tag{12.62}$$

Substituting Equations 12.61 and 12.62 into Equation 12.60 yields

$$\frac{JL_a}{K_t}\ddot{\varpi} + \frac{bL_a}{K_t}\dot{\varpi} + \frac{R_a J}{K_t}\dot{\varpi} + \frac{R_a b}{K_t}\varpi = v_a - K_e \varpi$$

$$\frac{JL_a}{K_t}\ddot{\varpi} + \frac{bL_a + R_a J}{K_t}\dot{\varpi} + \frac{R_a b + K_e K_t}{K_t}\varpi = v_a$$

$$s^2\frac{JL_a}{K_t}\varpi(s) + s\frac{bL_a + R_a J}{K_t}\varpi(s) + \frac{R_a b + K_e K_t}{K_t}\varpi(s) = v_a(s)$$

$$G(s) = \frac{\varpi(s)}{v_a(s)} = \frac{K_t}{JL_a s^2 + (bL_a + R_a J)s + (R_a b + K_e K_t)}$$

Assume that both the impedance L_a and the friction b are very low; that is, $L_a = b = 0$. Then the last equation can be rewritten as

$$G(s) = \frac{K_t}{R_a J s + K_e K_t} = \frac{1/K_e}{\dfrac{R_a J}{K_e K_t} s + 1}$$

Let $\tau = \dfrac{R_a J}{K_e K_t}$, $K = 1/K_e$, and the transfer function for the motor can written as

$$G(s) = \frac{K}{\tau s + 1}$$

This is the motor transfer function shown in Figure 8.32. τ is called the time constant of the motor. Further analysis of the system can be done using the information provided in Chapter 8.

12.10 SPINDLE

The spindle of a machine tool is a critical part of the machine. For most machine tools, the cutting tool is mounted to the spindle (in case of a lathe, the workpiece is attached to the spindle). It is used to deliver the cutting speed and cutting force. The spindle horsepower determines the material removal rate that a machine can deliver. The cutting speed is also limited by the rated spindle rpm. The run-out of the spindle assembly is also a major factor of machining error. A spindle is either driven directly or through a belt by a DC-servo motor. The motor control is no different than that used in axis motor control. The two major differences are the high power and high rpm of the spindle motor compared with the axis motor. A typical spindle motor is rated as 5–40 hp and spins below 8000 rpm. Spindles that rotate at more than 20,000 rpm are considered high-speed spindles. They are commonly used in the automobile industry for cutting aluminum parts for high-cutting-speed applications. For printed circuit board applications, due to the small size of the holes drilled, the spindle speed can be up to 100,000 rpm. As discussed in Chapter 6, the cutting speed is calculated by

$$V = \frac{\pi D N}{12}$$

where D is very small, and the spindle needs a high rpm (N) to generate the desired cutting speed. To drill a 0.02″-diameter hole at 200 fpm, the spindle rpm is $N = 12 \times 200/(\pi \times 0.02) = 38{,}197$.

Cutting tools for NC are mounted in a tool holder. The shank of the tool holder is inserted into the spindle. In order to have good surface contact to deliver the cutting power, both the shank and the spindle bore are tapered. Morse taper (Figure 12.37) is used as the standard. The greater the taper number, the larger is the diameter. It is obvious that higher-powered machines use a higher-number Morse taper on their spindles.

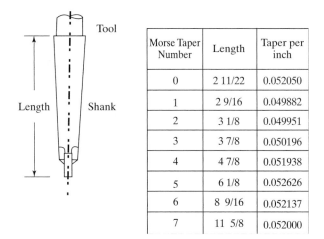

Morse Taper Number	Length	Taper per inch
0	2 11/22	0.052050
1	2 9/16	0.049882
2	3 1/8	0.049951
3	3 7/8	0.050196
4	4 7/8	0.051938
5	6 1/8	0.052626
6	8 9/16	0.052137
7	11 5/8	0.052000

Figure 12.37   Morse taper for cutting tools and spindles.

12.11 SUMMARY

In this chapter, the basic technologies used to build an NC machine were discussed. The purpose was to provide a general understanding of how NC machines work. It is important to understand the levels of controls required to achieve the final goal. The kinematics transformation converts the desired tool path in the Cartesian coordinate system into machine-specific axis control. For all axes, the desired axis motions are co-ordinated by an interpolator. Finally, the individual axis control uses a motor, lead-screw, and transducers. With this basic understanding of the machine tool structure and control, many machines encountered on the shop floor can be analyzed. The chapter also explained how to design a simple machine for a specific application. The addition-al knowledge of specific NC system components and the hardware–software interface can be found in handbooks and vendor's literature.

12.12 KEYWORDS

Accuracy and repeatability
Articulated machines
Backlash
Ball-bearing leadscrew
BLU
Cartesian machine
Closed loop
Continuous path
Control-loop unit (CLU)
Data-processing unit (DPU)
DC or AC servo motor
DDAs
Digital-to-analog converter
Five-axis machines
Forward kinematics

Hexapod machines
Incremental and absolute positioning
Interpolation
Inverse kinematics
Kinematics
Leadscrew
Machine-control unit (MCU)
Morse taper
Open loop
Pitch of the leadscrew
Point to point (PTP)
Position transducer
Spindle
Stewart platform
View transformation

12.13 REVIEW QUESTIONS

12.1   What are the major components of an NC machine system? Briefly describe the function of each.

12.2   What is point-to-point motion control? Contouring motion control? Describe the shapes of workpieces that can be machined by each of these control methods.

12.3   Write a computer program for the circular interpolation algorithm described in Section 12.7. The program should accept the following input: the location of the center point, the radius, the type of tolerance specification, and the maximum tolerance for the given tolerance type. The output should include the coordinates of all the interpolated points.

12.4   What are the major power sources for NC machines? Describe the advantages and disadvantages of each.

12.5   Define accuracy and repeatability.

12.6   Discuss the differences between an open-loop NC system and a closed-loop NC system.

12.7   What is a machining center? What is a turning center?

12.8   What are the sources of error in NC machining?

12.9   Modern CNC machines uses AC servo motors for axis control and DC servo motors for spindle drive control. What are the reasons behind such choices? What are the pros and cons of using an AC servo or a DC servo?

12.14 REVIEW PROBLEMS

12.1   Consider the four points p1, p2, p3, and p4, with coordinates $(10, 4, 9)$, $(29, 4, -10)$, $(-9, 21, 10)$, and $(4, 10, 5)$, respectively. It is desired to move the tool of an NC machine in a p1 $\rightarrow$ p2 $\rightarrow$ p3 $\rightarrow$ p4 sequence. Write down the corresponding X-, Y-, and Z-axis dimensions in an NC parts program if an incremental positioning system is used.

12.2   Given the data from Review Problem 12.1, what would be the X-, Y- and Z-axis dimensions for an absolute positioning system?

12.3   Two pulse dividers are used to control an X—Y table. One BLU equals 0.001 in. The input frequency is 1×10^4 Hz, and the word length is 10 bits. In order to move the table from $(0,0)$ to $(4,5)$ (both coordinates are in inches) at a speed of 8 ipm, what P values should be loaded into the X and Y pulse dividers? How many pulses should be sent to each axis?

12.4   A DC motor is connected directly to an external load. The moment of inertia of the load is 2 lb-m-s^2. Let $K_b = 0.9$ volt-S/rad, $R_a = 0.4$ ohm, $J_m = 0.4$ lb-m-s^2, $K_m = 8$ lb-in./A, and $f_m = f_1 = 0$. After applying 20 volts to the motor, what is the steady-state rotational speed of the motor? What is the time constant?

12.5   Design an NC system that yields a resolution of 0.0001 in. and a maximum feed speed (linear speed on the X–Y axis) of 100 fpm. Specify the leadscrew pitch, encoder resolution, pulse-divider word length, and pulse-divider input frequency.

12.6   Referring to Figure 12.35, derive $\omega(t)$ as a function of V_d. Compare the time constant and the steady-state angular speed of a system with a tachometer with a system without one, see Equation 12.44.

12.7   Write a computer program to simulate a DDA-based linear interpolator.

12.8   Write a computer program to simulate a DDA-based circular interpolator. (*Hint*: Use a loop to simulate the clock cycle. Use an IF statement to check whether a desired position has been reached.)

12.9 An NC machine uses a 200-pulse/rev. encoder and a l0-teeth/in. leadscrew for its axis control. What is the BLU of the machine?

12.10 An NC machine has a resolution (BLU) of 0.001 in. To move the cutter from coordinates (1,1) to (3,4) at 18 ipm, how many pulses and at what rate would the controller send to the X and Y servo motors?

12.11 Show a DDA-based linear interpolator for the NC machine described in Review Question 12.17. What p values should be loaded in the feed-rate control, and what p values should be loaded in the X and Y DDAs?

12.12 Design a circular interpolator (for the X and Y axes). Prove it works using a set of data you have chosen. You may do it by hand or with a computer. The interpolator can use only addition, subtraction, multiplication, and division operators. No trigonometric function is allowed.

12.13 Use a block diagram to show the structure of a closed-loop NC system (one axis only). Begin with the reference signal from the interpolator and include all system components (motor, transducers, and electronic and mechanical elements).

12.14 A 2-D position table is driven by stepping motors that have a 1.8° stepping angle. The leadscrews have 0.1″ pitch. What is the BLU of the system? In order to cut from (0,0) to (4,5) at 20 ipm, what should be the pulse rates and pulse counts for each motor?

12.15 A five-axis machine has the same configuration as the one shown in Figure 12.23. The machine design parameters are as follows:

$$l_t = 10 \quad \Delta x_{gm} = -10 \quad \Delta y_{gm} = 0 \quad \Delta z_{gm} = 30$$
$$\Delta x_{mr} = 30 \quad \Delta y_{mr} = 0 \quad \Delta z_{mr} = 0$$
$$\Delta x_{rb} = 5 \quad \Delta y_{rb} = 10 \quad \Delta z_{rb} = 5$$
$$\Delta x_{ba} = -5$$
$$\Delta x_{aw} = -2 \quad \Delta y_{aw} = -5 \quad \Delta z_{aw} = -2$$

Forward kinematics:

Let $x = 10$, $y = 5$, $z = 0$, $\beta = 15°$, $\alpha = 30°$. What is the tool-tip position with respect to the workpiece? In other words, what are the tool tip coordinates in the workpiece coordinate system?

Inverse kinematics:

The desired tool-tip position with respect to the workpiece is (1,4,0) and the tool orientation vector is [1,1,1]. What programming parameters should be used to control the machine? That is, what are x, y, z, β, and α?

12.16 Use the hexapod machine configuration given in Example 12.1 to find the strut lengths at $P_t = [1 \quad 2 \quad -50]^T$ and $P = [1 \quad 0 \quad -48]^T$. What is the tool length?

12.17 A new three-axis machine is invented by your professor. The machine design is shown in Figure 12.38.

The linear joints on the machine are controlled by a leadscrew and stepping-motor mechanism. The leadscrew has 0.1″ pitch and the stepping motor has 3.6° step angle. The rotational join is controlled by a DC-servo motor with a rotary encoder. The encoder has a resolution of 400 pulses/rev. The tool is 6″ long and 0.25″ diameter. The machine coordinate frame is set at the lower left corner on the machine table as shown.

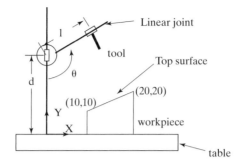

Figure 12.38

(a) In order for the flat-end milling tool to cut the top surface, what should be the d, l, and θ values at the start and the end positions?

(b) How many pulses do you need to send to the stepping motors and the servo motor in order to move the tool from the start position to the end position? Stepping motor 1 controls d, stepping motor 2 controls l, and the servo motor controls θ.

12.15 REFERENCES

Bedi, S., I. Ali, and N. Quan. "Advanced Interpolation Techniques for NC Machines," *ASME Journal of Engineering for Industry*, 115, 1993, 329–336.

Chou, J–J., and D. C. H. Yang. "On the Generation of Coordinated Motion of Five-Axis CNC/CMM Machines," *ASME Journal of Engineering for Industry*, 114, 1992, 15–22.

Electronic Association of America. *Axis and Motion Nomenclature for Numerically Controlled Machines*. Washington, DC: EIA. American National Standard, 1990.

Koren, Y. *Computer Control of Manufacturing Systems*. New York: McGraw-Hill, 1983.

Koren, Y., C–C. Lo, and M. Shpitalni. "CNC Interpolators: Algorithms and Analysis," *ASME Manufacturing Sciences and Engineering*, PED. 64, 1993, 83–92.

Lim, F. S., Y. S. Wong, and M. Rahman. "Circular Interpolators for Numerical Control: A Comparison of the Modified DDA Techniques and an LSI Interpolator," *Computers in Industry*, 19, 1992, 41–52.

Papaioannou, S. G. "Interpolation Algorithms for Numerical Control," *Computers in Industry*, 1, (1), 1979, 27–40.

Reintjes, T. F. *Numerical Control: Making a New Technology*. London: Oxford University Press, 1991.

Rolt, L. T. C. *A Short History of Machine Tools*. Cambridge, MA: The MIT Press, 1967.

Stewart, D. (1965) "A Platform with Six Degrees of Freedom," *Proceedings of the Institution of Mechanical Engineers*, Pt. I, 180 (15), 371–386.

Donmez, M. A., D. S. Blomquist, R. J. Hocken, C. R. Liu, and M. M. Barash, "A General Methodology for Machine Tool Accuracy Enhancement by Error Compensation," *Precision Engineering*, Publication No. 0141-6359/86/040187-10. 1986.

Donmez, M. A., C. Kang Lee, Richard Liu, and Moshe M. Barash. "A Real-Time Error Compensation System for a Computerized Numerical Control Turning Center," *Proceedings of the IEEE International Conference on Robotics and Automation*, San Francisco, CA, April 1986.

Chapter 13

Numeric-Control Programming

Objective

Building on Chapter 12, which covered the hardware aspects of numeric control, this chapter introduces the different levels of NC programming. CAD part programming and NC tool-path verification will be discussed. Mastery of these techniques does require some knowledge of the subject.

Outline

13.1 NC Part Programming
13.2 Manual Part Programming
13.3 Computer-Assisted Part Programming
13.4 CAD Part Programming
13.5 NC Cutter-Path Verification
13.6 Analytical Geometry for Part Programming
13.7 Summary
13.8 Keywords
13.9 Review Questions
13.10 References
Appendix: Computer-Assisted Part Programming Languages

13.1 NC PART PROGRAMMING

13.1.1 Coordinate Systems

In an NC system, each axis of motion is equipped with a separately controlled driving source that replaces the handwheel of a conventional machine. The driving source can

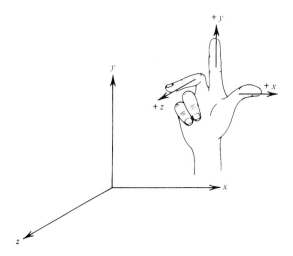

Figure 13.1   A right-hand coordinate system (Based on Y. Koren, *Computer Control of Manufacturing Systems*, McGraw-Hill, 1983. Reproduced with permission of McGraw-Hill Companies).

be a DC motor, a stepping motor, or a hydraulic actuator. The source selected is determined mainly based on the precision requirements of the machine.

The relative movement between tools and workpieces is achieved by the motion of the machine-tool slides. The three main axes of motion are referred to as the X, Y, and Z axes. The Z axis is perpendicular to both the X and Y axes in order to create a right-hand coordinate system, as shown in Figure 13.1. A positive motion in the Z direction moves the cutting tool away from the workpiece. This is detailed as follows:

Z Axis

1. On a workpiece-rotating machine, such as a lathe, the Z axis is parallel to the spindle, and the positive motion moves the tool away from the workpiece (Figure 13.2).
2. On a tool-rotating machine, such as a drilling or milling machine, the Z axis is parallel to the tool axis, and the positive motion moves the tool away from the workpiece (Figures 13.3 and 13.4).
3. On other machines, such as a press, a planing machine, or shearing machine, the Z axis is perpendicular to the tool set, and the positive motion increases the distance between the tool and the workpiece.

X Axis

1. On a lathe, the X axis is the direction of tool movement, and the positive motion moves the tool away from the workpiece.
2. On a horizontal milling machine, the X axis is parallel to the table.
3. On a vertical milling machine, the positive X axis points to the right when the programmer is facing the machine.

The Y axis is the axis left in a standard Cartesian coordinate system.

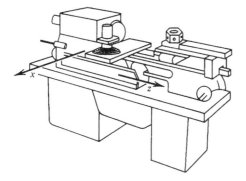

Figure 13.2 Coordinate system for a lathe (Based on Y. Koren, *Computer Control of Manufacturing Systems*, McGraw-Hill, 1983. Reproduced with permission of McGraw-Hill Companies).

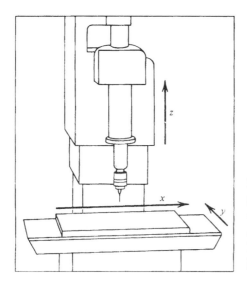

Figure 13.3 Coordinate system for a drill. (Based on Y. Koren, *Computer Control of Manufacturing Systems*, McGraw-Hill, 1983. Reproduced with permission of McGraw-Hill Companies).

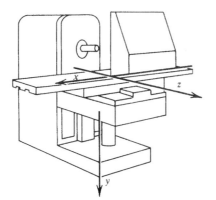

Figure 13.4 Coordinate system for a mill. (Based on Y. Koren, *Computer Control of Manufacturing Systems*, McGraw-Hill, 1983. Reproduced with permission of McGraw-Hill Companies).

13.1.2 NC Program Storage Media

During the numerical-control process, numerals and symbols representing the motion of the NC machine, as well as manual operating commands, are passed to the machine-control unit (MCU). Numerical data and symbols are represented by holes in punched tape or computer cards, magnetized domains on magnetic tape, or electronic impulses sent via computer networks. As long as the communication media are defined, an MCU can be designed to translate the information contained within the media and actuate servomechanisms to perform the required tasks.

The initiation of the NC process takes place by communication of data and symbols to the control unit. The ideal medium for communication and data-storage packs information into a dense, easily interpreted code that, after input to a reader, is sent to the MCU at high speeds.

There are several basic types of NC part program storage media: punched tape, magnetic tape, mylar tape, floppy disk, and a direct communication link [Pressman and Williams, 1979].

Floppy disks and a variety of data storage and transfer media are widely used in personal computers. CNC controllers are microcomputer-based devices, many of which use a standard personal-computer platform. Logically, floppy disks have been adopted for NC part program storage. Because several computer-assisted part programming software run on personal computers, data exchange is made easier, and database management is enhanced using standard computing platforms.

Modern CNC controllers provide several ways of transferring data. Perhaps the most typical data-communication methods used to transfer part program files is an RS-232C interface (see Chapter 10). An NC part program is stored in a file on a computer or a CNC controller. The file download (or upload) can be initiated by setting up a transfer mode on the CNC controller. On the other side of the communication cable is a computer that sends or receives data byte by byte. The operator must start and end the data-transfer process on both the CNC controller and the computer. Some machines use higher-level protocols to ensure an error-free data transfer. Two of the higher-level protocols used are Kermit and Xmodem. Kermit and Xmodem are widely accepted in the computer-to-computer telecommunication file-transfer process. These protocols allow the file transfer to be controlled by either the computer or the controller. The computer can send and retrieve data directly. Some machines also provide local-area networks (LAN) instead of serial communication. Ethernet and MAP are two technologies used. Some CNC controllers allow the entire controller function to be initiated from a remote computer through the data-communication network.

13.1.3 Symbolic Codes

A BCD (binary-coded decimal) or ASCII (American Standard Code for Information Interchange) code is frequently used in NC applications.

- BCD: An eight-track punched tape is one of the more common input media for NC systems. Hence, all data in the form of symbols, letters, and numbers must be representable by eight binary fields. The BCD code has been devised to satisfy this requirement. In a BCD code, the numerals 0 through 9 are specified using only the first four tracks, quantities 1, 2, 4, and 8. Note that the four numbers 1, 2, 4, and 8 are added together as needed to make all numbers from 1 to 15. Letters, symbols, and special instructions are indicated by using tracks 5 through 8 in combination with the numeral tracks. A complete BCD character set, based on EIA Standard RS-244A, is illustrated in Figure 13.5. Each BCD character must have an odd number of holes. By punching a parity bit along with all even bit strings, all characters have an odd number of holes. If an even number of holes is detected, it is by definition an error, and a parity check occurs. This simple method provides some protection from input errors resulting in part damage.

- ASCII: ASCII was formulated to standardize punched-tape codes regardless of the application [Pressman and Williams, 1979]. Hence, ASCII is used in computer and telecommunications as well as in NC applications. ASCII code was devised to support a large character set that includes uppercase and lowercase letters and additional special symbols not used in NC applications. Figure 13.5 illustrates the ASCII subset applicable to NC. Many new control systems now accept both BCD and ASCII codes. It is likely that the move toward ASCII standardization will progress as older NC equipment is replaced.

13.1.4 Tape Input Formats

The organization of words within blocks is called the tape format (EIA Standard RS-274) [Groover, 1980]. Four tape formats are used for NC input [Pressman and Williams, 1979]:

1. The fixed sequential format requires each NC block to be the same length and contain the same number of characters. This restriction enables the block to be divided into substrings corresponding to each of the NC data types. Because block length is invariant, all values must appear even if some types are not required.

2. The block-address format eliminates the need for specifying redundant information in subsequent NC blocks through the specification of a change code. The change code follows the block sequence number and indicates which values are to be changed relative to the preceding blocks. All data must contain a predefined number of digits in this format.

3. The tab sequential format derives its name because words are listed in a fixed sequence and separated by depressing the tab key (TAB) when typing the manuscript on a Flexowriter. Two or more tabs immediately following one another indicate that the data that would normally occupy the null locations are redundant and have been omitted. An example of tab sequential NC code is

```
T001 T01 T07500 T06250 T10000 T612 T718 T T EOB
T002 T T08725 T06750 T T T T EOB
T003 T T T T05000 T520 T620 T01 T EOB
```

where T represents a tab character.

4. The word-address format places a letter preceding each word and is used to identify the word type and to address the data to a particular location in the controller. The X prefix identifies an *X*-coordinate word, an S prefix identifies

Character	8	7	6	5	4		3	2	1	Character	8	7	6	5	4		3	2	1	Meaning
ISO code										EIA code										
0			o	o		•				0			o			•				Numeral 0
1	o		o	o		•			o	1						•			o	Numeral 1
2	o		o	o		•		o		2						•		o		Numeral 2
3			o	o		•		o	o	3				o		•		o	o	Numeral 3
4	o		o	o		•	o			4						•	o			Numeral 4
5			o	o		•	o		o	5				o		•	o		o	Numeral 5
6			o	o		•	o	o		6				o		•	o	o		Numeral 6
7	o		o	o		•	o	o	o	7						•	o	o	o	Numeral 7
8	o		o	o	o	•				8					o	•				Numeral 8
9			o	o	o	•			o	9				o	o	•			o	Numeral 9
A		o				•			o	a		o	o			•			o	Address A
B		o				•		o		b ?		o	o			•		o		Address B
C	o	o				•		o	o	c ?		o	o	o		•		o	o	Address C
D		o				•	o			d		o	o			•	o			Address D
E	o	o				•	o		o	e		o	o	o		•	o		o	Address E
F	o	o				•	o	o		f		o	o	o		•	o	o		Address F
G		o				•	o	o	o	g		o	o			•	o	o	o	Address G
H		o			o	•				h ?		o	o		o	•				Address H
I	o	o			o	•			o	i		o	o	o	o	•			o	Address I
J	o	o			o	•		o		j ?		o		o		•			o	Address J
K		o			o	•		o	o	k		o		o		•		o		Address K
L	o	o			o	•	o			l		o				•		o	o	Address L
M		o			o	•	o		o	m		o		o		•	o			Address M
N		o			o	•	o	o		n		o				•	o		o	Address N
O	o	o			o	•	o	o	o	o		o				•	o	o		Address O
P		o		o		•				p		o		o		•	o	o	o	Address P
Q	o	o		o		•			o	q		o		o	o	•				Address Q
R	o	o		o		•		o		r		o			o	•			o	Address R
S		o		o		•		o	o	s			o	o		•		o		Address S
T	o	o		o		•	o			t			o			•		o	o	Address T
U		o		o		•	o		o	u			o	o		•	o			Address U
V		o		o		•	o	o		v ?			o			•	o		o	Address V
W	o	o		o		•	o	o	o	w			o			•	o	o		Address W
X	o	o		o	o	•				x			o	o		•	o	o	o	Address X
Y		o		o	o	•			o	y ?			o	o	o	•				Address Y
Z		o		o	o	•		o		z			o		o	•			o	Address Z
DEL	o	o	o	o	o	•	o	o	o	Del	o	o	o	o	o	•	o	o	o	Delete (cancel an error punch).
NUL						•				Blank						•				Not punched. Can not be used in significant section in EIA code.
BS	o				o	•				BS			o		o	•		o		Back space
HT					o	•			o	Tab		o	o	o	o	•	o	o		Tabulator
LF or NL					o	•		o		CR or EOB	o					•				End of block
CR	o				o	•	o		o											Carriage return
SP	o		o			•				SP						•				Space
%	o		o			•	o		o	ER				o		•		o	o	Absolute rewind stop
(			o		o	•				(2-4-5)				o	o	•		o		Control out (a comment is started)
)	o		o		o	•			o	(2-4-7)		o			o	•		o		Control in (the end of a comment)
+			o		o	•		o	o	+		o	o	o		•				Positive sign
–			o		o	•	o		o	–		o				•				Negative sign
:			o	o	o	•		o												Colon
/	o		o		o	•	o	o	o	/			o	o		•			o	Optional block skip
.			o		o	•	o	o		.			o	o	o	•	o	o		Period (A decimal point)
#	o		o			•		o	o											Sharpe
$			o			•	o													Dollar sign
&	o		o			•	o	o		&				o		•	o	o	o	Ampersand
'			o			•	o	o	o											Apostrophe
*	o		o		o	•		o												Asterisk
,	o		o		o	•	o													Comma
;	o		o	o	o	•		o	o											Semicolon
<			o	o	o	•	o													Left angle bracket
=	o		o	o	o	•	o		o											Equal
>	o		o	o	o	•	o	o												Right angle bracket
?			o	o	o	•	o	o	o											Question mark
@	o	o				•														Commercial at mark
"			o			•		o												Quotation
[	o	o		o	o	•		o	o											Left brace
]	o	o		o	o	•	o		o											Right brace

Figure 13.5   EIA and ASCII codes for perforated tape used in NC applications (Courtesy of FANUC Ltd.).

spindle speed, and so on. The standard sequence of words in a block for a three-axis NC machine is

N word

G word

X word

Y word

Z word

F word

S word

T word

M word

EOB

A word-address NC code is

N001 G01 X07500 Y06250 Z10000 F612 S718 EOB

N002 X08752 Y06750 EOB

N003 Z05000F520 S620 M01 EOB

13.1.5 NC Words

A block of an NC part program consists of several words. A part program written in this data format is called a G-code program. A G-code program contains the following words:

$$N, G, X, Y, Z, A, B, C, I, J, K, F, S, T, R, M$$

Through these words, all NC control functions can be programmed. An EIA standard, RS-273, defines a set of standard codes. However, it also allows for the customization of certain codes. Even with this standard, there is still a wide variation of code formats. A program written for one controller often does not run on another. It is, therefore, essential to refer to the programming manual for the target machine before a program is written. In this section, before the meaning of each word is explained, we will first analyze the requirements of an NC control.

13.1.5.1. Basic Requirement of NC Machine Control.   To control a machine, it is necessary to begin by defining the coordinates of the tool motion. It is necessary to specify whether the motion is a positioning motion (rapid traverse) or a feed motion (cutting). The feed motion includes linear motion and circular motion. Linear motion requires the destination coordinates. When circular interpolation is used, the center of the circle must be given in addition to the destination. Before a cutting motion is called out, the spindle must be turned to the desired rpm and the feed speed must be specified. The spindle can rotate either clockwise or counterclockwise. Sometimes coolant is required in machining, and the coolant may be applied in flood or mist form. If an automatic tool changer is present, the next tool number has to be known to the controller before a tool can be changed to the machine spindle. The sequence to change the tool also needs to be specified. It is often desirable to aggregate a fixed sequence of operations such as drilling holes into a cycle. Using cycle codes can drastically reduce programming effort. Additional information is needed for specific cycle operations. Finally, there are other programming functions, such as units—inch or metric—positioning system—absolute or incremental, and so on. All of these activities can (and in some cases must) be controlled through the NC controller and related part program. These control functions and data requirements are summarized as follows:

(a) Preparatory functions: specify which units, which interpolators, which absolute or incremental programming, which circular interpolation plane, which cutter compensation, and so on.

(b) Coordinates: define three translational (and three rotational) axes.

(c) Machining parameters: specify feed and speed.

(d) Tool control: specifies tool diameter, next tool number, tool change, and so on.

(e) Cycle functions: specify drill cycle, ream cycle, bore cycle, mill cycle, and clearance plane.

(f) Coolant control: specify the coolant condition (i.e., coolant on–off, flood, and mist).

(g) Miscellaneous control: specifies all other control specifics (i.e., spindle on–off, tape rewind, spindle rotation direction, pallet change, clamps control, and so on).

(h) Interpolators: linear, circular interpolation, circle center, and so on.

These control functions are programmed through program words (codes).

13.1.5.2. NC Words.   A specific NC function may be programmed using an NC word or a combination of NC words. All functions can be programmed in one block of a program. Many CNC controllers allow several instances of the same "word" to be present in the same block. Thus, several functions can be included in one block. This is normally done by using a word-address format, which is the most popular format used in modern CNC controllers. The sequence of the words within one block is usually not important, except for the sequence number that must be the first word in the block. In order to make a program more readable, it is a good practice to follow a fixed sequence. Each word consists of a symbol and a numeral. The symbol is either N, G, X, Y, and so on. Numerals follow as data in a prespecified format. For example, the format for an X word might be "3.4," which means three digits before the decimal point and four digits after the decimal are used. The function of each NC word (code) and their applications are discussed in what follows:

> *N-code.* A part program block usually begins with an "N" word. The N word specifies the sequence number. It is used to identify the block within the program. It is especially useful for program editing. For example, when the format is "4," a proper sequence number would be

N0010

It is a good practice to program N values in increments of 10 or greater. This allows additional blocks to be inserted between two existing blocks.

G-code. The G-code is also called preparatory code or word. It is used to prepare the MCU for control functions. It indicates that a given control function is requested or

TABLE 13.1 G Codes

G00	Rapid traverse	G40	Cutter compensation—cancel
G01	Linear interpolation	G41	Cutter compensation—left
G02	Circular interpolation, CW	G42	Cutter compensation—right
G03	Circular interpolation, CCW	G70	Inch format
G04	Dwell	G71	Metric format
G08	Acceleration	G74	Full-circle programming off
G09	Deceleration	G75	Full-circle programming on
G17	X–Y plane	G80	Fixed-cycle cancel
G18	Z–X plane	G81–89	Fixed cycles
G19	Y–Z plane	G90	Absolute dimension program
		G91	Incremental dimension

that a certain unit or default should be taken. There are modal functions and nonmodal functions. Modal functions are those that do not change after they have been specified once, such as unit selection. Nonmodal functions are active in the block where they are specified. For example, circular interpolation is a nonmodal function. Some commonly used G-codes are listed in Table 13.1. Some of these functions will be explained next.

G00 is the rapid traverse code that makes the machine move at maximum speed. It is used for positioning motion. When G01, G02, or G03 are specified, the machine moves at the feed speed. G01 is linear interpolation; G02 and G03 are for circular interpolation. For circular interpolation, the tool destination and the circle center are programmed in one block (explained later). G04 (dwell) is used to stop the motion for a time specified in the block. G08 and G09 codes specify acceleration and deceleration, respectively. They are used to increase (decrease) the speed of motion (feed speed) exponentially to the desired speed. Before an abrupt turn, decelerate the tool. Rapid acceleration in the new direction may cause a tool to break. The best accuracy can be obtained with acceleration and deceleration codes on and set to lower values. Most NC controllers interpolate circles on only XY, YZ, and XZ planes. The interpolation plane can be selected using G17, G18, or G19. When a machine is equipped with thread-cutting capability (G33–G35), the part program must specify the proper way to cut the thread. Codes G40–G43 deal with cutter compensation. They simplify the cutter-center offset calculation. More details of cutter compensation are discussed later in Section 13.2. Most canned cycles are manufacturer-defined. They include drilling, peck drilling, spot drilling, milling, and profile-turning cycles. The machine-tool manufacturer may assign them to one of the nine G codes reserved for machine manufacturers (G81–G89). A user can also program the machine using either absolute (G90) or incremental (G91) coordinates. In the same program, the coordinate system can be changed. In order to simplify the presentation, most of the examples given in this chapter use absolute coordinates. Many controllers also allow the user to use either inch units (G70) or metric units (G71). Because hardwired NC circular interpolators work only in one quadrant and many CNC systems allow full-circle interpolation, a G74 code emulates NC circular interpolation for CNC controllers. G75 returns the CNC back to the full-circle circular interpolation mode.

X-, Y-, Z-, A-, B-, and C-Codes. These words provide the coordinate positions of the tool. X, Y, and Z define the three translational (Cartesian) axes of a machine. A, B, and C are used for the three rotational axes about the X, Y, and Z axes. For a three-axis machine, there can be only three translational axes. Most applications only require X, Y, and Z codes in part programs. However, for four-, five-, or six-axis machine tools, A, B, and C are also used. The coordinates may be specified in a decimal number (decimal programming) or integer number (BLU programming) format. For a controller with a data format of "3.4," to move the cutter to $(1.12, 2.275, 1.0)$, the codes are

```
X1.1200 Y2.2750 Z1.000
```

In BLU programming, the programmer may also need to specify leading zero(s), or trailing-zero formats. A leading-zero format means that zeros must be entered in the space proceeding the numeric value. In this format, the controller locates the decimal point by counting the digits from the beginning of a number. In trailing-zero format, it is reversed. The number specified is in the BLU unit. The data format "3.4" implies that a BLU equals 0.0001 in. (fourth decimal place). By using the data from the preceding example, the leading-zero program would be

```
X00112 Y002275 Z001
```

In the trailing-zero format, the program looks like

```
X11200 Y22750 Z10000
```

For circular motion, more information is needed. A circular arc is defined by the start and end points, the center, and the direction. Because the start point is always the current tool position, only the end point, the circle center, and the direction need to be specified. I, J, and K words are used to specify the center. Usually, circular interpolation works only on either X–Y, Y–Z, or X–Z planes. When interpolating a circular arc on the X–Y plane, the I word provides the X-coordinate value of the circle center and the J word provides the Y coordinate value. X and Y words specify the end point. Clockwise or counterclockwise motions are specified by the G-code (G02 versus G03). There are many variations in circular interpolation programming. Each NC controller vendor has its own form and format. Also, they can depend on the combination of absolute or incremental, full-circle on or off modes. The following example is based on absolute programming with full circle on for a hypothetical controller.

F-Code. The F-code specifies the feed rate of the tool motion. It is the relative speed between the cutting tool and the workpiece. It is typically specified in in./min. (ipm). From a machinability data handbook, feed is given in in./rev. (ipr). A conversion has to be done either by hand or on-board the controller. Some controllers offer a G-code that specifies the ipr programming mode. When the ipr programming mode is used, the tool diameter and the number of teeth must be specified by the operator. The F-code must be given before G01, G02, or G03 can be used. Feed speed can be changed frequently in a program, as needed. When an F-code is present in a block, it takes effect immediately. To specify a 6.00-ipm feed speed for the cutting motion in Figure 13.6, the program would read

```
N0100 G02 X7.000 Y2.000 I5.000 J2.000 F6.00
```

S-Code. The S-code is the cutting-speed code. Cutting speed is the specification of the relative surface speed of the cutting edge with respect to the workpiece. It is the result of the tool (or workpiece in turning) rotation. Therefore, it is programmed in rpm. The Machinability Data Handbook [Machinability Data Center, 1980] gives these

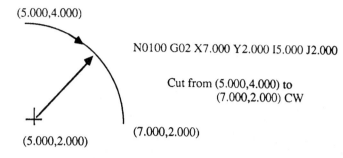

Figure 13.6   Cutting a circular arc.

values in surface feet per minute (sfpm), and conversion is required before programming is done. When a controller is equipped with an sfpm programming option, the operator must specify the tool diameter. The S-code is specified before the spindle is turned on. The S-code does not turn on the spindle. The spindle is turned on by an M-code. To specify a 1000-rpm spindle speed, the program block is

```
N0010 S1000
```

T-Code. The T-code is used to specify the tool number. It is used only when an automatic tool changer is present. It specifies the slot number on the tool magazine in which the next tool is located. Actual tool change does not occur until a tool-change M-code is specified.

R-Code. The R-code is used for cycle parameter. When a drill cycle is specified, the clearance height (R plane) must be given (see Figure 13.7). The R-code is used to specify this clearance height. In Figure 13.7, the drill cycle consists of five operations:

1. Rapid to location (1,2,2).
2. Rapid down to the R plane.
3. Feed to the *Z* point, the bottom of the hole.
4. Operation at the bottom of the hole, for example, dwelling.
5. Rapid or feed to either the R plane or the initial height.

The cycle may be programmed in one block, such as (cycle programming is vendor-specific)

```
N0010 G81 X1.000 Y2.000 Z0.000 R 1.300
```

M-Code. The M-code is called the miscellaneous word and is used to control miscellaneous functions of the machine. Such functions include turning the spindle on and off, starting and stopping the machine, turning the coolant on or off, changing the tool, and rewinding the program (tape) (Table 13.2). M00 and M01 both stop the machine in the middle of a program. M01 is effective only when the optional stop button on the control panel is pressed. The program can be resumed through the control panel. M02 marks the end of the program. M03 turns the spindle on (clockwise). The spindle rpm must be specified in the same line or in a previous line. M04 is similar to M03, except it turns the spindle on counterclockwise. M05 turns off the spindle. M06 signals the tool-change

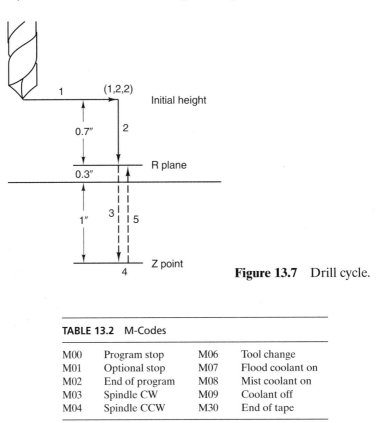

Figure 13.7 Drill cycle.

TABLE 13.2 M-Codes

M00	Program stop	M06	Tool change
M01	Optional stop	M07	Flood coolant on
M02	End of program	M08	Mist coolant on
M03	Spindle CW	M09	Coolant off
M04	Spindle CCW	M30	End of tape

operation. On a machine equipped with an automatic tool changer, it stops the spindle, retracts the spindle to the tool-change position, and then changes the tool to the one specified in the T-code. M07 and M08 turn on different modes of coolant. M09 turns off the coolant. M30 marks the end of the tape. It stops the spindle and rewinds the program (tape). On some controllers, more than one M-code is allowed in the same block.

13.2 MANUAL PART PROGRAMMING

13.2.1 Part Programs

In manual part programming, the machining instructions are recorded on a document, called a part-program manuscript (see Figure 13.8) by the part programmer. The manuscript is essentially an ordered list of program blocks. The manuscript is either entered as a computer file or punched on a paper tape. Each symbol on the manuscript, alphanumeric or special characters, corresponds to a perforation(s) on the tape (or magnetic bit pattern on a disk) and is referred to as a character. Each line of the manuscript is equivalent to a block on the punched tape and is followed by an EOB (end-of-block) character. When it is stored in a computer file, a tape-image format is used.

THE NC PART PROGRAMMING MANUSCRIPT

Part name _____					MANUSCRIPT CONTOURING PROGRAM			Prepared by _____ Date ____ Checked by _____ Date ____ Machine _____ Tape number _____					
Part number _____												
Sheet _____ _____												
Remarks _____												
n	g	x	y	z	i	j	k	f	s	t	m	REMARKS

Figure 13.8   NC part-program manuscript.

Because a part program records a sequence of tool motions and operations to produce the final part geometry, a process plan must be prepared with setups and fixtures before writing the program. The workpiece location and orientation, features (holes, slots, pockets) to be machined, and tools and cutting parameters used need to be determined. An example will illustrate how a part is programmed.

Example 13.1.

The part drawing shown in Figure 13.9 is to be machined from a 4-in. $\times$ 4-in. $\times$ 2-in. workpiece. The workpiece material is low-carbon steel. A hypothetical three-axis CNC machining center will be used for the process. The process plan for the part is as follows:

1. Set the lower-left bottom corner of the part as the machine zero point (floating-zero programming).
2. Clamp the workpiece in a vise.
3. Mill the slot with a $\frac{3}{4}$-in. four-flute flat, end mill made of carbide. From the Machinability Data Handbook, the recommended feed is 0.005 in./tooth/rev and the recommended cutting speed is 620 fpm.
4. Drill two holes with a 0.75-in.-diameter twist drill. Use 0.18-ipr feed and 100-fpm speed.

Figure 13.10 shows the setup, fixturing, and cutter path. Write a part program for the part.

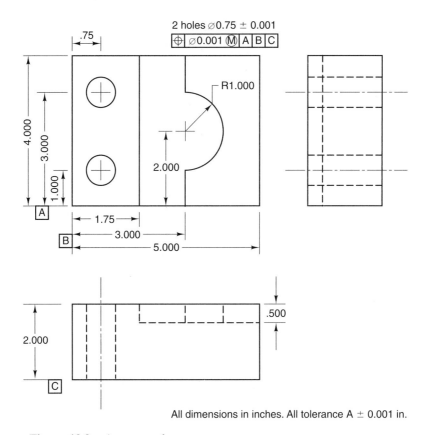

Figure 13.9   An example part.

Solution

The cutting parameters need be converted into rpm and ipm.

Milling:

$$\text{rpm} = \frac{12V}{\pi D} = \frac{12 \times 620 \text{ fpm}}{\pi \times 0.75 \text{ in.}} = 3157 \text{ rpm}$$

$$V_f = nf \text{ rpm} = 4 \text{ tpr} \times 0.005 \text{ iprpt} \times 3157 \text{ rpm} = 63 \text{ ipm}$$

Drilling:

$$\text{rpm} = \frac{12V}{\pi D} = \frac{12 \times 100 \text{ fpm}}{\pi \times 0.75 \text{ in}} = 509 \text{ rpm}$$

$$V_f = f \text{ rpm} = 0.018 \text{ ipr} \times 509 \text{ rpm} = 9.16 \text{ ipm}$$

For the milling operation, the cutter is smaller than the slot, and two passes are required.

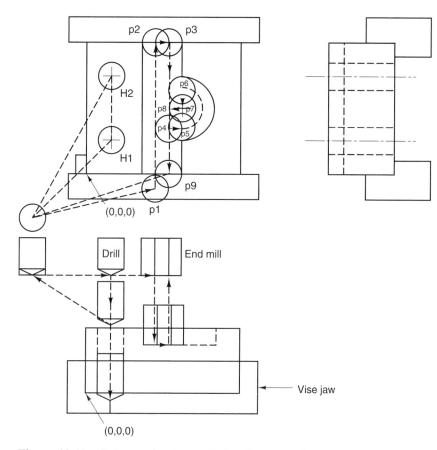

Figure 13.10   Setup and cutter path for the example part.

The cutter first moves to p1′ (the prime denotes the upper point). There must be clearance between the cutter and the workpiece, so that the cutter will not touch the workpiece during rapid positioning. We will use 0.1 in. as the clearance. The cutter then plunges down to p1, which is on the slot bottom level. Both p2 and p3 are outside the workpiece to ensure the slot edges are completely cut. The center of the cutter overhangs the edge by 0.1 in. The cutter moves to p4 from p3 to clear the slot. Note that p5 is the beginning of a circular interpolation and p6 is the end. From p6, the cutter moves to p7 to clear the center of the circular area, and then moves to p8. After the milling operation, a drill is installed in the spindle through an automatic tool change. The two holes are drilled using a drilling cycle.

The coordinates of each point (cutter location) are

p1′: $(1.75 + 0.375, -0.1 - 0.375, 400) = (2.125, -0.475, 4.000)$

p1: $(2.125, -0.475, 2.000 - 0.5000) = (2.125, -0.475, 1.500)$

p2: $(2.125, 4.000 + 0.100, 1.500) = (2.125, 4.100, 1.500)$

p3: $(3.000 - 0.375, 4.100, 1.500) = (2.625, 4.100, 1.500)$

Part program	Explanation
N0010 G70 G90 T08 M06	Set the machine to the inch format and absolute dimension programming.
N0020 G00 X2.125 Y-0.475 Z4.000 S3157	Rapid to p1'.
N0030 G01 Z1.500 F63 M03	Down feed to p1, spindle CW.
N0040 G01 Y4.100	Feed to p2.
N0050 G01 X2.625	To p3.
N0060 G01 Y.1375	To p4.
N0070 G01 X3.000	To p5.
N0080 G03 Y2.625 I3.000 J2.000	Circular interpolation to p6.
N0090 G01 Y2.000	To p7.
N0100 G01 X2.625	To p8.
N0110 G01 Y-0.100	To p9.
N0120 G00 Z4.000 T02 M05	To p9', spindle off, tool sign 2.
N0130 F9.16 S509 M06	Tool change, set new feed and speed.
N0140 G81 X0.750 Y1.000 Z-0.1 R2.100 M03	Drill hole 1.
N0150 G81 X0.750 Y3.000 Z-0.1 R2.100	Drill hole 2.
N0160 G00 X-1.000 Y-1.000 M30	Move to home position, stop the machine.

Figure 13.11   Part program for the part in Figure 13.9.

p4: $(2.625, 2.000 - 1.000 + 0.375, 1.500) = (2.625, 1.375, 1.500)$

p5: $(3.000, 2.000 - 1.000 + 0.375, 1.500) = (3.000, 1.375, 1.500)$

p6: $(3.000, 2.625, 1.500)$

p7: $(3.000, 2.000, 1.500)$

p8: $(2.625, 2.000, 1.500)$

p9: $(2.625, -0.100, 1.500)$

p9': $(2.625, -0.100, 4.000)$

Combining the information from the process plan and the cutter-location data, a part program can be written. The program for the example part is shown in Figure 13.11. A step-by-step explanation is presented on the right-hand side of the figure. The part program is verified using a program called Mac CNCS, and the results are shown in Figure 13.12. The result in Figure 13.12 is the same as that shown in Figure 13.10; thus, the program is correct. A three-dimensional view can also be found in Figure 13.13.

13.2.2 Tool-Radius Compensation

When machining a complicated workpiece, it is necessary to compensate (or position the tool further away from the desired cutting surface) in order to allow for the radius of the cutting tool. The geometry in Example 12.1 is relatively simple, and little calculation was required. However, when there are nonorthogonal lines or planes in

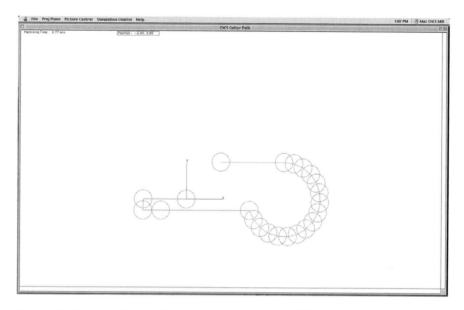

Figure 13.12   Verified cutter path (using Mac CNCS by T. C. Chang).

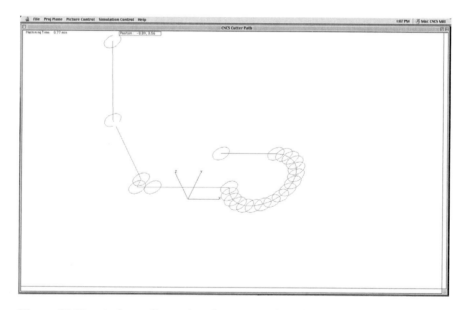

Figure 13.13   A three-dimensional cutter-path verification.

the drawing, the calculations can become more complicated. Also, when a slightly different size cutter is used instead of the programmed cutter, the program becomes invalid. In this case, a new program must be written. The tool-radius compensation feature of modern CNC machines can eliminate the tool-offset calculation for the finish cut. Tool-radius compensations make it possible to program directly from the drawing's measurements. The actual tool size is then "keyed into" the CNC controller prior to the operation. The CNC controller performs the offset calculations automatically. Tool-radius compensation only modifies the existing path. Because the number of roughing passes and their paths are determined by cutter size, tool-radius compensation cannot be used. Following are some typical tool-radius compensation functions:

- G40: cancel tool-radius compensation.
- G41: compensation—left; Figure 13.14(a). Assume that the cutter is on the left-hand side of the line. The direction is established by the tool-motion direction.
- G42: compensation—right; Figure 13.14(b).
- M96: additional block for external curves; Figure 13.14(c). The circular arc at the corner is inserted by the CNC controller.
- M97: go to the machining point when cutting external curves; Figure 13.14(d).

An example on how to use these functions follows.

Start of Compensation. If G41 (or G42) and G01 are in the same block, there will be a gradual effect of the compensation, as shown in Figure 13.15(a). This is known as a ramp-up block and takes place at block N0010:

N0010 G01 G42 X0.500 Y1.700
N0020 G01 X1.500

If G41 (or G42) and G01 are in blocks that are separate from X and Y, the compensation is effective from the start of the block, as shown in Figure 13.15(b):

N0010 G41
N0020 G01 X0.500 Y1.700
N0030 G01 X1.500

Inside Corner. When the cutter path determines the geometry of an inside corner, it stops at the inside cutting point, as shown in Figure 13.15(c):

N0010 G41
N0020 G01 X1.500 Y2.000
N0030 G01 X0.000 Y1.600

Use of M96 and M97. If a step is to be cut using a cutting tool that is larger than the height of the step, M97 must be used. If M96 is used, the cutter will roll over the corner and into the material [see Figure 13.15(d)]. The following code creates the error shown in Figure 13.15(d).

N0010 G41
N0020 G01 X1.000 Y1.000
N0030 G01 Y.800 M96
N0040 G01 X2.000

(a) G41   (b) G42

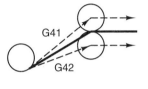

(c) M96   (d) M97

Figure 13.14   (a) Tool compensation—left, (b) tool compensation—right, (c) additional block for external curves, (d) go to the machining point when cutting external curves.

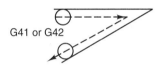

(a)   (b)

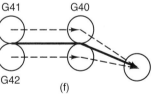

(c)   (d)

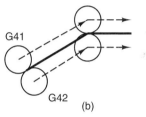

(e)   (f)

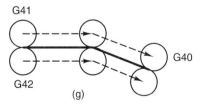

(g)

Figure 13.15   (a) Ramp on block, (b) tool compensation is effective from the start, (c) tool compensation at inside corner, (d) use of M96 improperly, (e) use of M97, (f) ramp off block, and (g) tool compensation is effective to the end point.

The correct program should be

    N0010 G41
    N0020 G01 X1.000 Y1.000
    N0030 G01 Y0.800 M97
    N0040 G01 X2.000

The result is shown in Figure 13.15(e).

Cancel Tool Compensation.   If G40 is in the same block as X and Y, there will be a gradual cancellation of compensation. This is known as a ramp-off block, as shown in Figure 13.15(f):

```
N0060 G40 X2.000 Y1.700 M02
```

If G40 is in a block following the last motion, the compensation is effective to the endpoint (2.000,1.700) (see Figure 13.15(g)):

```
N0060 X2.000 Y1.700
N0070 G40 M02
```

Example 13.2.

A 2.0-in. × 2.0-in. square is to be milled using a $\frac{1}{2}$-in. end milling cutter. The drawing is given in Figure 13.16. Write an NC part program to make the square.

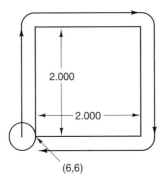

2.000

2.000

(6,6)

Figure 13.16   A 2 in. × 2 in. square is to be milled using a 0.5 in. end-milling cutter.

Solution

Let us set up the lower left corner of the square at (6.0,6.0). By using tool-radius compensation, the square can be produced by the program shown in Figure 13.17.

Part program	Explanation
N0010 G41 S1000 F5 M03	Begin compensation, set feed and speed, spindle on
N0020 G00 X6.000 Y6.000	Move to lower left corner
N0030 G01 Z-1.000	Plunge down the tool
N0040 Y8.000	Cut to upper left corner
N0050 X8.000	Cut to upper right corner
N0060 Y6.000	Cut to lower right corner
N0070 X6.000	Cut to lower left corner
N0080 Z1.000	Lift the tool
N0090 G40 M30	End compensation, stop the machine

Figure 13.17   Part program and explanation for Example 13.2.

13.3 COMPUTER-ASSISTED PART PROGRAMMING

In computer-assisted part programming, general-purpose computers are used as an aid in programming, and special-purpose, high-level programming languages perform the various calculations necessary to prepare the punched tape. The computer allows the economical programming of the machining of complex parts that could not be manually programmed economically. The part programmer's job is divided into two tasks: First, define the configuration of the workpiece in terms of basic geometric elements via points, lines, surfaces, circles, and so on. Second, direct the cutting tool to perform machining steps along these geometric elements. Programming languages that are capable of running on general-purpose computers have been developed. These languages are based on common English words and mathematical notations and are easy to use.

An NC processor is a computer application program that accepts as input user-oriented language statements that describe the NC operations to be performed. Figure 13.18 illustrates a generalized flow chart for most NC processors. Commonality among the many types of NC devices that exist is obtained by designing NC processors to produce a common interface code called CL (cutter-location) data. The CL data, in turn, are used as the input to another computer application program called a postprocessor, which produces the code for the particular NC device utilized. This output is normally in the form of punched tape for convenient storage and reading by the device's controller during the step-by-step execution of the operation.

There are two major classes of part-programming languages [Smith and Evans, 1977]:

1. *Machine-oriented languages* create tool paths by doing all the necessary calculations in one computer processing stage by directly computing the special coordinate-data format and the coding for speed and feed requirements.
2. *General-purpose languages* break down the computer processing into two stages, a processing stage and a postprocessing stage. The processing stage creates an intermediate set of data points called CL data. There are three steps in the processing stage.

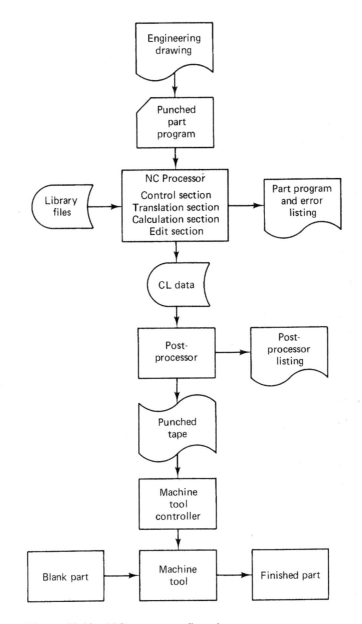

Figure 13.18   NC processor flowchart.

- *Translate input symbols.* This function translates symbolic inputs contained in the part program into a computer-usable form. It also establishes the link between the human operator and the computer.
- *Arithmetic calculation.* The arithmetic calculation unit performs geometric and trigonometric calculations required to generate the part surface.
- *Cutter offset calculations.* The cutter offset unit calculates the path of the center line of the cutter based on the part-outline information.

EIA RS-494B provides a new standard for computer numerical control that allows different machines to operate from the same input data. The name of this standard is "Standard for 32 Bit Binary CL Exchange (BCL) Input Format for Numerically Controlled Machines."

The BCL format represents NC machining input data as a series of records that are groups of 32-bit binary integer words. The content of these records closely parallels the content of CL data records defined in ANSI Standard X3.37 as prepared by ANSI Committee X3J7. The syntax and the semantics of records defined by this standard in most cases either conform to or are extensions of ANSI X3.37—1980 Programming Language APT. A BCL file is typically produced by a part-oriented postprocessor called a BCL converter. Its primary functions are to retrieve CL data records, select the relevant portions of the data, convert the floating-point data to 32-bit integer data, format data into records, and produce a BCL program file and program listing. It performs only machine-dependent functions. The BCL format structure is as follows:

FILE HEADER (optional)

PARTNO record (optional)

UNITS record

DATA record

DATA record

DATA record

END OF FILE record

Figure 13.19 illustrates a sample BCL program. In this program, corresponding APT commands are listed for cross-reference. The code is shown in hexadecimal.

At the postprocessing stage, the CL data are converted by the computer to make them specific for a particular machine-tool system and a punched tape (or floppy disk) is prepared. The output of the postprocessor is the NC tape (or the NC program on a floppy disk) written in an appropriate format for the machine in which it is to be used.

Since 1956, more than 100 NC part-programming languages have been developed. Some of them are special-purpose, machine-oriented languages. However, most of them are general-purpose. Some languages have stood the test of time, whereas many have not. We describe some of the more popular languages in the Appendix at the end of this chapter.

```
LINE  BCL code        interpretation          APT source statement
E.1                                           PARTNO BCL SAMPLE
      6415 0002  PARTNO (record no. 2)
      2042 434C  BCL
      2053 414D  SAM
      504C 4520  PLE
E.2                                           UNITS/INCHES
      9409 0004  UNITS (record no. 4)
      7FFF 012F  INCHES
E.3                                           SPINDL/RPM, 1200, CLW
      8407 0006  SPINDL (record no. 6)
      7FFF 004E  RPM
      0067 1600  12000000
      7FFF 003C  CLW
E.4                                           COOLNT/FLOOD
      8406 0008  COOLNT (record no. 8)
      7FFF 0059  FLOOD
E.5                                           RAPID
      8005 000A  RAPID (record no. 10)
E.6                                           GOTO/1,0,4.1
      9001 000C  GOTO (record no. 12)
      0000 2710  10000
      0000 0000  0
      0000 A028  41000
E.7                                           FEDRAT/FPM, 10
      83F1 000E  FEDRAT (record no. 14)
      7FFF 0142  FPM
      0001 86A0  100000
E.8                                           GOTO/1,0,1
      9001 0010  GOTO (record no. 14)
      0000 2710  10000
      0000 0000  0
      0000 2710  10000
E.9                                           RAPID
      8005 0012  RAPID (record no. 18)
E.10                                          GOTO/1.1,0,1
      9001 0014  GOTO (record no. 20)
      0000 2AF8  11000
      0000 0000  0
      0000 2710  10000
E.11                                          RAPID
      8005 0016  RAPID (record no. 22)
E.12                                          GOTO/1.1,0,4.1
      9001 0018  GOTO (record no. 24)
      0000 ZAF8  11000
      0000 0000  0
      0000 A028  41000
```

Figure 13.19   Listing of BCL codes.

```
LINE  BCL code        interpretation        APT source statement
E.13                                         RAPID
      8005 001A RAPID (record no. 26)
E.14                                         GOTO/.9,0,4.1
      9001 001C GOTO (record no. 28)
      0000 2328 9000
      0000 0000 0
      0000 A028 41000
E.15                                         THREAD/SCALE, 1000000,
      840C 001E THREAD (record no. 30)       LEAD, .05, TURN
      7FFF 0019 SCALE
      000F 4240 1000000
      7FFF 0146 LEAD
      0000 C350 50000
      7FFF 0050 TURN
E.16                                         THREAD/ON
      840C 0020 THREAD (record no. 32)
      7FFF 0047 ON
E.17                                         GOTO/.9,0,1.2
      9001 0022 GOTO                         (record no. 34)
      0000 2328 9000
      0000 0000 0
      0000 2EE0 12000
E.18                                         THREAD/OFF
      840C 0024 THREAD (record no. 36)
      7FFF 0048 OFF
E.19                                         RAPID
      8005 0026 RAPID (record no. 36)
E.20                                         GOTO/1.1,0,1.2
      9001 0028 GOTO (record no. 40)
      0000 2AF8 11000
      0000 0000 0
      0000 2EE0 12000
E.21                                         SPINDL/OFF
      8407 002A SPINDL (record no. 42)
      7FFF 0048 OFF
E.22                                         COOLNT/OFF
      8406 002C COOLNT (record no. 44)
      7FFF 0048 OFF
E.23                                         GOHOME
      8011 602E GOHOME (record no. 46)
E.24                                         END
      8001 0030 END (record no. 48)
E.25
      8000 0032 END OF FILE (record no. 50)
```

Figure 13.19   (*Continued*).

13.4 CAD PART PROGRAMMING

For a machine tool such as a mill or lathe, the part program describes the path that the cutter will follow, as well as the direction of rotation, rate of travel, and various auxiliary functions such as coolant flow. Traditionally, programs for NC machine tools have been created using one of the previously described methods; manual part programming or computer-assisted part programming. Simple programs are often created manually, perhaps with the aid of a calculator, and more complex programs are usually created using a computer and a part-programming language, such as APT. The manual method, while adequate for many simple point-to-point processes, requires the programmer to perform all calculations required to define the cutter-path geometry and can be time-consuming. Errors made by the programmer are often not discovered until the program is tested graphically or on the machine tool, and error correction can be cumbersome. Also, because most machine tools have their own language, the programmer is often working with different instruction sets, which further complicates part-program creation. The computer-assisted or part-programming language approach simplifies the process because the programmer uses the same language for each program, regardless of the target machine tool. The language processor, which translates the part-programming language (such as APT) into the NC machine tool's instructions, also performs most of the calculations needed to describe the cutter path. Errors, however, are often not discovered until the program is tested.

Although the computer-assisted approach offers advantages over the manual approach, both require the programmer to translate geometric information from one form (usually an engineering drawing) into another, which has a significant potential for errors in the process.

Creation of NC programs from CAD provides yet another option by allowing the part programmer to access the computer's computational capabilities via an interactive graphics display console. This allows geometry to be described in the form of points, lines, arcs, and so on, just as it is on an engineering drawing, rather than requiring a translation to a text-oriented notation. Use of a graphics display terminal also allows the system to display the resulting cutter-path geometry, allowing earlier verification of a program, which can avoid costly machine setups for program testing.

Most CAD/NC systems generally provide significant productivity gains. These systems allow the user to more rapidly define the geometry as well as to use powerful graphics display capabilities to quickly define, verify, and edit the actual cutter motion.

The part-programming operation typically starts with the receipt (by the manufacturing department) of a design in the form of a CAD/NC drawing or model. After a review by a production planner, the tool design/selection process is completed, often with the assistance of the CAD system.

The programmer, using the part's geometric description as created by the design department and, perhaps, other geometric data produced later by the manufacturing department (clamps, appliances, and so on), now begins the part-programming process. If the programmer was not using any CAD/NC system, the first step would be to define some or all of the part geometry. When using a CAD/NC system, this step is greatly simplified.

The programmer now generates a cutter path by selecting geometric elements with a digitizer (such as a light pen or mouse) attached to the display terminal. Various auxiliary and postprocessor commands are also entered at the terminal.

Once a cutter path is created, it can be replayed for verification. The computer assists the programmer by animating the entire path on the display terminal, showing the location of the cutter visually, and displaying the X, Y, and Z coordinates. Editing may be done interactively during the replay to correct errors or make changes to the program.

After verification, the various cutter passes are combined to form a program using a postprocessing command, which is provided by most of the CAD/NC systems. Information relating to the machine tool to be used is automatically extracted from the CAD/NC machine file on a storage medium such as magnetic disk and is merged with the program, which is then stored for processing by different postprocessors. Finally, the NC program is produced on punched tape or whatever medium is required by the NC machine tool.

Many manufacturers of discrete products have employed NC machine tools for decades. Whether it is a simple two-axis drill or a complex five-axis machining center, all NC machines require part programming. The preparation of paper tapes or tape images for NC machine tools is complex and time-consuming. It often takes 24 or even 500 hours to program and prepare a paper tape for a part on a machine tool [Knox, 1983]. Because part programmers have difficulty keeping up with the demand for new part programs, the tendency is to develop efficient procedures capable of replicating this human process and to install these procedures on a CAD/CAM system.

Several CAD/CAM systems, such as CADAM, Computervision, and CATIA, have the capability of generating NC machining instructions based on the geometric definition of a workpiece.

13.4.1 Computervision System

Computervision has a relatively powerful NC package, called NC Vision, for the generation of NC instructions. A rough description of the procedure to generate an NC tool path (face-milling operation) using NC Vision follows (the entire NC tool-path planning process is conducted in an interactive mode):

- Create a part drawing using CADD4.
- Select the NC planning function from the menu.
- Select a milling cutter from the tool library. The library contains a complete set of tools for turning, milling, drilling, countersinking, and several other operations.
- After entering the NC planning mode, the user answers a series of questions, such as tool geometry, stepover, approach type, retraction type, and tolerance. NC Vision then generates a tool path for the planned part.

NC Vision is able to assist the user in generating and verifying NC tool paths based on different algorithms.

13.4.2 CADAM System

The NC capability provided by CADAM is more limited than that of Computervision. For example, in milling a flat surface in CADAM, the user has to explicitly specify those points on which the tool stops or changes direction. CADAM then generates a tool path based on what the user has defined. As a result, the user is solely responsible for generating a "good" tool path.

13.4.3 CATIA System

CATIA utilizes a modeling scheme different from that used by CADAM and Computervision. Basically, CATIA is a CAD package that has the capability of generating NC tool paths internally. CATIA provides a utility function that allows the user to convert a CATIA file to a tool path using a sophisticated NC planning function.

13.4.4 NC Programmer

NC Programmer is an AutoCAD-based NC part-program generator. The NC instruction-planning capability of NC Programmer is relatively primitive. First, the user needs to create a part drawing using AutoCAD and export the drawing to an ASCII data file. Second, in the NC instruction-planning mode, the user has to explicitly "tell" NC Programmer the starting and the ending locations of each tool path using a pointing device such as a mouse. NC Programmer then generates an NC part program based entirely on the instructions by the user.

Despite the fact that many CAD/CAM systems provide the facility that is able to generate the NC tool path directly from CAD data, this area is far from fully developed. This can be addressed as follows:

1. *Geometric coverage.* The commercial CAD/NC packages are limited to very simple geometric shapes, such as 2-D peripheral cuts, pocket milling, and turned parts [Wang and Lin, 1987]. More complicated features demand human interpretation, which is time-consuming. In order to automate the CAD/NC interface, more sophisticated feature-recognition algorithms are required.
2. *Optimization of the NC tool path.* CAD/CAM systems are currently generating the cutter path for many NC operations. However, no mathematical model is available for creating an optimal path. By utilizing such a model, the minimum length of cut could be identified for all kinds of surface features. In the following section, a preliminary study on NC tool-path optimization, which was conducted exclusively for face-milling operations, is presented.

13.4.5 Interactive Part Programming

Interactive part programming can be done either on a CNC controller or on a personal computer. Because it is relatively easy to learn and inexpensive to purchase, many small shops opt for interactive part programming instead of a part-programming language or the more expansive CAD-based part programming. Interactive part programming is usually limited to $2\frac{1}{2}$ D parts.

Because CNC controllers are computer-based, some vendors add interactive or graphical part-programming functions to the controller. A machinist can enter the part program interactively on the controller. The part geometry can be entered using a keyboard and special function keys. After the part geometry is defined, the machinist can define the cutter and its path. The graphics display on the controller shows both the part geometry defined and the final tool path. The controller guides the user through each step, and little or no language syntax needs to be followed. Many advanced controllers provide this capability. One company, Hurco, produces controllers exclusively

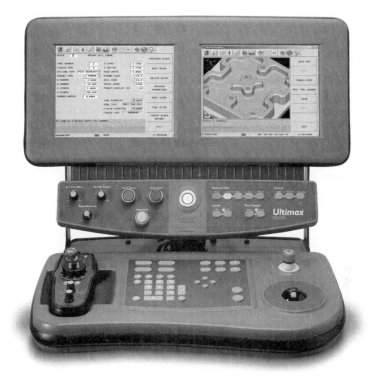

Figure 13.20   On controller programming-Hurco® UltiMax® 4. (Courtesy of Hurco Companies, Inc.).

using interactive part programming. Using dual displays, a programmer works on the geometric display in one window and observes the text on the other. The user is able to program the machine while the machine is actively cutting a workpiece. Figure 13.20 shows the Hurco CNC controller.

Another type of interactive part programming is graphical part-programming software such as SmartCAM. SmartCAM is capable of handling two-dimensional geometries. Part shape is defined using lines, arcs, chamfers, and so on, where geometric objects are selected from a menu. Endpoint coordinates of each geometric object are entered after the system prompt. The part geometry is assumed to be the 2-D profile swept in the Z direction. The geometry also can be imported from other CAD systems in DXF (see Chapter 4) format. The user can select menu items to change the direction of the cut, starting position, tool-path profile, and so on. The system is capable of generating cutter path for offset walls, simple pockets, pockets with islands, drill patterns, and turned profiles. Figure 13.21 shows the SmartCAM programming system.

13.4.6 Solid-Model-Based Cutter-Path Generation

Almost most of the cutter-path-generation packages use lines and arcs as the basic geometric entities (wire-frame model), some systems have begun to use solid models

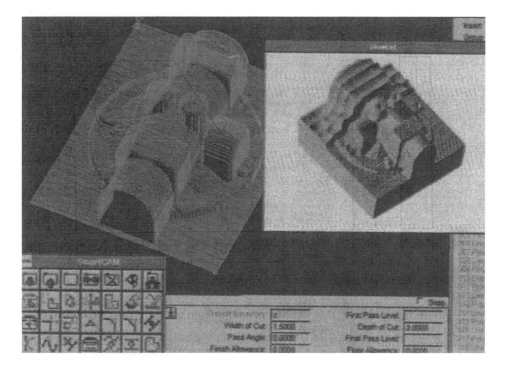

Figure 13.21   SmartCAM.

directly. When the cutter path is generated using a wire-frame model, there is no way of detecting interference between the tool and the part and the fixture. By using a solid model, the entire machining environment, including stock, part volume, tools, and fixtures, can be modeled and used to generate a collision-free cutter path. Gouge-avoidance algorithms can be used to check the entire tool and toolholder to ensure gouge-free machining. It is also possible to incorporate geometric reasoning rules and machining knowledge to select feeds, speeds, tools, and operation sequence. Machining knowledge is applied to each solid geometry feature to be machined. An optimal cutter-path algorithm can be written for each feature. Figure 13.22 shows a cutter-path simulation verifying a cutter path generated by the SDRC I-DEAS Master Series package. In the figure, it clearly shows the workpiece, fixtures, and the machine table. Any interference can be detected before the actual machining take place. Another example is a package called Strata by Spatial Technology. Strata is a feature-based NC cutter-path-generation package using an object-oriented solid modeler, ACIS. Strata contains optimum cutter-path-generation algorithms for many solid features.

13.5 NC CUTTER-PATH VERIFICATION

Before a part is machined, the part program needs to be verified. The purposes of verification are (1) to detect geometric errors of the cutter path, (2) to detect potential tool interference, and (3) to detect erroneous cutting conditions. Geometric error is the most common problem. The machined part geometry must agree with the design

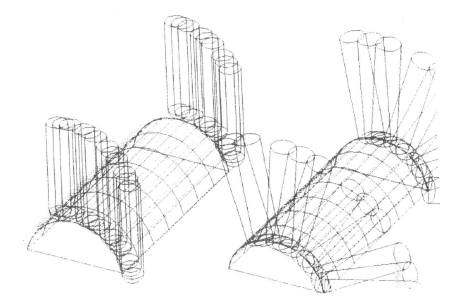

Figure 13.22   Solid-based machining from SDRC (Courtesy of UGS).

specification. When a tool follows the programmed cutter path, it may not produce the exact part surface the designer has specified. Although gross geometric errors do occur in some part programs (especially those written by less experienced part programmers), many errors occur because of local gauging and undercutting. These errors are more difficult to detect. Other types of errors are caused by the interference between the tool with the workpiece and the fixture elements. In three-axis machining, such errors often occur during rapid positioning motions. For five-axis machining, it can happen any time. It is difficult for the part programmer to detect the error during programming. The cutting condition is the specification of specific machining speed, feed, depth of cut, cutter geometry, and tool material on the workpiece material. The selection of tool and cutting parameters is experience-based. There is no guarantee that they are correct. In order to predict the cutting condition, the user must have an exact process model. Unfortunately, this is the area in which knowledge is least. In order to eliminate potential errors during actual machining, it is necessary that a part program be verified.

There are several ways to verify a part program. For instance, a dry cut can be made on the machine without the workpiece. A dry cut can detect gross programming mistakes, but not the detailed geometric ones. Another approach is to actually machine a prototype. Typically, a prototype part is machined in wax, machining plastic, wood, foam, or some other soft material. The actual geometry is then measured and compared with the design specification. Because material property is critical in determining the cutting condition, this approach can verify only the geometry.

When a part program is generated using a CAD-based system, a graphic output of the cutter path may be produced by the software. By visual inspection, cutter-path abnormalities may be detected. On a simple system, the cutter path is shown as a line drawing, sometimes overlayed on the part model. The exact final geometry is left for the user to imagine. In an advanced system, a solid model may be used to generate a realistic picture of the workpiece, tool, and the finished part. Real-time simulation of the cutting process can be displayed on screen. However, most simulations are purely geometry-based, where the cutting condition is not considered. One example of such a system is shown in Figure 13.23.

13.6 ANALYTICAL GEOMETRY FOR PART PROGRAMMING

13.6.1 Cutter-Center Location and Tool Offset

The following examples show the mathematical foundation of a computed cutter path. The method illustrated is implemented in MAPT. The method uses homogeneous coordinate transformation to convert a complex geometric program into an easier one.

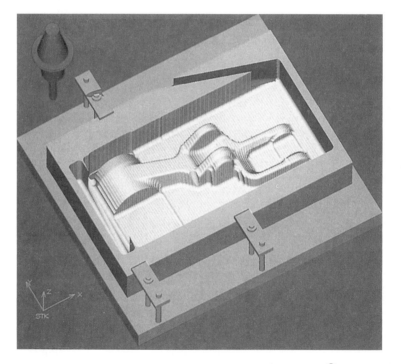

Figure 13.23 Cutter path verification using VERICUT®
(Courtesy CG Tech).

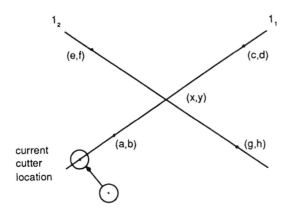

Figure 13.24 Calculating the cutter-center location.

Directions for a cutter motion and offset can be easily taken care of by this method. The idea may be extended to solve many other related problems.

Algebra provides the foundation for CL data generation. However, in implementing procedures to compute CL data, problems may occur. This is especially true when writing computer programs for the computational procedure. Consider the following example: Suppose that the goal is to find the cutter-center location after the following APT command is issued (see Figure 13.24):

$$\text{GO} \left\{ \begin{array}{c} \text{LFT} \\ \text{RGT} \end{array} \right\} /l_1, \left\{ \begin{array}{c} \text{TO} \\ \text{ON} \\ \text{PAST} \end{array} \right\}, l_2$$

By solving the following simultaneous equations, the X and Y coordinates can be obtained:

$$l_1: \frac{x - a}{y - b} = \frac{a - c}{b - d} \tag{13.5}$$

$$l_2: \frac{x - e}{y - f} = \frac{e - g}{f - h} \tag{13.6}$$

However, in reality, problems occur when

1. $b - d = 0$ or $f - h = 0$
2. l_1 / l_2
3. $b - d \rightarrow 0, a - c/b - d \rightarrow \infty$
4. It is impossible to tell whether the tool goes to the left- or right-hand side of l_1 in order to determine the tool offset.

The following section presents a new approach to finding the exact location of the cutter center.

First, examine an ideal situation, in which l_1 is on the Y axis and the current cutter location is at the origin of the X—Y coordinate system, as shown in Figure 13.25.

In this ideal case, the following procedure provides the solution for x and y:

1. If $|e - g| \leq \delta$, then there is no solution.
2. Otherwise,

$$x = 0$$

$$y = f + (-e)\frac{f - h}{e - g} \qquad (13.7)$$

Now it is time to transfer a pair of arbitrary lines l_1 and l_2 to this ideal situation. Find the angle, θ, between the Y axis and l_1 so that l_1 is lined up with the Y axis (see Figure 13.26). The next step is to rotate about the Z axis for θ.

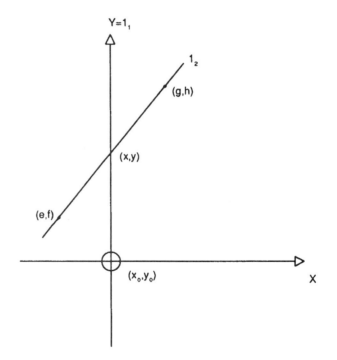

Figure 13.25 l_1 is lined up with the Y axis.

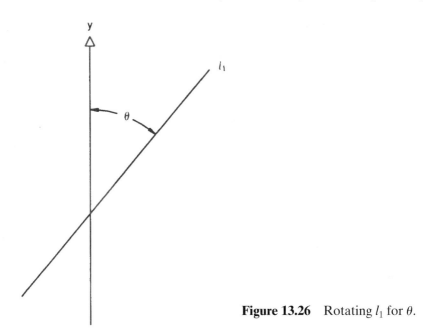

Figure 13.26   Rotating l_1 for θ.

The rotation transformation matrix is

$$RT_z(\theta) = \begin{bmatrix} \cos\theta & \sin\theta & 0 & 0 \\ -\sin\theta & \cos\theta & 0 & 0 \\ 0 & 0 & 1 & 0 \\ 0 & 0 & 0 & 1 \end{bmatrix} \tag{13.8}$$

Chapter 3 contains a more detailed description of coordinate transformation. The following is the solution for procedure for a general case:

1. Find θ.

2. Translate the current cutter location to the origin of the X–Y coordinate system. Rotate θ about the Z axis.

$$[e', f', 0, 1] = [e, f, 0, 1] \cdot \text{Tran}(a, b, c) \cdot RT_z(\theta)$$

$$[g', h', 0, 1] = [g, h, 0, 1] \cdot \text{Tran}(a, b, c) \cdot RT_z(\theta)$$

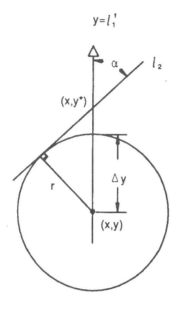

$y = l_1'$

Figure 13.27 Calculating tool offset.

3. If $|e' - g'| \leq \delta$, stop. (No solution is found.) δ is a very small number arbitrarily chosen by the user.

4. Find α for the tool-offset calculation (see Figure 13.27).

5. Calculate the tool offset:

$$x' = 0$$

$$y' = \left| \left(f' - e' \frac{f' - h'}{e' - g'} \right) \right| + (\text{flag}) \frac{r}{\sin \alpha} \qquad (13.9)$$

where

$$\text{flag} = \begin{cases} -1 & \text{for} & \text{TO} \\ 0 & \text{for} & \text{ON} \\ 1 & \text{for} & \text{PAST} \end{cases} l_2$$

6. Transform (x', y') back to the original coordinate system.

$$[x, y, 0, 1] = [x', y', 0, 1] \cdot RT_z(-\theta) \cdot \text{Tran}(-x_0, -y_0, 0) \qquad (13.10)$$

(x, y) is the solution.

The same procedure can be applied to find the cutter-center location for this command (see Figure 13.28).

$$\text{GO} \begin{Bmatrix} \text{LFT} \\ \text{RGT} \end{Bmatrix} / l_1, \begin{Bmatrix} \text{TO} \\ \text{ON} \\ \text{PAST} \end{Bmatrix}, c_1$$

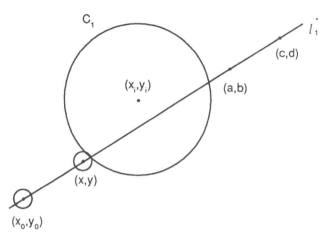

Figure 13.28   A different example for calculating the cutter-center location.

The procedure that finds the cutter-center location is as follows (see Figure 13.29):

1. Find θ, the angle between l_1 and the Y axis.
2. Translate (x_0, y_0) and rotate θ about the Z axis.

$$[x_i', y_i', 0, 1] = [x_i, y_i, 0, 1] \cdot \text{Tran}(x_0, y_0, 0) \cdot RT_z(\theta) \qquad (13.11)$$

3. $R' = R + (\text{flag}) \cdot r$
   where

$$\text{flag} = \left\{ \begin{array}{cc} 1 & \text{TO} \\ 0 & \text{ON} \\ -1 & \text{PAST} \end{array} \right\} c_1$$

If $|x_i'| > |R'|$, no solution is found.

4. Transform the solution back to the original coordinate system.

$$[x, y, 0, 1] = [x', y', 0, 1] \cdot RT_z(-\theta) \cdot \text{Tran}(-x_0, -y_0, 0) \qquad (13.12)$$

(x, y) is the solution.

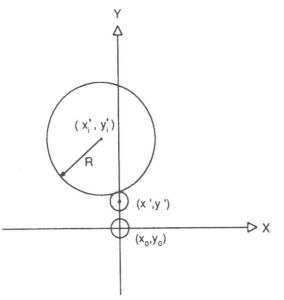

Figure 13.29 The tool path generated by a MAPT statement.

13.6.2 Computational Geometry

13.6.2.1. Parametric Representation. In preparing NC part programs, engineers have to deal with the geometry of many different objects. Under these circumstances, having the knowledge of computational geometry is helpful. The following basic formulas serve as a starting point for further study:

1. *Tangent of a curve.* The tangent of a curve $y = f(x)$ at point $p(x_1, y_1)$ is computed from the following equations (see Figure 13.30):

$$y = y_1 + f'(x_1)(x - x_1) \tag{13.13}$$

$$f'(x) = \frac{df}{dx}(x) \tag{13.14}$$

$$\frac{y - y_1}{x - x_1} = \frac{df}{dx} = \frac{dy}{dx} \tag{13.15}$$

2. *Tangent at point p.* The tangent at point p with parameter $v = v_1$ is calculated from the following equations (see Figure 13.30):

$$x = x(v_1 + \tau \dot{x}(v_1))$$
$$y = y(v_1 + \tau \dot{y}(v_1)) \tag{13.16}$$

Here,

$$\tau = \text{parameter for the new line}$$
$$\dot{x}(v_1) = dx/dt \text{ at } v = v_1$$
$$\dot{y}(v_1) = dy/dt \text{ at } v = v_1$$

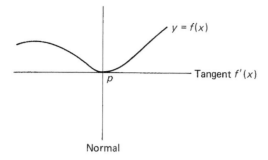

Figure 13.30   Tangent of a curve.

3. *Normal to a curve.* The normal to a curve is available from the following equations (see Figure 13.30):

$$x = x(v_1) - \tau \dot{y}(v_1)$$
$$y = y(v_1) - \tau \dot{x}(v_1) \tag{13.17}$$

4. *Parametric description of surfaces.* A plane passing through the point r_0 and containing the lines $\mathbf{r} = \mathbf{r}_0 + u\mathbf{r}_1$ and $\mathbf{r} = \mathbf{r}_0 + v\mathbf{n}_2$, where $\mathbf{n}_1$ and $\mathbf{n}_2$ are unit vectors, is shown in Figure 13.31.

5. *Tangent to a curve.* A tangent vector (unit vector) is defined as (see Figure 13.32)

$$T = \frac{d\mathbf{r}/du}{|d\mathbf{r}/du|} \tag{13.18}$$

where

$$\left| \frac{d\mathbf{r}}{du} \right| \neq 0 \tag{13.19}$$

A tangent line at $u = u_0$ can then be expressed as

$$\mathbf{r}_t = \mathbf{r}(u_0) + \lambda \mathbf{T}(u_0) \tag{13.20}$$

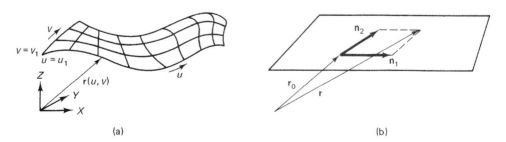

(a)   (b)

Figure 13.31   Parametric description of a surface.

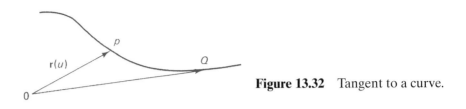

Figure 13.32   Tangent to a curve.

Example 13.3.

Find the tangent line for the following curve:

$$\mathbf{r} = u^3\mathbf{a} + u^2\mathbf{b} + u\mathbf{c} + \mathbf{d} \tag{13.21}$$

$$\frac{d\mathbf{r}}{du} = 3u^2\mathbf{a} + 2u\mathbf{b} + \mathbf{c} \tag{13.22}$$

$$u_0 = 0 \tag{13.23}$$

$$\frac{d\mathbf{r}}{du} = c \tag{13.24}$$

$$\mathbf{T}(0) = \frac{c}{|c|} \tag{13.25}$$

$$\mathbf{r}(0) = d \tag{13.26}$$

$$\mathbf{r}_t = \mathbf{d} + \frac{\lambda c}{|c|} \tag{13.27}$$

Therefore, the tangent line for

$$\mathbf{a} = \begin{bmatrix} 1 \\ 2 \end{bmatrix} \quad \mathbf{b} = \begin{bmatrix} 2 \\ 2 \end{bmatrix} \quad \mathbf{c} = \begin{bmatrix} 3 \\ 2 \end{bmatrix} \quad \mathbf{d} = \begin{bmatrix} 4 \\ 1 \end{bmatrix}$$

is

$$\mathbf{r}_t = \begin{bmatrix} 4 \\ 1 \end{bmatrix} + \frac{\lambda}{\sqrt{13}}\begin{bmatrix} 3 \\ 2 \end{bmatrix} \tag{13.28}$$

6. **Tool offset.** The tangent and normal vectors to a surface are calculated as follows (see Figure 13.33):

$$\mathbf{r} = \mathbf{r}(u, v) \tag{13.29}$$

At (u_0, v_0), $\partial\mathbf{r}/\partial u$ is a vector tangent to the curve $\mathbf{r} = \mathbf{r}(u, v_0)$, and $\partial\mathbf{r}/\partial v$ to the curve $\mathbf{r} = r(u_0, v)$. Given a unit normal vector

$$\mathbf{n} = \pm\frac{[\partial\mathbf{r}/\partial u \times \partial\mathbf{r}/\partial v]}{[\partial\mathbf{r}/\partial u \times \partial\mathbf{r}/\partial v]} \tag{13.30}$$

in cutting the surface using a ball end-mill, $\mathbf{r}$ is the cutter contact point. The normal vector passes through the center of the tool. Since the tool is aligned with the z axis, the tip of the tool is right under the center. At $u = u_0$ and $v = v_0$, the tool tip coordinate $\mathbf{r}'$ is

$$\mathbf{r}' = \mathbf{r}(u_0, v_0) + R(\mathbf{n} - \mathbf{u}) \tag{13.31}$$

where R is the tool radius, $u = \begin{bmatrix} 0 & 0 & 1 \end{bmatrix}$.

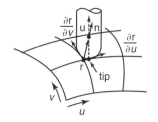

Figure 13.33 Tool offset.

13.6.2.2 Sculptured Surface Machining.   Sculptured surfaces are usually machined on three-, four-, or five-axis machines. To machine a sculptured surface with a three-axis Cartesian machine, a flat end-milling cutter is used to remove most of the material on the surface. A ball end-milling cutter is then used to finish the surface by approximating the surface using line segments. Because the entire spherical surface of the ball end-milling cutter can be used for cutting, it can cut curved surfaces. At each step is a cutter contact point. The surface normal vector at the cutter contact point passes through the cutter center. The cutter path is defined by the cutter center points. The resultant surface produced by the cutter must be within the tolerance specified on the design. The smaller the tolerance, the more cutter path steps have to be taken. After the machining, usually grinding and polishing operations are used to smooth the surface. Because grinding and polishing are slow and expensive, it is important that machining is done properly, so that a minimum amount of secondary processes will be needed.

There are several ways to generate finishing cutter paths for a sculptured surface: APT type cutting, Cartesian cutting, and isoparameter cutting. In APT-type cutting, the sculptured surface is defined as the part surface (PSIS; see Appendix 13.1). Newton's method is used to find the tool position iteratively. In Cartesian cutting, the tool path is generated using the Cartesian coordinates mapped onto the part surface. For example, the cutter moves first in the X direction and then sidesteps in the Y direction (Figure 13.34). Because the part surface is defined in the parameter domain (u, v), it is difficult to calculate the cutter contact point (given X and Y coordinates, find u and v). In isoparameter cutting, the cutter moves on u–v space either incrementing on u first and then on v, or vice versa. This section discusses only the isoparameter approach.

The algorithm to be discussed is not optimal, however it works well for simple surfaces. The objective of the cutter-path generation is to generate a minimum step cutter path that is within the tolerance limit. The control points describing the surface, the tool diameter, and the desired tolerance value are given in the figure.

Step 1. Select a direction u or v to move first. Set $u = 0$, $v = 0$, $i = 0$, and $j = 0$.

Step 2. Move along u (or v), and find step size Δu that satisfies the tolerance requirement (Figure 13.35). (The step-size algorithm will be presented later.)

$u_{i+1} = u_i + \Delta u$, $r(u_{i+1}, v_j)$ is the cutter contact point. Find the cutter center at

$$r_{CL} = r(u_{i+1}, v_j) + \frac{\text{dia}}{2} \frac{\mathbf{n}(u_{i+1}, v_j)}{|\mathbf{n}(u_{i+1}, v_j)|}$$

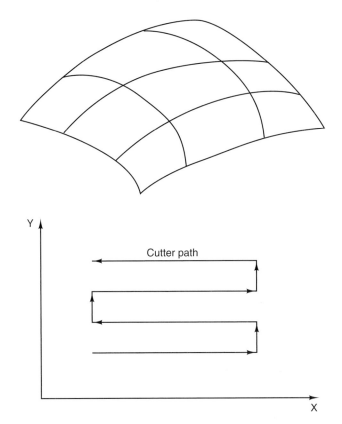

Figure 13.34 Cartesian cutting, the cutter path on the *X*–*Y* plane is mapped onto the part surface above.

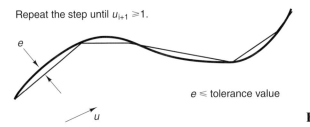

Repeat the step until $u_{i+1} \geqslant 1$.

$e \leqslant$ tolerance value

Figure 13.35 Forward step size.

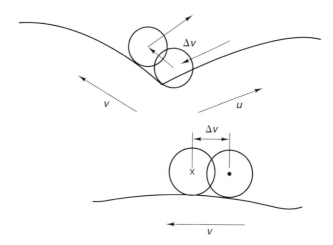

Figure 13.36 Step size.

where dia is the tool diameter, $\mathbf{n}(u_{i+1} \geq v_j)$ is the surface normal at $r(u_{i+1}, v_j)$.

Repeat the step until $u_{i+1} \geq 1$.

Step 3. Move (sidestep) Δv for the next curve. $v_{j+1} \geq v_j + \Delta v$. Make sure that the scallop height δ is less than tolerance (Figure 13.36). (The scallop-height calculation will be presented later.)

Step 4. Repeat steps 2 and 3 until the entire surface has been machined ($v_{j+1} \geq 1$). When j is an odd number, set the initial u_0 to 1 and $u_{i+1} = u_i - \Delta u$. This makes a zigzag cutter path.

Finding the Step Size. (Figure 13.37)

At constant v,* the curve $r(u) = r(u, v^*)$.

Set an initial Δu, and find the maximum cordal deviation. If d^* is greater than the tolerance, reduce Δu and try again until d^* is within the tolerance.

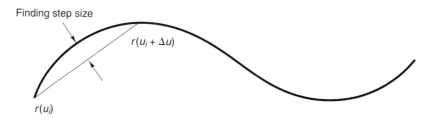

Finding step size

$r(u_i + \Delta u)$

$r(u_i)$

Figure 13.37 Step size.

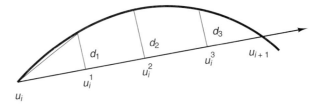

Figure 13.38 Error at each test point.

Because the exact error is difficult to find, a "quick-and-dirty" method may be used to get a tool path (Figure 13.38). The method is as follows:

(a) Set $\Delta u = \text{constant}$, $u_i = 0$
(b) $u_{i+1} = u_i + \Delta u$. If $u_{i+1} \geq 1$, $u_{i+1} = 1$.
(c) Find three points: $r(u_i^1)$, $r(u_i^2)$, and $r(u_i^3)$.

$$u_i^j = u_i + j \frac{\Delta u}{4}$$

$$d_j = \left| \frac{[r(u_i^j) - r(u_i)][r(u_{i+1}) - r(u_i)]}{|r(u_{i+1}) - r(u_i)|} \right|$$

Because this is a curve in a surface, $r(u_i)$ should be $r(u_i, v^*)$, where v^* is the current v value. On a two-dimensional map, the distance between a point and a line can also be calculated. Let the line equation be $ax + by + c = 0$. The point is located at (x_1, y_1). The distance is

$$d = \frac{ax_1 + by_1 + c}{\sqrt{a^2 + b^2}}$$

(d) If $\max(d_j) > $ tolerance, set $\Delta u = 0.75 \Delta u$ and repeat step (c); otherwise, set the next step at $r(u_{i+1})$.
(e) If $u_{i+1} = 1$, end; otherwise, go to step (a).

Find the Sidestep. At constant u^*, the curve $r(v) = r(u, v^*)$. Assume that the surface normals are the same at $r(v_i)$ and $r(v_{i+1})$ (Figure 13.39).

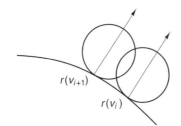

Figure 13.39 Tool at two adjacent paths.

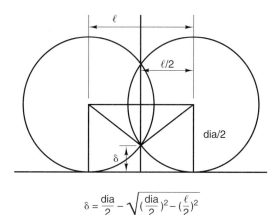

$$\delta = \frac{\text{dia}}{2} - \sqrt{(\frac{\text{dia}}{2})^2 - (\frac{\ell}{2})^2}$$

Figure 13.40   Approximate scallop height.

(a) Set Δv = constant.
(b) $v_{i+1} = v_i + \Delta v$
(c) $l = |r(v_{i+1}) - r(v_i)|$ (see Figure 13.40)
(d) $\delta = \text{dia}/2 - \sqrt{(\text{dia}/2)^2 - (l/2)^2}$
(e) If $\delta > 10$, set $\Delta v = 0.75\Delta v$ and go to step (b).
(f) Otherwise, v_{i+1} is the new step.

The preceding equation for calculating δ is approximate. There are several potential problems associated with the approach:

(a) The normal and curvature on the surface are changing. For the concave area, the estimated error is less than the actual error (Figure 13.41). For the convex region, the estimate is conservative (Figure 13.42).
(b) When the surface curvatures at two ends are very different the paths generated are either too close to each other or not close enough (Figure 13.43).

The procedure discussed before can find the "least-number-of-steps" solution. However, it is computationally expensive because of the search strategy employed.

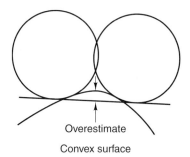

Overestimate

Convex surface

Figure 13.41   Overestimate on a convex surface.

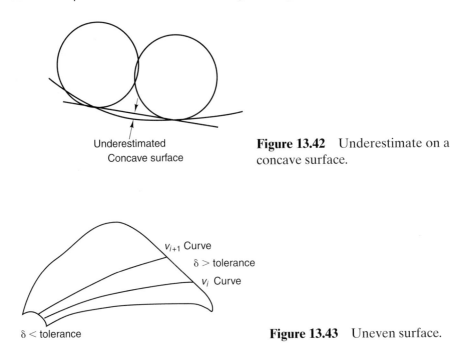

Underestimated
Concave surface

Figure 13.42   Underestimate on a concave surface.

v_{i+1} Curve

$\delta >$ tolerance

v_i Curve

$\delta <$ tolerance

Figure 13.43   Uneven surface.

In order to improve the computational efficiency, a quick estimate approach may be more desirable. Using the initial step size Δu, the error at u_i can be found. If the error is smaller than the tolerance output, then position and increment $u, u_{i+1} = u_i + \Delta u$. If the error is greater than the tolerance, reduce Δu and check again. Smaller steps may take more time than necessary, but the computation is greatly simplified.

13.7 SUMMARY

Numerical-control (NC) machining has changed the appearance of an entire industry in a way that nobody would have dreamed, NC machining has impacted almost every aspect of manufacturing—accuracy, repeatability, flexibility, and economics—in a positive and profound manner. The soul of NC machining, NC programming, is introduced in this chapter. Depending on the level of automation, NC programming can be divided into (1) manual programming, (2) computer-assisted programming, and (3) automated generation of NC code (CAD/CAM). All three levels of NC programming were discussed, along with analytical aspects of NC. Although NC programming is moving toward the direction of fully integrated CAD/CAM, a modern manufacturing engineer should be equipped with the knowledge of manual programming and computer-assisted part programming, as they will still exist on the shop floor for years to come.

13.8 KEYWORDS

APT	Miscellaneous control
ASCII	NC Words
BCD	N-code
BCL	Part programming
Block-address format	Post processor
BLU programming	Preparatory function
CL-data	R-code
Computer-assisted part programming	Sculptured surface machining
Cutter compensation	Tab sequential format
EIA	T-code
Fixed sequential format	Tool-radius compensation
G-code	Word-address format
Interpolation	X, Y, Z-code
M-code	

13.9 REVIEW QUESTIONS

13.1 Prepare an NC part program for making the slot shown in Figure 13.44. The cutter diameter is $\frac{3}{16}$ in. The start point is (0,0) at 2 in. to the left and 2 in. below the lower left corner of the part.

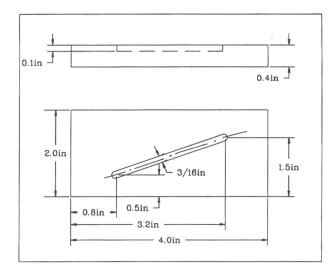

Figure 13.44 Part drawing for Review Question 13.1 (Courtesy of Terco).

13.2 Prepare an NC part program for the part shown in Figure 13.45 The dimensions given in the figure are in millimeters. The tool diameter is 20 mm.

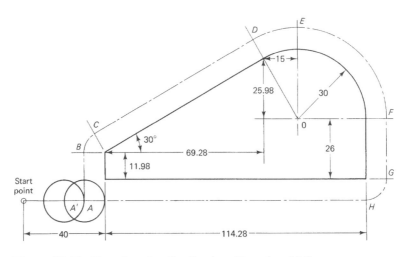

Figure 13.45 Part drawing for Review Question 13.2.

13.3 Prepare an NC part program for the part shown in Figure 13.46. The suggested cutting conditions are 6 ipm for the feed rate and 400 fpm for the speed. The cutter diameter is 5 in. The dimensions are all in inches.

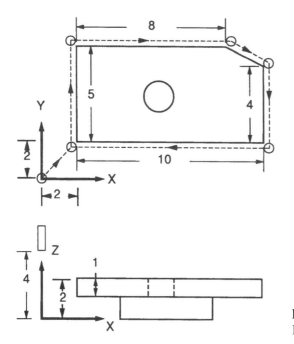

Figure 13.46 Part drawing for Review Question 13.3.

13.4 Prepare an NC part program for the part shown in Figure 13.47. The cutting conditions are the same as those given in Review Question 13.3. The dimensions are all in inches.

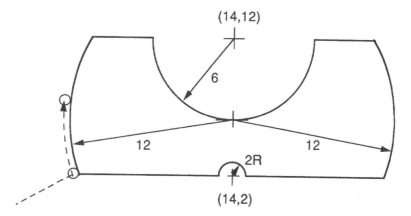

Figure 13.47   Part drawing for Review Question 13.4

13.5 Prepare an NC part program for making a ball with a diameter of 2.8 in. The ball will be made out of a 3-in. × 3-in. × 3-in. cubic blank.

13.6 Write an APT part program for the part shown in Figure 13.48. The cutter diameter is 1 in., the suggested feed rate is 8 ipm, and the spindle speed is 764 rpm. The start is 2 in. to the left and 2 in. below the corner of the part.

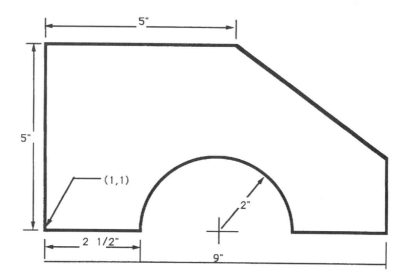

Figure 13.48   Part drawing for Review Question 13.6

13.7 Prepare an APT part program for the part shown in Figure 13.45.

13.8 Turned parts are machined on lathes. A lathe usually has two axes, X and Z. To turn a part from a cylinder down to the part profile, first program the roughing passes and then the finish pass. The roughing passes remove most of the materials from the workpiece. Only a small layer of finishing allowance is left on the workpiece. The finishing cut is close to the part profile. For the part shown in Figure 13.49, the roughing passes are shown in Figure 13.50. Assume that the cutter tip has a $\frac{1}{32}$-in. radius. Write a part program for the roughing and the finishing passes. The depth of cut for each roughing layer is 0.2 in. and the finishing allowance is 0.1 in.

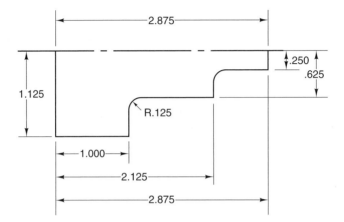

Figure 13.49 A turned part (half of the-cross section).

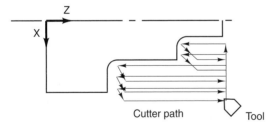

Figure 13.50 A turned part (half of the cross section).

13.9 The profile of a turned part is defined by a Bezier's curve. The four control points are (0,0), (3,6), (6, −2), and (8,1). Write a program for the finishing cut. Use $\frac{1}{32}$ in. as the tool-nose radius. The tolerance allowed is 0.001 in.

13.10 Develop an algorithm for turning a profile consisting of lines and curves. (See Figure 13.50.) The algorithm must generate both roughing and finishing cuts. Implement the algorithm in a computer language that you know.

13.11 Develop an algorithm to mill a rectangular pocket. Assume the corner radius is the same as the tool radius. Use the staircase approach for the pocketing.

13.12 Repeat Review Question 13.11 using the contouring approach.

13.13 Write a computer program to generate a cutter path for a Bezier's curve (Figure 13.51). The input consists of four control points, a cutter diameter, and tolerance. The output consists of a cutter path and graphics.

Cutter path 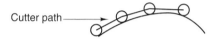   **Figure 13.51**   Bezier's curve cutting.

13.14 Write a computer program to generate cutter path for a Bezier's surface (Figure 13.52). The input consists of 16 control points, a tolerance = 0.1 in., and a direction of interpolation (t first or s first). The output consists of a cutter path and verification graphics. The tool consists of a $\frac{1}{2}$-in. ball-nose end mill, 1-in. maximum depth, and $\frac{1}{4}$-in. maximum cut depth/pass. Set (0,0,0) to be the center of the 4-in.-diameter workpiece (floating-zero, decimal programming), for example,

```
N0100 G00 X0.000 Y0.000 Z1.000
```

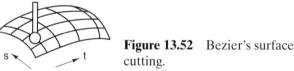   **Figure 13.52**   Bezier's surface cutting.

Show the derived equations, the flow chart, the source code listing, and three experimental results (data and graphics). The graphics must include both the surface plot and the cutter-path plot. Use the same control points and three different tolerances for the experiments. Use the algorithm introduced in the chapter to implement the program.

13.15 A rule surface is created by blending two curves, r_0 and r_1.

$$r(t, s) = r_0(t) + s[r_1(t) - r_0(t)]$$

where r_0 and r_1 can be any parametric curves. Write a program to generate a cutter path for a ruled surface.

13.10 REFERENCES

Childs, J. J. *Principles of Numerical Control.* New York: Industrial Press, 1982.

Choi, B. K. *Surface Modeling for CAD/CAM.* Amsterdam: Elsevier, 1991.

Ding, Q., and B. J. Davies. *Surface Engineering Geometry for Computer-Aided Design and Manufacture.* West Sussex, UK: Ellis Horwood, 1987.

Electronic Industries Association. *32-Bit Binary CL (BCL) and 7-Bit ASCII CL (ACL) Exchange Input Format for Numerically Controlled Machines.* Washington, DC: EIA, 1992.

Koren, Y. *Computer Control of Manufacturing Systems.* New York: McGraw-Hill, 1983.

Kral, I. H. *Numerical Control Programming in APT.* Englewood Cliffs, NJ: Prentice Hall, 1986.

Machinability Data Center. *Machining Data Handbook*, 3d ed. Cincinnati, OH: Metcut Research Associates, 1980.

Marciniak, K. *Geometric Modelling for Numerically Controlled Machining.* Oxford: Oxford University Press, 1992.

Pressman, R. S., and J.E. Williams. *Numerical Control and Computer-Aided Manufacturing.* Aerospace Industries Association, 1979.

Wang, H. P., and A. C. Lin. "Automated Generation of NC Part Programs for Turned Parts Based on 2-D Drawing Files," *International Journal of Advanced Manufacturing Technology*, 2(3), 1987, 23–36.

Wysk, R. A., T. C. Chang, and H. P. Wang. *Computer-Aided-Manufacturing PC-Application Software.* Albany, NY: Delmar Publishers, 1988.

Yellowley, I., A. Wong, and B. DeSmit. "The Economics of Peripheral Milling," in *Manufacturing Engineering Transactions.* Dearborn, MI: Society of Manufacturing Engineers, 1987, 388–394.

APPENDIX: COMPUTER-ASSISTED PART PROGRAMMING LANGUAGES

13.1 APT

APT is an acronym for Automatically Programmed Tool. Initially developed in 1956 at MIT, it is the most popular part-programming language in the United States [Childs, 1982]. Since that time, APT has continued to evolve, and the Illinois Institute of Technology Research Institute (IITRI) has continued the development and administrative responsibility for the fourth version of the software.

APT is the most powerful general-purpose part-programming system against which other systems are commonly compared and evaluated. Summary characteristics of APT are as follows [Illinois Institute of Technology Research Institute, 1967]:

- Three-dimensional unbounded surfaces and points are defined to represent the part to be made.
- Surfaces are defined in an *X–Y–Z* coordinate system chosen by the part programmer.
- In programming, the tool does all the moving; the part is stationary.
- The tool path is controlled by pairs of three-dimensional surfaces; other motions, not controlled by surfaces, are also possible.
- A series of short straight-line motions are calculated to represent curved tool paths (linear interpolation).
- The tool path is calculated so as to be within specified tolerances of the controlling surfaces.

- The X, Y, and Z coordinates of successive tool-end positions along the desired tool path are recorded as the general solution to the programming problem.
- Additional processing (postprocessing) of the tool-end coordinates generates the exact tape codes and format for a particular machine.

An APT processor interprets and computes the cutter path, whereas a postprocessor translates the cutter path to a format acceptable by a specific machine. There are five types of statements in the APT language:

1. *Identification statements*. These define a specific project.
2. *Geometry statements*. These define a scaler or geometric quantity.
3. *Motion statements*. These describe a cutter path, such as GOLFT.
4. *Postprocessor statements*. These define machining parameters such as feed, speed, coolant on–off and so on.
5. *Auxiliary statements*. These describe auxiliary machine-tool functions to identify the tool, part, tolerances, and so on.

Some of the APT geometry and motion statements are explained with examples as follows:

13.1.1 Geometry Statements

Geometry statements are used to define basic geometric entities such as points, lines, circles, and so on.

Points.   A point can be defined by the following:

1. The explicit specification of the X, Y, and Z coordinates of the point:

$$p_1 = \text{POINT}/x, y, z$$

2. The intersection of two lines (see Figure 13.21(a)):

$$p_2 = \text{POINT}/l_1, l_2$$

3. The center of a circle (see Figure 13.21(b)):

$$p_3 = \text{POINT}/\text{CENTER}, c_1$$

4. The intersection of a line and a circle (see Figure 13.53(b)):

$$p_4 = \text{POINT}/\text{YLARGE}, \text{INTOF}, \ l_1, c_1$$

$$p_5 = \text{POINT}/\text{XLARGE}, \text{INTOF}, \ l_1, c_1$$

5. The intersection of two circles (see Figure 13.53(c)):

$$p_6 = \text{POINT}/\text{YLARGE}, \text{INTOF}, \ c_1, c_2$$

$$p_7 = \text{POINT}/\text{XLARGE}, \text{INTOF}, \ c_1, c_2$$

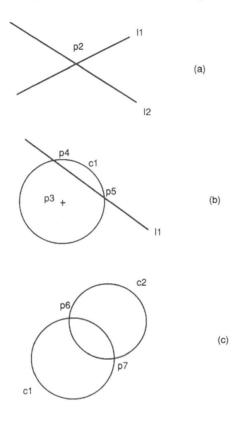

Figure 13.53 POINT statements.

Lines. A line can be defined by the following:

1. The specification of the coordinates of the two points explicitly connecting the line.

$$l_1 = \text{LINE}/x_1, y_1, z_1, x_2, y_2, z_2$$

2. The specification of the two points connecting the line (see Figure 13.54(a)):

$$l_2 = \text{LINE}/p_1, p_2$$

3. A point and a line (see Figure 13.54(b)):

$$l_3 = \text{LINE}/p_1, \text{PARLEL}, l_{20}$$

$$l_4 = \text{LINE}/p_1, \text{PERPTO}, l_{20}$$

4. A point and a circle (see Figure 13.54(c)):

$$l_5 = \text{LINE}/p_1, \text{LEFT}, \text{TANTO}, c_1$$

$$l_6 = \text{LINE}/p_1, \text{RIGHT}, \text{TANTO}, c_1$$

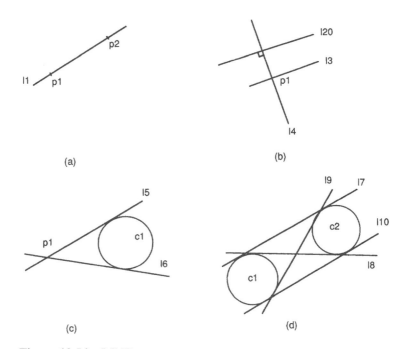

Figure 13.54   LINE statements.

5. Two circles (see Figure 13.54(d)):

$$l_7 = \text{LINE/LEFT, TANTO, } c_1, \text{LEFT, TANTO, } c_2$$

$$l_8 = \text{LINE/LEFT, TANTO, } c_1, \text{RIGHT, TANTO, } c_2$$

$$l_9 = \text{LINE/RIGHT, TANTO, } c_1, \text{LEFT, TANTO, } c_2$$

$$l_{10} = \text{LINE/RIGHT, TANTO, } c_1, \text{RIGHT, TANTO, } c_2$$

Circles.   A circle can be defined by the following:

1. The specification of the X, Y, and Z coordinates of the center and the radius of the circle:

$$c_1 = \text{CIRCLE/}x, y, z, R$$

2. The specification of the center point and a radius (see Figure 13.55(a)):

$$c_2 = \text{CIRCLE/CENTER, } p_1, \text{RADIUS, } R$$

3. A point and a line (see Figure 13.55(b)):

$$c_3 = \text{CIRCLE/CENTER, } p_1, \text{TANTO, } l_1$$

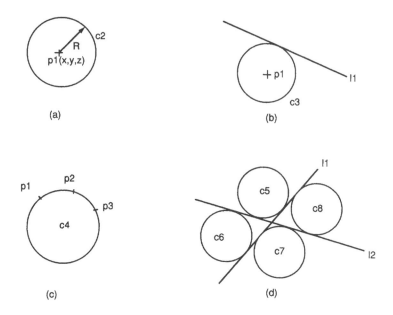

Figure 13.55   CIRCLE statements.

4. Three points (see Figure 13.55(c)):

$$c_4 = \text{CIRCLE}/p_1, p_2, p_3$$

5. Two lines and a radius (see Figure 13.55(d)):

$$c = \text{CIRCLE}/\left\{\begin{array}{l} \text{XSMALL} \\ \text{XLARGE} \\ \text{YSMALL} \\ \text{YLARGE} \end{array}\right\}, l_1, \left\{\begin{array}{l} \text{XSMALL} \\ \text{XLARGE} \\ \text{YSMALL} \\ \text{YLARGE} \end{array}\right\}, l_2, \text{RADIUS}, R$$

By using different combinations of the preceding statements, four circles can be specified.

13.1.2 Motion Statements

From.   The FROM statement is used to define the starting point of certain motion:

$$\text{FROM}/\left\{\begin{array}{c} p_1 \\ x, y, z \end{array}\right\}$$

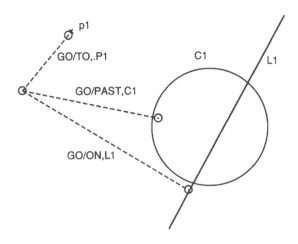

Figure 13.56   Motion statements—I

Go.   The GO statement is used to define the destination of certain motion (see Figure 13.56; the difference between GOTO/ and GO/TO is obvious):

$$GOTO/\left\{ \begin{array}{c} p_1 \\ x, y, z \end{array} \right\}$$

$$GOTO/\left\{ \begin{array}{c} TO \\ PAST \\ ON \end{array} \right\}, \left\{ \begin{array}{c} p_1 \\ l_1 \\ c_1 \end{array} \right\}$$

Go Directions.   These statements are used to define the path of certain tool motion in order to avoid ambiguity:

1. GOLFT and GORGT (see Figure 13.57(a)):

$$GOTO/l_1$$

$$GORGT/l_1, TO, \ l_2$$

2. GOFWD and GOBACK (see Figure 13.57(b)):

$$GOTO/l_1$$

$$GORGT/l_1, TO, \ l_2$$

$$GOFWD/l_1, PAST, c_1$$

3. GOUP and GODOWN (see Figure 13.57(c)):

$$GORGT/l_1, TO, l_2$$

$$GOUP/l_4, TO, \ l_3$$

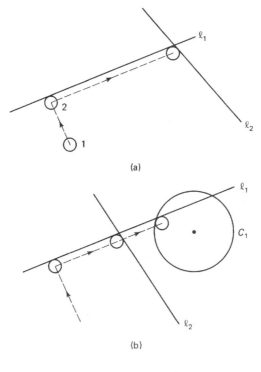

(a)

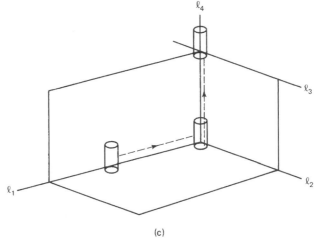

(b)

(c)

Figure 13.57 Motion statements—II.

There are two other important features provided in the APT language:

- Macros: Individual macros similar to FORTRAN subroutines can be created to add to APT program routines. A library of frequently used routines and definitions can be created as special macros.
- Loops: Individual sections of an APT program can be repeated until a specified result is obtained.

In addition to the basics just presented, APT also provides more geometric capabilities and advanced functions to handle complex parts. Other geometries supported by APT include the following:

- *Plane curves:* Ellipse, Hyperbola, GCONIC, and LCONIC.
  An ellipse is defined by its center, major, and minor axes, and the angle between the major axis and the $+X$ axis. A hyperbola is defined in the same way as an ellipse. GCONIC is a general conic section, which includes circles, ellipses, hyperbolas, and parabolas. LCONIC defines loft conics, which is interpolated from a set of points. The defining equations are as follows:

$$\frac{x^2}{a^2} + \frac{y^2}{b^2} = 1 \quad \text{(ellipse)} \tag{13.32}$$

$$\frac{x^2}{a^2} - \frac{y^2}{b^2} = 1 \quad \text{(hyperbola)} \tag{13.33}$$

$$ax^2 + by^2 + cz^2 + dx + ey + f = 0 \quad \text{(GCONIC)} \tag{13.34}$$

- *Surfaces:* Quadratic surfaces, cylinders, tabulated cylinders, surfaces of rotation, ruled surfaces, and polyconic surfaces.
  Quadratic surfaces are commonly used in engineering design. A quadratic surface has the following form:

$$ax^2 + by^2 + cz^2 + dyz + ezx + fxy + gx + hy + iz + j = 0 \tag{13.35}$$

The APT statement for a quadric surface is

QADRIC/a, b, c, d, e, f, g, h, i, j

A cylinder (CYLNDR) in APT can be defined in many ways. There is an infinitely long cylinder. A tabulated cylinder (TABCYL) is a surface produced by the linear sweep of a curve. The curve in a tabulated cylinder is defined by a set of points (spline curve). The sweep is defined by a vector or using one of the three axes.

Surfaces of rotation include cones, spheres, and toruses.

A ruled surface (RLDSRF) is blended together by two spline curves. If two spline curves are defined as $r_p r_2$ (each defined by a set of points), the ruled surface is

$$r = \lambda r_1 + (1 - \lambda)r_2 \tag{13.36}$$

where λ is a parameter, $0 \le \lambda \le 1$.

A polyconic surface (POLCON) defines a surface using a series of cross-sectional curves. It is an extension of a ruled surface. The cross sections of a polyconic surface are perpendicular to the longitudinal axis. Each section is approximated from a set of points by polynomial functions. The profile on the longitudinal axis is approximated by a polynomial function. Surfaces like aircraft wing profiles can be described by a polyconic surface.

- *Advanced functions:* Advanced functions include pattern cutting, pocket milling, and sculptured surface machining. The sculptured surface machining function is not typically available in APT. It is an add-on to the APT processor. The pattern-cutting function (PATERN) is especially useful in drilling hole patterns. Patterns can be either linear or circular. When combined with a set of modifiers for transformation and editing, it can describe very complex patterns.

The pocket milling function (POCKET) automatically generates roughing and finishing paths for milling a pocket. A pocket is defined by a set of points. APT produces a contouring pattern that begins from the center of the pocket to the pocket boundary. The tool motion follows the sequence in which the points are defined. The pocket geometry is limited to convex polygons. A nonconvex pocket must be split into several convex pockets by the programmer. Figure 13.58 shows an example of the POCKET function.

The PSIS statement is used to assign the part surface. The surface can be a planar surface defined by PLANE or one of the curved surfaces defined in the previous section. The APT processor uses an iterative method to locate the cutter onto the part surface while moving the cutter either along a line or in a pocket. The error created by the cutting is kept within the specified tolerance limit.

13.2 OTHER PART-PROGRAMMING LANGUAGES

AUTOSPOT is a 2-D part-programming language developed at IBM. This processor allows point-to-point operations in two dimensions only. This type of operation requires the movement of a cutter to a discrete position, enables the performance of a desired machining function at that position (i.e., drilling, boring, and so on), and then enables these steps to be repeated.

ADAPT (Adaptation of APT) was the first attempt to adapt more commonly used APT routines for smaller computers. It was developed at IBM under a U.S. Air Force contract. It was constructed in a modular manner, providing greater flexibility to the user wanting to add and delete routines. It has full two-dimensional and some limited three-axis capabilities. It has routines for curve fitting, inclined planes, polygonal pockets, and macro definitions.

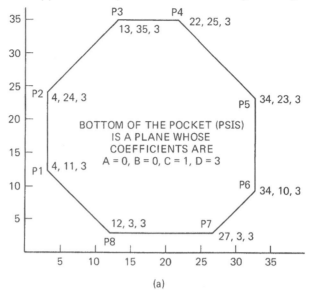

(a)

Pocket Part Program

```
 1 REMARK POCKET POLYGON COLLAPSE DEMONSTRATION TEST
 2 NOPOST $$ NO POSTPROCESSING FOR THIS TEST CASE
 3 CLPRNT $$ PRINT CUTTER CENTER DATA
 4 TOLER/.001 $$ TOLERANCE BAND
 5 $$ POINTS DEPINING POCKET PERIMETER
 6 P1 = POINT/4, 11, 3 $$ STARTING POINT OF POCKET DEFINITION
 7 P2 = POINT/4,24,3
 8 P3 = POINT/13,35,3
 9 P4 = POINT/22,35,3
10 P5 = POINT/34,23,3
11 P6 = POINT/34,10,3
12 P7 = POINT/27,3,3
13 P8 = POINT/12,3,3 $$ ENDING POINT OF POCKET DEFINITION
14 H = .01 $$ SCALLOP HEIGHT MAXIMUM
15 D = .38 $$ CONSTANT CUTTER DIAMETER
16 CR = .19 $$ CUTTER CORNER RADIUS
17 D2 = SQRTF((D * B) - B ** 2)$$ A BALL END MILL EFFECTIVE CUTTER RADIUS
18 CV = (4 * D2)/D $$ A MEASURE OF POCKETING CUT OFFSET
19 CUTTER/D, CR $$ BALL END MILL
20 FEDRAT/50 $$ MODAL FEED RATE
21 FROM/0,0 $$ STARTING CUTTER POSITION
22 GO TO/20,20,5 $$ MOVE CUTTER TOWARD AND OVER CENTER OF POCKET
23 PSIS/(PLANE/0,0,1,3) $$ BOTTOM PLANE OF POCKET
24 POCKET/D2,CV,CV,3,10,10,1,1,P1,P2,P3,P4,P5,P6,P7,P8 $$ STATEMENT
25 GO DLTA/0,0,2 $$ CLEARANCE POSITION OF CUTTER
26 GO TO/0,0,0 $$ END CUTTER POSITION
27 FINI $$ END OF PART PROGRAM
```

(b)

Figure 13.58 (a) A pocket to be machined using the IBM APT, and (b) the program listing for machining the pocket.

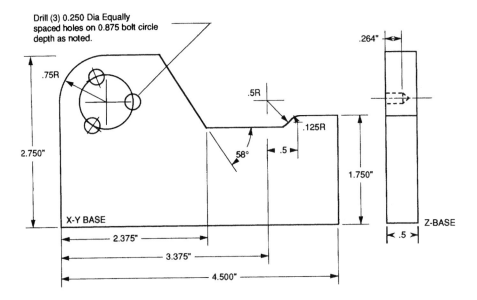

Figure 13.59   A part to be machined using the COMPACT II program of Figure 13.60.

UNIAPT was developed by the United Computing Corporation of Carson, California. It was the first successful attempt to handle the full power of APT on a small computer. Externally, it is completely compatible with APT; it differs only in the internal design of the processor.

EXAPT (Extended Subset of APT) was developed jointly in West Germany around 1964 by several technical universities to make APT more appropriate to European conditions.

AUTOSPOT (Automatic System for Positioning Tools) was developed at IBM for three-axis, point-to-point motion control around 1962. It was subsequently combined with ADAPT to provide an effective language for both point-to-point and continuous-path applications.

COMPACT was developed by Manufacturing Data Systems, Ann Arbor, Michigan, to simultaneously service multiple users from a remote computer over telephone lines. The COMPACT system converts its language statements to machine-control codes in a single computer iteration, thus eliminating the postprocessing stage completely [Smith and Evans, 1977]. COMPACT II, the latest version (together with APT), is the most popular part-programming language. It is also supported by most turnkey CAD/CAM systems. The COMPACT II program of Figure 13.58(b) was written to machine the part shown in Figure 13.59. The geometry describing the part and the machining motion generated by the program is illustrated in Figure 13.60.

In a COMPACT II program, a geometric entity is identified by its type and a number. For example, LN1 refers to line 1, PT1 refers to point 1. A "D" in front of a geometric name defines the geometry. To define line 1, DLN1 is used. In Figure 13.60, all the statements beginning with DLN, DPT, DCIR, are geometry definitions.

```
MACHINE MILL
IDENT DEMONSTRATION PROGRAM
SETUP
BASE, 3XB, 8YB, 6ZB
DLN1, XB
DLN2, 2. 75YB
DPT1, .75XB, 2YB, ZB
DCIR1, PT1, .75R
DPT2, 2.375XB, 1.5YB, ZB
DLN3.PT2, 58CW
DLN4, 1.5YB
DPT3, 3.5XB, 1.5+.5YB,ZB
DCIR2.PT3,.5R
DPT4, 3.5+.5XB, 1.75-.125YB.ZB
DCIR3, PT4,.125R
DLN5, 1.75 YB
DLN6. 4. 5XB
DLN7, YB
DCIR4, PT1,.875DIA
ATCHG, TOOL1, .5TD. 6GL, 250FPM, .01IPR
MOVE, PASTLN1.PASTLN7, .6ZB
CUT, -.03ZB
OCON, CIR1, CW.S(180) .F(270)
CUT,PARLN2.PASTLN3
CUT.PARLN3.TOLN4
ICON.CIR2.CCW.S (TANLN4), F(TANCIR3)
OCON, CIR3.CW.S (TANCIR2), F(TANLN5)
CUT, PASTLN6
CUT, PASTLN7
CUT, PASTLN1
ATCHG, TOOL2, .25TD, 6GL, 800RPM, 011PR, 118TPA
DRL3, CIR4, CW, SO, 5ZB, 3DP
END
```

Figure 13.60   A sample COMPACT II part program.

Tool-change and speed and feed specifications are done using ATCHG (automatic tool-change) and MTCHG (manual tool-change) statements. MOVE is used to rapid traverse to a geometry; it is similar to the GO statement in APT. CUT does the same, except feed motion is used. To move around a circular arc, three statements are used: OCON, ICON, and CONT, which are outer contouring, inner contouring, and contouring, respectively. The DRL statement is used for hole-pattern drilling.

   MAPT (Micro-APT) is a subset (a microcomputer version) of the APT language processor. Using MAPT, a manufacturing engineer specifies the geometry of a part to be machined, the motion of the tool (cutter), and the operations involved in producing the part. The MAPT system translates these instructions into numerical information that guides the machine tool to produce the part. Tool-center offsets are automatically calculated. Currently, MAPT is capable of three-axis programming.

   MAPT is an interpretive language. Like the full-scale APT language processor, MAPT contains four classes of statements. With MAPT, the user can compile a MAPT source program line by line. The final CL data and error messages, if any, are printed on the screen as the program compiles. The CL data can be plotted with user-defined geometry to verify the program. Paper tape or paper-tape-format data are generated by postprocessors for the target NC machine tools.

MAPT also contains a built-in screen editor with an online help facility. The user is able to obtain online assistance on both editor commands and MAPT language syntax. A MAPT source program can be created, compiled, edited, and then compiled again without leaving the MAPT working environment. A built-in graphics package allows the user to verify a program. It can also simulate the cutter path on a real-time basis. The verification plot of both the part geometry and tool path also can be attained on a digital plotter.

The MAPT program structure and syntax is presented in Figure 13.61. The details about the MAPT language are available in Wysk, Chang, and Wang [1988]. A newer version that contains most of the APT features described in this chapter is also available from the authors.

During the past few years, significant changes have occurred that affect the NC environment. These require a reassessment of the adequacy of current NC processors.

A program written in MAPT for the part shown in Figure 13.16 is

```
partno/example 10.2
fedrat/5.0
splindl/1000,cw
cutter/0.5
clprint/on
pt1 = point/6,6,0
pt2 = point/pt1,0,2,0
pt3 = point/pt2,2,0,0
l1 = line/pt1,p2
l2 = line/pt2,p3
l3 = line/pt3, parlel,l1
l4 = line/pt1,parlel,l2
go/to,l1
golft/l1,past,l2
gorgt/l2,past,l3
gorgt/l3,past,l4
gorgt/l4,past,l1
fini
```

In the program, all points and lines are defined with respect to pt1. The part location can be changed easily by redefining the coordinates of pt1. The statement "clprint/on" prints the data value of each geometric entity and cutter location. It is used for debugging purposes. The tool motion commands consist of one GO statement and four GOLFT/GORGT statements. A MAPT "word" can be written in either uppercase or lowercase.

PART PROGRAM STRUCTURE AND SYNTAX

The MAPT program has the following structure
— Part Identification
    SYNTAX:
        PARTNO/part identification

— Environment Description
        CUTTER/dia
        INTO/tolerance
        OUTTO/tolerance
        TOLT/tolerance
        CLPRINT/{ON, OFF}
        PSIS/{pl-name, OFF}

— Geometric Definitions
    Point
        p-name = POINT/x, y, z
        p-name = POINT/p-name, dx, dy, dz
        p-name = POINT/l-name, l-name
        p-name = POINT/CENTER, c-name
    Line
        l-name = LINE/xl, yl, zl, x2, y2, z2
        l-name = LINE/p-name, p-name
        l-name = LINE/p-name, {LEFT RIGHT}, c-name
        l-name = LINE/p-name, parlel, l-name
        l-name = LINE/p-name, angle
        l-name = LINE/{LEFT RIGHT}, c-name, {LEFT RIGHT}, c-name
    Circle
        c-name = CIRCLE/x, y, r
        c-name = CIRCLE/l-name, {LEFT RIGHT}, {FAR NEAR}, l-name, r
        c-name = CIRCLE/p-name, RADIUS, r
        c-name = CIRCLE/p-name, TAN, l-name
        c-name = CIRCLE/p-name, p-name, p-name
    Plane
        p-name = PLANE/a, b, c, d

— Motion Statements
    Positioning Statements
        SETPT/x, y, z
        GODLTA/x value, y-value, z-value
    Start-Up Motion
        GO/{TO ON PASS}, {p-name l-name c-name}
    Continuous Motion
        GORGT/{l-name c-name}, {TO ON PAST}, {p-name l-name c-name}
        GOLFT/{l-name c-name}, {TO ON PAST}, {p-name l-name c-name}

—Postprocessor Commands
        FEDRAT/feedrate
        MACHINE/machine type
        COOLNT/{ON OFF}
        RAPID/speed
        SELECT/TOOL, tool#
        SPINDL/{OFF rpm, {CLW CCLW}}
        TORCH/{ON OFF}

—Termination
        FINI/

Figure 13.61   MAPT program structure and syntax.

- Based on the experiences reported by NC users, new requirements have been identified and have to be satisfied. Among these are the following:

    — NC machine tools have grown in complexity and capability, and their new functions need to be supported via NC processors and resultant postprocessors.

    — Most of the worldwide NC users support the standardization of an NC language so that it provides a common Englishlike language based on APT syntax and language, with subsets and modularized features therein. (Both ANSI and ISO have committees addressing this subject.)

- Computer hardware and programming technology have evolved considerably. Among the areas directly affecting current NC processors are the following:

    — The architecture of current NC processors has inhibited their ability to take advantage of new advances in computer hardware that may improve price/performance (i.e., new disk devices, increased core capacity, and so on).

    — The advent of new operating systems enables applications that properly utilize their functions to be smaller, more efficient, and more reliable. The current NC processors were developed prior to the existence of elaborate operating-system facilities and cannot be conveniently altered to take advantage of these improvements.

    — Many improvements in programming technology have resulted in more efficient methods for performing algorithmic operations, especially for geometric functions. Current NC processors may only selectively utilize these new techniques, at a sacrifice of efficiency.

- The concept of computer-aided manufacturing (CAM) is evolving. CAM encompasses the spectrum of manufacturing applications and utilizes the computer to integrate plant operations. NC processors are a vital application with CAM, however, current processors do not lend themselves to this environment and would be extremely difficult to extend to do so. For example:

    — Large disk files containing data required by several CAM applications need to be utilized by NC processors (i.e., material files, tool files, and so on) for the automatic calculation of feeds and speeds at the user's option.

    — CAM applications lend themselves to be used in a terminal-oriented environment to increase efficiency and reliability, thereby improving flow time.

It is clear from this discussion that computer-aided part programming has received a great deal of attention over the past 20 years. The languages developed can be of tremendous advantage to the part programmer. With the advent of computer-aided design (CAD), the geometric information needed for part programming is already resident in the computer. The new generation of computer-aided part-programming systems is capable of automatically generating some limited part programs. Future systems should be able to do this more effectively.

Chapter 14

Rapid Prototyping

Objective

Chapter 14 provides an overview of existing rapid prototyping technologies and the basic principles that they use. The capabilities and limitations of each process are discussed, and general guidelines in part design and process selection are presented.

Outline

14.1 INTRODUCTION

The idea of building a three-dimensional object using two-dimensional cross sections is not a new one. Three-dimensional topographic maps have been built out of cut-out cardboard. The topographic curves are first traced onto the cardboard, and then they are cut. The cardboard layers are stacked and glued together to form the final 3-D map. However, this process was not able to handle arbitrary geometric shapes until very recently.

In the 1970s, Swanson, et al.[1,2] invented a means of using two intersecting radiation beams to produce three-dimensional objects in a liquid or other medium. However, without the solid model of the object and with the restriction of two radiation beams, the success of the approach was limited. But the ability to use solid models was not generally available until the 1980s. The two-radiation-beam idea allows the stationary liquid or glass to be penetrated by the beam and the intersection to cure the liquid or alter the glass structure. However, the beam intensity varies with the depth of penetration. In 1982, Charles Hull[3] invented the idea of using a UV laser to selectively cure a cross-sectional pattern of the object on the surface of a fluid medium; this is the stereolithography process. After each layer of the cross section is completed, the object is lowered to allow a thin layer of fluid to cover the completed cross section. Repeating the process for each thin cross-sectional layer, a 3-D solid object is built. The end result is a totally new process and industry. The first stereolithography machine, the 3D Systems SLA-1 (StereoLithography Apparatus), was sold in 1988. Since then, many machines based on the layered manufacturing concept have been invented and used in industry. This new technology has been described using several different names: rapid prototyping, layered manufacturing, 3-D fabrication, desktop manufacturing, automated fabrication, tool-less manufacturing, free-form fabrication, time-compressed manufacturing, etc.

The material used in stereolithography is a photopolymer. When monomer (building blocks of polymer) resins in the vat are activated by a UV laser, they link together to form long chains of molecules; thus, a polymer is formed. Early polymer resins were quite brittle and the quality of finish was poor. Objects made using the process were used only for concept models. In 1988, Ciba–Geigy initiated intensive research into developing new polymers. Now, many different types of polymers with much better properties are available for use in stereolithography applications.

What has made the stereolithography process a success is not only the use of a UV laser to cure layers of polymer, but also it is just as important to have the data necessary to drive the process. Since objects are built layer-by-layer and the layer thickness is small, it takes a large number of layers to build an object. For example, for a layer thickness of $0.005''$ and an object height of $5''$, the total number of layers is $5''/0.005'' = 1000$. Each of the 1000 cross sections has to be calculated and represented in order to control the laser to draw hatch patterns (to fill the cross-sectional areas). Such a task is far too complex and tedious for humans to do. Traditional 3-D wireframe or surface models lack the volume property; thus, they cannot generate correct cross sections. (See Chapters 4 and 5.) A solid model is the only usable representation.

It was not until the late 1980s that solid-modeling technology matured for practical use. The stereolithography format (STL) file used in most rapid prototyping machines is a simplified solid model consisting of only triangular faces (facets).

---

[1]Swanson, Wyn K., "Method, medium and apparatus for producing three-dimensional figure products," US Patent 4041476, August 9, 1977.

[2]Swanson, Wyn K. and Stephen D. Kremer, "Three dimensional systems," US Patent 4078229, March 7, 1978.

[3]Hull, Charles W., "Apparatus for production of three-dimensional objects by stereolithography," US Patent 4575330, March 11, 1986.

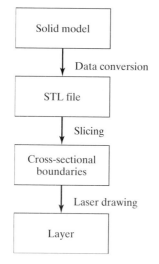

Figure 14.1   The stereolithography process.

The machine controller can slice the model efficiently (mathematically calculate the intersection between the model and equation $z = h_i$, where h_i is the z coordinate of layer i). Each slice consists of one or several closed-boundary piece-wise linear curves. Hatching algorithms are used to generate the desired laser drawing paths. Figure 14.1 shows this process.

Since the 1990s, many other rapid prototyping processes have been invented. Almost all of these processes are based on the same general concept, according to which the layer-build process (physical object creation) uses different chemistry and physics. As mentioned in the beginning of the chapter, one way to make a three-dimensional topological map is to cut pieces of cardboard and stack them together. They can be cut with either a computer-controlled laser or a knife. The stacking can be done either by hand or done automatically. The layer can also be produced by squeezing molten wax or plastic on the previous layer. Or a glue/binder can be spread on a layer of small, uniform-sized powder particles. As a matter of fact, particles of any materials could be used. Of course, if the laser is strong enough, powders could simply be sintered together instead of glued together.

Another variation of the process combines the concept of using powder and molten material by squeezing or spreading molten powders (droplets) onto a layer. A heat source, such as a high-powered laser, melts the powder before it touches the surface of the object. Another process is based on the process of welding layers of metal sheets together. The layers can be cut by machine or by wire-EDM (Electrical Discharge Machining). In this case, the positioning of the sheets can be a difficult problem. These processes are limited only by the designer's imagination. A lot of different build processes have been tried, and definitely more will come in the future. The success of a process is based on the cost, accuracy, speed, and material properties. So far, none of these processes can produce objects with material properties and accuracies that are comparable to those made by machining. This continues to be a major disadvantage of the rapid prototyping (RP) processes.

Early rapid-prototyping parts were used primarily for visualization of the general geometry. However, as material properties and build accuracy improved, parts made from the RP process have been used in some functional applications. The applications of rapid prototyping processes can be classified into three categories: form, fit, and function. *Form* implies that the prototype is used for visualization. Only the general shape is of interest. Early rapid-prototyping systems could produce objects for form applications only. *Fit* requires dimensional accuracy. Prototypes are used to check the assembly fitting. Dimensional accuracy and surface-finish requirements are much higher than those for form. Currently, a dimensional accuracy of $\pm 0.005''$ is achievable. Finally, *Function* means that the prototypes will be used for function checking, where a load is applied to the prototype as it would be applied to the real parts. To ensure success, the material property must be similar to that used in the final product. Prototypes built using starch powders and wax do not have the properties necessary for most functional applications. On the other hand, prototypes built using epoxy, nylon, or metal powders are much stronger. They are good enough for certain applications. However, none of them have achieved the strength of parts made from machining or even forming processes.

In the next section, several popular rapid prototyping processes are introduced. The modeling of rapid prototypes is discussed in the next section. Finally, the cost justification issues are discussed in the last section.

14.2 RAPID PROTOTYPING PROCESSES

Currently, there are quite a number of commercially available rapid-prototyping processes. Many more are being developed either in academic research laboratories or in industry. In this section, only the popular commercial systems will be introduced. All have been used by industry and are available to the general public. Proprietary systems and systems still being developed are omitted for good reasons: they are not generally available and may not be commercially viable.

14.2.1 Stereolithography Apparatus (SLA)

The stereolithography apparatus was the first commercial RP machine,[4] and it is still the most widely used rapid-prototyping technology. The stereolithography process itself was discussed in the introduction of this chapter; thus, it will be omitted here. However, the device itself and the materials used in the process will be further discussed.

Stereolithography builds plastic parts layer by layer by tracing a UV laser beam on the surface of liquid photopolymer in a vat. This class of materials, originally developed for the printing and packaging industries, quickly solidifies wherever the laser beam strikes the surface of the liquid. Once one layer is completely traced, it is lowered a small distance (i.e., layer thickness) into the vat and a second layer is traced right on top of the first. The self-adhesive property of the material causes the layers to bond to one another, and they eventually form a complete three-dimensional object after many such layers are deposited.

---

[4]3D Systems started by Charles Hull and others.

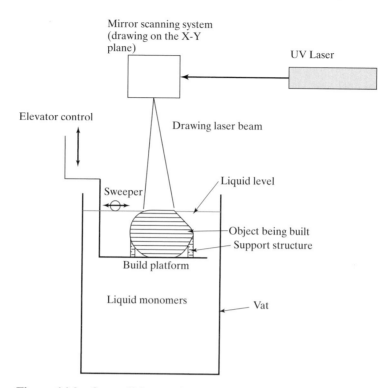

Figure 14.2   Stereolithography machine.

A stereolithography apparatus is illustrated in Figure 14.2. The size of the vat determines the size of the part that can be built. Before the process can start, the vat must be filled with liquid monomer (the building block for polymers). However, in a large vat, the cost of the monomer can be very high. A build platform has an elevator that tells it to move down one layer thickness at a time. After a layer has been drawn, the platform is lowered. Then, a sweeper levels the liquid before the laser hatching can start. This laser process is controlled by a scanner system with a rotating mirror. The mirror reflects the laser beam onto the desired location on the surface of the liquid. When there is an overhang, a support structure must be build to support the part. Of course, it is made of the same material, since there is only one monomer in the vat. However, the hatch pattern that draws the support is not as dense as that used for the part. Thus, it takes less time and less material to build the support. The supports are either manually or automatically designed. Upon completion of the fabrication process, the object is elevated from the vat and the supports are cut off.

The total build time is equal to the total drawing time plus the platform moving time and the sweeping time. Since thousands of layers may be needed to build a part, the build time could easily be several hours long.

This process is a polymerization process—the process of linking small molecules (monomers) into larger molecules (polymers) composed of many monomer units.

For example, consider the following example, which links vinyl-type monomers to the polymer on the right-hand side:

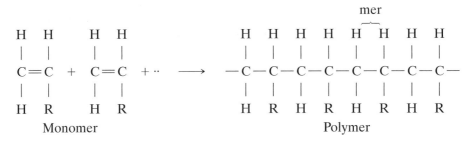

A commonly used material is acrylate. An acrylate monomer is a vinyl-type monomer with —COOH carbon–carbon double bond instead of the "R."

The polymerization of an acrylate monomer is an energetically favorable reaction. It generates a large amount of heat. Also, acrylate formulations can be stabilized to remain unreacted indefinitely at ambient temperature. A catalyst is required for the polymerization to proceed.

Since a catalyst is a free radical, it may be generated either thermally or photochemically. Stereolithography uses photons to general radical. Photoinitiators (I) absorb photons (hv) of the appropriate frequency to generate free radicals:

$$I + hv \rightarrow 2\,R\bullet$$

"$\bullet$": radical

In this reaction, h is Planck's constant, v is the frequency, and hv is the photon energy. The reaction proceeds as follows:

$$R\bullet + M \longrightarrow (R - M)\bullet$$

The macroradical creates a larger macroradical:

$$(M_n\bullet) + M \longrightarrow k_p\,(M_{n+1}\bullet)$$

Here, k_p is the reaction rate for polymerization and M denotes a mer. This equation shows how monomers are connected with the help of a radical. One radical can polymerize 1000 acrylate monomers.

Therefore, only low-power ultraviolet lasers are needed. But lasers require regular maintenance and tube replacement. Indeed, it can cost several thousand dollars per month to run the laser in the SLA machine, depending on the use.

Stereolithography is generally considered to provide a greater accuracy (±0.005 to ±0.010) and surface finish (200–300 μin.) than any rapid-prototyping technology can achieve. Over the years, a wide range of materials that mimic the properties of several engineering thermoplastics have been developed. Limited selective color changing materials for biomedical and other applications are available, and ceramic materials are currently being developed. The technology is also notable for the large object sizes that are possible. On the negative side, working with liquid materials can be messy, and parts often require a postcuring operation (the laser drawing process does not completely cure the polymer) in a separate ovenlike apparatus using UV light for a more complete cure and greater polymeric stability. The SLA systems can cost $100K to $800K and the resins can cost up to $300/gallon.

Several models of SLA machines are available from 3D Systems. At the time of this writing, all three machines use Nd:YVO$_4$ laser at a wavelength of 354.7 nm. The power output range is from 100 mW to 800 mW. The beam size is about 0.2–0.3 mm. The layer thickness is about 0.05mm. The laser drawing speed is about 5.0 m/sec. The maximum build volume is 508 × 508 × 584 mm^3. Up-to-date information can be found at the 3D System's website.

14.2.2 Selective Laser Sintering (SLS)

The selective laser-sintering process utilizes the capability of a laser to deliver heat at a focused spot to melt plastic powder. Instead of using photopolymer as in the SLA process, a layer of plastic powder is laid on the table, and a more powerful laser is used to draw (heat) the powder. Layer by layer, a 3-D part can be built.

A roller used to spread uniform-sized thermoplastic powder over the surface of a build cylinder. The cylinder's piston moves down one object-layer thickness to accommodate the new layer of powder. The powder-supply piston moves up incrementally to supply a measured quantity of powder for each layer. The excessive powder is swept off the build tank and recovered for later use. A laser beam then traces over the surface of this tightly compacted powder to selectively melt and bond it to form a layer of the object. The fabrication chamber is maintained at a temperature just below the melting point of the powder, so that heat from the laser need only elevate the temperature slightly—enough to cause sintering. This greatly speeds up the process. The process is repeated until the entire object is fabricated.

After the object is completely built, the piston is raised to elevate it. Excess powder is vacuumed away and a part is completed. No supports are required with this method since overhangs and undercuts are supported by the solid powder bed.

A variety of thermoplastic materials such as nylon, glass-filled nylon, thermoplastic elastomer, polyamide, and polystyrene are available. It is also possible to use

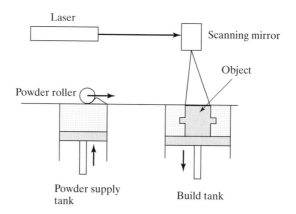

Figure 14.3   Selective laser-sintering device.

plastic-coated metal or ceramic powders. When metal or ceramic powders are used, the finished part is sintered in an oven and infiltrated with metals such as bronze. The surface finishes (300 to 400 μin.) and accuracy ($\pm 0.005''$ to $\pm 0.010''$) of SLS parts are not quite as good as with stereolithography, but the material properties can be quite close to those of the intrinsic materials. The build speed is about 1/8″ vertical inches per hour. A typical machine will cost $250K to $400K.

Several machines can be found on the market, including one from 3D Systems. This machine can build parts using thermoplastic, thermoplastic elastomer, metal, and composite. The machine uses a 25 W or 100 W CO^2 laser. The build volume is $370 \times 320 \times 445$ mm^3 ($14.5 \times 12.5 \times 17.5$ in^3).

EOS of Germany produces several machines. One of them is used to build sand cores or molds. The build volume is $720 \times 380 \times 380$ mm^3 and the speed is 2500 cm^3/h. Another machine is used to build steel molds. It has a build volume of $250 \times 250 \times 250$ mm^3. The layer thickness of this machine is 20–100 microns. It is designed to make accurate steel parts. The speed can reach up to 54 cm^3/h. To build steel parts, a much more powerful 200 W CO_2 laser is used. To make plastics, a machine that uses a 50 W CO^2 laser is available. It uses polyamide or polystyrene powders. The build volume is $700 \times 380 \times 580$ mm^3.

14.2.3 Inkjet-Based Processes

There are two versions of inkjet-based process. The first is based on thermoplastics, the second on photopolymers. The idea is to use the inkjet to print layers of liquid plastics (polymer) that are then solidified through temperature change (thermal-phase change) or UV light curing (photopolymer-phase change). Both processes are shown in Figure 14.4 and are examined next.

a. Thermal-Phase-Change Inkjets   Solidscape's inkjet machine uses a single jet for its plastic build material and a single jet for its waxlike support material, which is held in a melted liquid state in reservoirs. The liquids are fed to individual jetting

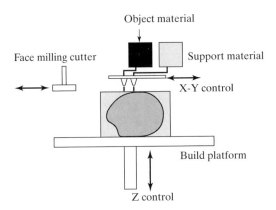

Figure 14.4   An inkjet machine.

heads, which then squirt tiny droplets of the materials as they are moved in X–Y fashion in the pattern required to form a layer of the object. The materials harden by rapidly dropping in temperature as they are deposited.

After an entire layer of the object is formed by jetting, a milling cutter flattens the surface and ensures a uniform thickness. The process is repeated to form the entire object. Following completion of the object, the wax support material is either melted or dissolved.

The most outstanding characteristic of the Solidscape system is its ability to produce extremely fine resolution and surface finishes, essentially equivalent to those produced by CNC machining. However, the technique is very slow for large objects. While the size of the machine and materials are office friendly, the noise of a milling head may be objectionable in an office environment. The materials selection is also very limited. The Solidscape machine has a build volume of $30.48 \times 15.24 \times 14.24$ mm^3 ($12 \times 6 \times 6$ in^3). The layer thickness is 0.013–0.076 mm (0.0005–0.0030 in.). The accuracy is $\pm 0.001''$ or 0.025 mm per inch in X, Y, and Z. The surface finish can achieve 32–63 μin (RMS).

Other manufacturers use considerably different inkjet techniques, but they all rely on squirting a build material in a liquid or melted state, which cools or otherwise hardens to form a solid on impact. 3D Systems has a similar system called the Thermojet™ printer. It has a build volume of $250 \times 190 \times 200$ mm^3 ($10 \times 7.5 \times 8$ in^3), and the resolution is $300 \times 400 \times 600$ dpi. It uses several hundred nozzles in a wide head configuration, as well as a hair-like matrix of build material to provide support for overhangs that can be easily brushed off once the object is complete. This machine is much faster than the Solidscape approach, but its surface finish or resolution is not as good.

All thermal-phase-change inkjets have material limitations and make fragile parts. Their applications range from concept models to precise casting patterns for industry and the arts, particularly jewelry.

b. Photopolymer-Phase-Change Inkjets

The photopolymer-phase-change inkjet process replaces the laser with the inkjet. The major advantage of the inkjet is its low initial and maintenance costs. Due to the low cost, it is also possible to use an array of inkjets instead of a single inkjet. This increases the printing speed. The object's photopolymer material and the support material are both squirted out of the inkjet nozzle. An inexpensive UV flood lamp mounted on the inkjet head cures the photopolymer. (The material changes from liquid phase to solid phase.) The support material is later washed away.

In early 2000, Objet Geometries introduced its first machine based on **PolyJet™** technology. It is a potentially promising replacement for stereolithography. Both the low initial system price (approximately $65K) and specifications that are similar to laser-based stereolithography systems (costing ten times as much) make this technology an important one. The **PolyJet™**-based machine has a resolution of 600×300 dpi and a layer thickness of 16 microns ($0.0006''$). The **Eden™** 333 machine has a build volume of $336 \times 336 \times 200$ mm^3 ($13.2 \times 12.8 \times 7.9$ in^3). The machine is also marketed by Stratasys.

14.2.4 3-D printing

Three-dimensional printing was developed at MIT under a National Science Foundation grant. Several companies licensed this technology. A typical 3-D-printing machine is illustrated in Figure 14.5.

The process starts by depositing a layer of powdered object material at the top of a fabrication chamber. To accomplish this, a measured quantity of powder is first dispensed from a similar supply chamber by moving a piston up incrementally. The sweeper blade then distributes and compresses the powder at the top of the fabrication chamber. The inkjet head subsequently deposits a liquid adhesive onto the layer of the powder in a two-dimensional pattern to form a layer of the object.

Once a layer is completed, the fabrication piston moves down by the thickness of a layer. The process is repeated until the entire object is formed within the powder bed. After completion, the object is elevated and the extra powder is vacuumed away, leaving a "green" object. No external supports are required during fabrication, since the powder bed supports overhangs. The green object can be strengthened in two ways: first, by infiltrating it with wax; second by coating it with epoxy, elastomeric urethane, or cyanoacrylate. This increases the material strength or elastic property (using elastomerica urethane).

Three-dimensional printing offers the advantages of speedy fabrication (1 vertical inch per hour) and low materials cost. In fact, it is probably the fastest of all the RP methods. Recently, color output has also become available. However, there are limitations on resolution ($\pm0.020''$ accuracy), surface finish, part fragility, and available materials. The closest competitor to this process is probably fused deposition modeling.

The 3-D printing process is represented by Z Corporation's 3-D printers. At the time of this writing, the largest build volume was $20'' \times 24'' \times 16''$. The most popular materials used in Z Corporation's machines are starch and plaster. The normal layer thickness is $0.003''-0.010''$, depending on the powder used.

Soligen developed a variation of the process. Parts are built out of ceramic powder, which is fused together by a liquid binder from an inkjet-printer head. Soligen calls this process *Direct Shell Production Casting*. The available materials include aluminum, magnesium, copper, brass, bronze, cast iron, stainless steel, super alloy, and so on. The build volume is $14'' \times 18'' \times 14''$ in^3. For parts smaller than $1''$, the accuracy is $\pm0.021''$ per inch. For parts greater than $6''$, the accuracy is $\pm0.003''$ per inch.

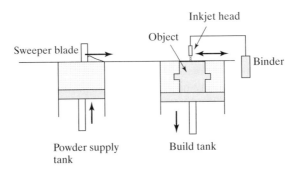

Figure 14.5   3-D printing.

Although 3-D printing is relatively inexpensive, there are costs beyond the material cost. The printer heads tend to clog if the machine is not used regularly, and they must be replaced ($75/each). The parts tend to be delicate; strengthening them requires infiltration with CA sealer. The material cost is very low ($1/inch^3); however, the sealer is expensive ($25 per 50 gram bottle), and replacement printer heads are expensive, too. The powder also has a shelf life of a few months.

14.2.5 Laminated-Object Manufacturing (LOM)

A laminated-object manufacturing machine is shown in Figure 14.6. Profiles of object cross sections are cut from paper or another material by means of a laser (Figure 14.7). The paper is unwound from a feed roll onto the stack. Then it is bonded to the previous layer with the use of a heated roller, which melts a plastic coating onto the bottom side of the paper. The profiles are then traced by an optics system that is mounted to an X–Y stage.

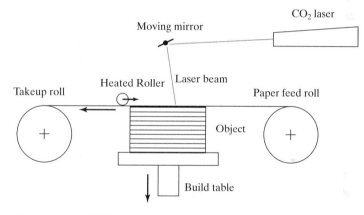

Figure 14.6   LOM Process.

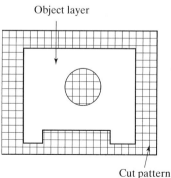

Figure 14.7   Cut layer.

After the layer cutting is complete, excess paper is cut away to separate the layer from the web. Waste paper is wound on a take-up roll. The method is self-supporting for overhangs and undercuts. Areas of cross sections are flagged for removal in the final object and are heavily crosshatched with the laser to facilitate their removal. However, it can be time consuming to remove extra material for some geometry.

Variations on this method have been developed by many companies and research groups. In general, the finish, accuracy ($\pm 0.015''$), and stability of paper objects are of a lower quality than that of materials used with other RP methods. However, the material costs are very low; the objects look and feel like wood, and they can be worked and finished in the same manner. The build process is also relatively fast (1/4″ vertically per hour). This has fostered applications such as pattern making for sand castings. While there are limitations on materials, work has been done with plastics, composites, ceramics and metals. Some of these materials are available on a limited commercial basis.

The principal commercial provider of LOM systems, Helisys, ceased operation in 2000, so this product is currently sold through Cubic Tech. The higher-end machine has a build volume of $813 \times 559 \times 508$ mm^3 ($32 \times 22 \times 20$ in^3). Thus far it is, the largest build volume among all machines. The layer thickness is 0.003″–0.008″, and the material can be paper, plastic, or composite. A 50 W CO_2 laser is used to cut the material.

Another product is Kira's Paper Lamination Technology (PLT). It uses a knife instead of a laser to cut each layer and applies adhesive to bond layers using the xerographic process. The machine has a similar layer thickness as the Cubic Tech machines (when adding the paper plus adhesive thickness). The build volume is $400 \times 280 \times 300$ mm^3 ($15.7 \times 11 \times 11.8$ in^3).

14.2.6 Fused-Deposition Modeling (FDM)

FDM (Figure 14.8) is the second most widely used rapid-prototyping technology. (The first is stereolithography.) A plastic filament is unwound from a coil and supplies material to an extrusion nozzle. The nozzle is heated to melt the plastic and has a mechanism which allows the flow of the melted plastic to be turned on and off. The nozzle is mounted to a mechanical stage, which can be moved either horizontally or vertically.

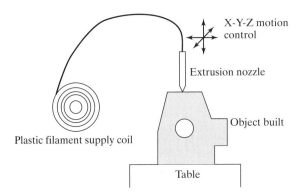

Figure 14.8 Fused-deposition modeling device.

As the nozzle is moved over the table in the required geometry, it deposits a thin bead of extruded plastic to form each layer. The plastic hardens immediately and bonds to the layer below. The entire system is contained within a chamber, which is held at a temperature just below the melting point of the plastic.

Several materials are available for the process, including acrylonitrile-butadiene-styrene (ABS) and simple wax. ABS offers good strength; recently, polycarbonate and polysulfone materials have been introduced to further extend the capabilities of this method (in terms of strength and temperature range). Support structures are fabricated for overhanging geometries and are later removed by breaking them away from the object. A water-soluble support material, which can simply be washed away, is also available.

The FDM method is office-friendly and quiet. Fused-deposition modeling (FDM) is fairly fast for small parts (a few cubic inches) or those that have tall, thin form factors. It can be very slow for parts with wide cross sections, however. The finish of the parts produced with this method has greatly improved over the past few years, but still is not quite as good as that produced by stereolithography. The accuracy is limited to typically 0.010″. The closest competitor to the FDM process is probably three-dimensional printing.

The Stratasys FDM machine can build objects using ABS plastics and polycarbonate. Several models are available; the largest has a build volume of $600 \times 500 \times 600$ mm^3 ($16 \times 14 \times 16$ in^3). The layer thickness is 0.007″–0.013″. The Solidscape (formerly Sanders) machine may produce objects using thermoplastics or wax. It has a build volume of $304 \times 152 \times 152$ mm^3 ($12 \times 6 \times 6$ in^3). The layer thickness is 0.0005″–0.007″.

14.2.7 Laser-Engineered Net Shaping (LENS)

The LENS process (Figure 14.9) builds a metal object using molten metal droplets. Metal is delivered in the powder form through a nozzle. As soon as the metal powder

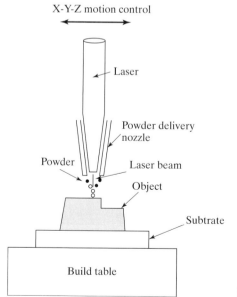

Figure 14.9   Laser-engineered net shaping.

leaves the nozzle, a powerful laser beam melts it into a liquid droplet. The deposit of this droplet is controlled by positioning the nozzle in the three-dimensional space. Initially, the droplets are deposited on a substrate. The subsequent droplets are fused onto the previous layer. The process uses commercially available metal powders, thus reducing the material cost. To deliver the power needed for melting the metal, a 1000 W Nd:YAG laser is used. The machine has a build volume of $18 \times 18 \times 42$ in^3. The build speed is 0.5 in^3/hr. Feature sizes of 0.030″ and greater can be made.

14.3 DESIGN MODELING FOR RAPID PROTOTYPING

Rapid prototyping (RP) is a process that takes a solid model and converts it into an artifact. Before the physical process can begin, the design model must be exported to STL or SLC format. Since STL format was developed first, it has been adopted by most RP machine manufacturers. This section introduces both the STL and SLC formats. The slicing of the model into cross sections will also be discussed.

14.3.1 The STL File

An STL file has the extension "STL" on its file name. The purpose of using STL format is to make the solid model as simple as possible, so the RP machine can slice it in a timely fashion. An STL model is a tessellated triangular B-Rep. It is consists of only triangular planar faces (facets) (Figure 14.10). Any surface, either planar or curved, is approximated using triangular facets. It contains only geometric entities; everything else (such as attributes, dimensions, and tolerances, etc.) is removed during the export process.

A triangular facet is the simplest representation since all faces are planar and bounded exactly by three linear edges. There is no hole allowed in the facet; thus, there is only one outer loop. Since all the edges are linear, there is no need to define them. A facet is defined by facet normal and the coordinates of the three vertices.

There are two STL file formats: ASCII and binary. ASCII can be read by humans; however, it is inefficient in data storage because its file size is much larger than binary. Regardless of whether the file is in ASCII or binary format, the data are identical. In practice, before importing an STL file, the user must specify whether it is ASCII or binary. In exporting a solid model to the STL file (except for debugging purposes), it is advisable always to output it in binary format.

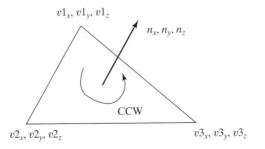

Figure 14.10   A facet.

Consider an ASCII STL file that stores information pertaining to a solid. Since a solid consists of many facets, each facet is defined by a normal and three vertices. The facet section repeats for each facet in the solid, as follows:

```
solid name
facet normal nᵢ   nⱼ   nₖ
             outer loop
                        vertex v1ₓ   v1_y   v1_z
                        vertex v2ₓ   v2_y   v2_z
                        vertex v3ₓ   v3_y   v3_z
             endloop
          endfacet
endsolid name
```

Example 14.1.

The cube shown in Figure 14.11 is represented by an STL file. Only the facet with the normal vector is represented in this example. Other facets are not captured in the STL file, which looks like this:

```
solid example_cube
facet normal 1.0    0.0    0.0
          outer loop
                     vertex 1.0    0.0    0.0
                     vertex 1.0    1.0    1.0
                     vertex 1.0    0.0    1.0
          endloop
       endfacet
       ... (next facet)
       endsolid example_cube
```

The syntax of a binary version of the STL file is

```
<Binary STL file>::=<STL file entity name><facet number N><facet info>
<STL file entity name>::=<80 bytes>
<facet number N>::=<4 bytes long integer>
<facet info>::=<facet normal><facet vertices><2 bytes spaces>
<facet normal>::=<lx, ly, lz, float, 12 bytes>
<facet vertices>::=<x1, y1, z1, x2, y2, z2, …,float, 36 bytes>
```

In this format, "::=" defines the term on the left-hand side with the terms on the right-hand side.

Example 14.2.

The same cube in Figure 14.11 is represented in a binary STL file. The file is

|<- Repeat for each facet -->|

Example_cube	12	1.0 0.0 0.0	1.0 0.0 0.0	1.1 1.1 1.1	1.0 0.0 1.0
80 bytes	4 bytes	12 bytes	12 bytes	12 bytes	12 bytes

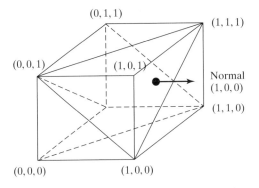

Figure 14.11 Triangular tessellation of a unit cube.

In the preceding example, ASCII representation is used in each field to show the content. However, in an actual binary file, the data in each field are represented in binary. They are not easily readable by humans.

While the file entity name and the facet number in the file fields appear only once, the facet information field repeats for each facet.

The face normal points to the outside of the object, and the order of the vertices is used for a redundancy check. The vertices are listed in a particular order, when the right-hand rule is applied, it will point the thumb outward, so the order is counterclockwise. Inconsistent inside–outside information is checked before a model is sliced.

Another rule used in the STL file is the vertex-to-vertex rule. Each facet shares exactly two vertices with each of its three adjacent facets. Figure 14.12 shows an invalid case.

Three consistency rules can be used to check the model:

- The number of faces must be even.
- The number of edges must be a multiple of three.
- Two times the number of edges must equal three times the number of faces.

STL file errors Almost all STL models were produced by converting CAD solid models using algorithms built into the CAD package. Unfortunately, algorithms are not always correct. Often, due to the error in the algorithm or the numerical data

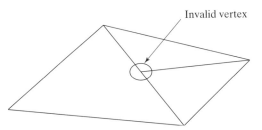

Figure 14.12 Invalid vertex in an STL file.

round-off, an STL model becomes incorrect. An incorrect STL model will cause the slicing process to fail; then, the part cannot be built. It is important to check the validity of the model before it is sent to the machine for building. In this section, some common errors are discussed.

A. Topology error An STL model must be a manifold model, as discussed in Section 5.5.1. In slicing a *non*manifold model, the cross section may not be closed. An open cross section is said to leak. The process used to create the layer does not know where to stop; thus it leaks. A manifold model must satisfy Euler's equation (Equation 5.63), which, for a single solid, is

$$F - E + V = 2$$

where

F = number of faces
E = number of edges
V = number of vertices

However, STL allows multiple solids to be present in the same file. A modified Euler's equation for a multiple solid is

$$F - E + V = 2B \tag{14.1}$$

where B = number of separate solid bodies.

Commonly found errors are as follows:

- *Intersecting facets.* A poor-quality Boolean operator algorithm may generate intersecting facets.
- *Mismatched adjacent edges.* The triangulation of two adjacent faces creates mismatched facets. In Figure 14.13(b), the original adjacent edge is a curve, so the triangulation on the upper side of the edge does not match the triangulation on the lower side of the edge.

B. Degenerated facets During the conversion to STL, the facets sometimes degenerate into an edge or a vertex. These errors are caused by a rounding error. When the facets are very narrow and small, the truncation of the digits may make the facet

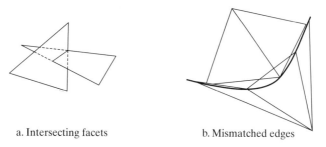

a. Intersecting facets   b. Mismatched edges

Figure 14.13   Common manifold errors.

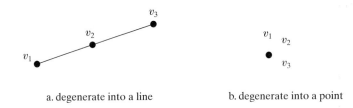

a. degenerate into a line   b. degenerate into a point

Figure 14.14   Degenerated facet errors.

into a line or a point. These errors are less serious during slicing. Correct cross sections can still be generated. However, they may cause other applications, such as support generation, to fail. The two types of degenerated-facet errors are presented in Figure 14.14.

C. Model errors   These errors are not generated during the STL conversion process, but rather inherited from the incorrect solid models. These errors may cause inconvenience to the RP process.

14.3.2 The SLC format

While the STL model is a simplified B-rep model, an SLC model consists of $2\frac{1}{2}$-D contours. The contours are actually the sliced cross sections defined on the X–Y plane. Each slice has a specific z height and one or more contours. Each contour is represented by either an interior or an exterior closed-boundary polyline. SLC data can be generated by slicing the CAD solid model or can be taken directly from systems that produce data arranged in layers, such as CT-scanners.

A boundary is a closed polyline (Figure 14.15(a)) representing interior or exterior solid material. An exterior boundary has its polyline listed in counterclockwise order. The solid material is inside the polyline. An interior boundary has its polyline listed in clockwise order and the solid material is outside the polyline.

Consider a sample SLC file that contains two layers. The first layer has two contours, each with five vertices. The second layer has only one contour, which is defined

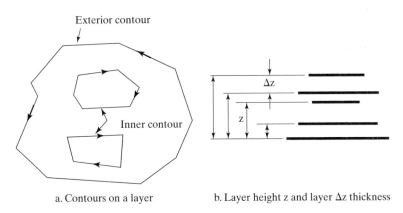

a. Contours on a layer   b. Layer height z and layer Δz thickness

Figure 14.15   SLC layers and contours.

by four vertices. In the following example, the bold characters should be typed as is. The italic characters should be substituted with their corresponding values:

```
Slice V0
FILE(filename)
SLICES()
Z z dz              #  Layer 1, z = absolute z of layer,
x y                 #  dz = layer thickness
x y                 #  Vertex
x y                 #  n vertices, each represented by (x, y)
x y
x y
C                   #  End of a contour
x y                 #  New contour
x y
x y
x y
x y
C
Z z dz              #  Begin a new layer
x y                 #  New contour on a new level...
x y
x y
x y
C                   #  End of a contour.
END
```

It is obvious that, when an SLC file is used, there is no need to slice the model. However, there will be a large file size.

14.3.3 Model slicing

The STL model is sliced by parallel slice planes $z = h_k$, where h_k is the height of slice k. The sliced layers are at a constant thickness $\Delta h = h_{k+1} - h_k$. The slicing algorithm for the STL file is rather simple, and the algorithm is divided into two major steps. During the first step, the intersecting lines are found and saved in a list. During step two, the intersecting lines are joined together to form contours. This is shown in a simple algorithm:

Step 1. Find the intersecting lines

The STL model used in the algorithm consists of m facets $f_1, f_2, \ldots, f_n$. Facet f_i consists of three vertices v_1^i, v_2^i, v_3^i. Vertex v_j^i has the coordinates $v_j^i.x, v_j^i.y, v_j^i.z$.

(a) Sort all the facets

- For $\forall f_i$, sort three vertices in order of decreasing z-coordinates ($v_j^i.z$).
- Sort all facets in decreasing order based on $v_1^i.z$, $v_2^i.z$, and $v_2^i.z$, in that order.

(b) Find the intersections

$\quad$ For $i = 1$ to m

$\quad\quad\quad$ If $v_3^i.z \leq h_k$ & $v_1^i.z \geq h_k$, then find_intersecting_points.

$\quad$ End

(c) Finding intersecting points

One of the intersecting points is between v_1^i and v_3^i. Find the intersection between lines: $z = h_k$ and $v_1^i v_3^i$.

If $v_2^i.z \le h_k$, find the intersection between lines $z = h_k$ and $\overline{v_1^i v_2^i}$.

Else the intersection is between lines $z = h_k$ and $\overline{v_2^i v_3^i}$.

An intersecting line L_l between two intersecting points p_1^l, p_2^l is stored in a list $IL = L_1 \rightarrow L_2 \rightarrow \cdots \rightarrow L_n$, where n is the number of lines.

Example 14.3.

Four facets are being sliced by $z = h_k$, as shown in Figure 14.16. First, the vertices in all facets are sorted on the basis of their z-coordinates. The sorted result is shown by the j index of the vertices, $v_1^i \rightarrow v_2^i \rightarrow v_3^i$. Then the facets are sorted. The facet with the highest z-coordinate is on the top of the list. The order is $f_1 \rightarrow f_2 \rightarrow f_3 \rightarrow f_4$. Now the condition $v_3^i.z \le h_k$ & $v_1^i.z \ge h_k$ is checked. In this example, only f_2 satisfies the condition.

The intersections are between v_1^2, v_3^2 and v_1^2, v_2^2. Intersections between the line and the slicing plane can easily be found.

Step 2. Contour development

1. For each line in list IL, sort two points on the basis of their x-coordinate values.
2. Sort the lines in list IL according to $p_1^l.x$ values, where $IL = L_1 \rightarrow L_2 \rightarrow \cdots \rightarrow L_n$ and n equals the number of lines.
3. Let $i = 1$ and $m = 1$.
4. While IL is not empty, remove L_1 from IL and add it to the contour segment
   m: $C_m = C_m \longrightarrow L_1$, where $j = 1$.
5. If $p_2^{1,m} == p_1^i$, remove L_j from IL and add it to the contour segment m:
   $C_m = C_m \longrightarrow L_j$.

   Else if $p_2^{1,m} == p_2^i$, reverse p_2^i and p_1^i, remove L_j from IL, and add it to the contour segment m: $C_m = C_m \rightarrow L_j$.

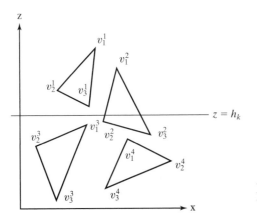

Figure 14.16 Finding intersections between the slicing plane and facets.

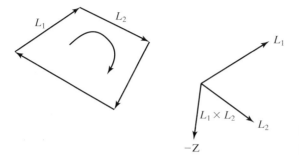

Figure 14.17 Clockwise path and cross product.

Else $j = j + 1$. If the end of the list IL has not been reached, repeat this step.

Else, if the $p_2^{k,m} == p_1^{1,m}$ contour m is completed, then $m = m + 1$. Go to step d.

Else there is an error: The contour is not closed.

Check all contours C_m, and make sure that the edges in C_1 are in the counterclockwise orientation and the edges in the other contours are in the clockwise orientation.

The orientation of the edges can be checked by using the cross product of two adjacent edges. Figure 14.17 is a closed contour in the clockwise orientation. Select the first two lines in the contour: L_1, L_2. The cross product is a vector pointing downward. (Use the right-hand rule.) It can easily be checked by verifying the z-coordinate of the cross-product vector. If the z-coordinate is negative, the vector points downward.

Example 14.4.

The edge list IL has 11 lines, as shown in Figure 14.18. They are already sorted properly. The process begins with adding L_1 to contour C_1. Since $p_2^1 \neq p_1^2$, check the next line in the list, $p_2^1 == p_1^3$. Then remove L_3 from the list and add it to $C_1 = L_1 \rightarrow L_3$. Continue the process, adding L_{10}, L_{11}, L_8, L_6 and L_2 to $C_1.C_1 = L_1 \rightarrow L_3 \rightarrow L_{10} \rightarrow L_{11} \rightarrow L_8 \rightarrow L_6 \rightarrow L_2$. Finally, when $p_1^2 == p_1^1$, contour C_1 is completed.

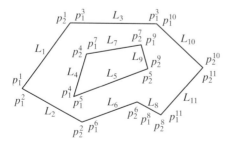

Figure 14.18 Contour development.

The next line in the list is L_4. Add L_4 to $C_2 : C_2 = L_4$. Repeat the preceding process to find that $C_2 = L_4 \rightarrow L_7 \rightarrow L_9 \rightarrow L_5$. The edges in C_1 need to be reversed in order to have a counterclockwise order.

Example 14.5.

A cell phone is designed with the use of a solid model (Figure 14.19), and the model is converted into an STL file (Figure 14.20). Finally, the STL file is sliced into 144 slices. Figure 14.21 shows three consecutive slices of the model.

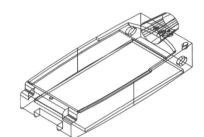

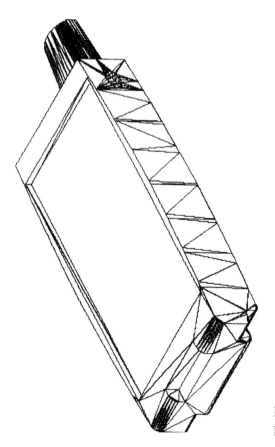

Figure 14.19   The solid model of a cell phone.

Figure 14.20   The STL file of the phone.

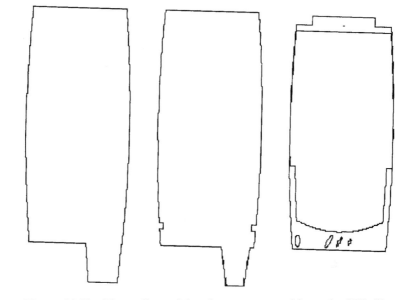

Figure 14.21   Three slices of the phone generated from the STL file.

14.4 CHOOSING A SYSTEM

As discussed in Section 2, there are quite a few rapid prototyping technologies and machines available. Choosing one is not an easy decision. To select a machine, first consider the purpose of the prototype. Prototypes can be classified into three categories: form, fit, and function. The form prototype conveys the design appearance or the general size and shape of the part. A form prototype does not need to have good dimensional control or a good surface finish. The material strength is not an issue either. The fit prototype is intended for checking the assembly of a design with other parts. It must be built at a tight tolerance and with a good surface finish. Finally, the function prototype is a prototype that can truly replace a "real" part. It is used to test the performance under the designed conditions. The dimensional accuracy must be high, and the material property must match the actual product material.

Based on this description, it is clear that the prototypes built on a 3-D printer using starch can only be form prototypes. The SLA process can satisfy both the form and fit requirements, but not the function requirements. A LENS prototype has the potential of being a function prototype for most mechanical-engineering applications. After the prototype's purpose has been determined, several technologies can be eliminated from the selection process.

Material properties cannot be ignored even for form prototypes. Each machine offers a limited set of materials. Typical material strength can be found in Table 14.1. It is necessary to check whether the material is strong enough for the intended use.

The economics of the process is always an important factor in determining the selection of a machine. A traditional project-selection model can be used to compare

TABLE 14.1 Material Properties

Process	Material	Tensile Strength MPa
SLA	Acrylate	24.2
	Epoxy	58.3
	AccuGen™	67
SLS	Polycarbonate	23
	Fine nylon	36
	Glass-filled nylon	49
LOM	Paper	
3-D printing	Starch	3
	Plaster	10
Traditional		
Manufacturing	Aluminum	90
	Copper	220
	Steel	485–1750
	Stainless steel	510

different alternatives. However, if the true cost of a rapid prototype machine is not used, the economics model will lead to the wrong conclusion. Thus, some of the important cost factors should be carefully considered in the analysis:

(a) *Machine cost and depreciation.* The machine will initially cost $30K to $800K.

(b) *Room preparation.* While some machines can be used in offices, others need special room preparation before installation. The electrical, air, structural, exhaust, and hazardous-chemical disposal requirements must be met.

(c) *Electricity consumption.* Many machines, especially those which use high-power lasers, can consume a lot of power. The electricity cost must be included.

(d) *Maintenance.* Usually, the annual maintenance contract costs 10% of the purchase price. Then the cost of replacement parts must be added. Lasers are good for about 5000 hours, and a replacement laser can cost more than $40K. The software maintenance also costs several thousand dollars a year.

(e) *Cleaning and finishing parts.* The prototypes that were just finished by the machine are not ready to be used. Postprocessing steps include curing, infiltration, cleaning, and finishing (e.g., removal of supports, minor surface machining, etc.); these steps are usually necessary.

(f) *Material costs.* Photopolymers can cost a few hundred dollars a gallon. In a large machine, the initial cost of filling the vat can cost $45K. It is also necessary to keep an inventory worth several thousand dollars. The photopolymers degrade in the vat and must be replaced regularly. For processes using metal powders, the shelf life is long. However, 20%–50% of the powders are wasted during the build process. (They are usually swept off the table and not recovered.) Powders made of starch and plaster definitely have a limited shelf life.

(g) *Staffing cost.* A staff is required to run and maintain the machines, clean the parts, and provide advice on design limitations.

There are also other factors, such as the technical support provided by the vendor. In the past decade, the RP industry has seen its share of failures. Since it is a new technology, many vendors are start-up companies. They may not be in the market for long. An orphaned machine will definitely not receive the proper support needed to keep it running.

In sum, machine selection is based on needs, capability, and economic justification. Each factor must be studied carefully before a decision can be made.

14.5 SUMMARY

Since its invention, RP has been successfully applied in many applications. The accuracy and strength of objects built by the process keep improving. In addition, the machines are widely available in industry, research institutes, and universities. For those who cannot justify purchasing a machine due to the lack of volume, many service bureaus are ready to help build parts at a cost.

However, RP is not a replacement for all manufacturing processes. It is a wonderful complement to traditional manufacturing processes. It provides both design and manufacturing engineers an alternative to build prototypes. For certain applications, RP can produce function prototypes and even final products (one-of-a-kind products). For other applications, it is good tool to help visualize designs and speed up design iterations. This chapter discussed the basic RP technologies and provided an overview of existing devices. Since RP is a young technology, many more ideas and devices are being developed all the time, so it is important to keep browsing manufacturing magazines and the Internet for up-to-date information.

14.6 KEYWORDS

Contour development	Photopolymer
Fit	Photopolymer phase-change inkjets
Form	Selective laser sintering
Function	SLC model
Fused-deposition modeling	Stereolithography
Laminated-object manufacturing	STL file
Laser-engineered net shaping (LENS)	Thermal phase-change inkjets
Layer thickness	Three-dimensional printing
Model slicing	UV laser
Monomer	

14.7 REVIEW QUESTIONS

14.1 What are the fundamental principles used in the stereolithography process?

14.2 Can 3-D wire-frame models be used in generating STL or SLC files?

14.3 What is a photopolymer? How does it work?

14.4 Why do some RP processes use high-powered CO_2 lasers and some use lower-powered UV lasers?

14.5 What are the advantages of photopolymer-phase change inkjets over stereolithography?

14.6 What is a thermal phase-change inkjet?

14.7 What are the differences between 3-D printing and the selective laser-sintering processes?

14.8 What kinds of material property can be expected from a part built using the LOM process?

14.9 What materials can be used by the fused-deposition modeling process?

14.10 What are the differences between ASCII and binary STL files?

14.11 Why are STL files used instead of general B-Rep models as the input to RP machines?

14.12 What are the pros and cons of using SLC files instead of STL files?

14.13 Why is facet normal defined in the STL file?

14.14 What are different types of errors found in STL files? What are the sources of those errors?

14.8 REVIEW PROBLEMS

14.1 A function prototype is needed for a hypothetical product. The material must be stronger than 50 MPa. The accuracy requirement is 0.010″, and the surface finish requirement is 60 μinch. Which processes would you recommend and why?

14.2 A $10 \times 8 \times 5$ in³ rectangular block is being built on an RP machine. The drawing line width is 0.008″. Two adjacent lines are drawn with 20% overlap and the layer thickness is 0.005″. The drawing speed is 500 inches per second. The elevator positioning time and the sweeping time are 1 second and 2 seconds, respectively. Compare the build times for setting the block at a height equal to 10″ and at a height equal to 5″.

14.3 A tetrahedron is defined by four triangular facets (Figure 14.22). Convert the model into an ASCII STL file. Use the algorithm discussed in this chapter to slice the STL model at $z = 4$. Write step-by-step instructions to show how the slicing operation is done.

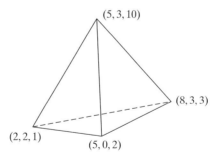

Figure 14.22   A tetrahedron.

14.9 REFERENCES

Beaman, Joseph J. *Solid Freeform Fabrication: A New Direction in Manufacturing with Research and Applications in Thermal Laser Processing.* Dordrecht: Kluwer Academic Publishers, 1997.

Binnard, M., *Design by Composition for Rapid Prototyping.* Boston: Kluwer Academic, 1999.

Chua, C. K., and K. F. Leong. *Rapid Prototyping: Principles & Applications in Manufacturing.* New York: J. Wiley, 1997.

Chua, C. K., K. F. Leong, and C. S. Lim. *Rapid Prototyping: Principles and Applications*. Singapore: World Scientific, 2003.

Hull, C. W. "Apparatus for production of three-dimensional objects by stereolithography," US Patent 4575330, March 11, 1986.

Jacobs, P. F. *Stereolithography and Other RP&M Technologies: From Rapid Prototyping to Rapid Tooling*. Dearborn, MI: ASME Press, 1996.

Jacobs, P. F. and D. T. Reid. *Rapid Prototyping & Manufacturing: Fundamentals of Stereolithography*. Dearborn, MI: Society of Manufacturing Engineers in cooperation with the Computer and Automated Systems Association of SME, 1992.

Kochan, D., *Solid Freeform Manufacturing: Advanced Rapid Prototyping*. Amsterdam: Elsevier, 1993.

Lu, L., J. Y. H. Fuh, and Y. S. Wong. *Laser-Induced Materials and Processes forRapid Prototyping*. Boston: Kluwer Academic Publishers, 2001.

Pham, D. T. and S. S. Dimov. *Rapid Manufacturing: The Technologies and Applications of Rapid Prototyping and Rapid Tooling*. London: Springer, 2001.

Swanson, W. K. and S. D. Kremer. "Three dimensional systems," US Patent 4078229, March 7, 1978.

Swanson, W. K. "Method, medium and apparatus for producing three-dimensional figure product," US Patent 4041476, August 9, 1977

3D Systems: http://www.3dsystems.com/

Cubic Technologies: http://www.cubictechnologies.com/

Envision Tech (Germany): http://www.envisiontec.de/

EOS: http://www.eos-gmbh.de/

Kira (Japan): http://www.kiracorp.co.jp/kira/rp/pr-E.htm

Object Geometries (Israel): http://www.objet.co.il/home.asp

Optomec Design: http://www.optomec.com/

Solid Scape: http://www.solid-scape.com/

Solidimension (Isarel): http://www.solidimension.com/

Soligen: http://www.soligen.com/

Stratasys: http://www.stratasys.com/

Z-Corp: http://www.zcorp.com/

Chapter 15

Industrial Robotics

Objective

Chapter 15 provides information about robotics and the industrial use of robots. The chapter demonstrates how industrial robots serve as an integral part of modern manufacturing systems. It also discusses the economic use of robots in manufacturing systems.

Outline

15.1 INTRODUCTION

15.1.1 Background

With a pressing need to upgrade productivity, manufacturing industries are turning more and more toward computer-based flexible or programmable automation. Currently, many automated manufacturing tasks are carried out by hardwired automation designed to perform predetermined functions. The inflexibility of these handwired machines has led to a broad interest in the use of industrial robots. An industrial robot is a general-purpose computer-controlled manipulator consisting of several rigid links connected in series by revolute or prismatic joints. Typically, one end of the robotic arm is attached to a supporting base, and the other end is free and equipped with a gripper or a tool to manipulate objects or perform assembly or fabrication tasks.

The concept of robotics, although not referred to by that term until relatively recently, has captured human imagination for centuries. One of the first automatic animals—a wooden bird that could fly—was built by Plato's friend, Archytas of Tarentum, who lived between 400 and 350 B.C. In the second century B.C. Hero of Alexandria described in his book, *De Automatis,* a mechanical theater with robotlike figures that danced and marched in temple ceremonies. In 1921, Karel Capek, the Czech playwright, novelist, and essayist, wrote the satirical drama *R.U.R.* (Rossum's Universal Robots), which introduced the word "robot" into the English language. The playwright coined the word to mean "forced labor;" the machines in his play resembled people, but worked twice as hard.

Generally, in manufacturing, robots, compared to humans, yield more consistent quality, yield more predictable output, and are more reliable. Compared with automated machines, the robots' flexibility becomes more useful than the rigidity of hardwired automation. As previously mentioned, a robot is essentially an arm fixed to a base on which it can move. The accurate and flexible characteristics of motion depend on the sophistication of the robot's control system. In contrast, a hardwired automated machine is inflexible and generally has no redundant degrees of freedom to allow it to process products outside its narrow focus. Figure 15.1 shows that a robot can be

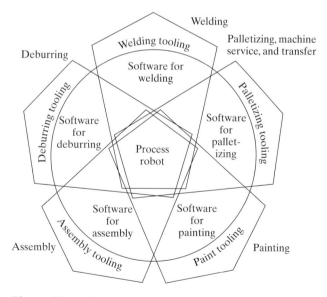

Figure 15.1   The use of appropriate software and hardware allows a standard robot to be customized to a specific task.

reconfigured (through special software and hardware accessories) to perform a wide range of tasks. Therefore, it is generally considered economically sensible to install a robot whose optional features are relatively low-priced when compared to the robot's base price, rather than a specialized machine for each task.

However, the robot is a generalized device; it has many redundant links and features, resulting in a slower process time and less accurate motions than a machine designed for specific tasks. Robots also have limited dexterity when compared with humans. If the tasks performed are in an uncertain environment, sophisticated sensors and controls are needed. The complexity of the control algorithm and high hardware and software costs may make the robotic application undesirable. Therefore, it is necessary to know the capabilities and limitations of robots before it is used in an application. The reprogrammable characteristic of robots, which makes them so attractive in a flexible manufacturing environment, actually may be a negative factor due to the high cost of program development. Since the introduction of industrial robots in the 1960s, there have been many successful applications reported. However, at the same time, many failures also occurred. Thus, it is important to recognize that the use of robotics is sound for the type of problem best suited for it. The technology is far from making a robot as capable as the one described in Capek's play. This chapter discusses the capabilities and limitations of robots and shows how to build an application. Some theoretical background on kinematics is also included. These materials provide a general background in industrial robotics.

15.1.2 Classification of Robots

The Robotics Institute of America defines a robot as "a reprogrammable multifunctional manipulator designed to move material, parts, tools, or other specialized devices through variable programmed motions for the performance of a variety of tasks." Robots like NC machines can be powered by electrical motors, hydraulic systems, or pneumatic systems. Controls for these devices can be either open or closed loop. In fact, many robotic developments have evolved from the NC industry, and many of the manufacturers of NC machines and NC controllers also manufacture robots and robot controllers.

Robots have been used in industry since 1965. Today, industrial robots are widely used in manufacturing and assembly tasks such as material handling, stock selection, welding, parts assembly, product inspection, paint spraying, machine loading and unloading, foundry application, space and undersea exploration, and in handling hazardous and toxic materials. They can also work in complete darkness, in poor light, in excessive heat and in noisy and confined environments.

A physical robot is normally composed of a main frame (or arm) with a wrist and some tooling (usually some type of gripper) at the end of the frame. An auxiliary power system (i.e., a hydraulic power source) may also be included with the robot. A controller with some type of teach pendant, joystick, or keyboard is also part of the system.

Robots are usually characterized by the design of the mechanical system. Generally, there are six recognizable robot configurations: (1) Cartesian, (2) gantry, (3) cylindrical, (4) spherical, (5) articulated, and (6) SCARA.

1. *Cartesian robots.* A robot whose main frame consists of three linear axes (joints) is called a Cartesian robot and is shown in Figure 15.2. The Cartesian robot derives

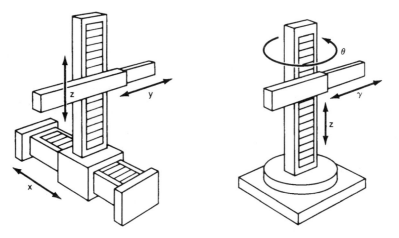

Figure 15.2   Rectilinear or Cartesian robot (From *Robotics in Practice* by Joseph F. Engelberger. Published by Kogan Page, 1980. Reproduced by permission of the publisher).

its name from the coordinate system. Travel normally takes place linearly in three axes. The X-, Y-, and Z-coordinates of a Cartesian robot can be easily derived from the equations

$$X = a \tag{15.1}$$
$$Y = b \tag{15.2}$$
$$Z = c \tag{15.3}$$

where a, b, and c are the joint variables.

2. *Gantry robots.* A gantry robot is a type of Cartesian robot whose structure resembles a gantry. This structure is used to minimize deflection along each axis (see Figure 15.3). Many large robots are of this type. The X, Y, and Z coordinates of a gantry robot can be derived using the same set of equations used for the Cartesian robot.

3. *Cylindrical robots.* A cylindrical robot has two linear axes and one rotary axis. The robot derives its name from the operating envelope—the space in which a robot operates—that is created by moving the axes from limit to limit (see Figure 15.4). The X-, Y-, and Z-coordinates of a cylindrical robot can be obtained from the equations

$$X = a \cos \alpha \tag{15.4}$$
$$Y = a \sin \alpha \tag{15.5}$$
$$Z = c \tag{15.6}$$

where α, a, and c are the joint variables.

4. *Spherical robots.* A spherical robot has one linear axis and two rotary axes (see Figure 15.5). Spherical robots are used in a variety of industrial tasks, such as

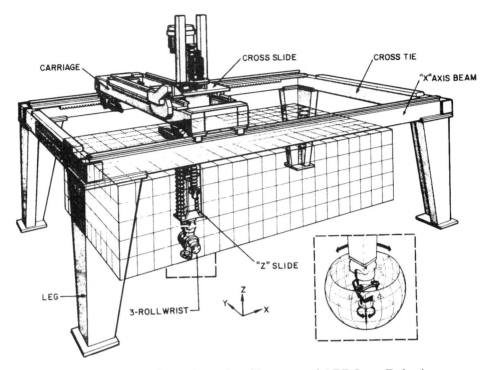

Figure 15.3 Gantry-configuration robot (Courtesy of ABB Inc.– Robotics, Automotive & Manufacturing Group).

welding and material handling. The X-, Y-, and Z-coordinates of a spherical robot can be obtained from the equations

$$X = a \cos \alpha \cos \beta \qquad (15.7)$$

$$Y = a \sin \alpha \cos \beta \qquad (15.8)$$

$$Z = a \sin \beta \qquad (15.9)$$

where a, α, and β are the joint variables.

5. *Articulated robots.* An articulated robot has three rotational axes connecting three rigid links and a base, as shown in Figure 15.6. An articulated robot is frequently called an anthropomorphic arm because it closely resembles a human arm. The first joint above the base is referred to as the shoulder. The shoulder joint is connected to the upper arm, which is connected at the elbow joint. Articulated robots are suitable for a wide variety of industrial tasks, ranging from welding to assembly. As with other types of robots, the X-, Y-, and Z-coordinates of an articulated robot can be obtained from the equations

$$X = [l_1 \cos \beta + l_2 \cos (\beta + \gamma)] \cos \alpha \qquad (15.10)$$

$$Y = [l_1 \cos \beta + l_2 \cos (\beta + \gamma)] \sin \alpha \qquad (15.11)$$

$$Z = l_1 \sin \beta + l_2 \sin (\beta + \gamma) \qquad (15.12)$$

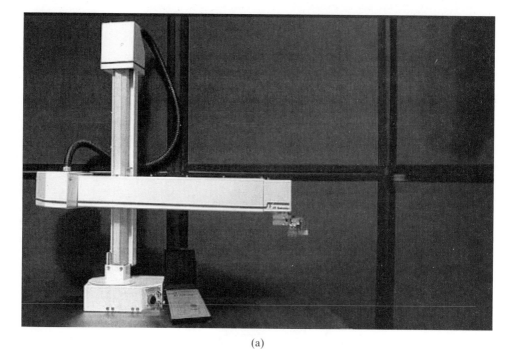

(a)

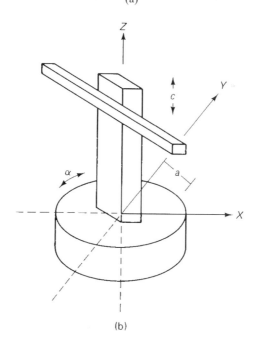

(b)

Figure 15.4   Joint variables of a cylindrical robot (Courtesy of ABB Inc.– Robotics, Automotive & Manufacturing Group).

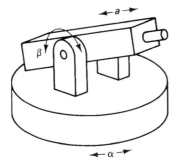

Figure 15.5 Joint variables of a spherical robot.

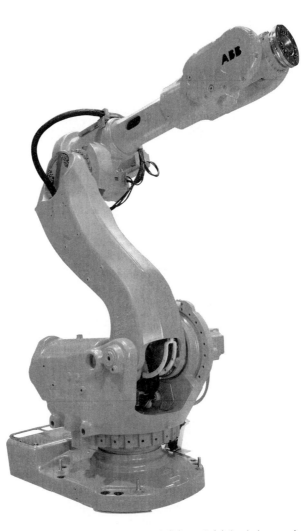

Figure 15.6 (a) An articulated robot and (b) and (c) its joint variables (Courtesy of ABB Inc.– Robotics, Automotive & Manufacturing Group).

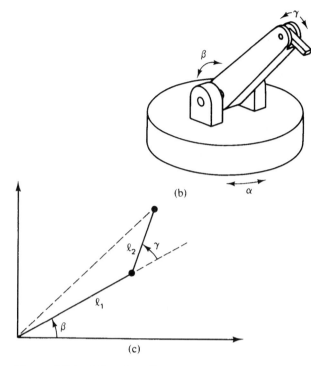

Figure 15.6 (*Continued*).

where α, β, are γ are the joint variables, and l_1 and l_2 are the lengths of the lower arm and upper arm, respectively.

6. *SCARA robots.* One style of robot that has become quite popular is a combination of the articulated arm and the cylindrical robot. The robot has more than three axes and is called a SCARA robot. It is used widely in electronic assembly. As illustrated in Figure 15.7, the rotary axes are mounted vertically rather than horizontally. This configuration minimizes the robot's deflection when it carries an object while moving at a programmed speed.

From the standpoint of the type of control, robots can also be classified as follows:

1. *Point-to-point (PTP) control.* A PTP robot (also called a bang–bang robot) is able to move from one specified point to another, but cannot stop at arbitrary points not previously designed. It is the simplest and least expensive type of control. Stopping points are often just mechanical stops that must be adjusted from operation to operation. This type of robot is also called a modular robot, because a PTP robot is typically constructed out of some building blocks (modules). The control is normally done using an electromechanical sequencer or a programmable controller. It is usually driven by compressed air. This type of robot is fast, accurate, and inexpensive.

Figure 15.7   SCARA configuration robot. (Courtesy of Adept Technology, Inc.)

2. *Continuous-path control.* A robot with continuous-path control is able to stop at any specified number of points along a path. However, if no stop is specified, the robotic arm may not stay on a straight line or a constant curved path between specified points. Every point must be explicitly stored in the robot's memory.

3. *Controlled-path (computed trajectory) control.* Control equipment on controlled-path robots can generate straight lines, circles, interpolated curves, and other paths with high accuracy. Paths can be specified in geometric or algebraic terms in some of these robots. Good accuracy can be obtained at any point along the path. Only the start and finish points and the path definition function are required for control.

15.1.3 Workspace Envelopes

An important concept in robotic applications is the workspace envelope of a robot. It is constrained by its mechanical systems configuration. Each joint of a robot has a limit of motion range. By combining all the limits, a constrained space can be defined. A

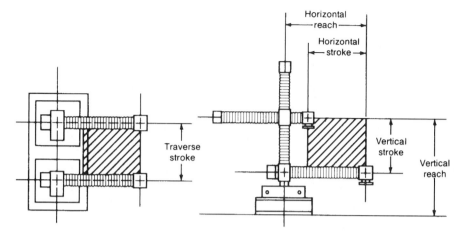

Figure 15.8   Work envelope: Cartesian coordinate robot. (Based on Critchlow, *Introduction to Robotics*, 1st Edition, © 1985. Reprinted by permission of Pearson Education Inc., Upper Saddle River, NJ.)

workspace envelope is defined as all the points in the surrounding space that can be reached by the robot or the mounting point for the end effector or tool. The area reachable by the end effector itself is usually not considered part of the workspace envelope. A clear understanding of the workspace envelope of a robot to be used is important because all interaction with other machines, parts, and processes only takes place within this volume of space.

Figures 15.8 to 15.11 show the workspace envelopes of Cartesian, cylindrical, spherical, and articulated robots, respectively. Because the workspace envelope is

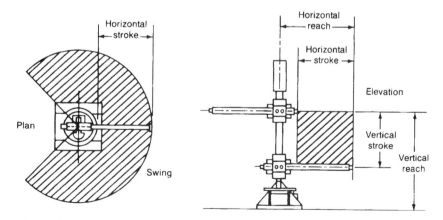

Figure 15.9   Work envelope: cylindrical coordinate robot. (Based on Critchlow, *Introduction to Robotics*, 1st Edition, © 1985. Reprinted by permission of Pearson Education Inc., Upper Saddle River, NJ.)

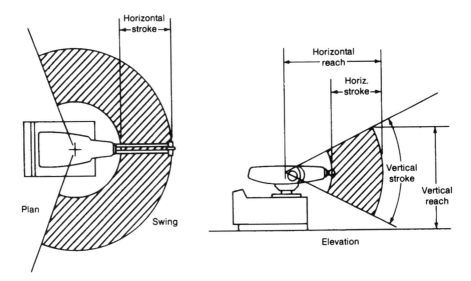

Figure 15.10   Work envelope: spherical or polar coordinate robot (Based on Critchlow, *Introduction to Robotics*, 1st Edition, © 1985. Reprinted by permission of Pearson Education Inc., Upper Saddle River, NJ.)

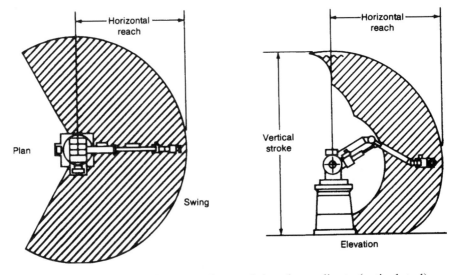

Figure 15.11   Work envelope: revolute or jointed coordinate (articulated) robot. (Based on Critchlow, *Introduction to Robotics*, 1st Edition, © 1985. Reprinted by permission of Pearson Education Inc., Upper Saddle River, NJ.)

defined by the joint motion range, given the robot configuration and range limits, the workspace envelope can be derived. To derive the workspace envelope, robot kinematic equations, which are discussed in a later section, are used. Robot manufacturers also usually provide drawings of workspace envelopes similar to those shown in Figures 15.8 to 15.11.

To design a robot application, the application workspace must be within the envelope. However, as discussed before, this envelope is defined by the end of the arm and does not take into consideration what tool may be attached to the arm. Also, it only defines the position or points that can be reached. The possible orientations of the tool at each point have to be found separately. Because the tool orientation is affected not only by the base's three axes, but also by the wrist joints and tool length, a kinematic equation including all these variables is needed.

Example 15.1.

In a proposed manufacturing cell, a robot is used to load/unload two machines M/C 1 and M/C 2. The machines, feeder, buffer, and finished-part rack dimensions are shown in Figure 15.12. Design a layout of the cell.

Solution

A circular layout is adopted, because it maintains a relatively constant distance between each device and the robot. The selected robot can reach all of them. Figure 15.13 shows a proposed design. The robot workspace envelope is a circular area with a 7-ft diameter and a 270° sweep angle.

Example 15.2.

Given the inside radius, r_1, and the outside radius, r_2, of the footprint of a cylindrical robot workspace (see Figure 15.14), calculate the maximum rectangular area that the robot can reach while fitting a maximum workbench in the robot workspace.

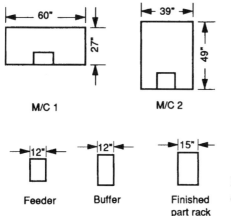

Figure 15.12   Manufacturing cell device.

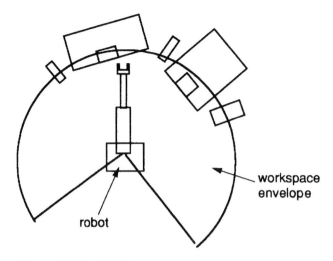

Figure 15.13 Cell layout.

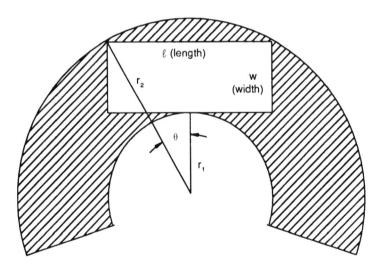

Figure 15.14 Footprint of a cylindrical robot workspace.

Solution

From Figure 15.14,

$$w = r_2 \cos \theta - r_1$$
$$l = 2r_2 \sin \theta$$
$$A = lw$$
$$= 2r_2 \sin \theta (r_2 \cos \theta - r_1)$$
$$\frac{dA}{d\theta} = \frac{d}{d\theta}(2r_2^2 \sin \theta \cos \theta - 2r_1 r_2 \sin \theta)$$
$$= 2r_2^2 \cos 2\theta - 2r_1 r_2 \cos \theta$$
$$= 0$$
$$r_2 \cos 2\theta - r_1 r_2 \cos \theta = 0$$
$$r_2(2 \cos^2 \theta - 1) - r_1 \cos \theta = 0$$

Let $\cos \theta = x$:

$$2r_2 x^2 - r_1 x - r_2 = 0$$
$$x = \frac{r_1 \pm \sqrt{r_1^2 + 8r_2^2}}{4r_2}$$
$$\theta = \cos^{-1}\left(\frac{r_1 \pm \sqrt{r_1^2 + 8r_2^2}}{4r_2}\right)$$

For example, if r_1 is 10 and r_2 is 100, the maximum rectangular area that the robot can reach is 8611.70.

15.1.4 Accuracy and Repeatability

Two very important terms in describing machine characteristics are accuracy and repeatability. The accuracy of a linear axis is one-half the control resolution plus the mechanical error (see Section 12.4.1). In a machine tool or a Cartesian coordinate robot where all three base axes are linear, the theoretical accuracy can be considered uniform throughout the entire workspace envelope. (Realistically, nothing is truly

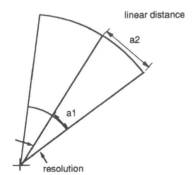

Figure 15.15   Resolution of a rotational axis.

uniform; however, in an ideal condition, they are.) Robots of other configurations employ one or several rotational axes. The control resolution is on the angle of rotation. The accuracy is normally defined as the error in the Cartesian space. The linear error due to the resolution of a rotational axis, as shown in Figure 15.15, is a variable; $a_1 \neq a_2$. When combining the effect of several rotational and linear axes, the term accuracy no longer can be used.

Robots are usually characterized by their repeatability. Repeatability is a statistical term. It does not describe the error with respect to absolute coordinates; instead, it describes how a point or position is repeated. Figure 12.11 shows positional error and repeatability on a two-dimensional plane. Although a target is always missed by a large margin, if the same error is repeated, then the repeatability is high and the accuracy is poor. Robot repeatability is normally measured in thousandths of an inch. Positional accuracy depends on the position in the workspace envelope. For this reason, it is difficult to do robot off-line programming without using sensors.

15.2 POWER SOURCES, ACTUATORS, AND TRANSDUCERS

The actuators and transducers for building industrial robots are essentially the same as those for building NC machinery. The major power sources for robots are hydraulic, pneumatic, and electric.

Hydraulics can deliver large forces, so it is commonly used on large robots that have to move large payloads. Pneumatics is used on those robots whose payload requirements are low, but that require high-speed movement. Electric-powered robots provide precise and quiet motion. This type of robot is usually used for assembly work demanding fine movement. Most robots are powered by electric motors. Detailed discussions of these power sources are found in Chapter 12.

15.3 ROBOTIC SENSORS

To perform some of the tasks presently done by humans, robots must be able to sense both their internal and external states (the environment). A sensor is a measurement device that can detect characteristics through some form of interaction with the characteristics. Only rudimentary sensors are currently applied to robots on factory floors; this reduces the flexibility, accuracy, and repeatability of robots. However, with the newly developed sensors, especially visual sensors, much more accurate and intelligent robots are expected.

Sensors for robotics can be classified in different ways, such as contact or noncontact, internal sensing versus external sensing, passive versus active sensing, and so on. In the following paragraphs, some typical robotic sensors are introduced.

A range sensor measures the distance from a reference point to a set of points in the scene. Humans can estimate range values based on visual data by perceptual processes that include stereopsis and comparison of image sizes and classified projected views of world-object models. Basic optical range-sensing schemes use the method of illumination (passive or active) and the method of range computation. Range can be sensed with a pair of TV cameras or sonar transmitters and receivers. Range sensing based on triangulation has a drawback: it misses the data of points not seen from both positions of the transmitters. This problem can be reduced, but not eliminated, by using additional cameras.

A proximity sensor senses and indicates the presence of an object within a fixed space near the sensor without physical contact. Different commercially available proximity sensors are suitable for different applications. A common robotic proximity sensor consists of a light-emitting-diode (LED) transmitter and a photodiode receiver. The major drawback of this sensor stems from the dependency of the received signal on the reflectance and orientation of the intruding object. The drawback can be overcome by replacing proximity sensors with range sensors.

As its name implies, an acoustic sensor senses and interprets acoustic waves in gases, liquids, or solids. The level of sophistication of sensor interpretation varies among existing acoustic sensors; it can range from a primitive detection of the presence of acoustic waves, to a frequency analysis of acoustic waves, to the recognition of isolated words in continuous speech.

A touch sensor senses and indicates a physical contact between the object carrying the sensor and another object. The simplest touch sensor is a microswitch. Basically, touch sensors are used to stop the motion of a robot when its end-effector makes contact with an object.

A force sensor measures the three components of the force and three components of the torque acting between two objects. In particular, a robot-wrist force sensor measures the components of force and torque between the last link of the robot and its end-effector by transducing the deflection of the sensor's compliant sections, which results from the applied force and torque.

Human workers effectively use their ability to sense the presence and outline of an object with the sense of touch. Researchers are also developing artificial tactile sensors for robots. Whereas vision may guide the robot arm through many manufacturing operations, it is the sense of touch that will allow the robot to perform delicate gripping and assembly. Tactile sensors will provide position data for contacting parts more accurately than that provided by vision.

15.4 ROBOT GRIPPERS

In order for a robot to perform an assembly task, it must be equipped with an application-dependent device, often called an end-effector. An end-effector can be either a tool (e.g., a screwdriver) or a grasping device, commonly called a gripper. In line with industrial practice, the term gripper is used to mean a grasping device, tool, or any other end-effector.

Robots are usually specified without grippers because grippers are very task- and environment-dependent. In other words, the selection of a robot gripper depends on the nature of the task to be performed. Consequently, grippers are normally part of the customizing package along with other auxiliary equipment and installation.

It is usually assumed that a gripper has no independent degrees of freedom, as it is anticipated that all degrees of freedom are provided by other robot elements. It has been recognized, however, that certain devices (e.g., autoscrewdrivers), because of their inherent degrees of freedom, duplicate certain capabilities of a robot.

The tasks performed during operation often require precise movements of the objects being handled or assembled. There are two alternative methods of achieving precise motions robotically. The first is to use a very sophisticated robot and a simple gripper. The advantage is that the gripper does not have to be technically complex; hence, its inherent reliability will be high and its cost low. The disadvantage is that the investment on the robot is significant. The second alternative is to use a simple robot and a sophisticated gripper. The main drawback is that the gripper has to be specially designed for the task.

Although a gripper action is essentially one of opening or closing, both the design of action and the structure of the gripper are very task-specific, though not always product-specific. Grippers have many actions: (1) they can have parallel motion; (2) they can make a scissors motion; (3) one jaw may be fixed and the other moving; and (4) they can be sprung open or closed. It is generally recognized that there are five classifications of grippers:

1. Mechanical clamping—the most common mechanism, whereby pneumatic or hydraulic devices apply a surface pressure on a component.
2. Magnetic clamping—applying electromagnetism to hold the component.
3. Vacuum clamping—applying negative pressure to components so that they adhere to the grippers.
4. Piercing grippers—puncturing the component to lift it. The technique is only used where slight damage to the component is acceptable.
5. Adhesive grippers—used for components that do not permit any of the preceding methods. These grippers use sticky tape to pick up and hold the component.

15.5 ROBOT SAFETY

A robotic system is an integration of robots, machines, computerized information channels, and humans, no element of which can be considered perfect or immune from eventual failure and malfunction. The proximity of humans to the robots allows the risk of mutual damage, resulting in the formulation of safety guidelines that indicate how the conditions of conflict can be minimized. The high productivity levels associated with robotic systems can be only realized if all the system elements are functioning safely and reliably.

However, until definitive regulations are imposed by law, attempting to determine the safety hazards of a robotic assembly system is best done on a piecemeal basis, whereby each element is analyzed for risk. The relationships between elements are known on a quantitative or qualitative basis. Therefore, the risk factors can be transferred from one element to the others.

There are four groups of humans at risk from direct personal injury from a robot:

1. *Programmers.* A robot programmer using any one of the previously mentioned programming methods is in direct contact with the robot. This closeness with the robot's work envelope, with its inherent danger of injury, distinguishes robotics from any other form of automation.
2. *Maintenance engineers.* A maintenance engineer is at risk from much the same dangers as programmers, with the added risk of electrocution. Also, because maintenance procedures often require that safety interlocks be disconnected, the inherent risk of injury is greater.
3. *Casual observers.* To the casual observer, robots are often seen standing still, apparently doing nothing, for long periods of time. The programmer, of course,

would know whether or not these pauses are intentional: the robot may be performing a programmed delay or waiting. However, if, as is usually the case, the assembly robot is not rigidly guarded, then a casual observer may move toward a seemingly stationary robot and be injured when it continues its operation.

4. *Others outside the assumed danger zone.* Even though a robot has a known maximum work envelope, the risk of injury is not limited to encounters within this envelope. If components manipulated by the robot are not properly secured, then it is possible for them to fly out of the grippers and strike personnel well outside the assumed danger zone of the robot.

In a practical sense, safety procedures and devices allow the authorized entry of humans into a robot's work envelope with a minimal risk of injury. Hardware devices and sensors monitor all anticipated reasonable access to a robot's work envelope.

Physical safeguards are many and varied. They include the following:

1. simple contact switches
2. restrained keys
3. pressure mats
4. infrared light beams
5. vision systems
6. flashing red lights within a work zone indicating that an apparently stationary robot is activated, but awaiting an input, or performing a time-delayed operation.

15.6 ROBOT PROGRAMMING

Programming conventional robots normally takes one of three forms: (1) walking through, or pendant, teaching, (2) lead-through teaching, or (3) textual language programming. Each has advantages and disadvantages depending on the application being considered.

Pendant programming uses a teach pendant to instruct a robot to move in working space. A teach pendant (see Figure 15.16) is a device equipped with switches and dials used to control the robot's movements to and from the desired points in the space. These points are recorded into memory for subsequent playback. This is the most common programming method for playback robots.

Lead-through programming is for continuous-path playback robots. In pendant programming, to program a straight-line path between two points, the teach pendant could teach the robot the locations of the points. The trajectory between the two points to be followed is computed by the robot controller. In lead-through programming, on the other hand, the programmer simply physically moves the robot through the required sequence of motions. The robot controller records the position and speed as the programmer leads the robot through the operation.

Textual language programming methods use an English-like language to establish the logical sequence of a work cycle. To input the program instructions, a computer terminal is used. To define the location of various points in the workplace, a teach pendant might be used. When a textual language program is entered without a teach pendant defining locations in the program, it is often called offline programming.

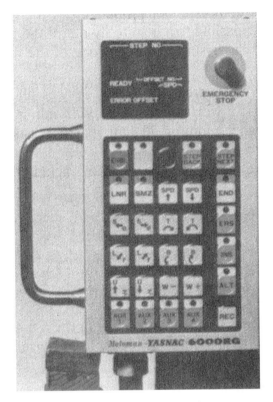

Figure 15.16   A teach pendant (Courtesy of MOTOMAN).

15.6.1 Robot Programming Languages

All robots are programmed with some kind of programming language. These programming languages command the robots to move to certain locations, to signal outputs, and to read inputs. They are created and evolve along with the robot controller technology developments. According to robot generations, robot control and their programming has evolved over the years in the following stages:

1. *Robot playback* generates a sequence of recorded instructions, such as those of a robot used in spray or arc welding. These robots commonly have open-loop control.
2. *Robots controlled by sensors* have control in the loopback of manipulated movements and make decisions based on data collected by sensors.
3. *Robots controlled by vision* are programmed to perform certain tasks based on feedback from a vision system.
4. *Robots that are adaptable* can automatically reprogram their own actions on the basis of the data collected by their sensors.
5. *Robots with artificial intelligence* use the techniques of artificial intelligence to make their own decisions and solve problems.

Correspondingly, the robot programming languages fall into three basic categories:

1. *Specialized robot languages.* These languages have been developed specifically for robots. The commands found in these languages are mostly motion commands with minimal logic statements available. Most of the early robot languages were of this type, and many still exist today. VAL is an example of this category.

2. *Robot libraries for a new general-purpose language.* These libraries were established by first creating a new general-purpose programming language and then adding robot-specific commands to it. These languages are generally more capable than a specialized language, since they tend to have better logic-testing capabilities. Fanuc's KAREL language is an example.

3. *Robot libraries for an existing computer language.* Languages in these libraries are developed by creating extensions to popular computer programming languages. Consequently, these languages closely resemble traditional computer programming languages, and benefit from the power of existing, widely used languages.

The specialized robot languages have been characterized by "teaching-by-showing" programming techniques [Sciavicco and Siciliano, 2001]. The operator guides the controller or manipulator manually or via a teach pendant along the desired motion path. During motion execution, the data read by the joint position transducers are stored; thus, they can be utilized later as references for the joint drive servos. In this way, the mechanical structure can immediately execute the motion taught by a direct acquisition. Typically, this programming technique can solve applications like spot welding, spray painting, and simple palletizing.

This programming environment does not allow the implementation of logic conditioning and queuing. It could not deal with complex logic strategy and multiple sensor data. The setup of a working program obviously requires the robot to be available to the operator at the time of teaching; thus, the robot itself must be taken offline to be programmed.

Robot programming codes that are based on an existing computer language are becoming a widespread trend. They use the open architecture of the robot controller, which is driven by the Internet's technology and armed with the object-oriented programming methodology.

As robot workcells become more complex, the robot must perform more complicated moves and interact with more sensors and other peripheral devices. The sensors include thermal couples, velocity and acceleration transducers, and video cameras. These peripheral devices allow the user to connect a robot controller to an enterprise-wide network for logging data, backing up programs monitoring and controlling robots remotely, or other communications tasks. Obviously, the specialized languages often have a limited ability to handle such sophisticated tasks. On the other hand, when the robot programming codes are based on an existing computer language, they are able to use subroutines and do logic testing; thus, they provide very English-like commands to deal with complicated controls.

This current development trend is a result of the open-architecture, PC-based robot controllers. These controllers work with virtually any robot and give the user increased flexibility and capabilities over the standard OEM controllers. In fact, they usually use prevailing operation software as their operating systems. This allows the operator to run third-party software, such as statistical process controls (SPC), management execution systems (MES), enterprise resource planning (ERP) systems, or custom software, along with robot programming languages. The open-architecture and PC-based robot controllers provide platforms for the robot library to use an existing computer language.

Most factories have robots from multiple manufacturers; thus, multiple languages are running on these control systems. This causes a dilemma: Robot programmers must be proficient in more than one language, or they must specialize in certain languages. The situation has caused an outcry for a common language that can be used on any type of robot. A language that is based on a common existing computer language would provide an effective solution to master a robot controller quickly.

One example of the open-architecture industrial robotic control system is based on Windows NT. It has the architecture of a single-CPU structure and can run on general industrial computers [Zhang et al., 2003]. Thanks to the Windows NT's operating system and RTX real-time extension environment, the control system has a friendly graphical user interface, as well as good real-time performance. To promote the application of the robot system, C is used as the basic programming language of the robot system. A series of robot application programming interfaces (API) is provided to help users develop their own robot applications easily and flexibly.

An example of a robot programming language that has an existing computer language is RobotScript™ [Lapham, 1999]. It was developed in conjunction with the development of the Universal Robot Controller™ (URC). The URC is an open-architecture robot controller that is PC-based and uses Windows NT as its operating system. RobotScript™ was developed on the basis of Microsoft Visual Basic scripting language. The syntax of the Visual Basic Family is very easy to learn. The commands are similar to English, and they allow for variable names up to 255 characters. This makes the completed programs easy to follow, and consequently, easy to debug. For example, instead of using a command like MOVL (a Motoman Inform 2 command) to command a linear move, RobotScript uses MoveLinearTo. When a command is less cryptic, it is easier for someone to learn. (See Table 15.1.)

Another popular computer language is also used for robots: the object-oriented programming language called Java [Bergin, 2002]. As an example, consider the typical UrRobot. Its specification interface in the robot programming language is shown in Figure 15.17.

The UrRobot class has several methods to control robot movements, such as move(), turnLeft(), and send alarms, such as putBeeper(). In Figure 15.17, { ... } indicates

TABLE 15.1 Comparison of Robot Language Syntax

Description	Inform 2	KAREL	RobotScript
Joint motion	MOVJ	$MOTYPE = JOINT MOVE TO	MoveJointTo
Linear motion	MOVL	$MOTYPE = LINEAR MOVE TO	MoveLinearTo
Turn on output	DOUT OT = (12) ON	DOUT[12] = ON	SetDigitalOutput 12, 1
Addition	Add 112 113	X = 112 + 113	X = 112 + 113

```
class UrRobot
{
        void move() {...}
        void turnOff() {...}
        void turnLeft() {...}
        void pickBeeper() {...}
        void putBeeper() {...}
        ..
}
```

Figure 15.17 Sample program for UrRobot.

that some coding is not being shown. These are the descriptions of how an UrRobot would carry out each of these instructions.

A sample task for an UrRobot might be to start at the origin, facing east, and then walk three blocks east to a corner known to have a beeper, pick up the beeper, and turn off on that corner.

A program to accomplish this task is written in pure Java programming style and is shown in Figure 15.18. In this program, the robot is called Karel.

There are other applications of advanced computer languages in robot programming. An XML-based language, RoboML, is designed to serve as a common language for robot programming, and it allows communication between the operator and the robot controller through the Internet [Makatchev and Tso, 2000]. Special effort is given to deal with the communication between the real-time robot operations and the user-interface agents via a communication channel of uncertain quality. Advanced

```
Package kareltherobot;
public class Tester implements Directions
{
        public static void main(String [] args)
        {
                UrRobot Karel = new UrRobot(1, 2, East, 0);
                Karel.move();
                Karel.move();
                Karel.pickBeeper();
                Karel.move();
                Karel.turnLeft();
                Karel.move();
                Karel.move();
                Karel.putBeeper();
                Karel.move();
                Karel.urnoff();
        }
}
```

Figure 15.18 Java style of the robot program.

multimedia and distributed programming languages, such as Java and Java3D, make it possible for robots to be accessed through the Internet. The operator can specify a task to a 3-D model in a virtual-robot environment, and commands will be sent to the real robot for production [Marin et al., 2002].

15.7 ROBOT APPLICATIONS

Robotics has rapidly moved from theory to application during the past three decades, primarily due to the need for improved productivity and quality.

One of the key features of robots is their versatility. A programmable robot used in conjunction with a variety of end-effectors can be programmed to perform specific tasks. It can later be reprogrammed and refitted to adapt to process or production line variations or changes. The robot offers an excellent means of utilizing high-technology to make a given manufacturing operation more profitable and competitive.

Robots are used today primarily for welding, painting, assembly, machine loading, and foundry activities. The number of robots used for welding accounts for about 50% of the applications in the United States, primarily because the automotive industry is the major user of robots. The sharp visibility given to the automotive industry's robotic applications and its declared intention to even more aggressively increase the installation rates have made that industry a major focus for robot builders.

In the U.S. metalworking industry, machine loading and unloading appears to be the biggest single area of the application of robots. Assembly is another extremely important application area for robots.

Arc welding by robots is a growing application, but a rudimentary vision for seam tracking needs to be developed before its full potential is realized. Also coming into its own is the use of robots for stamping; the robot's lack of speed has been the deterrent up to now, but that is expected to improve.

One manufacturing process that owes much of its current growth to robotics is investment casting. The series of steps in patternmaking can now be performed easily and consistently. The assembly robot is getting intense attention in the United States and Japan. In assembly, the driving impetus is economics, and the multishift capabilities of robots may revolutionize one-shift assembly operations. Die casting, injection molding, heat treating, and glass handling are other application areas of robotics.

15.7.1 Principles for Robot Applications

There are four principles for robot applications, as follows [Heginbotham et al., 1973]:

- If the conditions of dimension and state at the workplace repeat without significant unstructured variability, then a simple robotic solution is possible. For this principle, two categories of problems are considered: (1) first-order problems—position variations; (2) second-order problems—position and orientation variations; and repeatability of the gripping device.
- If the machine can be deceived into thinking that it is handling the same thing when it is not, then the third-order robot assembly or handling problem can be reduced to a second-order problem. For this principle, smart engineering design or complex sensory systems are required.

- For all changes of state or dimension (either structure or unstructured) that are imposed on a robot installation, the robot behavior and the workplace must be capable of change. It is important not to try to use robots to imitate human functions.
- Recognize and exploit or create robot-compatible mechanical states or environments. Thus, eliminate or reduce the need for human skill and judgment and/or the drudgery of a function by using robot-compatible mechanical states or environments.

15.7.2 Application Planning

The planning for a robot application involves several stages of consideration:

1. *Workpiece analysis and evaluation:* a thorough study of what is to be produced, including parts breakdown, process plans, and accessibility analysis.
2. *Establishment of alternative methods for automation:* the layout of the workplace and the interlink of workplaces are major issues.
3. *Selection of optimum methods:* simulation and graphical emulation are common tools at this stage.
4. *Search for solutions to implement the selected methods:* the selection of the robot and the selection of peripheral devices such as the parts feeder, gripper, and fixtures.
5. *Cost analysis:* this is discussed in the next section.
6. *Measures taken to implement the overall method solution:* popular tools for project management, such as PERT, CPM, and Gantt charts, are commonly used.

15.8 ECONOMIC CONSIDERATIONS OF ROBOTIC SYSTEMS

Several U.S. industries have shown good foresight by investing in high-tech production equipment. Some of these investments include robots for welding and assembly. However, as with any capital-intensive production investment, these systems must be economically justified before they can be purchased. Installation, however, does not end the economic justification considerations of a system. As products are redesigned, a different mix of parts enters the production system, and a new problem must be resolved: What parts can be produced economically with robotic equipment?

The economical consideration for robot applications includes the following:

- facility and equipment
- operation and maintenance
- product
- parameters of analysis

15.8.1 Facility and Equipment

This category concerns (1) the robot(s), associated tooling, and spare parts; (2) taxes and tax consequences, such as investment tax credits and tax savings due to depreciation;

(3) energy requirements; (4) space requirements; (5) safety equipment; (6) programming; and (7) compatibility of current equipment with the robot(s).

15.8.2 Operation and Maintenance

In the category of operation and maintenance, key issues are (1) operating costs; (2) direct cost of illness, absenteeism, insurance, and injuries; (3) training costs; (4) supervision costs; (5) retooling and setup costs for batch processing; and (6) maintenance costs.

15.8.3 Product

This category focuses on the cost issues associated with the product. Those issues are (1) changes in product design; (2) material costs such as raw material and in-process inventory; (3) production rates (productivity, scrap rate, defective items); (4) handling and reworking of defective products; and (5) costs due to undetected defective product released to the customer (loss of good will and complaints).

15.8.4 Parameters of Analysis

The parameters for cost analysis involve (1) income tax rates (federal, state, and local); (2) engineering and consulting costs not considered in the previous categories, and (3) costs of capital (discount rate).

15.9 ROBOT KINEMATICS AND DYNAMICS

Robot arm kinematics involves the analytical study of the geometry of motion of a robotic arm with respect to a fixed reference coordinate system without regard to the forces/momenta that cause the motion. In other words, robot kinematics deals with the analytical description of the spatial displacement of the robot as a function of time, in particular, the relations between the joint-variable space and the position and orientation of the end-effector of a robot arm.

There are two fundamental problems in robot-arm kinematics. The first is usually referred to as the direct (or forward) kinematics problem and the second is the inverse kinematics problem. If the locations of all of the joints and links of a robot arm are known, it is possible to compute the location of the end of the arm. This is defined as the direct kinematics problem. The inverse kinematics problem is to determine the necessary positions of the joints and links in order to move the end of the robot arm to a desired position and orientation in space.

Vector and matrix algebra are used to develop a systematic and generalized approach to describe and represent the locations of the links of a robot arm with respect to a fixed reference frame. Since the links of a robot arm can rotate and/or translate with respect to a reference (world) coordinate frame, a body-attached (joint) coordinate frame is established along the joint axis for each link. In general, the direct kinematics problem reduces to finding a transformation matrix that relates joint coordinates to world coordinates. Matrix manipulation, discussed in Chapter 5, applies here.

Computer-based robots are usually servo controlled in the joint-variable space, whereas objects to be manipulated are usually identified in the world or part coordinate system. In order to control the position and orientation of the end-effector of a robot to reach the target object, the inverse kinematics solution is necessary to obtain the correct joint angle. In other words, given the position and orientation of the end-effector of a six-axis arm and its joint and link parameters, it is possible to find the corresponding joint angles of the robot so that the end-effector can be positioned as desired.

15.9.1 Basics

15.9.1.1. Translation and Rotation.
Translation of a vector from one coordinate system to another is usually accomplished by multiplying a matrix that defines the linear translations of two coordinate systems on the three major axes.

Rotation is accomplished by rotation matrices in homogeneous coordinates. Matrices performing linear translations or rotations about the X, Y, or Z axes by an angle θ are given in Chapter 5 (object transformation).

The transformation matrices used are T_{to}, T_{xo}, T_{yo}, T_{zo}. To transform a point P, use the formula

$$P' = PT$$

where P and P' are row vectors $[x\ y\ z\ 1]$ and $[x'\ y'\ z'\ 1]$.

Example 15.3.

Suppose there is a vector, $\mathbf{A} = 5i + 2j + 4k$, in space. Find the new vector that results from rotating $\mathbf{A}$ through $90°$ around the $\mathbf{Z}$ axis, $-90°$ around the $\mathbf{X}$ axis, and translating through the vector value $2i + 2j + 2k$.

Solution

The three transformation matrices are

$$T_{zo} = \begin{bmatrix} 0 & 1 & 0 & 0 \\ -1 & 0 & 0 & 0 \\ 0 & 0 & 1 & 0 \\ 0 & 0 & 0 & 1 \end{bmatrix}$$

$$T_{xo} = \begin{bmatrix} 1 & 0 & 0 & 0 \\ 0 & 0 & 1 & 0 \\ 0 & -1 & 0 & 0 \\ 0 & 0 & 0 & 1 \end{bmatrix}$$

$$T_{to} = \begin{bmatrix} 1 & 0 & 0 & 0 \\ 0 & 1 & 0 & 0 \\ 0 & 0 & 1 & 0 \\ 2 & 2 & 2 & 1 \end{bmatrix}$$

$$A' = A\,T_{zo}\,T_{xo}\,T_{to}$$

$$= \begin{bmatrix} 5 & 2 & 4 & 1 \end{bmatrix} \begin{bmatrix} 0 & 0 & 1 & 0 \\ -1 & 0 & 0 & 0 \\ 0 & -1 & 0 & 0 \\ 2 & 2 & 2 & 1 \end{bmatrix}$$

$$= \begin{bmatrix} 0 & -2 & 7 & 1 \end{bmatrix}$$

The new vector is

$$0i - 2j + 7k$$

15.9.2 Robot Coordinate Systems

Use the information learned to construct a series of coordinate systems for robot elements. First, look at a robot lower arm (Figure 15.19). Construct two coordinate systems for the arm: One relates to the robot waist and the other relates to the upper arm. These coordinate systems are called X–Y–Z and U–V–W, as shown in the figure. How can the transformation matrix $[L]$ be obtained? It translates the location of a point from one coordinate system to the other. By observing the figure, this transformation is a linear translation. The distance between the two coordinate frames is S. The generic form of a translation matrix is

$$T_{to} = \begin{bmatrix} 1 & 0 & 0 & 0 \\ 0 & 1 & 0 & 0 \\ 0 & 0 & 1 & 0 \\ \Delta x & \Delta y & \Delta z & 1 \end{bmatrix}$$

In this case, $\Delta_x = s$, $\Delta_y = \Delta_z = 0$; therefore, the translation matrix relating U–V–W to X–Y–Z is

$$T = \begin{bmatrix} 1 & 0 & 0 & 0 \\ 0 & 1 & 0 & 0 \\ 0 & 0 & 1 & 0 \\ S & 0 & 0 & 1 \end{bmatrix}$$

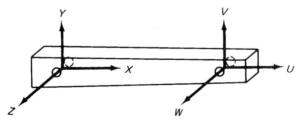

Figure 15.19   Lower arm of a robot.

In other words, a point P in U–V–W can be transformed to a point in X–Y–Z as follows:

$$P_{X-Y-Z} = P_{U-V-W}T$$

For example, if P_{u-v-w} is $[1\ 0\ 2\ 1]$, P_{x-y-z} can be found from the equation

$$P_{X-Y-Z} = [1 \quad 0 \quad 2 \quad 1]\begin{bmatrix} 1 & 0 & 0 & 0 \\ 0 & 1 & 0 & 0 \\ 0 & 0 & 1 & 0 \\ S & 0 & 0 & 1 \end{bmatrix}$$

$$= [1 + S \quad 0 \quad 2 \quad 1]$$

Another important transformation usually occurs on a robot waist, as shown in Figure 15.20. Construct two coordinate systems for the robot body, and let these two systems be $(X$–Y–$Z)_1$ and $(U$–V–$W)_1$. By observing the figure, it becomes obvious that the relation between $(X$–Y–$Z)_1$ and $(U$–V–$W)_1$ includes two transformations: (1) rotate 90° about the X_1-axis, and (2) translate h along the Z_1-axis. Based on the discussion in the previous section, the matrices are formulated as follows:

$$T_{xo} = \begin{bmatrix} 1 & 0 & 0 & 0 \\ 0 & 0 & 1 & 0 \\ 0 & -1 & 0 & 0 \\ 0 & 0 & 0 & 1 \end{bmatrix}$$

$$T_t = \begin{bmatrix} 1 & 0 & 0 & 0 \\ 0 & 1 & 0 & 0 \\ 0 & 0 & 1 & 0 \\ 0 & 0 & h & 1 \end{bmatrix}$$

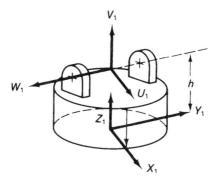

Figure 15.20   A robot base.

The transformation matrix can be obtained by multiplying these two matrices (be careful with the sequence). The matrix is

$$T = T_{xo}T_t = \begin{bmatrix} 1 & 0 & 0 & 0 \\ 0 & 0 & 1 & 0 \\ 0 & -1 & 0 & 0 \\ 0 & 0 & h & 1 \end{bmatrix}$$

How can the transformation matrix be validated? Since the determinant of a rigid-body transformation is 1, the determinant of T, $\det[T]$, is calculated as follows:

$$\begin{bmatrix} 1 & 0 & 0 & 1 & 0 \\ 0 & 0 & 1 & 0 & 0 \\ 0 & -1 & 0 & 0 & -1 \end{bmatrix}$$

Then $\det[T] = 0 - (-1) = 1$. If $\det[x]$ does not equal 1, the transformation is deformed. If $\det[x]$ equals -1, this means a mirror transformation.

15.9.3 Generic Formulas for Coordinate Transformation

Up to now, this chapter has examined several coordinate systems for robot elements. Now, consider a complete robot system. The following generic matrix representation describes two types of matrix:

L is a link matrix describing location of joints on a body

J is a joint matrix describing relative location or orientation across joints

and

L must be a translation transformation matrix

J could be a translation or a rotation-transformation matrix

The most general form of coordinate transformation from the end-of-tool to a global coordinate system is as follows:

$$O = L_{TOOL}J_Y L_5 J_R L_4 J_P L_3 J_E L_2 J_S L_1 J_W L_B T \qquad (15.13)$$

where X_G, Y_G, Z_G and X_T, Y_T, Z_T are global and tool coordinate system representations, respectively.

T is a transformation matrix from the base coordinate to a global coordinate

L_B is a base-body-link transformation matrix describing where the first (waist) joint is located on a robot

J_W is a transformation matrix for the waist joint

L_1 is a waist-link transformation matrix describing where the second (shoulder) joint is located

J_S is a transformation matrix for the shoulder joint

L_2 is a lower-arm-link transformation matrix describing where the third (elbow) joint is located

J_E is a transformation matrix for the elbow joint

L_3 is an upper-arm-link transformation matrix describing where the fourth (wrist pitch) joint is located

J_P is a transformation matrix for the wrist-pitch joint

L_4 is a wrist-pitch link transformation matrix describing where the roll joint is located

J_R is a transformation matrix for the roll joint

L_5 is a roll-joint transformation matrix describing where the yaw joint is located

J_Y is a transformation matrix for the yaw joint

L_{TOOL} is the end-of-tool-link transformation matrix describing where the end-of-tool point is located.

This section will establish the coordinate systems for the robot. The following rules are to be applied in the construction of such coordinate systems:

1. Always use the right-hand coordinate system.
2. The Z and W axes must be aligned with the axis of motion.
3. Construct two coordinate systems for each body.
4. Construct two coordinate systems for each joint.

Figure 15.21 illustrates an articulated robot as four (4) separate rigid bodies: base, waist, lower arm, and upper arm. The coordinate representations are also specified in the figure. Based on the general form of coordinate transformation and the given robot configuration, the transformation matrices are

$$
T = \begin{bmatrix} 1 & 0 & 0 & 0 \\ 0 & 1 & 0 & 0 \\ 0 & 0 & 1 & 0 \\ x_t & y_t & z_t & 1 \end{bmatrix}
$$

$$
L_B = \begin{bmatrix} 1 & 0 & 0 & 0 \\ 0 & 1 & 0 & 0 \\ 0 & 0 & 1 & 0 \\ 0 & 0 & t_b & 1 \end{bmatrix}
$$

$$
J_w = \begin{bmatrix} c\alpha & s\alpha & 0 & 0 \\ -s\alpha & c\alpha & 0 & 0 \\ 0 & 0 & 1 & 0 \\ 0 & 0 & 0 & 1 \end{bmatrix}
$$

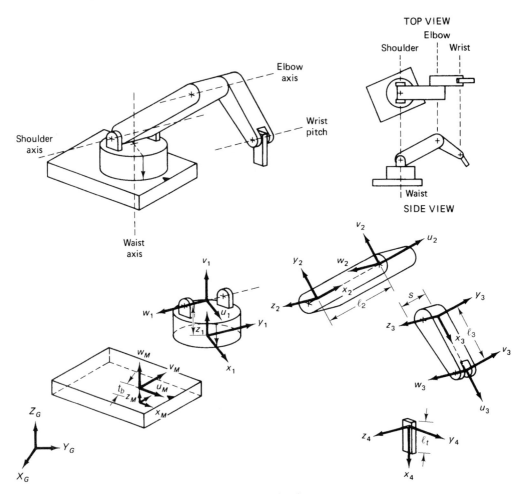

Figure 15.21 Links and joints of an articulated robot.

$$L_1 = \begin{bmatrix} 1 & 0 & 0 & 0 \\ 0 & 0 & 1 & 0 \\ 0 & -1 & 0 & 0 \\ 0 & 0 & t_w & 1 \end{bmatrix}$$

$$J_s = \begin{bmatrix} c\beta & s\beta & 0 & 0 \\ -s\beta & c\beta & 0 & 0 \\ 0 & 0 & 1 & 0 \\ 0 & 0 & 0 & 1 \end{bmatrix}$$

$$L_2 = \begin{bmatrix} 1 & 0 & 0 & 0 \\ 0 & 1 & 0 & 0 \\ 0 & 0 & 1 & 0 \\ l_2 & 0 & 0 & 1 \end{bmatrix}$$

$$J_E = \begin{bmatrix} c\gamma & s\gamma & 0 & 0 \\ -s\gamma & c\gamma & 0 & 0 \\ 0 & 0 & 1 & 0 \\ 0 & 0 & -s & 1 \end{bmatrix}$$

$$L_3 = \begin{bmatrix} 1 & 0 & 0 & 0 \\ 0 & 1 & 0 & 0 \\ 0 & 0 & 1 & 0 \\ l_3 & 0 & 0 & 1 \end{bmatrix}$$

$$J_P = \begin{bmatrix} C\eta & S\eta & 0 & 0 \\ -S\eta & C\eta & 0 & 0 \\ 0 & 0 & 1 & 0 \\ 0 & 0 & 0 & 1 \end{bmatrix}$$

$$L_{\text{TOOL}} = \begin{bmatrix} 1 & 0 & 0 & 0 \\ 0 & 1 & 0 & 0 \\ 0 & 0 & 1 & 0 \\ l_1 & 0 & 0 & 1 \end{bmatrix}$$

where

$C\alpha = \cos \alpha \qquad C\beta = \cos \beta \qquad Cr = \cos r \qquad C\eta = \cos \eta$

$S\alpha = \sin \alpha \qquad S\beta = \sin \beta \qquad Sr = \sin r \qquad S\eta = \sin \eta$

α = waist-rotation angle

β = shoulder-rotation angle

γ = elbow-rotation angle

x_l = X translation between the base and global origin

y_t = Y translation between the base and global origin

z_t = Z translation between the base and global origin

t_b = base thickness

t_w = waist thickness

l_2 = length of lower arm

l_3 = length of upper arm

s = offset on Z at the elbow joint

l_t = length of tool

η = pitch of angle

The final transformation matrix from the end of the tool coordinate system to the global coordinate system is

$$O = L_{\text{TOOL}} J_P L_3 J_E L_2 J_S L_1 J_W L_B T$$

Example 15.4.

Use the same robot shown in Figure 15.21. Suppose that the waist joint rotates 90°, the shoulder joint rotates 90°, the elbow joint rotates 0°, and the X, Y, and Z translations between the base and the global origin are 10, 10, and 0, respectively. Assume that the base thickness is 5, the waist thickness is 10, the offset on Z is 2, the length of the lower arm is 12, and the length of the upper arm is 8. Also, suppose that the location of an arbitrary point is $(5, 5, 5)$ with respect to the $(U-V-W)_3$ coordinate system. What is the location of the point with respect to the global coordinate system?

Solution

Based on the data given in the problem description, the transformation matrices are

$$[T] = \begin{bmatrix} 1 & 0 & 0 & 0 \\ 0 & 1 & 0 & 0 \\ 0 & 0 & 1 & 0 \\ 0 & 0 & 0 & 1 \end{bmatrix}$$

$$[L_B] = \begin{bmatrix} 1 & 0 & 0 & 0 \\ 0 & 1 & 0 & 0 \\ 0 & 0 & 1 & 0 \\ 0 & 0 & 5 & 1 \end{bmatrix}$$

$$[J_W] = \begin{bmatrix} 0 & 1 & 0 & 0 \\ -1 & 0 & 0 & 0 \\ 0 & 0 & 1 & 0 \\ 0 & 0 & 0 & 1 \end{bmatrix}$$

$$[L_1] = \begin{bmatrix} 1 & 0 & 0 & 0 \\ 0 & 0 & 1 & 0 \\ 0 & -1 & 0 & 10 \end{bmatrix}$$

$$[J_S] = \begin{bmatrix} 0 & 1 & 0 & 0 \\ -1 & 0 & 0 & 0 \\ 0 & 0 & 1 & 0 \\ 0 & 0 & 0 & 1 \end{bmatrix}$$

$$[L_2] = B\begin{bmatrix} 1 & 0 & 0 & 0 \\ 0 & 1 & 0 & 0 \\ 0 & 0 & 1 & 0 \\ 12 & 0 & 0 & 1 \end{bmatrix}$$

$$[J_E] = \begin{bmatrix} 1 & 0 & 0 & 0 \\ 0 & 1 & 0 & 0 \\ 0 & 0 & 1 & 0 \\ 0 & 0 & -2 & 1 \end{bmatrix}$$

$$[L_3] = \begin{bmatrix} 1 & 0 & 0 & 0 \\ 0 & 1 & 0 & 0 \\ 0 & 0 & 1 & 0 \\ 8 & 0 & 0 & 1 \end{bmatrix}$$

The transformation matrix relating the tool coordinate frame to the global frame $[O]$ is

$$O = \begin{bmatrix} 0 & 0 & 1 & 0 \\ 0 & -1 & 0 & 0 \\ 1 & 0 & 0 & 0 \\ 8 & 10 & 35 & 1 \end{bmatrix}$$

$$\begin{bmatrix} 5 & 5 & 5 & 1 \end{bmatrix} \begin{bmatrix} 0 & 0 & 1 & 0 \\ 0 & -1 & 0 & 0 \\ 1 & 0 & 0 & 0 \\ 8 & 10 & 35 & 1 \end{bmatrix} = \begin{bmatrix} 13 & 5 & 40 & 1 \end{bmatrix}$$

Multiplying the matrices and the point location $(5, 5, 5)$ yields the coordinates of the point with respect to the global coordinates system: $(13, 5, 40)$.

15.10 ROBOT-ARM DYNAMICS

Robot-arm dynamics, on the other hand, deals with the mathematical formulation of the equations of robot-arm motion. Specifically, dynamics is concerned with the use of information about the loads on a robot arm to adjust the servo operation to achieve optimum performance. The information includes inertia, friction, gravity, velocity, and acceleration. The dynamic equations of motion of an arm are a set of mathematical equations describing the dynamic behavior of the manipulator. Such mathematical formulation is useful for computer simulation of the robot-arm motion, the design of suitable control equations for a robot arm, and the evaluation of the kinematic design and structure of the robot [Fu, Gonzalez, and Lee, 1987].

This section derives the dynamics for the $\theta-r$ robot arm. Figure 15.22 shows the $\theta-r$ robot and its schematic representation. This robot arm includes three parts: a fixed-length body, an extended part, and a gripper. Let the mass of the fixed-length body be m_1, as shown in the figure. The extended part and the load are modeled as mass m_2. The Cartesian location of mass m_1 is

$$x_1 = r_1 \cos \theta \tag{15.14}$$

$$y_1 = r_1 \sin \theta \tag{15.15}$$

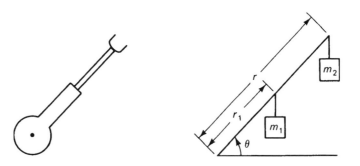

Figure 15.22   Schematic of a $\theta-r$ arm.

These equations are differentiated with respect to time to obtain the Cartesian velocities:

$$\dot{x}_1 = -r_1 \sin \theta \dot{\theta} \tag{15.16}$$

$$\dot{y}_1 = r_1 \cos \theta \dot{\theta} \tag{15.17}$$

The magnitude of the velocity vector can be expressed as follows:

$$V_1^2 = \dot{x}_1^2 + \dot{y}_1^2$$

$$= r_1^2 \sin^2 \theta \dot{\theta}^2 + r_1^2 \cos^2 \theta \dot{\theta}^2$$

$$= r_1^2 \dot{\theta}^2 (\sin^2 \theta + \cos^2 \theta)$$

$$= r_1^2 \dot{\theta}^2 \tag{15.18}$$

The kinetic energy of mass m_1 is

$$K_1 = 0.5 m_1 (r_1^2 \dot{\theta}^2) \tag{15.19}$$

Similarly, the kinetic energy of mass m_2 is

$$K_2 = 0.5 m_2 (\dot{r}^2 + r^2 \dot{\theta}^2) \tag{15.20}$$

The total kinetic energy of the robot arm is

$$K = K_1 + K_2$$

$$= 0.5 m_1 (r_1^2 \dot{\theta}^2) + 0.5 m_2 (\dot{r}^2 + r^2 \dot{\theta}^2) \tag{15.21}$$

The potential energy is

$$P = m_1 g r_1 \sin \theta + m_2 g r \sin \theta \tag{15.22}$$

where g is acceleration due to gravity.

The Lagrangian energy for this θ–r arm is

$$L = 0.5 m_1 (r_1^2 \dot{\theta}^2) + 0.5 m_2 (\dot{r}^2 + r^2 \dot{\theta}^2) - m_1 g r_1 \sin \theta - m_2 g r \sin \theta \tag{15.23}$$

The torque about the rotational actuator is

$$T_\theta = m_1 r_1^2 \ddot{\theta} + m_2 r^2 \ddot{\theta} + 2 m_2 r \dot{r} \dot{\theta} + g \cos \theta (m_1 r_1 + m_2 r_2) \tag{15.24}$$

The force applied by the linear actuator is

$$F_r = m_2 \ddot{r} - m_2 r \dot{\theta}^2 + m_2 g \sin \theta \tag{15.25}$$

These basic dynamics formulas impart useful information about the θ–r arm, especially the following applications:

- *Robot-arm design.* A robot-arm designer may want to enter the geometry of a proposed arm design along with estimates of masses, loads, and so on, and simulate the dynamic performance of the arm.

- *Path planning.* Basic path-control techniques provide a robot programmer with a tool to plan the desired path for a robot. However, as the robot moves, and speeds and accelerations increase, kinetic effects may result in an unexpected deviation from the planned path. Path simulation that considers the dynamic model can be used to develop worst-case estimates of path deviations at high speeds.

- *Real-time control.* It is known that no single choice of servo gains is appropriate to provide the best performance of a robot. With the dynamic model of the arm, there is the potential to attain such optimal control because the interaction of the joints can now be described.

In sum, the knowledge of kinematics and dynamics allows the control of an arm actuator to accomplish a desired task following a desired path. Trajectory planning and motion control are of considerable interest and importance, as these issues involve the degree of automation and intelligence of the robot.

15.11 COMPUTER VISION

Computer vision has become an indispensable part of an "intelligent" robotic system. As is true with humans, vision endows a robot with a sophisticated sensing mechanism that allows the machine to respond to its environment in an intelligent and flexible manner. The use of vision and other sensing schemes is motivated by the continuing need to increase the flexibility and scope of applications of robotic systems. Although proximity, touch, and force sensing play a significant role in the improvement of robot performance, vision is recognized as the most powerful robot sensory capability.

Robot vision may be defined as the process of extracting, characterizing, and interpreting information from images of a three-dimensional world. This process, also commonly referred to as machine or computer vision, may be subdivided into six principal areas: (1) sensing, (2) preprocessing, (3) segmentation, (4) description, (5) recognition, and (6) interpretation. It is convenient to group these various areas according to the sophistication involved in their implementation. The three levels of processing are low-, medium-, and high-level vision. Although there are no clear-cut boundaries between these subdivisions, they do provide a useful framework for categorizing the various processes that are inherent components of a machine-vision system.

In this section, robot vision is divided into three fundamental tasks: image transformation, image analysis, and image understanding. Image transformation involves the conversion of light images to electrical signals that can be used by a computer. Once a light image is transformed to an electronic image, it may be analyzed to extract such image information as object edges, regions, boundaries, color, and texture. This process is called image analysis. The last and most difficult process in robot vision is that once the image is analyzed, a vision system must interpret what the image represents in terms of information about its environment. This is called image understanding.

15.11.1 Image Transformation

Image transformation is the process of electronically digitizing light images using image devices. An image device is the front end of a vision system, which acts as an image transducer to convert light energy to electrical energy. In humans, the image device is the eye. In a vision system, the image device is a camera, photodiode array, charge-coupled device (CCD) array, or charge-injection device (CID) array.

The output of an image device is a continuous analog signal that is proportional to the amount of light reflected from an image. In order to analyze the image with a computer, the analog signals must be converted and stored in digital form. To this end, a rectangular image array is divided into small regions called picture elements, or pixels. Figure 15.23 illustrates the idea. With photodiodes or CCD arrays, the number of pixels equals the number of photodiodes or CCD devices. The pixel arrangement provides a sampling grid for an analog-to-digital (A/D) converter. At each pixel, the analog signal is sampled and converted to a digital value. With an 8-bit A/D converter, the converted pixel value will range from 0 for white to 255 for black. Different shades of gray are represented by values between these two extremes. This is why the term *gray level* is often used in conjunction with the converted values. As the pixels are converted, the respective gray-level values are stored in a memory matrix, which is called a picture matrix.

15.11.2 Image Analysis

A computer needs to locate the edges of an object in order to construct drawings of the object within a scene. Line drawings provide a basis for image understanding, as they define the shapes of objects that make up a scene. Thus, the basic reason for edge detection is that edges lead to line drawings, which lead to shapes, which lead to image understanding. This is illustrated in Figure 15.24.

15.11.2.1. Edge Detection. The edges are usually represented by the points that exhibit the greatest difference in gray-level values within a smoothed picture matrix. Look at the graphs of Figure 15.25. Figure 15.25(a) is a gray-level intensity function

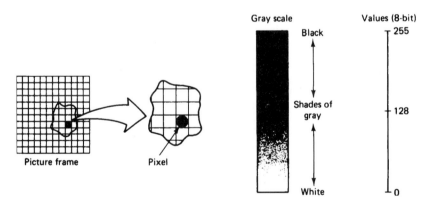

Figure 15.23 A picture frame is divided into picture elements, called pixels, for conversion to gray-scale values (Andrew C. Staugaard, Jr., *Robotics and AI: An Introduction to Applied Machine Intelligence,* © 1987, p. 207. Reprinted by permission of Prentice Hall.)

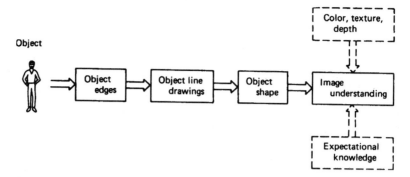

Figure 15.24   Edges lead to line drawings, which lead to shapes, which lead to image understanding (Andrew C. Staugaard, Jr., *Robotics and AI: An Introduction to Applied Machine Intelligence,* Prentice Hall, 1987, p. 209.)

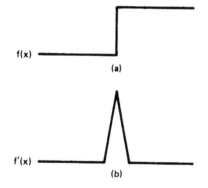

Figure 15.25   (a) An ideal step edge has (b) a first derivative approaching infinity (Andrew C. Staugaard, Jr., *Robotics and AI: An Introduction to Applied Machine Intelligence,* Prentice Hall, 1987, p. 212.)

and Figure 15.25(b) is the slope function of the intensity function. Calculus shows that the slope of a step edge approaches infinity, as illustrated in the figure. Using this idea, all that is necessary is to calculate the first derivative between adjacent gray-scale values, which is usually called the gradient. The technique is called pixel differentiation.

A pixel differentiator must operate on the digital gray-level picture matrix stored in memory. The obvious question at this point is how to differentiate a digital image. The Roberts cross operator provides a good approximation of the first derivative, or gradient, of a digital image. The Roberts cross operator is defined as

$$R(m, n) = \{[i(m + 1, n + 1) - i(m, n)]^2 + [i(m, n + 1) \\ - i(m + 1, n)]^2\}^{1/2} \tag{15.26}$$

where $i(m, n)$ is the image intensity of pixel (m, n).

As illustrated in Figure 15.26, the operator is applied to diagonal pixels within a 2×2 window of a single picture matrix. The following example shows the procedure.

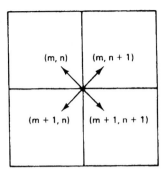

Figure 15.26   A 2 × 2 window needed for the Roberts cross operator (Andrew C. Staugaard, Jr., *Robotics and AI: An Introduction to Applied Machine Intelligence,* Prentice Hall, 1987, p. 213.)

Example 15.5.

Given the gray-level values of the following picture matrix, construct the gradient matrix:

$$\begin{bmatrix} 9 & 9 & 9 & 3 \\ 9 & 7 & 5 & 3 \\ 9 & 5 & 4 & 3 \\ 3 & 3 & 3 & 3 \end{bmatrix}$$

Solution

By applying the Roberts cross operator to each 2 × 2 pixel window in the matrix, the gradient matrix is obtained:

$$\begin{bmatrix} 2.0 & 4.5 & 6.3 & x \\ 4.5 & 3.0 & 2.2 & x \\ 6.3 & 2.2 & 1.0 & x \\ x & x & x & x \end{bmatrix}$$

15.11.2.2. Thresholding.   The function of thresholding is to decide which elements of the differentiated picture matrix should be considered as edge candidates. Edges are found by applying the Roberts cross operator to each intensity value and comparing the resulting gradient approximation, $R(m, n)$, to a threshold level, T. An edge is present if the gradient is greater than T. It is needless to say that the selection of the threshold level is very important. During the thresholding operation, the differentiated (gradient) matrix is converted to a binary picture matrix as follows:

If the Roberts operator value exceeds the threshold level for a given pixel, that matrix element is set to a value of 1; otherwise, a value of 0 is assigned.

Example 15.6.

Using the gradient matrix obtained in Example 15.5, construct a binary matrix using a threshold level of 4.

Solution

A pixel value is set to 1 if the Roberts operator value is greater than 4; otherwise, a 0 is given to that pixel. Thus, the matrix from Example 15.5 becomes

$$\begin{bmatrix} 0 & 1 & 1 & x \\ 1 & 0 & 0 & x \\ 1 & 0 & 0 & x \\ x & x & x & x \end{bmatrix}$$

Lines then might be identified from the binary matrix that is thresholded. Some popular techniques for finding lines from an edge-point matrix are tracking, model matching, and template matching.

15.11.3 Image Understanding

The final task of robot vision is to interpret the information obtained during the image-analysis process. This is called image understanding, or machine perception.

Most image-understanding research is centered around the "blocks world." The blocks world assumes that real-world images can be broken down and described by 2-D rectangular and triangular solids. Several AI-based image-understanding programs, which can interpret real-world images, have been successfully written under this blocks-world assumption.

15.12 SUMMARY

This chapter covers useful material of different technologies that go into a modern robotic system. It is an introduction for engineering students, technology students, and practicing engineers who are interested in obtaining some knowledge of industrial robotic systems.

With that in mind, this chapter presented some fundamental information such as robot classification, robot kinematics and dynamics, sensory systems, power sources, and robot grippers.

Interest has been growing in enhancing a robotic system by interfacing it to computer-vision equipment. Because of this, it is necessary to discuss some basics about computer vision: image transformation, processing, and understanding.

The application aspects of industrial robots are extremely important from a manufacturer's point of view. Therefore, the chapter discussed robot programming, safety issues, application planning, and economical considerations of robotic systems.

15.13 KEYWORDS

Articulated robot

Cartesian robot

Coordinate system

Cylindrical robot

Gantry robot

Gripper

Kinematics and dymanics

Lead-through robot

SCARA robot

Spherical robot

Textual-language programming

Walking through teaching

Workspace envelopes

15.14 REVIEW PROBLEMS

15.1 Using R = revolute and P = prismatic, kinematically describe the following basic robot geometries; (a) Cartesian, (b) cylindrical, (c) spherical, (d) articulated, and (e) SCARA.

15.2 The three primary DOF of a robot are used to achieve the correct _____ of the end-effector in space, whereas the three wrist DOF are needed to achieve part _____ in space.

15.3 Find the inverse kinematic solution for the cylindrical robot arm in Figure 15.4. That is, find the expression for α, a, and c as functions of X, Y, and Z.

15.4 Find the inverse kinematic solution for the spherical robot arm in Figure 15.5. In other words, find the expressions for α, β, and a as functions of X, Y, and Z.

15.5 Find the inverse kinematic solution for the articulated robot arm in Figure 15.6. In other words, find the expressions for α, β, and γ as functions of l_1, l_2, X, Y, and Z.

15.6 As viewed from above, a SCARA robot has the configuration shown in Figure 15.27).

(a) Find the homogeneous transformation relating the gripper position and orientation to the base.

(b) Assuming θ_3 is 0, find θ_1 and θ_2 to position the gripper at coordinates $(1, 1)$.

(c) Position the gripper at $(1, 1)$ with the orientation of the gripper being vertical as shown. Set up the equations that would be required to find angles θ_1, θ_2, and θ_3.

(d) Find θ_1, θ_2, and θ_3.

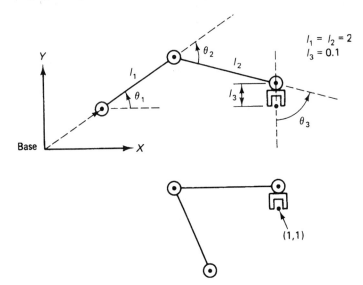

Figure 15.27   A SCARA robot.

15.7 An articulated robot has a vertical waist axis, a horizontal shoulder axis, and a horizontal elbow axis, as shown in Figure 15.28. The shoulder axis is 6 in. above and 4 in. forward of the base of the waist axis. The shoulder-to-elbow distance is 13 in. and the elbow-to-wrist-link distance is 5 in. The waist-axis joint limits allow $\pm 90°$ motion centered on the X_M axis. The shoulder-axis joint limits allow 0 to $+90°$ motion up from the horizontal. The elbow-axis joint limits allow $\pm 135°$ motion centered about the center line of the shoulder-to-elbow link.

(a) Calculate the X_M–Y_M–Z_M coordinates of the wrist center for the robot position shown in the figure.

(b) Sketch the outer envelope of this robot's work space. Show both top and side views.

(c) What is the transformation matrix describing the relative pose of the waist and shoulder axes using the X_1–Y_1–Z_1 and U_1–V_1–W_1 coordinate systems, as shown in Figure 15.28.

15.8 You have been asked to prepare specifications for robot selection for two different applications. As a preliminary step, you have to indicate the recommended drive system, the control type, and the programming capabilities. Indicate your selections and briefly explain your choices.

(a) Palletizing and depalletizing of boxes on pallets. Average box weight is 3 pounds. The boxes are placed in seven layers, with 20 boxes in each layer.

(b) Spray painting of car bodies in a closed paint booth. The brand of paint used contains flammable materials.

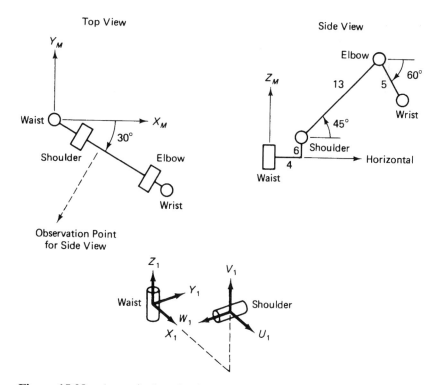

Figure 15.28   An articulated robot.

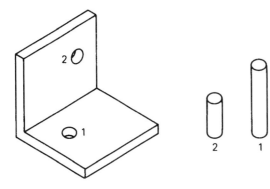

Figure 15.29 Assembly of a bracket and two pegs.

15.9 A robot is considered for assembly of two different pegs into the bracket shown in Figure 15.29. It is suggested that the use of a RCC (remote center of compliance) device will help to successfully implement the robot for this application. Do you agree with this suggestion? Comment on how you would implement the suggested device, and indicate any problems related to such implementation.

15.10 A cylindrical robot was selected to pick 80-mm $\times$ 100-mm boxes from a conveyor line and place the boxes on a pallet, as shown in Figure 15.30. Specify the degrees of freedom that will be needed on the wrist to perform this task. There is a price that will be paid for each degree of freedom on the wrist—do not overspecify!

15.11 Why is an accumulator required in a hydraulic robotic system?

15.12 Find a repetitive task that a robot might perform. Choose a robot and design the related tooling for the application.

15.13 Determine if the design developed in Review Problem 15.12 is economic for the robot.

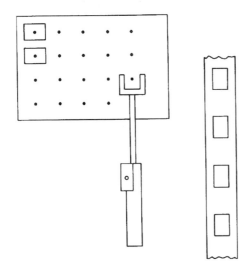

Figure 15.30 Palletizing boxes using a cylindrical robot.

15.15 REFERENCES

Anon. *Robotics Industrial Directory.* Conroe, TX: Technical Database Corporation, 1984a.

———. *The Specifications and Applications of Industrial Robots in Japan.* Tokyo: Japan Industrial Robot Association, 1984b.

Asada, H., and J. E. Slotine. *Robot Analysis and Control.* New York: John Wiley, 1986.

Ayres, R. U., and S. M. Miller. *Robotics: Applications and Social Implications.* Cambridge, MA: Ballinger, 1983.

Balestrino, A., G. De Maria, L. Sciavicco, and B. Siciliano. "An Algorithmic Approach to Coordinate Transformation for Robotic Manipulators," *Advanced Robotics,* 2, 1988, 315–404.

Chiu, S. L. "Task Compatibility of Manipulator Postures," *International Journal of Robotics Research,* 7(5), 1988, 13–21.

Craig, J. J. *Introduction to Robotics: Mechanics and Control,* 2d ed. Reading, MA: Addison-Wesley, 1989.

Critchlow, A. *Introduction to Robotics.* New York: Macmillan, 1985.

Csakvary, T. "Planning Robot Applications in Assembly," in S.Y. Nof, Ed., *The Handbook of Industrial Robotics.* New York: John Wiley, 1985, 1054–1083.

Dorf, R. C. *Robotics and Automated Manufacturing.* Reston, VA: Reston Publishing, 1983.

Engelberger, J. F. *Robotics in Practice.* New York: AMACOM, 1980.

Fu, K. S., R. C. Gonzalez, and C. S. G. Lee. *Robotics: Control, Sensing, Vision, and Intelligence.* New York: McGraw-Hill, 1987.

Groover, M. P. *Industrial Robotics: Technology, Programming, and Applications.* New York: McGraw-Hill, 1986.

Hartenberg, R. S., and J. Dentavit. *Kinematic Synthesis of Linkages.* New York: McGraw-Hill, 1964.

Heginbotham, W. B., A. Pugh, C. J. Page, D. W. Gatehouse, and P. W. Kitchen. "The Nottingham SIRCH Assembly Robot," in *Proceedings of the First CIRT.* Bedford, UK: IPS, 1973, 129–142.

Klein, C. A., and C. H. Huang. "Review of Pseudoinverse Control for Use with Kinematically Redundant Manipulators," *IEEE Transactions of Systems, Man, and Cybernetics,* 13, 1983, 245–250.

Kohli, D., and A. H. Soni. "Kinematic Analysis of Spatial Mechanisms via Successive Screw Displacements," *Journal of Engineering for Industry,* New York: ASME, 2B, 1963, 739–747.

Lapham, J., "RobotScript™: The Introduction of a Universal Robot Programming Language," *Industrial Robot,* 26(1), 1999, 17–25.

Lee, C. S. G. "Robot Arm Kinematics. Dynamics. and Control," *Computer,* 15(12), 1982, 62–80.

———. "Robot Arm Kinematics and Dynamics," in *Advances in Automation and Robotics: Theory and Applications.* Greenwich, CT: JAI Press, 1985.

Liegeois, A. "Automatic Supervisory Control of the Configuration and Behavior of Multibody Mechanisms," *IEEE Transactions of Systems, Man, and Cybernetics,* New York: IEEE, 7, 1977, 868–871.

Lin, S. K. "Singularity of a Nonlinear Feedback Control Scheme for Robots," *IEEE Transactions for Systems, Man, and Cybernetics,* 19(1), 1989, 134–139.

Luh, J. Y. S., M. W. Walker, and R. P. C. Paul. "Resolved-Acceleration Control of Mechanical Manipulators," *IEEE Transactions on Automatic Control,* 25(3), 1980, 468–474.

Maciejewski, A. A., and C. A. Klein. "Obstacle Avoidance for Kinematically Redundant Manipulators in Dynamically Varying Environments," *International Journal of Robotics Research,* 4(3), 1985, 109–117.

Makatchev, M. and S. K. Tso. "Human-Robot Interface Using Agents Communicating in an XML-Based Markup Language," *Robot and Human Communication—Proceedings of the IEEE International Workshop*, 2000, 270–275.

Marin, R., P. J. Sanz, and J. S. Sanchez, "A Very High Level Interface to Teleoperate a Robot Via Web Including Augmented Reality," *Proceedings—IEEE International Conference on Robotics and Automation*, 3, 2002, 2725–2730.

Marsh, P. "American's Factories Race to Automation," *New Scientist,* June 25, 1981, 845–847.

McCormick, D. (1982). "Making Points with Robot Assembly," *Design Engineering,* 1981, August, 24–28.

Meacham, J. "TIG Welding and Robotics," *Robotics Age,* March 1981, 28–31.

Milenkovic, V., and B. Huang. "Kinematics of Major Robot Linkages," in the *Proceedings of the 13th International Symposium on Industrial Robots.* Chicago, 1983, 16–31 to 16–47.

Nakamura, Y., H. Hanafusa, and T. Yoshikawa. "Task-Priority Based Redundancy Control of Robot Manipulators," *International Journal of Robotics Research,* 6(2), 1987, 3–15.

————. *Advanced Robotics: Redundancy and Optimization.* Reading, MA: Addison-Wesley, 1991.

Nakamura, Y., and H. Hanafusa. "Inverse Kinematic Solutions with Singularity Robustness for Robot Manipulator Control," *ASME Journal of Dynamic Systems, Measurement, and Control,* 108, 1986, 163–171.

————. "Optimal Redundancy Control of Robot Manipulators." *International Journal of Robotics Research,* 6(1), 1987, 32–42.

Nevins, J. L., and D. E. Whitney. "Computer-Controlled Assembly," *Scientific American,* 238(2), 1978, 62–74.

Nof, S. Y. (Ed). *Handbook of Industrial Robotics.* New York: John Wiley, 1985.

Nof, S.Y., and E. L. Fisher. (1982) "Analysis of Robot Work Characteristics," *Industrial Robot,* September 1982, 166–177.

Orin, D. E., R. B. McGhee, M. Vukobratovic, and G. Hartoch. "Kinematic and Kinematic Analysis of Open-Chain Linkages Utilizing Newton–Euler Methods," *Mathematical Bioscience,* 43, 1979, 107–130.

Orin, D. E., and W. W. Schrader. "Efficient Computation of the Jacobian for Robot Manipulators," *International Journal of Robotics Research,* 3(4), 1984, 66–75.

Ottinger, L. V. "A Plant Search for Possible Robot Applications," *Industrial Engineering,* December 1981, 26.

————. "Evaluating Potential Robot Applications in a System Context," *Industrial Engineering,* January 1985, 80–87.

Owen, T. *Assembly with Robots.* Englewood Cliffs, NJ: Prentice Hall, 1985.

Parent, M., and C. Laurgeau. *Robot Technology: Logic and Programming.* Englewood Cliffs, NJ: Prentice Hall, 1985.

Paul, R. P., B. E. Shimano, and G. Mayer. Kinematic Control Equations for Simple Manipulators," *IEEE Transactions on Systems, Man and Cybernetics,* SMC-11(6), 1981, 449–455.

Penington, R. A., E. L. Fisher, and S. Y. Nof. "Survey of Industrial Characteristics: General Distributions, Trends, and Correlations," in Ed., *Robotics and Material Flow.* Amsterdam: Elsevier, 1986, 37–54.

PERA Robots. *A Further Survey of Robots and Their Current Applications in Industry,* Report 337. London: Melton Mowbray, 1983.

Pieper, D. L. *The Kinematics of Manipulators Under Computer Control,* Artificial Intelligence Project Memo No. 72, Palo Alto, CA: Stanford University Computer Science Department, 1968.

Robot Institute of America. *Worldwide Robotics Survey and Directory.* Dearborn, MI: RIA, 1982.

Romeo, G., and A. Camera. "Robots for Flexible Assembly Systems," *Robotics Today,* Fall 1980, 23–43.

Salisbury, J. K., and J. J. Craig. "Articulated Hands: Force Control and Kinematic Issues," *International Journal of Robotics Research,* 1(1), 1982, 4–17.

Saveriano, J. W. "Industrial Robots Today and Tomorrow," *Robotics Age,* 1980, 4–17.

Sciavicco, L. and B. Siciliano. *Modeling and Control of Robotic Manipulation.* New York: McGraw-Hill, 1996.

Sciavicco, L. and B. Siciliano. *Modeling and Control of Robot Manipulators,* 2d Ed., London: Springer-Verlag, 2001.

————. "A Solution Algorithm to the Inverse Kinematic Problem for Redundant Manipulators," *IEEE Journal of Robotics and Automation,* 4(4), 1988, 403–410.

Siciliano, B. *Algoritmi di Soluzione al Problema Cinematico Inverso per Robot di Manipolazione* [in Italian], Tesi di Dottorato di Ricerca, Universita degli Studi di Napoli, 1986.

————. "Kinematic Control of Redundant Robot Manipulators: A Tutorial," *Journal of Intelligent and Robotic Systems,* 3, 1990, 201–212.

Spong, M. W., and M. Vidyasagar. *Robot Dynamics and Control.* New York: John Wiley, 1989.

Stadler, W. (1995). *Analytical Robotics and Mechanics.* New York: McGraw-Hill, 1989.

Staugaard, Jr., A. C. *Robotics and AI: An Introduction to Applied Machine Intelligence.* Englewood Cliffs, NJ: Prentice Hall, 1987.

Suh, C. H., and C. W. Radcliffe. *Kinematics and Mechanisms Design.* New York: John Wiley, 1978.

Tanner, W. *Industrial Robots.* Dearborne, MI: Society of Manufacturing Engineers, 1978.

Thompson, T. "Robots for Assembly," *Assembly Engineering,* July 1981, 32–36.

Wampler, C. W. "Manipulator Inverse Kinematic Solutions Based on Damped Least-Squares Solutions," *IEEE Transactions on Systems, Man, and Cybernetics,* 16(1), 1986, 93–101.

Warnecke, H. J., and R.D. Schraft. *Industrial Robots: Applications Experience.* Bedford, UK: I.P.S. Publications, 1982, 202–206.

Warnecke H. J., R. D. Schraft, and U. Schmidt-Streier. Computer-Aided Planning of Industrial Robot Application in Workpiece Handling," in *Proceedings of the Eleventh International Symposium on Industrial Robots.* Tokyo: Society of Biomechanisms of Japan and Japan Industrial Robot Association, 1981, 349–3S9.

Whitney, D. E. "Resolved Motion Rate Control of Manipulators and Human Prostheses," *IEEE Transactions of Man-Machine Systems,* 10, 1969, 47–53.

Wysk, R. A., and P. I. Pheffenberger. A Computerized Time Cycle Sheet for Welding Robot Justification," in *Proceedings of the 1986 Spring IIE Conference.* Dallas: IIE, 1986.

Yoshikawa, T. "Manipulability of Robotic Mechanisms," *International Journal of Robotics Research,* 4(2), 1985, 3–9.

————. *Foundations of Robotics.* Cambridge, MA: MIT Press, 1990.

Yuan, J.S.C. "Closed-Loop Manipulator Control Using Quaternion Feedback," *IEEE Journal of Robotics and Automation,* 4(4), 1988, 434–440.

Yuan, M. S. C., and R. Freudenstein. "Kinematics Analysis of Spatial Mechanisms by Means of Screw Coordinates." *Transactions of the American Society of Mechanical Engineers, Journal of Engineering Industry,* 93(1), 1971, 61–73.

Zhang, G.-L., Y. Fu, R.-Q. Yang, and W.-J. Zhang. "Robotic Real-Time Control System with Open Architecture Based on Windows NT," *Journal of Shanghai Jiaotong University,* 37(5) May 2003, 724–728.

Vector Algebra

This appendix introduces some basic properties of vectors.

A.1 COMPONENTS OF A VECTOR

Let $\mathbf{A}$ be a vector and $\mathbf{i}, \mathbf{j}$, and $\mathbf{k}$ be unit vectors in the directions of the positive X-, Y-, and Z-axes. Then

$$\mathbf{A} = A_1\mathbf{i} + A_2\mathbf{j} + A_3\mathbf{k}$$

where A_1, A_2, and A_3 are the components of $\mathbf{A}$. (See Figure A.1.)

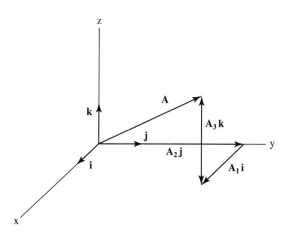

Figure A.1

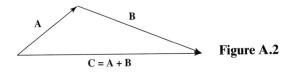

Figure A.2

$$C = A + B$$

A.2 MAGNITUDE OF A VECTOR

The magnitude of a vector is

$$|\mathbf{A}| = \sqrt{A_1^2 + A_2^2 + A_3^2}$$

A.3 MULTIPLICATION OF A VECTOR BY A SCALAR

If m is a real number (scalar) and $\mathbf{A}$ is a vector, then $m\mathbf{A}$ is a vector whose magnitude is $|m|$ times $|\mathbf{A}|$.

A.4 SUM OF VECTORS

$\mathbf{C}$ is the sum of two vectors, $\mathbf{A}$ and $\mathbf{B}$. See Figure A.2.

A.5 UNIT VECTORS

A unit vector is a vector with unit magnitude. If $\mathbf{A}$ is a vector, then a unit vector in the direction of $\mathbf{A}$ is $\mathbf{A}/|\mathbf{A}|$.

A.6 LAWS OF VECTOR ALGEBRA

Commutative law of addition: $\qquad \mathbf{A} + \mathbf{B} = \mathbf{B} + \mathbf{A}$
Associative law of addition: $\qquad \mathbf{A} + (\mathbf{B} + \mathbf{C}) = (\mathbf{A} + \mathbf{B}) + \mathbf{C}$
Associative law of scalar multiplication: $\quad m(n\mathbf{A}) = (mn)\mathbf{A} = n(m\mathbf{A})$
Distributive law: $\qquad (m + n)\mathbf{A} = m\mathbf{A} + n\mathbf{A}$
Distributive law: $\qquad m(\mathbf{A} + \mathbf{B}) = m\mathbf{A} + m\mathbf{B}$

A.7 DOT OR SCALAR PRODUCT

The dot product, or scalar product, of two vectors $\mathbf{A}$ and $\mathbf{B}$ is

$$\mathbf{A} \cdot \mathbf{B} = |A||B| \cos \theta \qquad 0 \le \theta \le \pi$$

where θ is the angle between $\mathbf{A}$ and $\mathbf{B}$. (See Figure A.3.)

Figure A.3

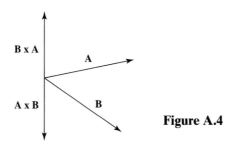

Figure A.4

$$\mathbf{A} \cdot \mathbf{B} = \mathbf{B} \cdot \mathbf{A}$$
$$\mathbf{A} \cdot (\mathbf{B} + \mathbf{C}) = \mathbf{A} \cdot \mathbf{B} + \mathbf{A} \cdot \mathbf{C}$$
$$\mathbf{A} \cdot \mathbf{B} = A_1B_1 + A_2B_2 + A_3B_3$$

A.8 CROSS OR VECTOR PRODUCT

The cross product, or vector product, of two vectors **A** and **B** is

$$\mathbf{A} \times \mathbf{B} = |A||B|(\sin \theta)\mathbf{u} \qquad \le \theta \le \pi$$

where θ is the angle between **A** and **B**, and **u** is a unit vector perpendicular to the plane of **A** and **B**. (See Figure A.4.) The direction of **u** is determined by the right-hand rule. Mathematically,

$$\mathbf{A} \times \mathbf{B} = \begin{bmatrix} \mathbf{i} & \mathbf{j} & \mathbf{k} \\ A_1 & A_2 & A_3 \\ B_1 & B_2 & B_3 \end{bmatrix}$$
$$= (A_2B_3 - A_3B_2)\mathbf{i} + (A_3B_1 - A_1B_3)\mathbf{j} + (A_1B_2 - A_2B_1)\mathbf{k}$$
$$\mathbf{A} \times \mathbf{B} = -\mathbf{B} \times \mathbf{A}$$
$$\mathbf{A} \times (\mathbf{B} + \mathbf{C}) = \mathbf{A} \times \mathbf{B} + \mathbf{A} \times \mathbf{C}$$
$$|\mathbf{A} \times \mathbf{B}| = \text{area of parallelogram having sides } \mathbf{A} \text{ and } \mathbf{B}$$

Appendix B

Transfer Functions and Block Diagrams

A transfer function is defined as the ratio of the output to the input of a system in the Laplace domain. A block diagram is a way to represent a system. In Figure B.1, a block diagram shows a component of a system.

B.1 TRANSFER FUNCTIONS

The transfer function of the system shown in Figure B.1 is

$$T(s) = \frac{O(s)}{I(s)}$$

where $O(s)$ is the output in the s domain, and $I(s)$ is the input in the s domain. This relation is valid under the assumption that all initial conditions are zero. When there are multiple inputs, $I_1(s), I_2(s), \ldots$,

$$I(s) = \sum_{i=1}^{n} a_i I(s)$$

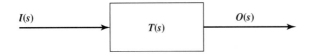

$I(s)$ $T(s)$ $O(s)$

Figure B.1 A block diagram of a simple system.

where a_i is the proportion of each input, and

$$O(s) = T(s)I(s)$$

$$= \sum_{i=1}^{n} a_i I(s)T(s)$$

$$= \sum_{i=1}^{n} a_i O(s)$$

B.2 BLOCK DIAGRAMS

Other than the simple block-diagram component shown in Figure B.1, there is another commonly used component—the error detector (Figure B.2). Depending on the corrective action, an error detector can be a sum operator or a difference operator. It is also sometimes called a comparator.

B.2.1 Block-Diagram Algebra

A complex block diagram can be written in simple form. In this section, several rules are given.

B.2.1.1. A Series of Blocks.   The system shown in Figure B.3 can be simplified into the diagram of Figure B.4.

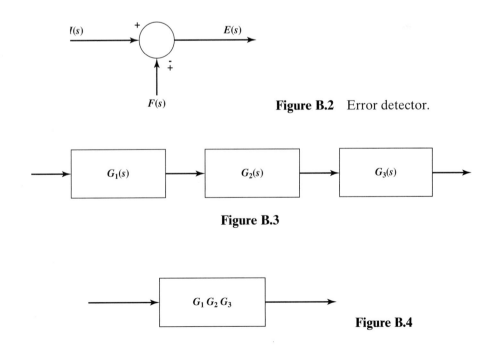

Figure B.2   Error detector.

Figure B.3

Figure B.4

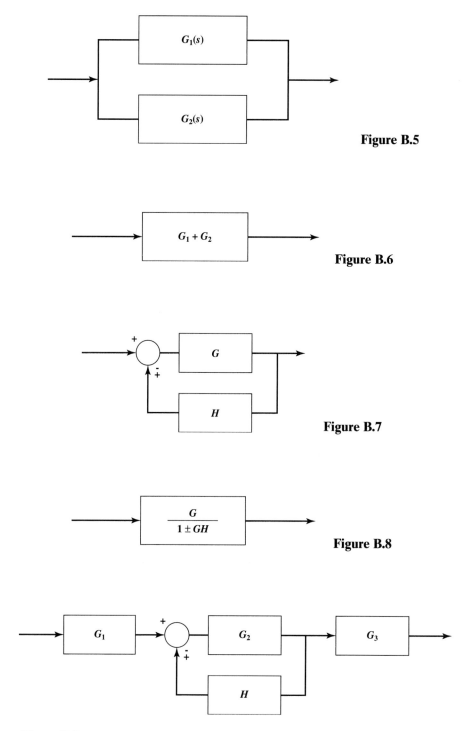

Figure B.5

Figure B.6

Figure B.7

Figure B.8

Figure B.9

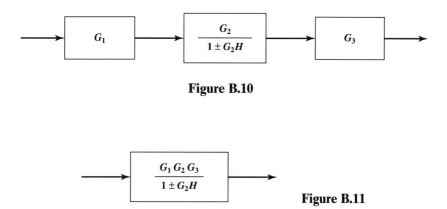

Figure B.10

Figure B.11

B.2.1.2. Parallel Blocks. The transfer function of the equivalent system (Figure B.5) is the sum of the original system (Figure B.6).

B.2.1.3. System with Feedback. The system of Figure B.7 can be simplified into the equivalent system of Figure B.8.

Example B.1.

Simplify the block diagram of Figure B.9.

Solution

First, the feedback loop is simplified. The result is a system in series (Figure B.10).

Finally, the three blocks are combined into one block (Figure B.11). The equivalent transfer function of the system is

$$\frac{G_1 G_2 G3}{1 \pm G_2 H}$$

Appendix C

Laplace Transforms

C.1 DEFINITION

The definition of the Laplace transform of $F(t)$ is

$$L\{F(t)\} = \int_0^\infty e^{-st} F(t)\, dt = f(s)$$

where L is the Laplace transform operator.

C.2 DEFINITION OF THE INVERSE LAPLACE TRANSFORM

The definition of the inverse Laplace transform of $f(s)$ is

$$F(t) = L^{-1}\{f(s)\}$$

$$= \frac{1}{2\pi i} \int_{c-i\infty}^{c+i\infty} e^{st} f(s)\, ds$$

C.3 GENERAL PROPERTIES

The general properties of the Laplace transform are

$F(t)$	$t(s)$
$aF_1(t) + bF_2(t)$	$af_1(s) + bf_2(s)$
$aF(at)$	$f\left(\dfrac{s}{a}\right)$

$F(t)$	$f(s)$
$F'(t)$	$sf(s) - F(0)$
$F''(t)$	$s^2 f(s) - sf(0) - F'(0)$
$\int_0^1 F(u)\, du$	$\dfrac{1}{s} f(s)$

C.4 LAPLACE TRANSFORMS

The following table gives a summary of commonly used Laplace transforms:

$F(t)$	$f(s)$
1	$\dfrac{1}{s}$
t	$\dfrac{1}{s^2}$
e^{at}	$\dfrac{1}{s - a}$
te^{at}	$\dfrac{1}{(s - a)^2}$
$\dfrac{\sin \omega t}{\omega}$	$\dfrac{1}{s^2 + \omega^2}$
$\cos \omega t$	$\dfrac{s}{s^2 + \omega^2}$
$\dfrac{e^{at}\sin \omega t}{\omega}$	$\dfrac{1}{(s - a)^2 + \omega^2}$
$e^{at}\cos \omega t$	$\dfrac{s - a}{(s - a)^2 + \omega^2}$
$\dfrac{1}{\sqrt{1 - \zeta^2}} e^{-\zeta \omega t}\sin \omega \sqrt{1 - \zeta^2}\, t$	$\dfrac{1}{s^2 + 2\zeta \omega s + \omega^2}$

Example C.1. Solve the first-order differential equation

$$\frac{dx}{dt} + 3x = 5t$$

where $x(t) = 0$.

Solution

$$L\left\{ \frac{dx}{dt} + 3x \right\} = L\{5t\} \qquad (\text{C.1})$$

$$sL\{x\} - x(0) + 3L\{x\} = \frac{5}{s^2} \tag{C.2}$$

By using the given initial condition, Equation (C.2) can be rewritten as

$$(s + 3)L\{x\} = \frac{5}{s^2} \tag{C.3}$$

so

$$L\{x\} = \frac{5}{s^2(s + 3)} \tag{C.4}$$

The partial-fraction expansion gives

$$\frac{5}{s^2(s + 3)} = \frac{A}{s + 3} + \frac{B}{s} + \frac{C}{s^2} \tag{C.5}$$

Solving Equation (C.5) yields

$$C = \frac{5}{3}$$

$$B = -\frac{5}{9}$$

$$A = \frac{5}{9}$$

Therefore,

$$L\{x\} = \frac{5/9}{s + 3} - \frac{5/9}{s} + \frac{5/3}{s^2} \tag{C.6}$$

Applying the inverse transformation results in

$$L^{-1}\{x\} = L\left\{\frac{5/9}{s + 3} - \frac{5/9}{s} + \frac{5/3}{s^2}\right\} \tag{C.7}$$

$$x(t) = \frac{5}{9}e^{-3t} - \frac{5}{9} + \frac{5}{3}t \tag{C.8}$$

Rewriting Equation (C.8) gives

$$x(t) = \frac{5}{9}(e^{-3t} - 1) + \frac{5}{3}t \tag{C.9}$$

Appendix D

Z-Transforms

D.1 DEFINITION

The Z-transform of a function $f(t)$ is defined as

$$Z[f(t)] = \sum_{n=0}^{\infty} f(t)z^{-n}$$

$$f(t) = f(0)\delta(t) + f(T)\delta(t - T) + f(2T)\delta(t - 2T) + \cdots$$

$$= f_0\delta(t) + f_1\delta(t - T) + f_2\delta(t - 2T) + \cdots$$

where $\delta(t)$ is a unit-impulse function.

D.2 DEFINITION OF THE INVERSE *Z*-TRANSFORM

$$f(t) = Z^{-1}[F(z)]$$

D.3 GENERAL PROPERTIES

The Z-transform $Z[f(t)] = F(z)$ satisfies a number of important properties, including the following:

linearity

$$Z[a_n + b_n] = Z[a_n] + Z[b_n]$$

translation

$$Z[f_{n-k}] = z^{-k}Z[f_n]$$

$$Z[f_{n+1}] = zZ[f_n] - zf_0$$

$$Z[f_{n+2}] = z^2Z[f_n] - z^2f_0 - zf_1$$

$$Z[f_{n+k}] = z^m Z[f_n] - \sum_{r=0}^{m-1} z^{k-r} f_{rt}$$

scaling

$$Z[a^n f_n] = F(z/a)$$

multiplication and division

$$Z[af_n] = aZ[f_n]$$

$$Z[f_n/a] = Z[f_n]/a$$

multiplication by powers of n

$$Z[n^k f_n] = (-1)^k \left(z \frac{d}{dz} \right)^k F(z)$$

$$Z[n^{-1} f_n] = -\int_0^z \frac{F(z)}{z} dz$$

D.4 *Z*-TRANSFORM TABLE

	f(t)	$Z[f(t)]$
1	$\delta(t)$ delta function	1
2	$1(t)$ step function	$\dfrac{z}{z-1}$
3	t ramp function	$\dfrac{Tz}{(z-1)^2}$
4	t^2 parabolic	$\dfrac{T^2 z(z+1)}{(z-1)^3}$
5	n	$\dfrac{z}{(z-1)^2}$
6	n^2	$\dfrac{z(z+1)}{(z-1)^3}$
7	n^3	$\dfrac{z(z^2 + 4z + 1)}{(z-1)^4}$

8	a^n power	$\dfrac{z}{z-a}$
9	na^n	$\dfrac{az}{(z-a)^2}$
10	e^{-at}	$\dfrac{z}{z-e^{-aT}}$
11	te^{-at}	$\dfrac{Tze^{-aT}}{(z-e^{-aT})^2}$
12	$\sin(\alpha n)$	$\dfrac{z\sin\alpha}{1-2z\cos\alpha+z^2}$
13	$\cos(\alpha n)$	$\dfrac{z(z-\cos\alpha)}{1-2z\cos\alpha+z^2}$

Example D.1.

Use the Z-transform to solve the following difference equations:

$$8x_{k+2} - 6x_{k+1} + x_k = 0, \quad x_1 = 1$$

Using the linearity and multiplication properties, taking the Z-transform, and ignoring initial conditions that are zero yields

$$8z^2 X(z) - 8zx_1 - 6zX(z) + X(z) = 0$$

Solving for $X(z)$ and expanding $X(z)/z$ in partial fractions gives

$$\frac{X(z)}{z} = \frac{8}{8z^2 - 6z + 1} = \frac{8}{2z-1} - \frac{16}{4z-1}$$

$X(z)$ simplifies to

$$X(z) = \frac{4z}{z-1/2} - \frac{4z}{z-1/4}$$

and the inverse Z-transform (table entry 8) now gives the solution for x_k:

$$x_k = 4[(1/2)^k - (1/4)^k]$$

Appendix E

Numbering Systems

E.1 BINARY NUMBER

A binary number consists of only two digits: 0 and 1. Any integer number can be represented by these two digits, for example, 10100110_2. To convert a binary number into decimal, use the following example:

$$
\begin{array}{ccccccccc}
1 & 0 & 1 & 0 & 0 & 1 & 1 & 0 \\
2^7 & 2^6 & 2^5 & 2^4 & 2^3 & 2^2 & 2^1 & 2^0 \\
\hline
128 & + 0 & + 32 & + 0 & + 0 & + 4 & + 2 & + 0 & = 166_{10}
\end{array}
$$

The weight of each digit is 2 to the power of the digit position minus 1. In a digital system, such as a computer, the state is either on or off, so a binary number can represent the state of a set of components. For example, computer memory is grouped into an eight-bit unit, called a byte. Each bit can be either on or off (1 or 0). A byte, therefore, stores an eight-digit binary number.

To convert a decimal number into binary, divide the decimal number by 2 repeatedly until the remainder is either 0 or 1.

$$25_{10}$$

remainder

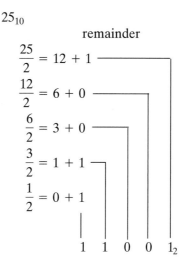

$$\frac{25}{2} = 12 + 1$$

$$\frac{12}{2} = 6 + 0$$

$$\frac{6}{2} = 3 + 0$$

$$\frac{3}{2} = 1 + 1$$

$$\frac{1}{2} = 0 + 1$$

$$1 \quad 1 \quad 0 \quad 0 \quad 1_2$$

E.2 OCTAL NUMBER

An octal number is based on eight digits: 0, 1, 2, 3, 4, 5, 6, and 7. Any integer number can be represented by these eight digits, for example, 374_8. Because the base, 8, is a power of 2, an octal number can easily be translated into binary:

3	7	4	Octal number
8^2	8^1	8^0	Octal weight of each digit
$2^2\, 2^1\, 2^0$	$2^2\, 2^1\, 2^0$	$2^2\, 2^1\, 2^0$	Binary weight of each digit
0 1 1	1 1 1	1 0 0	Binary representation

To do the conversion, each digit is converted into binary. The binary equivalent is the concatenation of the separately converted binary numbers. To convert into a decimal number, the preceding approach is again used:

3	7	4	Octal number
8^2	8^1	8^0	Octal weight of each digit

$$3 \times 64 + \quad 7 \times 8 + \quad 4 = 252_{10}$$

To convert a decimal number into an octal number:

$$25_{10}$$

remainder

$$\frac{25}{8} = 3 + 1$$

$$\frac{3}{8} = 0 + 3$$

$$3 \quad 1_8$$

E.3 HEXADECIMAL NUMBER

A hexadecimal number (hex) is based on 16 digits and/or symbols: 0, 1, 2, 3, 4, 5, 6, 7, 8, 9, A, B, C, D, E, and F. Any integer number is represented by these eight digits, for example, $2AF_{16}$. Because the base, 16, is 2 to the power of 4, each digit of a hexadecimal-number represents four bits. A byte can be conveniently represented by two hexadecimal digits, a concise representation. The conversion from hexadecimal to binary is also straightforward:

2	A	F	Hexadecimal number
16^2	16^1	16^0	Hex weight of each digit
$2^3\, 2^2\, 2^1\, 2^0$	$2^3\, 2^2\, 2^1\, 2^0$	$2^3\, 2^2\, 2^1\, 2^0$	Binary weight of each digit
0 0 1 0	1 0 1 1	1 1 1 1	Binary representation

To do the conversion, each digit is converted into binary. The binary equivalent is the concatenation of the separately converted binary numbers. To convert into a decimalnumber, the preceding approach is again used:

2	A	F	Hex number
16^2	16^1	16^0	Hex weight of each digit

$$2 \times 256 + \quad 10 \times 16 \quad + \quad 15 = 687_{10}$$

To convert a decimal number into a hexadecimal number, use the following logic:

$$125_{10}$$

remainder

$$\frac{125}{16} = 7 + 13$$

$$\frac{7}{16} = 0 + 7$$

$$7\ D_{16}$$

E.4 BINARY-CODED-DECIMAL (BCD) NUMBER

One way to represent a decimal number is to represent individual decimal digits in binary separately. For example, the decimal number 25 is represented as

2	5
010	101

Typical applications of BCD codes include data entry (time, volume, weights, and so on) via thumbwheel switches, data display via seven-segment displays, and input from absolute encoders.

E.5 BINARY ARITHMETIC

ADDITION

$$0 + 0 = 0$$
$$1 + 0 = 1$$
$$1 + 1 = 10$$

Example:

$$
\begin{array}{r}
1\ 0\ 1\ 1\ 0 \\
+\ \ 1\ 0\ 0\ 1\ 1 \\
\hline
1\ 0\ 1\ 0\ 0\ 1_2
\end{array}
$$

SUBTRACTION

$$1 - 0 = 1$$
$$1 - 1 = 0$$
$$0 - 1 = 1 \text{ borrow } 1$$

Example:

$$
\begin{array}{r}
1\ 0\ 1\ 1\ 0 \\
-\ 1\ 0\ 0\ 1\ 1 \\
\hline
0\ 0\ 0\ 1\ 1_2
\end{array}
$$

MULTIPLICATION

$$0 \times 0 = 0$$
$$1 \times 0 = 1$$
$$1 \times 1 = 1$$

Example:

$$
\begin{array}{r}
1\ 0\ 1\ 1\ 0 \\
\times\ \ \ \ \ \ \ \ 1\ 1 \\
\hline
1\ 0\ 1\ 1\ 0 \\
1\ 0\ 1\ 1\ 0 \\
\hline
1\ 0\ 0\ 0\ 0\ 1\ 0
\end{array}
$$

Multiplication is basically a series of shift operations, followed by adding the results together.

E.6 SIGNED AND UNSIGNED BINARY NUMBER

A binary number can be signed or unsigned. In a signed binary number, the most significant bit (leftmost) stores the sign. A negative binary number is stored in the *2's-complement form*. To obtain a 2's-complement form, all 1's are first converted to 0's and 0's to 1's. The result is called the 1's complement. By adding 1 to the 1's-complement, the 2's-complement form is obtained.

ORIGINAL NUMBER	**1's-COMPLEMENT**	**2's-COMPLEMENT**
01001101	$10110010 + 1 =$	10110011
77_{10}		-77_{10}

An eight-bit signed binary number has the following range:

10000000	... 11111111	00000000	00000001	... 01111111	
-128_{10}	... -1_{10}	0_{10}		1_{10} ...	127_{10}

MATLAB Fundamentals

MATLAB is based on matrix manipulations.

Basic commands:

help "command": get help.
clc: clear the work session.
clg: clear graphic window.
clear: clear workspace, reset variables.

MATLAB commands are shown in boldface:

>> pi
ans =
   3.1416
>> **A = [1, 2, 3];**

A ";" at the end of the line suppresses the output. Without it, MATLAB will print the vector out.
A "%" indicates that the text which follows is a comment.

MATLAB is case sensitive

>> C = [−1, 0, 0; 1, 1, 0; 1, −1, 0; 0, 0, 2];   % "," separate element in a row. ";" separate rows.
>> C
C =
   −1   0   0
    1   1   0
    1   −1   0
    0   0   2
>> 5*C

```
ans  =
   −5     0     0
    5     5     0
    5    −5     0
    0     0    10
```

Indexing:

```
>> C(1, 1)
ans  =
  −1
```

C(i,j): element i,j.
C(1,:): Row 1

```
>> C(1, :)
ans  =
  −1   0   0
```

C(:,1): Column 1.

```
>> C(:, 1)
ans  =
  −1
```

```
1
1
0
```

Series:

```
>> x = 1:8
x  =
   1   2   3   4   5   6   7   8
>> 1:2:8 % 1 to 8 at step of 2
ans  =
   1   3   5   7
```

Partial matrix:

```
>> y = C(:, 2:3)    %Column 2 and 3
y  =
    0   0
    1   0
   −1   0
    0   2
>> z = C(3:4,1:2)    %intersection of row 3−4 and column 1−2
z  =
    1  −1
    0   0
```

Special matrices:

```
>> eye(3)
```

```
ans  =
    1   0   0
    0   1   0
    0   0   1
>> ones(3,2)
ans  =
    1   1
    1   1
    1   1
>> zeros(1,2)
ans  =
    0 0
```

Scalar Operations

$+, -, *, /, \backslash, \wedge$ "$\backslash$" is left division, $a \backslash b = b/a$

Array Operations

Element-by-element $C(1, 1) = a(1)*b(1);$
Two matrices $C = A.*B;$ where the operator is "*".

Common Functions

$abs(x), sqrt(x),$
round(x): round to the nearest integer.
fix(x): round down toward zero
floor(x): round down to the nearest integer toward $-\infty$
ceil(x): round up
sign(x)
rem(x,y): remainder of x/y
$exp(x)$: e^x
log(x): natural log
log10(x): log base 10
sin,cos, tan, asin, acos, atan, atn2
sqrt(x): square root

Specialized Functions

cross(a,b): cross product (outer product) of two vectors
dot(a,b):   dot product (inner product) of two vectors

X-Y plot

plot(x,y), . . .   % ", . . . shows a continuation of the command,
title('title of plot"), . . . %the rest of the command will be interpreted
xlabel("label on x"), . . . % at the same time.
ylabel("label on y"), . . .
grid %add grid to the plot.

Function – M-file

```
% r = ferguson(p, n)
% 3-D Ferguson's curve, return curve as points
% p = [x0 x1 tx0 tx1; y0 y1 ty0 ty1; z0 z1 tz0 tz1]
%   x0 x1 two terminal points,
%   tx0 tx1 two tangents
% n: number of output points on the curve, n > 3
% r: contains 3-D curve points
% to plot the curve on 2-D, plot(r(1,:),r(2,:))
%
function r = ferguson(p,n)
m = 1/(n - 1);
t = 0:m:1;
A = [2 -3 0 1; -2 3 0 0; 1 -2 1 0; 1 -1 0 0];
tt = [t.^3; t.^2; t; ones(size(t))];
r = p*A*tt;
```

help

```
>> help ferguson
```
r = ferguson(p, n)
3D Ferguson's curve, return curve as points
p = [x0 x1 tx0 tx1; y0 y1 ty0 ty1; z0 z1 tz0 tz1]
x0 x1 two terminal points,
tx0 tx1 two tangents
n: number of output points on the curve, n > 3
r: contains 3D curve points
to plot the curve on 2D, plot (r(1,:),r(2,:))

run the function

```
ferguson(p,n);
```

Matrix manipulation

Transpose

```
A = B'
```

Solving simultaneous equations

$$3x_1 + 2x_2 - x_3 = 10$$
$$-x_1 + 3x_2 + 2x_3 = 5$$
$$x_1 - x_2 - x_3 = -1$$
$$AX = B$$
$$X = A \backslash B$$

```
>> A = [3, 2, -1; -1, 3, 2; 1, -1, -1];
>> B = [10, 5, -1]';        % B is a column vector, use transpose operator.
>> x = A\B
```

X =
       − 2.0000
        5,0000
       − 6.0000

Matrix Inverse

C = inv(A);
The previous equation can be solved by
$A^{-1}X = A^{-1}$
≫ **inv(A)*B**
ans =
       − 2.0000
        5.0000
       − 6.0000

Cubic Spline

3rd degree polynomial passing through a set of points.
spline(x,y,newx);
x,y are row vectors defining a set of points.
newx can be a single x value or a set of values.
The function returns corresponding y values.
≫ **x = [0, 1, 2, 3, 4, 5];**
≫ **y = [0.0, 20.0, 60.0, 68, 77, 110];**
≫ **newx = 0:0.1:5;**
≫ **newy = spline(x, y, newx)**
newy =
Columns 1 through 8
        0   − 1.0470   − 1.2160   − 0.5823   0.7787   2.7917   5.3813   8.4723
Columns 9 through 16
   11.9893   15.8570   20.0000   24.3430   28.8107   33.3277   37.8187   42.2083
Columns 17 through 24
   46.4213   50.3823   54.0160   57.2470   60.0000   62.2230   63.9573   65.2677
Columns 25 through 32
   66.2187   66.8750   67.3013   67.5623   67.7227   67.8470   68.0000   68.2380
Columns 33 through 40
   68.5840   69.0527   69.6587   70.4167   71.3413   72.4473   73.7493   75.2620
Columns 41 through 48
   77.0000   78.9780   81.2107   83.7127   86.4987   89.5833   92.9813   96.7073
Columns 49 through 51
  100.7760   105.2020   110.0000

Polynomial functions

$f(x) = 3x^4 - 0.5x^3 + x - 5.2$
a = [3, 0.5, 0, 1, −5.2];
f = polyval(a, x);   % set the x value to find the corresponding f(x).
                         % x can be a range.

Polynomial multiplication and division.
a, b: coefficient vector of a function.
g = conv(a, b);   % multiple two functions.
g is the coefficient vector for the resultant function.
a = deconv(g, b);   % get a back by dividing function g by function b.
Roots of a function
r = roots(g);   % r contain all the roots.
To find the root of the function a, use
         r = root(a);
Computes the coefficients of a polynomial from roots:
A = ploy([−1, 1, 3]); % three roots are −1, 1 and 3

Numerical Integration

$K = \int_a^b f(x)\,dx$
K = quad('function',a,b); % "function" is the name of the function.
         K = quad('velocity', 0, 1);
         % M-file function
         function v = velocity(r)
         % velocity This function calculate the velocity
         v = r.*(1 − r/1).^(1/8);

Differentiation

$f'(x_k) = \dfrac{f(x_{k+1}) - f(x_k)}{x_{k+1} - x_k}$
diff(f); % differences between adjacent values in a vector.
first derivative is: diff(f)./diff(x);
Let $f(x) = x^5 - 3x^4 - 11x^2 - 10$
         x = 1:.1:3;
         f = x.^5 − 3*x.^4 − 11*x.^2 − 10;
         df = diff(f)./diff(x);

3-D Graphics

Plot 3-D Example
         t = linspace(0, 10*pi);
         plot 3(sin(t),cost(t),t)
         xlabel('sin(t)'),ylabel('cos(t)'),zlabel('t')
         text(0,0,0,'Origin')
         grid on
         title ('Helix')
Mesh Plot
         mesh (x,y,z); % x,y,z contain 3-D data points;
Waterfall
         waterfall (x,y,z); % similar to mesh, except plots only in x direction
Surface Plot
         surf(x,y,z); % shaded between mesh

shading flat; % flat shading
shading interp;
surfnorm (x,y,z); % surface plot with normal vectors
Contour Plot
contour (x,y,z,10); % contour with 10 lines
contour 3(x,y,z10); % contour in 3-D
pcolor (x,y,z); % pseudocolor plot on contour

Control Flow

For loop
for x = array
        (commands)
end
for n = 1:10
        x(n) = sin(n*pi/10);
end
for m = 10:−1:1
        y(n) = sin(n*pi/10);
end

While loop
n = 1;
while (n + 1) > 1
        n = n/2;
        x = x + 1;
end

If-Else-End

if n > 5
        x = 1;
elseif n > 4
        x = 0;
else
        x = −1;
end

Switch-Case
switch n
case 1
        x = 1;
case 2
        x = 2;
case 3
        x = 3;
otherwise
        x = −1;
end;

```
switch name
      case {'john', 'jack')
               x  =  1;
      case 'david'
               x  =  − 1;
      end;
```

Index

A

A-code, 495–496
Abuse/misuse (product failure), 47
AC exciting signal, 338
AC servo motor, 345
Acceleration sensors, 340–341
Access Point (AP), 430
Accuracy, 43–44, 337, 448–449, 593–594
  and repeatability, 449
  and stereolithography, 558
ACIS, 517
Acoustic sensor, 595
Acrylonitrile-butadienestyrene (ABS), and FDM, 565
Actuators, 594
Adaptable robots, 598
Adhesive grippers, 596
Alberti, Leon Battista, 19
Allowed tolerance, 58
AlohaNet, 429
Alumina, on high-speed steel (HSS) tools, 235
Aluminum alloys, machinability, 244–245
Aluminum–magnesium alloys, machinability, 245
Aluminum–zinc–magnesium alloys, machinability, 245
American National Standards Institute (ANSI),
    dimensioning standards, 24
Analog input/output, 367
Analytical curves, 147–150
  degree of the polynomial, 149
  ellipse, 147–148
  explicit polynomial form, 147
  hyperbola, 147–148
  implicit polynomial form, 147–148
  parabola, 147–148
Analytical geometry, for part programming, 518–532
  computational geometry, 523–532
    parametric representation, 523–527
    sculptured surface machining, 527–532
  cutter-center location and tool offset, 518–523

Analytical surfaces, 162–167
  nonparametric surfaces, 163–164
  parametric surfaces, 164–167
AND logic, 375–376
Angle plates, 257
Angularity, 66, 93
Annotations, adding, 107
ANSI Standard X3.37, 509
ANSI Y14.5-1973, 55
Application layer, OSI model, 424
Applicon, 98
APT, 509, 512
Archytas of Tarentum, 581
Area, 106
Arithmetic average (AA), 29
Arithmetic-mean surface finish, 194
Armature, 342
Articulated robots, 584–587
Artificial intelligence (AI), 9
  robots with, 598
Artificial tactile sensors, development of, 595
ASCII (American Standard Code for Information
    Interchange), 490–491
  form, 404–406
    code chart, 406
    I/O, 367
    as STL file format, 566–568
ASME Y14.5, 54, 86
ASME Y14.5M-1994, 285
Assembly features, 121
Assembly robots, 602
Assignment statements, 381
ATN line, 412
AutoCAD, 99–100, 172
Automated manufacturing cells, 9
Automatic feature recognition, 121
Automatically Programmed Tool (APT), 98
Automation, 353–354